Physics of Nonlinear Transport in Semiconductors

NATO ADVANCED STUDY INSTITUTES SERIES

A series of edited volumes comprising multifaceted studies of contemporary scientific issues by some of the best scientific minds in the world, assembled in cooperation with NATO Scientific Affairs Division.

Series B: Physics

RECENT VOLUMES IN THIS SERIES

Volume 42 – Electrons in Disordered Metals and at Metallic Surfaces
edited by P. Phariseau, B. L. Györffy, and L. Scheire

Volume 43 – Recent Advances in Group Theory and Their Application to Spectroscopy
edited by John C. Donini

Volume 44 – Recent Developments in Gravitation – *Cargèse* 1978
edited by Maurice Lévy and S. Deser

Volume 45 – Common Problems in Low- and Medium-Energy Nuclear Physics
edited by B. Castel, B. Goulard, and F. C. Khanna

Volume 46 – Nondestructive Evaluation of Semiconductor Materials and Devices
edited by Jay N. Zemel

Volume 47 – Site Characterization and Aggregation of Implanted Atoms in Materials
edited by A. Perez and R. Coussement

Volume 48 – Electron and Magnetization Densities in Molecules and Crystals
edited by P. Becker

Volume 49 – New Phenomena in Lepton-Hadron Physics
edited by Dietrich E. C. Fries and Julius Wess

Volume 50 – Ordering in Strongly Fluctuating Condensed Matter Systems
edited by Tormod Riste

Volume 51 – Phase Transitions in Surface Films
edited by J. G. Dash and J. Ruvalds

Volume 52 – Physics of Nonlinear Transport in Semiconductors
edited by David K. Ferry, J. R. Barker, and C. Jacoboni

This series is published by an international board of publishers in conjunction with NATO Scientific Affairs Division

A Life Sciences B Physics	Plenum Publishing Corporation London and New York
C Mathematical and Physical Sciences	D. Reidel Publishing Company Dordrecht and Boston
D Behavioral and Social Sciences	Sijthoff International Publishing Company Leiden
E Applied Sciences	Noordhoff International Publishing Leiden

Physics of Nonlinear Transport in Semiconductors

Edited by
David K. Ferry
Colorado State University
Ft. Collins, Colorado

J. R. Barker
Warwick University
Coventry, England

and

C. Jacoboni
University of Modena
Modena, Italy

PLENUM PRESS • NEW YORK AND LONDON
Published in cooperation with NATO Scientific Affairs Division

Library of Congress Cataloging in Publication Data

Nato Advanced Study Institute on Physics of Nonlinear Electron Transport, Urbino, Italy, 1979.
Physics of nonlinear transport in semiconductors.

(NATO advanced study institutes series: Series B, Physics; v. 52)
"Published in cooperation with NATO Scientific Affairs Division."
Includes index.
1. Hot carriers–Addresses, essays, lectures. 2. Transport theory–Addresses, essays, lectures. 3. Nonlinear theories–Addresses, essays, lectures. 4. Semiconductors–Addresses, essays, lectures. I. Ferry, David K. II. Barker, John Robert. III. Jacoboni, C. IV. North Atlantic Treaty Organization. Division of Scientific Affairs. V. Title. VI. Series.
QC611.6.H67N37 1979 537.6'22 79-28383
ISBN 0-306-40356-0

Proceedings of the NATO Advanced Study Institute on Physics of Nonlinear Electron Transport, held at Sogesta Conference Centre, Urbino, Italy, July 16–26, 1979

A Division of Plenum Publishing Corporation
227 West 17th Street, New York, N.Y. 10011

Printed in the United States of America

Preface

The area of high field transport in semiconductors has been of interest since the early studies of dielectric breakdown in various materials. It really emerged as a sub-discipline of semiconductor physics in the early 1960's, following the discovery of substantial deviations from Ohm's law at high electric fields. Since that time, it has become a major area of importance in solid state electronics as semiconductor devices have operated at higher frequencies and higher powers. It has become apparent since the Modena Conference on Hot Electrons in 1973, that the area of hot electrons has extended well beyond the concept of semi-classical electrons (or holes) in homogeneous semiconductor materials. This was exemplified by the broad range of papers presented at the International Conference on Hot Electrons in Semiconductors, held in Denton, Texas, in 1977.

Hot electron physics has progressed from a limited phenomenological science to a full-fledged experimental and precision theoretical science. The conceptual base and subsequent applications have been widened and underpinned by the development of *ab initio* nonlinear quantum transport theory which complements and identifies the limitations of the traditional semi-classical Boltzmann-Bloch picture. Such diverse areas as large polarons, pico-second laser excitation, quantum magneto-transport, sub-three dimensional systems, and of course device dynamics all have been shown to be strongly interactive with more classical hot electron pictures. From this diverse group of exciting areas and from the controversies that arose during the Denton conference, it became evident that a summer school which coupled these new views with the more classical concepts would be invaluable to young researchers who were working in this and overlapping fields. Such a summer school also offered a good opportunity to reflect on the earlier developed picture of hot electron phenomena within the new perspectives afforded by modern and more specialized experimental and theoretical techniques.

The approach taken was to couple lectures treating the classical hot electron concepts with the modern approaches and lectures on these new diverse areas of carrier excitation. Moreover, these lectures were coupled with adequate discussion periods (both formal and informal) and with specialized seminars that highlighted current research, new directions, and unresolved problems. This type of approach provided a deep, basic total picture of the area through such cross-fertilization. An example of the new directional thrusts is provided by the role of quantum transport theory. Whereas, in the earlier days of high field transport, the quantum approach was relegated to the somewhat esoteric problem of transport in very-high magnetic fields, it has now assumed a leading role in the thrust of physics of ultrafast electronic processes that occur in size-constrained devices in very-large-scale-integrated (VLSI) circuits and in picosecond laser studies of semiconductor materials. Moreover, parameters such as electron-phonon coupling constants are no longer treated as phenomenological parameters but have been worked out from quantum mechanical first principles. These views were utilized to structure the lecture series so as to treat the basic fundamentals in a consistent manner prior to moving into the diverse areas of application of high-field transport. The seminars were especially picked to highlight the changing directions and thrusts of applications of high-field transport.

The Directors want to express their deep appreciation to the various lecturers, seminar speakers, and students. Special thanks go to Ms. Susan Daugaard for manuscript preparation.

D. K. Ferry
J. R. Barker
Ft. Collins, Colorado
18 October 1979

Contents

AN INTRODUCTION TO NONLINEAR TRANSPORT IN SEMICONDUCTORS

C. Jacoboni

Instituto di Fisica
Università de Modena
Italy

NONLINEAR TRANSPORT

In this brief introduction to the Advanced Nato Institute on the "Physics of Nonlinear Transport in Semiconductors" an outline will be given which should help the reader to follow the lectures given during the Institute by the various lecturers and to see how the diverse topics tie together.

Let us first of all recall that nonlinear transport deals traditionally with the problems which arise when the electric field is applied to a material, so intense that the current density deviates from the linear response. This subject is of particular interest both from the point of view of basic research, since it greatly contributes to the knowledge of the dynamics of electrons inside the crystals, and from the point of view of solid-state applications, since in most of the modern integrated microelectronics the necessary applied voltages imply electric fields large enough to enter the realm of nonlinear transport.

Most of the lectures which follow will be devoted to the basic physics of nonlinear transport; some discussion on the role of hot carriers in semiconductor devices are contained in H. Grubin's lecture and seminar.

HOT ELECTRONS

The presence of an electric field $\vec{E}$ inside a crystal modifies the free paths of the electrons between two successive scattering events by adding, with respect to the path without the field, a velocity component in the direction of the field:(+)

$$\vec{v}^{(i)} = \vec{v}_o^{(i)} + (q\vec{E}/m)t^{(i)} \quad (1)$$

where $\vec{v}^{(i)}$ is the instantaneous velocity of the i-th electron, $t^{(i)}$ is the time elapsed after the last scattering event of the i-th electron, $\vec{v}_o^{(i)}$ is the initial value of $\vec{v}^{(i)}$ at $t^{(i)} = 0$, and m and q are the electron mass and charge, respectively.

The drift velocity of the electron gas is obtained by averaging (1) over all the electrons at a given time t, corresponding to a different value of $t^{(i)}$ for each electron. If there is no accumulation of the effect of the electric field on the initial velocities $\vec{v}_o^{(i)}$ of the successive free electron paths, all these initial velocities reproduce the electron distribution function in the absence of the field. The mean value of the electron energy differs from its zero-field value only by the amount related to the energy gained during each single flight. As long as this energy is negligible with respect to the electron energy, the distribution of the $t^{(i)}$'s at any given time t is independent of the field since the electron scattering probabilities are functions of the electron energies. The drift velocity, obtained by averaging (1) over all electrons, is therefore linear in the electric field, according to Ohm's law.

If, on the contrary, the field is sufficiently high, the energy gained by the electrons during each single flight is not negligible; the electrons cannot fully dissipate this energy at each scattering event, and the effect of the field accumulates on the initial velocities $\vec{v}_o^{(i)}$; the distribution of the $t^{(i)}$'s also depends upon the field, and the linearity of Ohm's law is lost. The mean energy $\langle\varepsilon\rangle$ of the carriers then increases and would continue to increase up to breakdown of the sample were it not for scattering mechanisms whose dissipation capacity increases at increasing electron energies. These scattering mechanisms establish a new stationary state in which the average electron energy is higher than at equilibrium. This is the origin of the expression "*hot electron effects*" applied to nonlinear transport phenomena in semiconductors.

The concept of hot electrons is often associated with an electron temperature T_e, higher than the lattice temperature T. It is now known that this idea does not correctly describe reality since

(+) For simplicity a uniform system is considered here for electrons in a spherical and parabolic band.

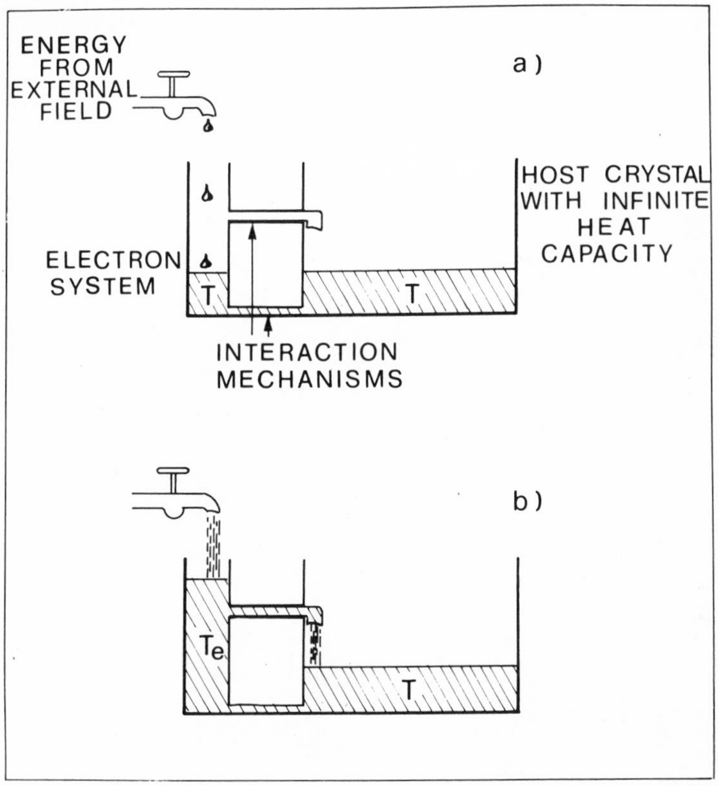

Fig. 1. Hydrodynamical analogue of the concept of hot electrons. Energy is transferred from the external electric field to the electron system and from the electron system to the host crystal. In the low-field case (a) some interaction mechanisms are capable of maintaining the temperature of the electron system equal to that of the thermal bath. At high fields (b) the energy supplied to the electrons is higher, and a new stationary state is attained by an increased efficiency of the scattering mechanisms already active at low fields and, sometimes, by the onset of new mechanisms.

the electron distribution function does not always allow an unambiguous definition of an electron temperature; that is, it is not of a Maxwellian type. It must be recognized, however, that the concept of an electron temperature has greatly helped the understanding of nonlinear transport and it still gives a useful terminology for a heuristic investigation of these problems. Figure 1 shows a hydraulic analogue of the concept of hot electrons which we will see again in the lecture by Hess.

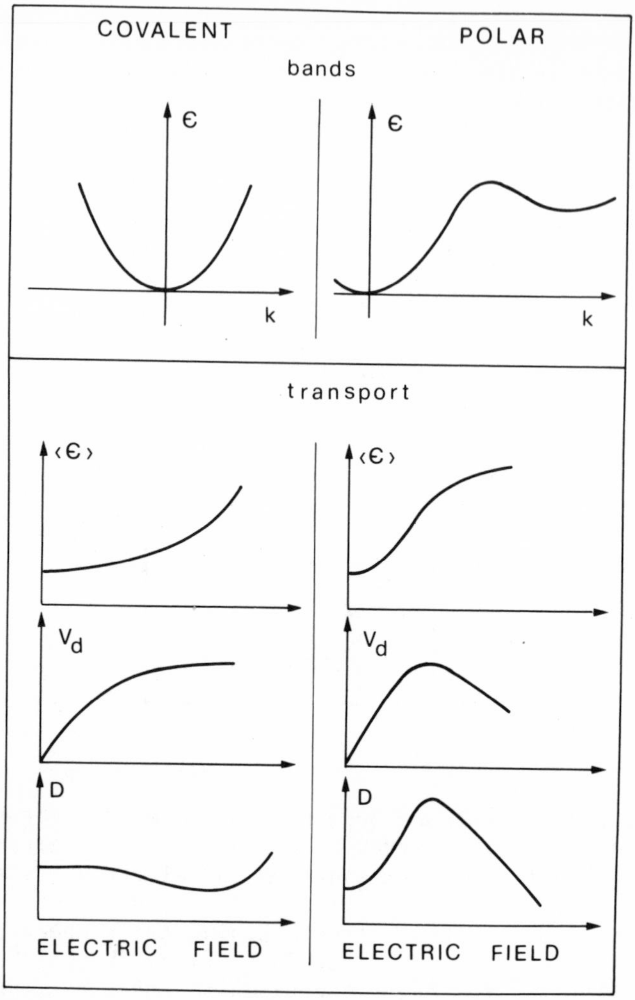

Fig. 2. A general physical picture of hot-electron effects (see text).

A GENERAL PHYSICAL PICTURE OF HOT-ELECTRON EFFECTS

A general picture of hot-electron effects in pure semiconductors can be given as follows (see Fig. 2). Let us first divide all semiconductors into two categories depending on whether the most effective scattering mechanisms are due to electrostatic fields (polar crystals) or to a deformation potential (covalent semiconductors). We then consider that as the external field increases above the Ohmic region, the mean energy $\langle\varepsilon\rangle$ of the electrons in steady-state conditions increases.

For covalent materials the scattering mechanisms become more effective as the electron energy increases so that a) the mobility $\mu(E)$ decreases as the field E increases, and b) steady state is attained with a small energy rise above thermal equilibrium: the electron mean energy increases slowly with field. Eventually, upper valleys are reached; this fact, however, does not happen in Si since the upper valleys are so high in energy that breakdown occurs first.

For polar materials the efficiency of the scattering mechanisms does not effectively increase (and may even decrease). Thus the electron mean energy increases very rapidly to reach a steady state condition as the applied field increases, while the mobility does not decrease. Above a threshold field for polar runaway, $\langle\varepsilon\rangle$ would increase indefinitely, giving rise to dielectric breakdown. Electrons then populate the upper valleys in the band and the populations of these valleys increase very rapidly as E increases. At this point the new mechanism of non-equivalent intervalley scattering comes into play. Since this is very effective in dissipating both energy and momentum, above threshold for polar run-away the energy increase with field is slowed down, and the drift velocity decreases giving rise to the well known phenomenon of negative differential mobility at the basis of the Gunn effect (see Fig. 2).

If we now consider the diffusion of electrons, we must remember that at high fields the Einstein relation does not hold. However, we may consider a very rough generalization of that equation (see also Nougier's lecture):

$$D = \frac{\mu(E)\,^2/_3\,\langle\varepsilon\rangle}{q}$$

where D is the diffusion coefficient, and, by taking into account the considerations made above on $\mu(E)$ and $\langle\varepsilon\rangle$, we obtain the general behaviour shown in the lowest part of Fig. 2. It is determined, in general, by the competing effects of reducing $\mu(E)$ and increasing $\langle\varepsilon\rangle$ as field increases. In covalent semiconductors the two effects are quite balanced, with the former effect prevailing slightly, so that D decreases slowly with increasing field. In polar semiconductors as we have seen, we first have a sharp increase of $\langle\varepsilon\rangle$ not accompanied by a decreased mobility, so that D increases very rapidly with E. Then we saw that the increase in energy is very much slowed while the mobility decreases very rapidly: a sharp reduction of D follows. For holes similar considerations should hold except for the absence of upper valleys. However, the experimental investigation of hot holes in polar materials is still scarce, and general conclusions cannot be drawn.

The particular problem of the validity of the Einstein relation at high fields will be discussed in some detail by Nougier in his

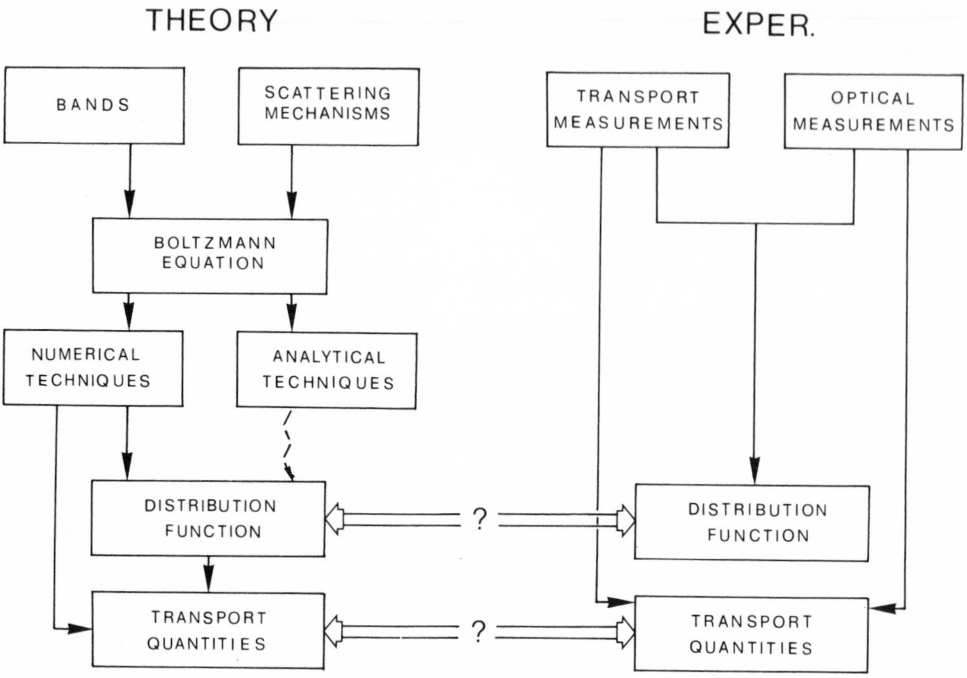

Fig. 3. Block diagram of electron transport investigations.

lecture. The phenomenology of hot carriers in semiconductors will be presented in Hess' lectures, and we shall see that most of the hot-electron effects can be described as details of the above general picture.

THE GENERAL FRAMEWORK

Figure 3 shows a block diagram of the general framework of electron transport investigations. In theoretical analyses, we start from the knowledge of the band structure of the material of interest and of the scattering processes which electrons undergo within the crystal. The fundamentals of band structure theory will be discussed in Calandra's lectures, while the analysis of scattering processes will be presented in the lectures of Vogl, Hearn, and Kocevar. It is then possible to write down the Boltzmann transport equation or its equivalent quantum equation for the density matrix when quantum transport is to be considered. These equations will be discussed in the lectures of Ferry and Barker.

With the electronic distribution function $f(\vec{k})$, which can in principle be obtained as solution of the transport equation, we may describe charge transport completely. However, this equation cannot be solved analytically in any case of practical interest without dramatic approximations, while "exact" numerical solutions can now be obtained with the modern powerful computers. In Monte Carlo simulation, which is the most popular of these numerical methods, transport parameters as mobility, mean energy, etc., can be obtained directly, without going through the determination of the distribution function, which is in general very time-consuming and not directly comparable with experimental results (with some exceptions, which will be discussed in Bauer's lectures).

On the experimental side the two major classes of investigations deal with transport measurements and optical measurements. Perhaps the most interesting results can be obtained when the two techniques are combined together. From these measurements, which will be discussed in the lectures of Bauer and Reggiani, quantities such as conductivity, mobility, diffusivity, optical absorption coefficients, energy relaxation times, etc., are determined. Then, a comparison of these transport quantities as determined by theory and experiment constitutes, as is usually the case in science, a test for the correctness of our physical model of charge transport in semiconductors.

In our case it is difficult to devise special measurements for checking particular details of the model used since, in general, transport quantities are obtained as averages over many physical processes. It is then difficult to determine from transport investigations e.g. a single coupling constant of the electron-phonon interaction or a single feature of the band structure. In order to get good confidence in the proposed model, it is then important to extend the investigations over the maximum possible ranges of temperature, field strength, frequency, etc.

When an intense magnetic field is applied, the quantization of the Landau orbits requires one to treat transport theory in a quantum framework, and the study of the resonant interaction of electrons and phonons when the energy separation of the Landau levels equals the energy of optical or intervalley phonons has recently shed new light on nonlinear transport in semiconductors. Galvano-magnetic effects will be discussed by Nicholas and Portal and by Calecki.

But the most recent developments of hot-electron physics is that associated with the ultrafast optical excitation and probe technique. Optical excitation of carriers in semiconductors has always been a useful technique to study electron transport in semiconductors, as will be described in the lectures of Ulbrich and of Hearn. But in particular, with the recent picosecond optical techniques it has become possible to monitor the fast dynamics of hot carriers, as we shall hear in the lectures of Smirl.

Seminars will be presented to introduce participants to some of the most recent developments of the physics and applications of nonlinear transport in semiconductors. They must be considered as complementary to the main lectures of this Advanced Study Institute.

PHENOMENOLOGICAL PHYSICS OF HOT CARRIERS IN SEMICONDUCTORS*

K. Hess

Department of Electrical Engineering
and Coordinated Science Laboratory
University of Illinois, Urbana

INTRODUCTION

Preparing a review on a topic developing as rapidly as nonlinear transport is a complicated problem. The material should be comprehensive for those new in the field and at the same time interesting in the sense that the basic principles are covered in the most efficient form. Also future applications should be pointed out whenever possible. In the past, the largest impetus to hot electron research came from device applications such as transistor characteristics, interface transport, and above all the Gunn effect. It is easy enough to guess that the largest impact in the future will come from very large scale integration (VLSI), which necessarily involves high electric fields, and from studies of multilayer heterojunction structures. It is not clear which of the hot electron effects will be important for these applications. Current device concepts are based on drift, diffusion, and generation-recombination. Therefore, I will concentrate in this review on the effects of hot electrons arising from these three basic mechanisms. It is my intention not only to give an introduction to hot carrier effects, but also to supply a complete "recipe" on how hot carrier effects can be calculated within the so-called carrier temperature model. In addition, I will discuss the cases of the largest deviations from this model.

This review is divided into 4 major parts:

(i) Deviations from Ohm's law in semiconductors are reviewed and the formal reason for the deviations is established.

*Supported by the Joint Service Electronics Program (U.S. Army, U.S. Navy, U.S. Air Force) under Contract DAAG-16-78-C-0016.

Then the microscopic reason is described, which is the acceleration of the charge carriers by electric fields to energies above the thermal equilibrium value. Then connecting the average energy <E> with a carrier temperature T_c by assuming Boltzmann statistics, the carrier temperature model is developed.

(ii) The consequences of a carrier temperature T_c larger than the lattice temperature T_L on drift, diffusion, and generation recombination are described. This part is divided into two subsections based on the consequences of the elevated electron temperature on the energy dependent scattering rate (e.g., saturation velocity, sticking probability) and changes in the effective mass m^* at higher energies (Gunn effect, equivalent valley repopulation).

(iii) The influence of a small magnetic field on hot electron phenomena is calculated. Here, a two-dimensional model applicable to the magnetoconduction in MOS transistors is treated.

(iv) In a final section, several hot electron effects are outlined with special emphasis on VLSI and multilayer heterojunction structures.

References for additional reading to each section are given at the end of the article.

THE CARRIER TEMPERATURE MODEL

Progress toward detailed understanding of high field effects began in 1951 with the experiments of Ryder and Shockley, who measured the nonlinear current-voltage characteristic of semiconductors. These measurements showed substantial deviations from Ohm's law as shown in Figure 1. Writing the current density as

$$j = \sigma F, \tag{1a}$$

or more generally,

$$j_i = \sum_j \sigma_{ij} F_j, \tag{1b}$$

Ohm's law states that σ_{ij} is independent of the electric field F.

As we see in Figure 1, the conductivity of semiconductors is strongly dependent on the electric field. We therefore have to replace (1) by

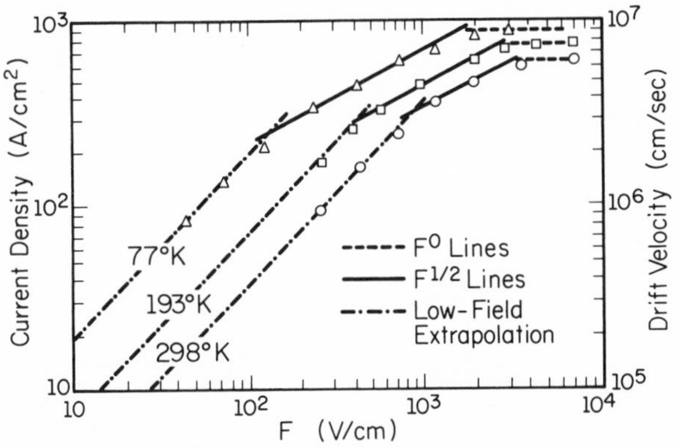

Fig. 1. Current density in n-Ge as a function of electric field intensity. [After Ryder (1953)].

$$j = \sigma(1 + \sum_{n=1}^{\infty} \beta_n F^{2n})\ F, \tag{2a}$$

or more generally by

$$j_i = \sum_j \sigma_{ij} F_j + \sum_{j,k,\ell} \beta_{ijk\ell} F_j F_k F_\ell + \ldots . \tag{2b}$$

Here σ_{ij} and $\beta_{ijk\ell}$ are independent of F, and σ_{ij} represents the zero field conductivity. We can immediately draw two important conclusions from the above series expansions:

(i) For small electric fields, the higher order terms are negligible which leads to Ohm's law. Since in metals Ohm's law is generally valid, the electric field strength involved must be small. Indeed we typically have in metals: $j \lesssim 10^3$ A/cm^2 and $\sigma \approx 10^6 (\Omega\text{cm})^{-1}$ which gives $F \lesssim 10^{-3}$ V/cm. But the small electric field is not the only reason for the validity of Ohm's law in metals. Heinrich and Jantsch (1976) showed that Ohm's law is valid up to $j \approx 10^9$ A/cm^2 in copper, corresponding to $E = 4 \times 10^3$ V/cm (see Fig. 2). The reason for this is that the conductivity of metals is mainly a function of the Fermi energy, which is practically independent of F. We will elaborate on this later. The typical field strengths in semiconductor devices are large. The typical dimension of a transistor in a very large scale integrated circuit (VLSI) is 10^{-4} cm. If one applies one Volt to this transistor, one obtains $F = 10^4$ V/cm. As we will see, the conductivity of semiconductors depends strongly on the average energy of the charge carriers, which in turn is a function of F. Therefore a large field dependence can be expected.

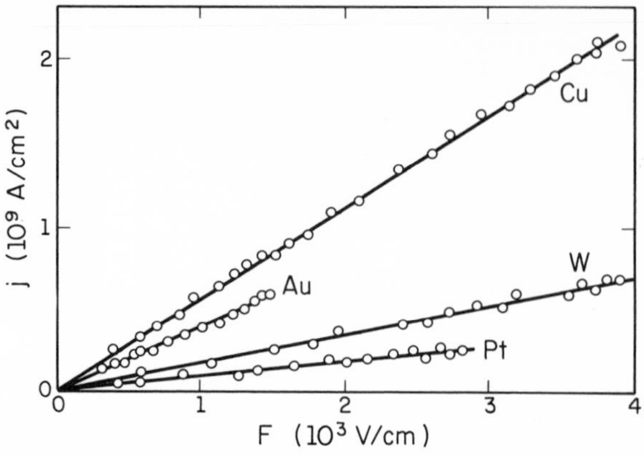

Fig. 2. Current density versus field strength for copper, platinum, gold and tungsten. [After Heinrich and Jantsch (1976)].

(ii) Assuming a certain crystal symmetry, we see that for most of the common semiconductors σ is diagonal. Using the point group of the zinc blende or diamond lattice, one can show that σ is scalar. The quantity β, however, is of higher rank and has off-diagonal terms even for cubic symmetry. This means that the high-field conductivity will be anisotropic, or the current will not generally have the same direction of the electric field. Indeed this effect was observed by Sasaki, Shibuya, and Mizuguchi (1958) and will be discussed later.

Let us now discuss the microscopic reason for the deviations from Ohms law. I have indicated already that one of the main reasons for this in semiconductors is the rise of the average energy of the electrons. If an electric field is applied, then the power input to the crystal is σF^2 (if σ is scalar); that means that per electron or hole we have a power input of $\sigma/n\ F^2 = e\mu F^2$ where μ is the mobility. Using typical values for metals ($F \leq 10^{-3}$ V/cm) we have $e\mu F^2 \leq 10^{-24}$ Watts which is very small as compared to other characteristic powers such as kT/s or the energy of the optical phonons $\hbar\omega_{opt}$/s. In semiconductor devices, on the other hand, we have typically $F \lesssim 10^4$ V/cm and higher mobilities, which give $e\mu F^2 \lesssim 10^{-7}$ Watts. Thus the electrons can gain a large amount of energy.

To establish a steady state the electrons must lose this energy. The main energy loss of electrons is due to the excitation of lattice vibrations. The hydrodynamic analog of the power balance is shown in Figure 3. The energy supplied to the carrier gas raises the carrier temperature to a value T_c which is above the lattice temperature T_L. Under steady state conditions, the average power transferred to the lattice $\langle dE/dt \rangle$ is exactly equal to the gain rate $e\mu F^2$. The theory of $\langle dE/dt \rangle$ is outlined in the Appendix. For the emission of optical

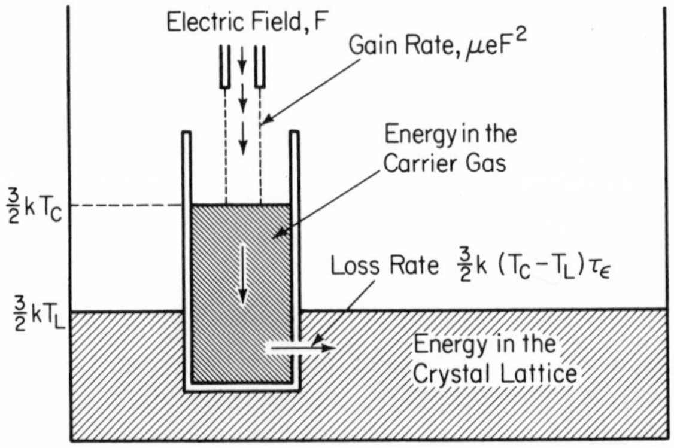

Fig. 3. Schematic representation of the carrier energy balance indicating the energy gain from an applied electric field F and the energy loss to the crystal lattice. For $F \neq 0$ the mean carrier energy $(3/2)kT_c$ is always larger than the mean thermal energy $((3/2)kT_L)$.

phonons, the above can be derived in a very simple manner. One simply has to calculate the scattering rate of absorption of optical phonons $1/\tau_{abs}$ and the rate of emission $1/\tau_{em}$. Then the energy loss is $h\omega_0\ (1/\tau_{abs} - 1/\tau_{em})$, which has to be averaged over all energies. This then gives $\langle dE/dt \rangle$. A crude estimate for $\langle dE/dt \rangle$ in nonpolar semiconductors at high temperatures ($T_c \gtrsim 300$ K) can be obtained from

$$\langle \frac{dE}{dt} \rangle = (1 - T_c/T_L)\sqrt{T_L/T_c} \times 10^{-8} \text{ Watts}, \qquad (3)$$

which is valid for scattering by optical phonons. At very low T_c however, only acoustic phonons can be emitted, which gives a much smaller rate of energy loss. References to exact formulas are given in the Appendix. Using (3) and the typical value of 3×10^{-8} Watts for $e\mu F^2$ in small devices we have $T_c/T_L \approx 10$. This means that the electrons in a device operating at 300 K can be as hot as 3000 K. Of course for $F = 0$ we have $T_c = T_L$.

The model just developed has the following implications: In order to describe the energy of the electrons by a temperature we have to assume that the momentum gained in the direction of the electric field is randomized by collisions of the electrons with each other and with other particles. A temperature can be defined in the ordinary sense only if the energy distribution function is close to the Fermi-distribution or the Maxwell-Boltzmann (MB) distribution, which is

$$f_o(E) = \exp(E_F - E)/kT_c \tag{4}$$

The drift term f_1 of the distribution function which is responsible for the current has to be small (and odd in $\vec{k}$), thus

$$f = f_o + f_1, \tag{5a}$$

$$f_1 << f_o, \tag{5b}$$

where f_o is indeed a MB distribution for dominant carrier-carrier collisions and elastic scattering mechanisms with an energy-independent scattering rate. However, scattering by ionized impurities as well as optical and acoustical phonons "distort" the distribution function whenever carrier-carrier collisions are less frequent than collisions with the other scattering agents.

Therefore f_o as given in (4) is a good approximation only at high carrier densities n. Typically $n \geq 10^{17}/cm^3$ is required. Table 1 shows the carrier densities which are usually encountered in semiconductor devices. The table reflects to some extent the hot electron research topics, which concentrate on the distribution function for the carrier-transfer devices and reverse-biased diodes, and on other things (size quantization, diffusion, nonstationary transfer, etc.) for MOS transistors, CCD's etc.

Table 1. Typical carrier densities in semiconductor devices and the applicability of the carrier temperature model.

Device	Density cm^{-3}	T_c Model
MOS	10^{18}	excellent
Transfer Devices (Gunn effect)	10^{15}	not good
Bipolar (bulk)	$> 10^{16}$	good
Bipolar (space charge region)	10^{8}	bad
Impatts (avalanche region)	$\leq 10^{16}$	good

The deviations from a MB distribution follow from the dependence of the scattering rates on the momentum $\hbar\vec{q}$ and energy $\hbar\omega$ exchanged between carrier and scattering agent. In the following, the most important scattering mechanisms and the corresponding deviations from a MB distribution (Fermi distribution) are discussed. The scattering rate per unit time is obtained from the matrix element $M_{\vec{k},\vec{k}'}$ for a transition from $\vec{k}$ to $\vec{k}'$, and from Fermi's golden rule

$$\frac{1}{\tau} = \frac{2\pi}{\hbar} \sum_{\vec{k}'} |M_{\vec{k},\vec{k}'}|^2 \delta(E_{\vec{k}'} - E_{\vec{k}} \pm \hbar\omega) \quad (6a)$$

The momentum relaxation time τ_m is obtained from a similar formula

$$\frac{1}{\tau_m} = \frac{2\pi}{\hbar} \sum_{\vec{k}'} |M_{\vec{k},\vec{k}'}|^2 \delta(E_{\vec{k}'} - E_{\vec{k}} \pm \hbar\omega)[1 - \cos(\theta)], \quad (6b)$$

where θ is the angle between $\vec{k}$ and $\vec{k}'$ and the $\pm$ sign stands for emission or absorption, respectively.

Acoustic Phonons

Scattering by acoustic phonons can be treated as elastic in most cases. For quasi two-dimensional carrier systems, this mechanism is independent of energy and f_o is given by (4). In three-dimensions, it leads to deviations from a MB distribution. The average value $\langle E\rangle$ of the carrier energy, however, does not change much which makes (for three-dimensions)

$$\langle E\rangle = \frac{3}{2} k T_c, \quad (7)$$

and the electron temperature model is a good approximation.

Ionized Impurity Scattering

Ionized impurities scatter the low energy electrons most efficiently, but do not strongly interact with high energy electrons. The scattering is elastic and leads to f_o close to a MB distribution. Again, (7) is a good approximation.

Optical Phonons, Scattering via Deformation Potential

In this case, the matrix element M is independent of q and momentum is randomized. Energy, however, is not conserved in the collisions and the probability of phonon emission is very large for carriers having energies above $\hbar\omega_o$. At intermediate electric-field strengths and a low carrier-carrier scattering rate, the field cannot supply enough carriers above $\hbar\omega_o$ and the carriers exceeding $\hbar\omega_o$ are scattered back to zero energy. This leads to a "depletion" of electrons for energies larger than $\hbar\omega_o$. Quantitatively the effect can be treated in the following way: Assuming $f = f_o + f_1$, the Boltzmann equation gives in the relaxation time approximation

$$f_1 = -\frac{e\hbar\tau}{m^*}\frac{df_o}{dE}(\vec{F}\cdot\vec{k}) = g(E)\cdot|\vec{k}|\cos\alpha, \qquad (8a)$$

and

$$\left(\frac{\partial f_o}{\partial t}\right)_c = \frac{e}{\hbar}F\,[g(E) + (2E/3)dg(E)/dE]. \qquad (8b)$$

Equation (8) is valid in a coordinate system where the constant energy surfaces are spheres. α is the angle between $\vec{k}$ and $\vec{F}$ and the factor 3/2 comes from an average over the energy. For two-dimensional systems, this factor has to be replaced by 1. $\left(\frac{\partial f_o}{\partial t}\right)_c$ is the collision operator. For intermediate electric fields, emission of phonons dominates over absorption. We then have

$$\left(\frac{\partial f_o}{\partial t}\right)_c \approx K\sqrt{\frac{E-\hbar\omega_o}{E}}\,(f_o(E-\hbar\omega_o)\exp(\hbar\omega_o/kT_L) - f_o(E)), \qquad (9)$$

where K is a constant [see E. M. Conwell (1967)]. For quasi-two-dimensional systems, the term $\sqrt{(E-\hbar\omega_o)/E}$ (density of states) is also replaced by a constant [K. Hess and C. T. Sah (1974)]. Equation (9) represents the dominant term in the Boltzmann equation for $E \geq \hbar\omega_o$ and intermediate $\vec{F}$. For $E < \hbar\omega_o$ optical phonons cannot be emitted and the Boltzmann equation can be solved easily. Using the trial solution $f_o(X) = e^{-X}(\xi(X)\cdot F^2 + i)$, where $X = E/k\,T_L$ one obtains

$$\xi(X) = \ell nX + \text{const}, \text{ for } X \leq X_o = \hbar\omega_o/k\,T_L, \qquad (10)$$

for three-dimensions and acoustic-phonon scattering below the energy $\hbar\omega_o$. As mentioned above, for $X > X_o$, (9) is the leading term in the Boltzmann equation and we can treat all other terms as a perturbation (or neglect them) which gives

$$\left(\frac{\partial f_o}{\partial t}\right) \approx 0, \qquad (11a)$$

for $X > X_o$, and

$$\xi(X - X_o) = \xi(X). \qquad (11b)$$

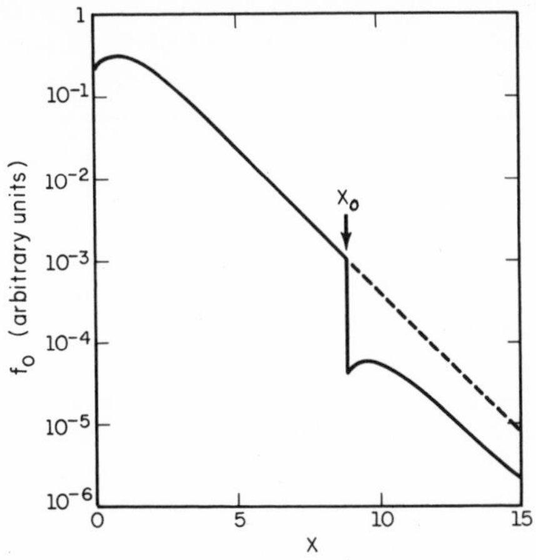

Fig. 4. Spherical part of the distribution function for acoustical plus optical scattering in a quasi-two-dimensional system.

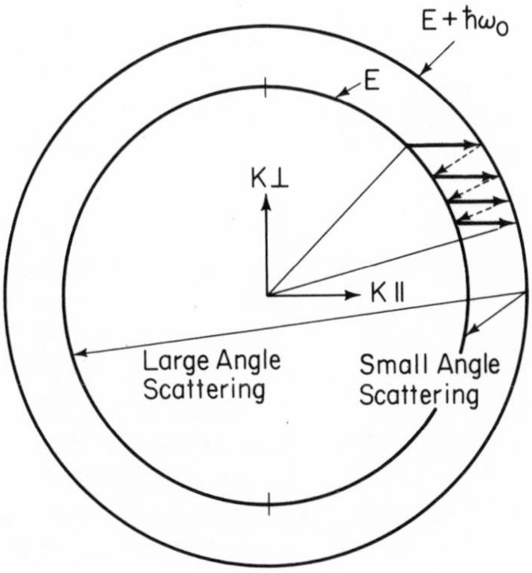

Fig. 5. Schematic representation of focusing effect on distribution of electric field. An electron is accelerated in k space in the direction of the field F, but scatters between energy surfaces with a probability which is symmetric about the radial to the initial state. For the purpose of illustration, we have assumed that the scattering takes place when the electron reaches the upper surface and that the scattering occurs to the center of the distribution of scattering probability. [After Dumke (1967)].

This means ξ is periodic in X with period X_o. A distribution function calculated in this way for a two-dimensional system is shown in Figure 4. Note the distinct kink at $X = X_o$.

Polar Optical Scattering

The polar-optical scattering mechanism is the "worst enemy" of the MB distribution, since $M \propto 1/q^2$ which means that scattering takes place with little momentum exchange (small q) and large energy exchange. As a consequence of the small q scattering, momentum is not randomized, which leads to a focusing of the electron velocities in the direction of the electric field in addition to kinks at multiples of $\hbar\omega_o$. The focusing effect is shown in Figure 5.

These deviations from the MB distribution are extremely important for the energy loss. Since only carriers above $\hbar\omega_o$ can lose energy and since the number of carriers is reduced above $\hbar\omega_o$, *the energy loss is strongly overestimated by assuming a MB distribution.* This overestimation can amount to a factor of 10 and above. The quantities depending strongly on $\langle dE/dt \rangle$ are therefore strongly changed from their Maxwellian values. Examples for these quantities are: β, defined by $\sigma = (\text{const.})(1 + \beta F^2)$, and the energy relaxation time τ_ε, which is the time constant responsible for the decrease of the average energy to the thermal equilibrium value. τ_ε is defined as [K. Hess and C. T. Sah (1975)]

$$\tau_\varepsilon = (\langle E \rangle - 3/2\ k\ T_L)/\langle dE/dt \rangle. \tag{12}$$

Experimental results of τ_ε in silicon measured at fields $F \leq 40$ V/cm are shown in Figure 6. Values of τ_ε calculated by using a MB distribution function are 100 times smaller. Using a distribution function as shown in Fig. 4 one obtains the right order of magnitude. The values of τ_ε shown in Fig. 6 are typical for all semiconductors. Note, however, the strong increase of τ_ε at low temperatures. At 4.2 K, τ_ε is typically 10^{-7}s. Other important hot electron parameters do not depend as sensitively on the distribution function and can be calculated with the electron temperature model within a factor of 2 or so.

DRIFT, DIFFUSION AND GENERATION-RECOMBINATION OF HOT ELECTRONS

By assuming that f_o is a MB distribution at carrier temperature T_c and that f_1 is given by (8a) the Boltzmann equation reduces to a series of balance equations. These equations can be obtained by multiplying the Boltzmann equation by the quantity under consideration (velocity, energy etc.) and integrating over all $\vec{k}$-space. Multiplying by $\vec{k}$ and integrating one obtains:

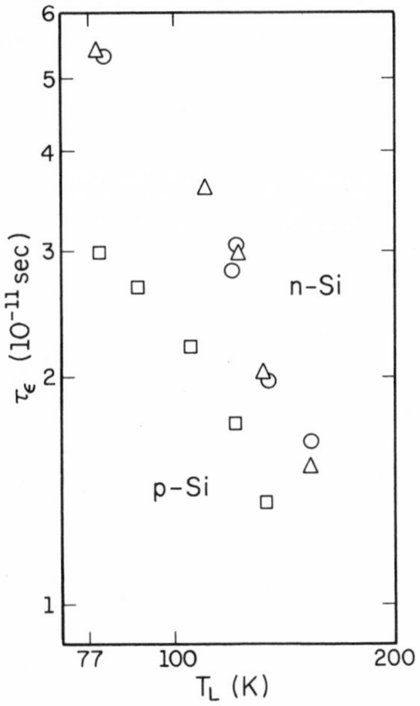

Fig. 6. Energy relaxation time τ_ε vs. lattice temperature for small electric fields ($T_c \approx T_L$) OΔ . . . n Si, $\square$ . . . p Si.

$$\langle\tau_m\rangle \frac{dj}{dt} + j = \sigma(T_c)F + e\,\nabla_r D(T_c)n, \tag{13}$$

where $\langle\tau_m\rangle$ is the average momentum relaxation time. To arrive at (13), one has to assume that $\langle\tau_m\rangle$ is independent of $\vec{r}$. Therefore the doping must be homogeneous or impurity scattering has to be unimportant. The diffusion constant is given by

$$D(T_c) = \mu(T_c)\,kT_c/e. \tag{14}$$

Equation (13) gives immediately the hot electron effects on drift and diffusion.

The Average Value of τ_m and μ as a Function of T_c

$\langle\tau_m\rangle$ as a function of T_L and T_c is given in the Appendix for most of the important scattering mechanisms. In all these formulas the effective mass is assumed to be constant. Of course, the effective mass can also depend on the energy of the electrons. This gives very important effects (Gunn effect) which will be treated below. $\langle\tau_m\rangle$ can in general be a function of both T_L and T_c because

T_L enters the equilibrium phonon occupation number N_q which is given by

$$N_q = 1/[\exp(\hbar\omega/kT_L) - 1]. \tag{15}$$

Of course when the electrons are very hot and emit many phonons N_q is also changed. I will discuss some prominent examples of disturbed phonon distributions in the last section. N_q is perturbed strongly if the phonon generation rate exceeds N_q/τ_{ph} where N_q is given by (15) and τ_{ph} is the phonon relaxation time, which strongly depends on temperature and varies within wide limits. (See the chapters on photoexcitation of hot electrons in this volume). For most practical cases N_q can be approximated by (15).

As can be seen in the Appendix, the functional dependence of $\langle\tau_m\rangle$ on T_c is generally complicated. It is different for the various kinds of phonon scattering. Scattering via the deformation potential is illustrated in Fig. 7b. The phonons change the distance of neighboring atoms and therefore the conduction and valence band

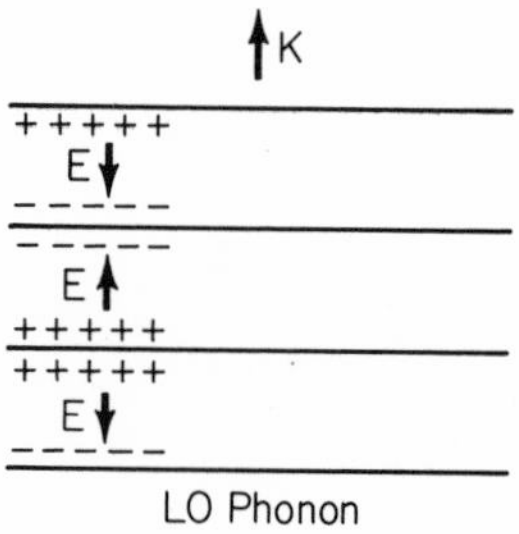

Fig. 7a. Electric field generated by polar phonons.

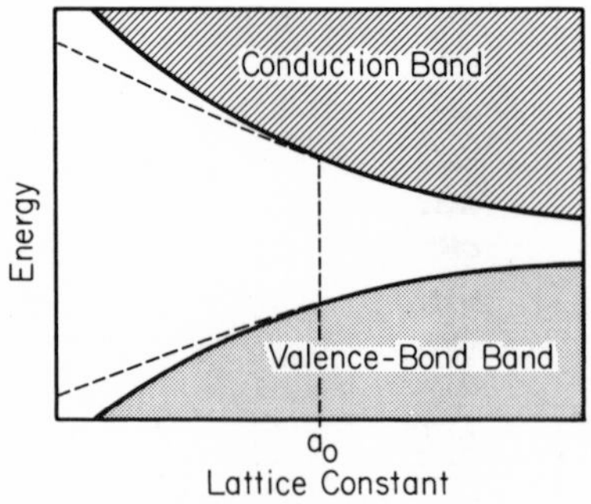

Fig. 7b. Dependence of the conduction and valence band energy on the distance between the atoms of a solid.

energies which gives rise to scattering. For $kT_c >> \hbar\omega_o$, one finds

$$<\tau_m> \propto 1/\sqrt{T_c}, \tag{16}$$

which means that at high electric fields the conductivity will decrease. Though (16) is only valid for rather high electron temperatures, it is not too bad an approximation to write

$$\sigma = \text{const.} \sqrt{T_L/T_c} , \tag{17a}$$

if deformation potential scattering dominates. This means σ decreases with the electric field F since T_c increases with F.

Impurity scattering and scattering by the electric fields associated with the polar phonons (Fig. 7a) leads to a conductivity increase, assuming, of course, that the carrier concentration and m* stay constant. The behavior of the electric current as a function of F is summarized for the various cases in Fig. 8. The diffusion coefficient as a function of electric field is shown in Fig. 9.

The values in Figs. 8,9 are for silicon at very high electric fields. Therefore μ decreases strongly (deformation potential scattering) and T_c increases. The two effects compensate each other in this case, giving an approximately constant value for D. The limitations of the electron temperature model are also clearly shown by the anisotropy of D. The electron temperature model gives an isotropic D. This measured anisotropy, which is also obtained by Monte Carlo calculations, is a consequence of the fact that the energy is not completely randomized. Let me say a few words about the electron temperature model and Monte Carlo calculations. At meetings one often hears the accusation: "Well your investigations sound interesting but they might be totally wrong as you used the electron temperature model. Monte Carlo would certainly give a different answer." In many instances this comment is justified. For high carrier concentrations, however, as they are found in MOS devices, one could exchange "electron temperature model" and "Monte Carlo" in the above statement. At present a MC calculation including e - e scattering and ionized impurity scattering at densities above 10^{18}cm^{-3} is hard to perform, but both scattering mechanisms will be very effective. To finish this section we calculate the current-voltage characteristic for the simple case of deformation potential scattering with the mobility (or conductivity) given by equation (17a) and the energy loss given by equation (3). Using the steady state condition (see Fig. 3 and the end of this chapter), we have

$$e\mu F^2 = <dE/dt> \tag{17b}$$

Using (17a,b) and (3), one obtains

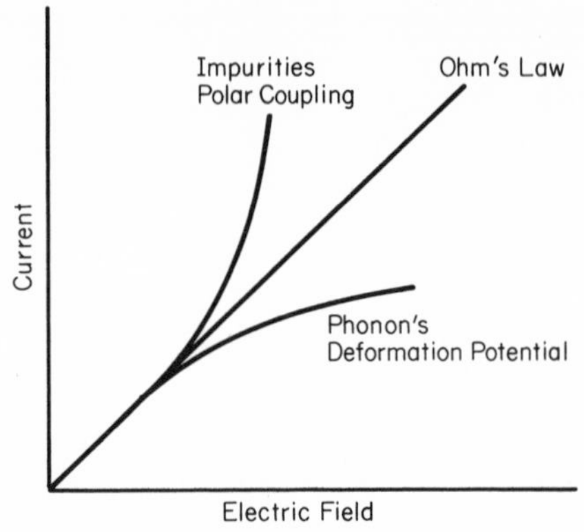

Fig. 8. Current density vs. electric field strength for various scattering mechanisms.

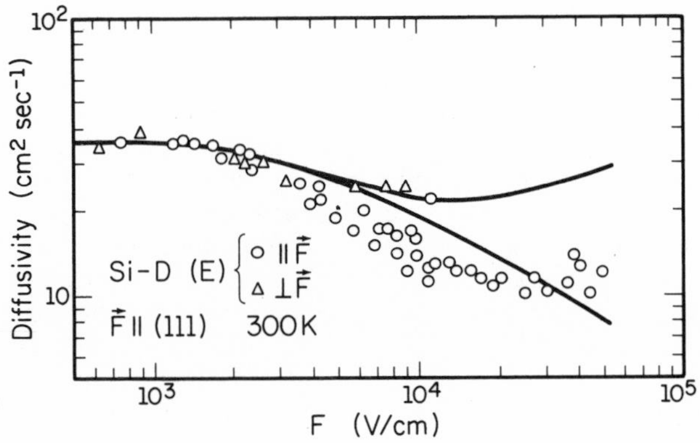

Fig. 9. Diffusion coefficients of electrons in Si at room temperature as a function of field applied parallel to a (111) crystallographic direction. Circles and crosses are found experimentally; the full lines are obtained by Monte Carlo calculations. [After Jacoboni, et al. (1977)].

$$\frac{T_c}{T_L} = 1 + e\mu_o F^2/10^{-8} \text{ Watts.} \tag{17c}$$

Here μ_o is the mobility at F = 0 and $\mu = \mu_o\sqrt{T_L/T_c}$. Since $j = ne\mu F$, this completes the calculation of the current-voltage characteristic.

Generation-Recombination Rates as a Function of F and T_c

As the electrons (holes) acquire very high kinetic energies they can lose their energy to an electron in the valence band, which then makes a transition to the conduction band. (This is exactly the inverse Auger process, where an electron makes a transition from the conduction band to the valence band and gives the energy to another electron in the conduction band). In this way the density of electrons in the conduction band increases, which leads in the case of a reverse biased p-n junction to avalanche breakdown. The corresponding increase in the reverse current is shown in Fig. 10 [Goetzberger, et al. (1963)]. The parameter of the curves is the lattice temperature. Theories for impact ionization have been given by Wolff (1954), Shockley (1961), and Baraff (1962). In these theories the ionization coefficient $\alpha_{n,p}$ (n for electrons, p for holes) is calculated. $\alpha_{n,p}$ is defined by

$$\frac{\partial I_{n,p}}{\partial x} = \alpha_{n,p} I_{n,p}, \tag{17}$$

where $I_{n,p}$ is the current due to the designated carrier.

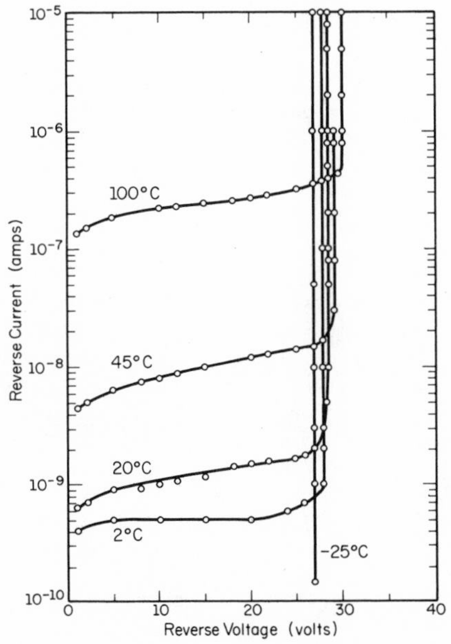

Fig. 10. Avalanche breakdown of a reverse biased silicon p-n junction [Goetzberger, et al., (1963)].

Since only electrons above the energy E_G of the band gap can contribute to impact ionization, (energy and momentum has to be conserved) only the high energy tail of the distribution function contributes to $\alpha_{n,p}$. Wolff (1954) assumed $f = f_0 + f_1$ with $f_1 << f_0$ and neglected all scattering mechanisms except scattering by optical phonons. The resulting f_0 showed a marked kink above E_1, the energy necessary for impact ionization. This is caused by the effect described in (8,11). Shockley (1961) realized that the assumption $f_1 << f_0$ is rather poor and assumed a peaked distribution function with large f_1. Baraff (1962) finally calculated the exact distribution function by solving Boltzmann's equation neglecting electron-electron scattering. This is very good in a reverse biased diode ($n \leq 10^9$ cm^{-3}) but it is certainly not a valid approach in Impatts or CCD's ($n \geq 10^{16}$ cm^{-3}).

Baraff's results depend on essentially three parameters: the mean free path of the electron for predominant optical phonon scattering, given by λ_{opt}, E_I (typically $E_I \approx 1.5\ E_G$), and the energy

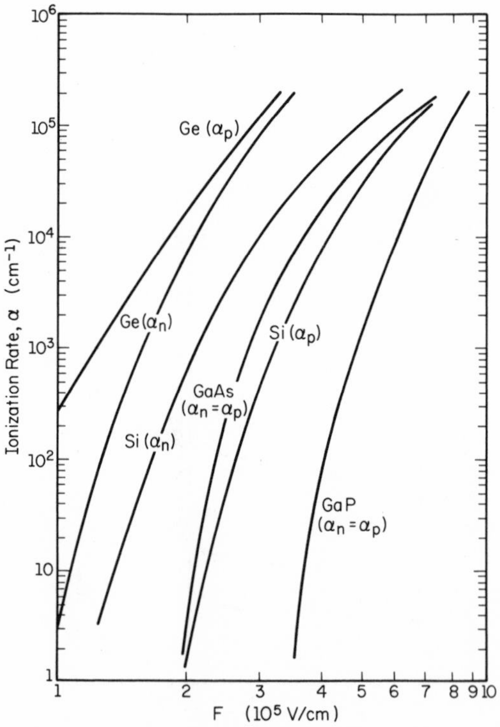

Fig. 11. Measured ionization coefficient for avalanche multiplication vs. electric field for Ge, Si, GaAs, and GaP. [See S. M. Sze (1969) and references therein].

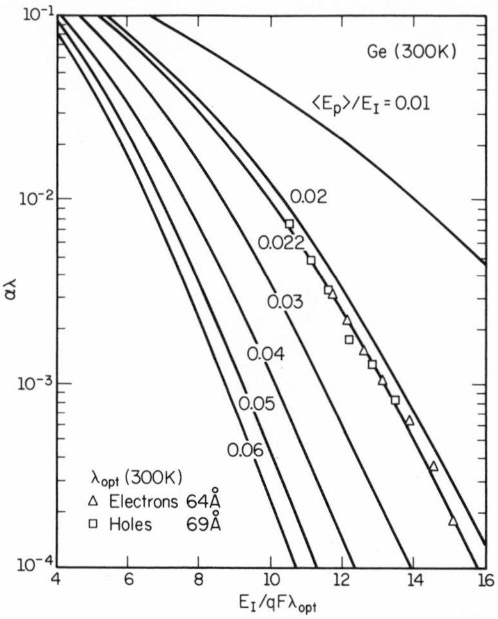

Fig. 12. Baraff's plot - product of ionization rate and optical phonon mean free path vs. X_1 [see S. M. Sze (1969) and references therein].

of the optical phonons $\hbar\omega_{opt}$. His results for $\alpha_{n,p}$ are shown in Fig. 11 and Fig. 12. Note that $\alpha_{n,p}\lambda_{opt}$ can be calculated from very simple polynomial expansions in terms of $(E_I/eF)\lambda_{opt} \equiv X_1$

$$\alpha_{n,p}\lambda_{opt} \simeq \exp\left(\sum_{n=1}^{3} r_n X_1^n\right) \tag{18}$$

The r_n are given in S. M. Sze (1969).

All this concerns band-to-band generation-recombination (GR). Trapping levels (deep lying impurity states) also play an important role in GR processes. Two parameters describe GR processes of electrons (holes) involving traps. These are the capture rate $c_{n,p}$ and the emission rate $e_{n,p}$ [see C. T. Sah (1976)].

Hot electrons are important also for $c_{n,p}$ and $e_{n,p}$ (for $e_{n,p}$ it is F instead of T_c which is important). Two examples are given below:

(i) Capture by the Lax cascade process [M. Lax (1960)]. Figure 13a shows the potential energy in the vicinity of a charged defect,

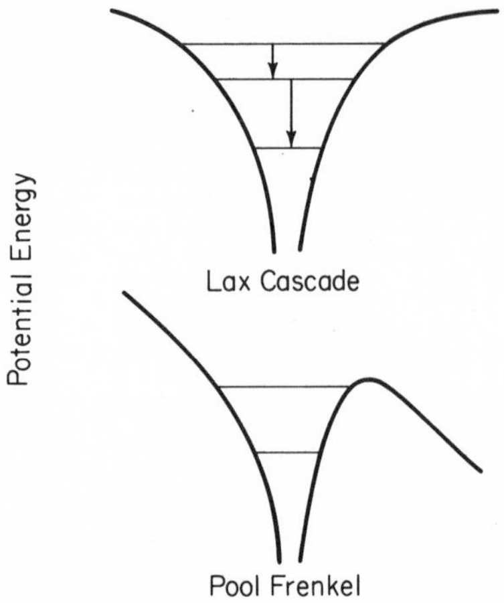

Fig. 13. a) The Lax cascade process b) The Pool-Frenkel effect

with the ground state and several excited states present. In the Lax cascade process an electron (hole) is captured in an excited state where it has a certain sticking probability, then the electron "cascades" down by emission of phonons. If the electrons are hot the sticking probability is reduced. This gives a dependence on T_c as $c_{n,p} \propto T_c^{-3/2}$.

Even without heating the electrons, the electric field F has an effect on c_n and e_n, since it lowers the barrier which the electron has to overcome in order to escape. As a consequence the emission rate which depends exponentially (activated process) on the energy of the trap level is changed (Fig. 13b). This is the well-known Pool-Frenkel effect. In passing we note also that the Lax cascade process is changed by this effect as the excited states are wiped out by high values of $F(\geq 10^5 V/cm)$.

(ii) Dependence of the Effective Mass on T_c (Intervalley Repopulation). The $E(\vec{k})$ relation for the conduction band of GaAs is shown in Fig. 14. At Γ the electrons have a very small effective mass of $m^* = 0.067\ m_o$. At higher energies, however, there are satellite minima with very large effective mass. If the electrons are heated by a high electric field and $3/2\ kT_c$ approaches 0.36 eV they can be scattered to the satellite "valleys" and their mass increases. As a consequence the current density decreases, which leads to the current voltage characteristic shown in Fig. 15. Note that at the time of this calculation it was unknown that the L minima are lower than the X minima. Of crucial importance for the occurrence of the negative differential resistance is the rapid increase of T_c with the electric field, which is typical for polar-optical scattering. Since the polar scattering probability decreases with increasing T_c,

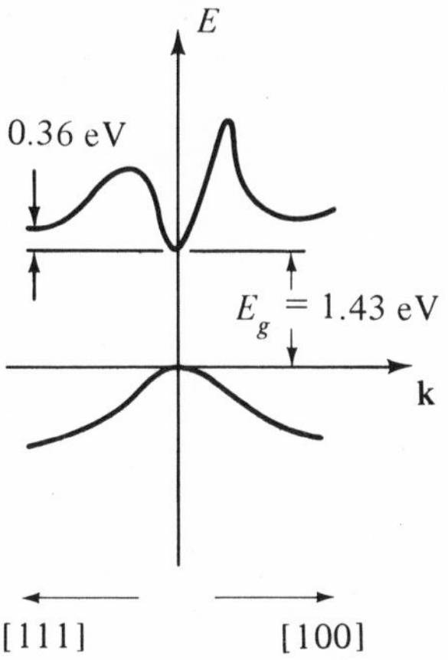

Fig. 14. $E(\vec{k})$ relation for the GaAs conduction band.

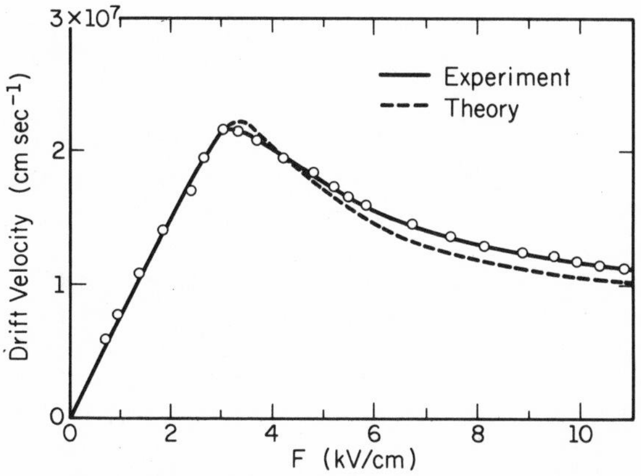

Fig. 15. Theoretical velocity-field characteristic of GaAs. [After Butcher and Fawcett (1966) and experimental results. [After Ruch and Kino (1967)].

the electron temperature runs away at a critical field. This was realized by Hilsum (1962) in his original publication and is shown in Fig. 16. The NDR was also predicted by Ridley and Watkins (1961) and finally observed by Gunn (1963). Gunn observed current oscillations in n-GaAs and InP at frequencies between 0.47 and 6.5 GH_z by applying electric fields of more than 3 kV/cm. The reason for the oscillations is that an operating point within the negative differential resistivity region is not stable. At an average magnitude F of the electric field two regions of length l_1 and l_2 with fields F_1, F_2 are formed such that the total voltage across the sample is given by $V = F_1 l_1 + F_2 l_2$. This is shown schematically in Fig. 17.

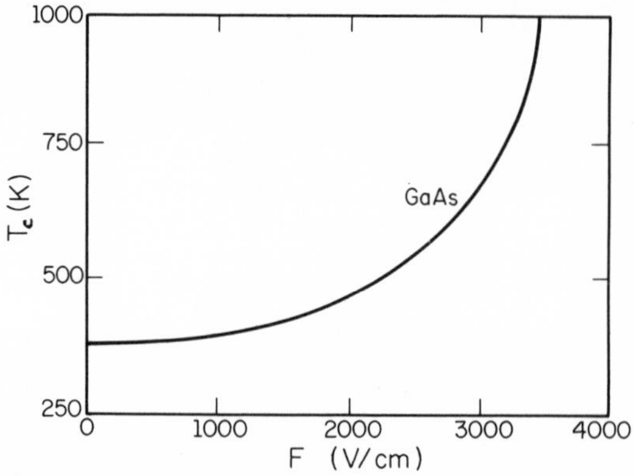

Fig. 16. Temperature-field dependence for GaAs at 300 °K. [After Hilsum (1962)].

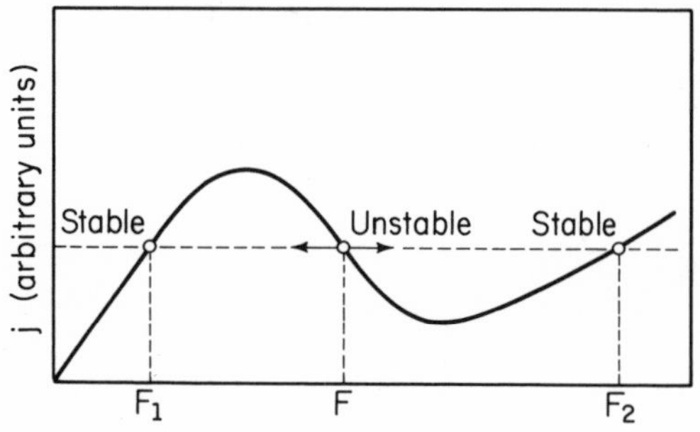

Fig. 17. Instability in the region of negative differential conductivity. [From Seeger (1973)].

The formation of the high and low field domains (Gunn domains) is not an instantaneous process but occurs typically within the dielectric relaxation time $\tau_D = \varepsilon\varepsilon_o/(en\mu)$. The high field domains travel with the drift velocity v_d through the semiconductor and vanish after hitting the anode. Then, the process of domain formation starts again, which leads to the microwave oscillations observed by Gunn. The travelling and growth of a domain is shown in Fig. 18. Of course a domain can only form if the time of travelling L/v_d is much larger than the formation time τ_D which leads to the requirement $nL > \varepsilon\varepsilon_o v_d/e\mu$. Various modes of operation of Gunn devices are possible depending on the value of nL and Lf where f is the operation frequency. A discussion of this is given by S. M. Sze (1969).

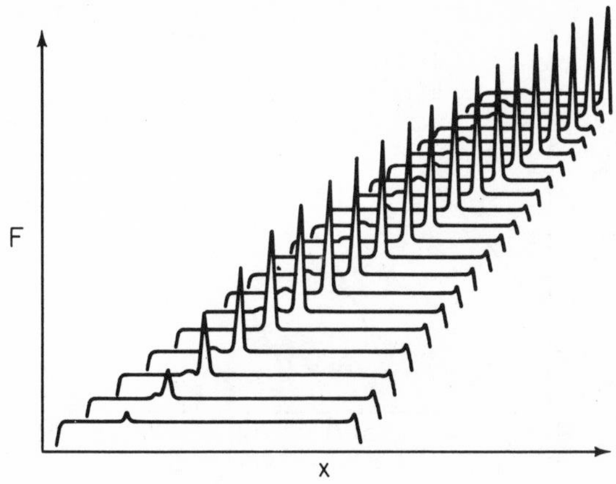

Fig. 18. A high-field domain nucleates at an inhomogeneity of doping and moves to the anode of a "long" sample as it grows. [After McCumber and Chynoweth (1966)].

Here we have considered a case with satellite valleys higher in energy. The repopulation of carriers in equivalent valleys (like the X valleys in silicon) also has a pronounced influence on the current voltage characteristic. The mechanism is shown in Fig. 19. The electrons in the valley with the small mass in the field direction are heated easier than the electrons in valleys having a larger mass

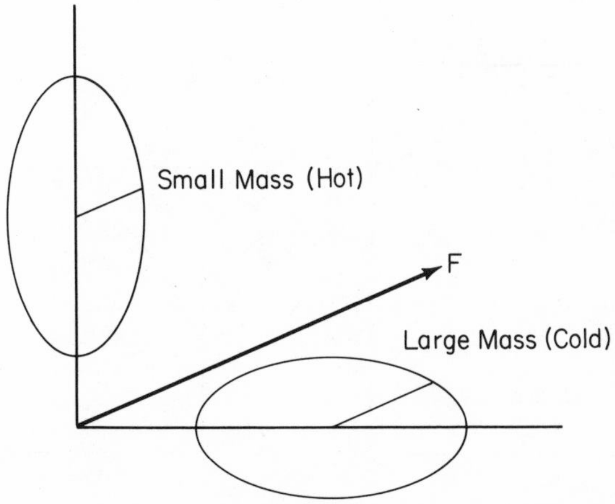

Fig. 19. Heating of charge carriers in different valleys. The valleys with the small effective mass in field direction are heated strongest.

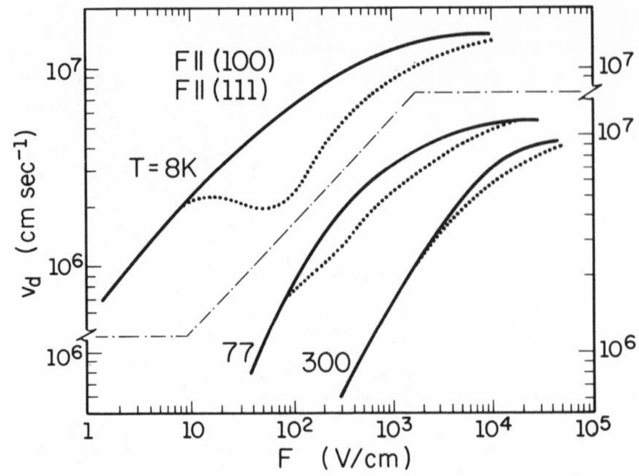

Fig. 20. Experimental results of electron drift velocity as a function of the electric field applied parallel to (111) and (100) crystallographic directions at several temperatures. [See the review of Jacoboni, et al., (1977)].

in the direction of the electric field. The hot electrons emit phonons at a higher rate and therefore are scattered to cooler valleys. This leads to a repopulation of electrons among the equivalent valleys. The colder valleys with the large mass in field direction are filled. Therefore the current drops below its ohmic value. In some cases a negative differential resistance can occur, which is shown in Fig. 20. The repopulation also has the effect that current and field

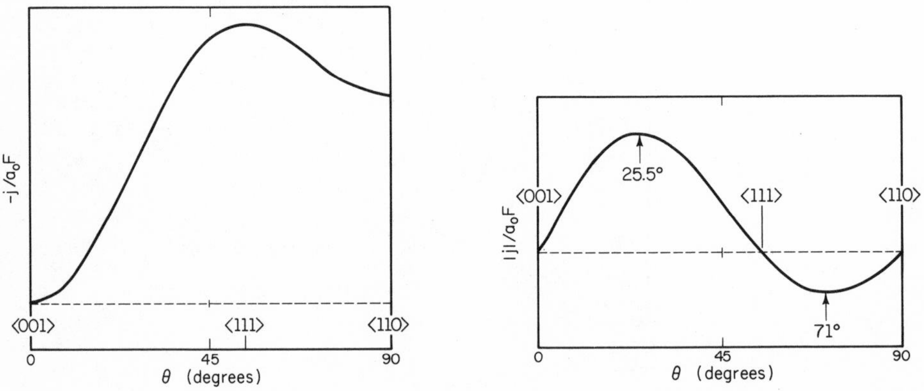

Fig. 21. Longitudinal (a) and transverse (b) Sasaki-Shibuya current components as a function of the angle between field strengt and the <001> direction. The numerical data are valid for warm electrons. [From Seeger, (1973)].

F have in general different directions. The longitudinal and transverse current components measured by Sasaki, et al., (1958) are shown in Fig. 21. Here the field was applied in the longitudinal direction.

Deviations from Ohms law are also caused by nonparabolicity, i.e. deviations of the $E(\vec{k})$ curve from the "normal" parabolic behaviour. Since the velocity $\vec{v}$ is given by

$$\hbar \vec{v} = \vec{\nabla}_{\vec{k}} E(\vec{k}), \tag{19}$$

the conductivity can be calculated from

$$\sigma_{\ell i} = -\frac{e^2}{4\pi^3\hbar^2} \int \tau(E) \frac{\partial f_o}{\partial E} \frac{\partial E}{\partial k_\ell} \frac{\partial E}{\partial k_i} d\vec{k}. \tag{20}$$

Figure 22 shows the influence of the nonparabolicity of the silicon valence band on the mobility.

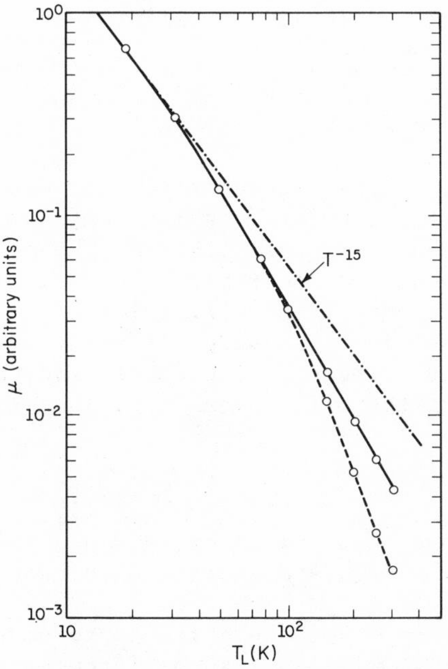

Fig. 22. Mobility vs. lattice temperature for p-Si. The full line is calculated for acoustic phonon scattering including the nonparabolicity. The dashed line includes also optical deformation potential scattering. The $T_L^{-1.5}$ curve corresponds to acoustic phonon scattering and parabolic bands. [After Asche and Borzeskowski (1970)].

The effect of nonparabolicity, is important in p-silicon, but even more important in semiconductors with small energy gap like InSb or PbTe. It causes a decrease of the mobility with both T_L and T_c.

To finish this section, I would like to show how to calculate the current voltage characteristic for a band with several valleys within the electron temperature model. It was indicated before that the electron temperature can be calculated from the balance of energy going in the electron gas and the energy lost by emission of phonons i.e. from

$$e\vec{F}^{(i)}\,\hat{\mu}^{(i)}\,\vec{F}^{(i)} = \langle\frac{dE}{dt}\rangle^{(i)}, \tag{21}$$

where $\hat{\mu}$ is the mobility tensor (see below). The superscript (i) stands for the i-th valley. The particle numbers in the different valleys can be calculated from the steady state condition

$$\Sigma_{j\neq i}\left(\frac{\partial n}{\partial t}\right)_{j\rightarrow i} = \Sigma_{j\neq i}\left(\frac{\partial n}{\partial t}\right)_{i\rightarrow j}. \tag{22}$$

The transfer rate of carriers from valley j to i is given by

$$\left(\frac{\partial n}{\partial t}\right)_{i\rightarrow j} = -\langle\frac{1}{\tau^{i\rightarrow j}}\rangle\, n^{(i)} \tag{23}$$

Here the intervally scattering rates $\langle 1/\tau^{i-j}\rangle$ are similar to the scattering rates derived before. A more extensive discussion of these rates was given by Conwell (1967). The effective mass in the valleys is not necessarily isotropic. For silicon and germanium, the surfaces of constant energy are ellipsoids of revolution (corresponding to a longitudinal and transverse effective mass of $m_\ell = 0.9\ m_o$ and $m_t = 0.19\ m_o$ for Silicon). The current in the valley (i) in the coordinate system of the ellipsoidal axis is then given by

$$j_\alpha^{(i)} = n^{(i)}\,\mu_\alpha^{(i)}\,F_\alpha^{(i)}, \tag{24}$$

where $\alpha = 1, 2, 3$ are vector indices, and $\mu_\alpha^{(i)}$ are the components of the mobility tensor which is diagonal in this coordinate system.

Thus the calculation of the current proceeds as follows: (i) $T_c^{(i)}$ is calculated from (21) using the formulas for μ and $\langle dE/dt\rangle$ as given in the Appendix, (ii) $n^{(i)}$ is calculated from (22,23), and (iii) the current is calculated from (24), transformed into a specific coordinate system, and the $j^{(i)}$ are summed. The whole procedure only applies for homogeneous semiconductors and electric fields. If $\vec{F}$ depends on the space coordinates T_c also becomes space dependent. Then the electron heat conduction and the electron drift can supply and remove energy. Therefore, the power balance equation (21) has to be replaced by

$$\frac{\partial n\langle E\rangle}{\partial t} = \vec{j}\cdot\vec{F} + n\langle\frac{dE}{dt}\rangle + \frac{\partial}{\partial x}\{\frac{\partial T_e}{\partial x} + \frac{j}{e}\frac{\langle\tau E^2\rangle}{\langle\tau E\rangle}\}. \tag{25}$$

This equation is valid for a spherical single valley band and was written in a one-dimensional form for simplicity. κ is the thermal conductivity of the electron given by

$$\kappa = 2\ kD. \tag{26}$$

For deformation potential scattering ($\tau_m = 1/\sqrt{E}$) one obtains:

$$\langle\tau_m E^2\rangle/\langle\tau_m E\rangle = 2\ kT_c \tag{27}$$

The effect of the thermal conduction of the free carriers is to broaden the T_c curves. For an Impatt device, (21) would yield electron temperatures of several thousand degrees in the avalanche zone but only room temperature about one micrometer from this zone. This, of course, is inconceivable and the elaborate form (25) has to be used to include heat conduction effects.

THE INFLUENCE OF CLASSICAL MAGNETIC FIELDS ON HOT ELECTRONS

When a magnetic field is applied, the field term in the Boltzmann equation becomes

$$-\frac{e}{h}(\vec{F} + \vec{v}\times\vec{B})\cdot\nabla_{\vec{k}} f. \tag{28}$$

This term makes an explicit solution of the Boltzmann equation complicated. The method of moments, which lead to the equation for the current and the power balance equation is not applicable anymore, because the magnetoresistance, which we would like to calculate is essentially a statistical effect. We therefore use a slightly different procedure. To simplify the notation we consider a two-dimensional electron gas to which a perpendicular magnetic field is applied. The calculation is not academic, as it might seem, but directly applicable to the magnetoconductivity of a MOS transistor. The lines of constant energy (surfaces in 3 dimensions) are assumed to be ellipses and we calculate in the system of the main axes. As before, we assume f_o to be a Maxwell-Boltzmann distribution at carrier temperature T_c. Multiplying the Boltzmann equation by the wave vector and integrating over a line of constant energy (over the polar angle) one obtains:

$$\frac{dj_x^{(i)}}{\tau_x} = E \frac{e}{m_x} \frac{F_x}{kT_c} dn^{(i)} + \frac{e}{m_x} \sum_{y,z} \delta_{xyz} B_z dj_y \tag{29}$$

and an analogous equation for $dj_y^{(i)}$. The superscript (i) indicates a specific valley. The differential currents and carrier concentrations are the contribution of a line of constant energy. The exact definitions are

$$dj_\alpha^{(i)} = \frac{e}{\sqrt{m_\alpha}} \int_0^{2\pi} k_\alpha \ f \ d\phi \quad \alpha = x,y \tag{30}$$

and

$$dn^{(i)} = \int_0^{2\pi} f \ d\phi \ . \tag{31}$$

We now assume energy loss to acoustic phonons only (which is a valid assumption at very low T_c and T_L). The energy loss has then a particularly simple form

$$\langle \frac{dE}{dt} \rangle_{ac} = k(T_c - T_L)/\tau_\varepsilon, \tag{32}$$

where the so-called energy relaxation time τ_ε is independent of T_c, which means that it is a genuine relaxation time. (In three dimensions, τ_ε always depends on T_c). τ_ε is of the order of 10^{-9} s for silicon inversion layers.

The momentum relaxation time we assume to be due to ionized impurity scattering [K. Hess, (1975)] as

$$\tau_\alpha = C_\alpha \ E \tag{33}$$

where C_α is a constant. Using (32) and (21) to determine T_c and (33) and (29) to determine μ_α, one obtains

$$j_x^{(i)} = en^{(i)} \ \mu_x^{(i)} \ F_x(1 - 3 \ B_z^2 \mu_x \mu_y) \tag{34}$$

and

$$T_c \simeq \frac{T_L}{1-a} \ [1 - \frac{3ab(\mu_x^o \ B_z)^2}{(1-a)^3}] \ . \tag{35}$$

Here we have used

$$\mu_\alpha = 2e\, C_\alpha\, kT_c/m_\alpha, \tag{36a}$$

$$a = 2e^2\, F_x^2\, C_x\, \tau_\varepsilon/m_x, \tag{36b}$$

$$b = C_y\, m_x/(C_x\, m_y) \tag{36c}$$

and μ_α is the zero field mobility ($T_c \to T_L$). From these equations we can deduce the important consequence that the magnetic field "cools" the electron gas as it reduces T_c [see (35)]. The electric field of course heats the electron gas, which is shown by the factor 1-a in the denominator of (35). The singularity of T_c at a-1 = 0 is unphysical since at high T_c, scattering by optical phonons is also important and leads to different relations. The Hall field can also be easily obtained from the above equations

$$\frac{F_y}{F_x} = \frac{3}{2}\, B_x\, \mu_x. \tag{37}$$

Measurements and calculations for bulk silicon have been performed by Heinrich and Kriechbaum (1970). The inclusion of the repopulation between different valleys makes hot-galvanomagnetic effects very complicated to calculate. The basic effect of the magnetic field, however, is the cooling of the electron gas.

In passing I would like to emphasize also that geometrical effects play an important role when a magnetic field is applied, as the electric field becomes inhomogeneous. [H. Heinrich, W. Jantsch and J. Rozenbergs (1975)].

HOT ELECTRONS IN SEMICONDUCTOR DEVICES AND LAYERED STRUCTURES

Though hot-electron-transfer devices have a much larger importance than some of the hot-electron device effects described below, I excluded the transfer devices from this chapter, since they are treated by others in following chapters. This section is divided into subsections on drift, diffusion, generation-recombination, hot electron thermionic emission and disturbances of the phonon distribution function.

Hot Electron Drift

Hot electrons are important for the current voltage characteristics of MOS transistors with very small size. Figure 23 shows the

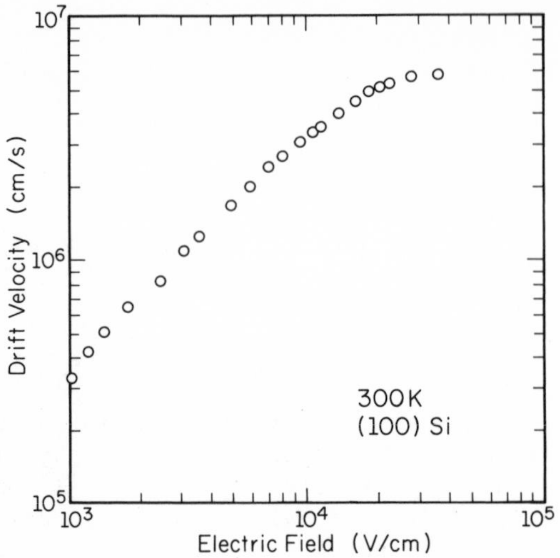

Fig. 23. Velocity-field curve measured by Fang and Fowler (1970) for $n_s = 6.6 \times 10^{12}/cm^2$.

current voltage characteristics of a MOS transistor measured by Fang and Fowler (1970). Ferry (1976) was able to achieve quite good agreement using a theory which included size quantization and three subbands. However, some of the features of this current voltage characteristic are still under discussion. One is the relatively low saturation velocity which is not reflected by the theory. This discrepancy might be caused by an oversimplification of the interface roughness scattering. Also, it was recently shown that even the polar phonons in the SiO_2 interact strongly with the electrons at high carrier temperatures T_c [K. Hess and P. Vogl, (1979)]. A more extensive review on this subject was given previously by Hess (1978) and by Ferry (1978). High electric fields occur also in charge coupled devices (CCD's). CCD's are formed by an array of MOS capacitors. Charge can be moved between these capacitors by applying a clock voltage to the metal gates. The applications of CCD's are numerous for imaging and signal processing. Figure 24 shows the transfer of charge from one gate to the other. The fraction of residual charge under the gate from which charge is transferred is shown vs. time. Curve one is calculated assuming Ohmic conductivity, while curve two takes into account current saturation by the hot electrons. The result is self-explanatory. Hot electrons always limit the current and impose a limit on the device speed since the drift velocity drops below its ohmic value at high electric fields.

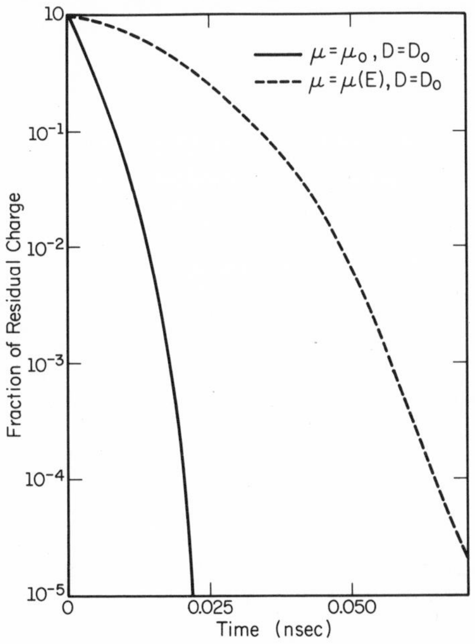

Fig. 24. Fractional residual charge under the gate for $n = 2.5 \times 10^{11}/cm^2$. Curve 1: $\mu = \mu_o$, $D = D_o$. Curve 2: $\mu = (E)$, $D = D_o$.

Hot Electron Diffusion

Hot electron diffusion is important to a lesser extent in devices when compared to hot electron drift. Therefore I would like to point out only a few features. The power input in the electron gas is equal to the product j.F (Eq. 25). Since the diffusion and drift currents can compensate each other, there exists the possibility of having very high electric fields present without any (average) heating of the charge carriers, as the net current is zero. This is the case in a p-n junction without bias when the built-in field is high. Therefore, as soon as diffusion becomes important the mobility and conductivity are no longer unique functions of the electric field, as indicated in (1) and (2) but a unique function only of the carrier temperature T_c. T_c in turn is a function of j.F, not of F alone.

This makes calculations very complicated. A simplified approach was given recently by Hess and Sah (1978). Note that hot electron diffusion is usually not included in device modeling because of the large numerical difficulties and in some cases because of the wrong assumption that the diffusion constant decreases with the electric field as the mobility does.

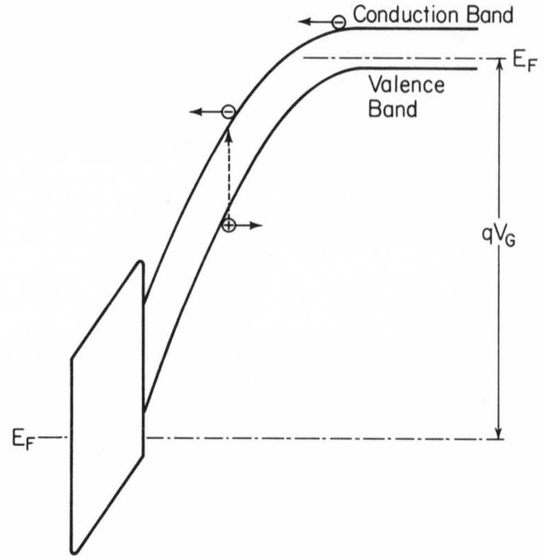

Fig. 25. Band structure of a MOS device at high gate and substrate voltage [After Ning (1978)].

Generation-Recombination

The electric fields in MOS devices can be high both in a parallel direction as well as perpendicular to the semiconductor-insulator interface. The electric field perpendicular to the interface of a MOS transistor with high gate and substrate voltage is shown in Fig. 25. Electrons which are thermally generated in the bulk of the semiconductor are accellerated by the high electric field. On their way down they multiply by impact ionization. Finally some are collected at the minimum of the potential well but some are emitted in the insulator (SiO_2), where they are trapped. This trapping gives rise to device instabilities since the potential distribution at the interface is changed. This effect was discussed by Ning (1978) and seems to be fairly important for long-term device stability.

High electric fields parallel to the interface can also limit the device performance. As an example we will estimate the ultimate speed of a CCD. The geometry of gates in a CCD is shown in Fig. 26. The electric fields are highest between neighboring gates and lowest beneath the middle of the gates (see Fig. 27). To achieve ultimate transfer speed one wants to have the lowest electric field, F_ℓ, larger than the field at which the drift velocity (current) saturates. This is typically a field of 10^4 V/cm. On the other hand the highest electric field F_n (between the gates) must not exceed the field at which impact ionization becomes important. This field is typically 10^5 V/cm as can be seen from Fig. 12. The lowest and highest electric field are connected by the approximate relation

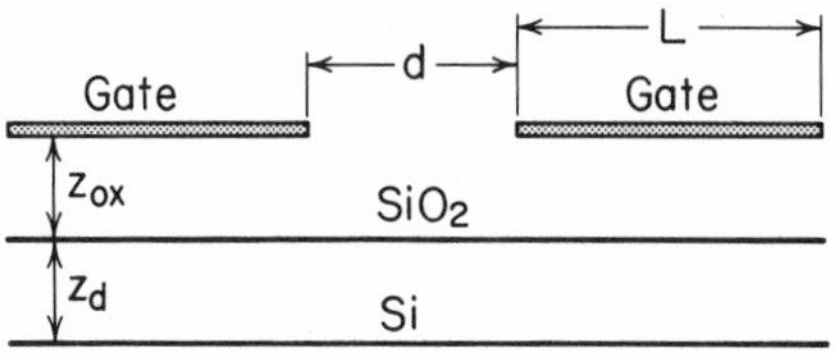

Fig. 26. Typical CCD geometry with non-overlapping gates.

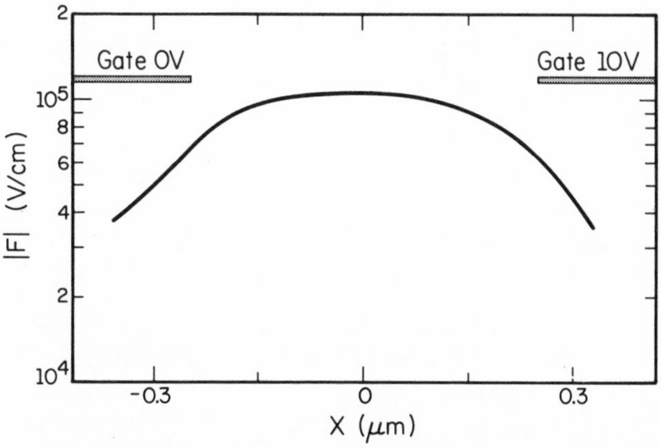

Fig. 27. Electric field under the gates of CCD's.

$$F_\ell \approx 6.5 \frac{d\, Z_{ox}\, F_h}{L^2} \qquad (38)$$

This relation was found by computer simulations. Typical values are $d \approx Z_{ox} \approx 10^{-5}$ cm. Therefore ultimate transfer speed is only possible for $L \leq 8 \times 10^{-5}$ cm, which gives $F_\ell = 10^4$ V/cm when $F_h = 10^5$ V/cm. CCD's with a longer gatelength are slower. This is only valid for surface channel CCD's. For buried channel CCD's, one has to use different relations.

As indicated earlier the capture and emission rates of deep lying impurity levels (recombination centers) are also changed by high electric fields. This influences the hot electron device noise and reverse leakage currents etc. However, only minor research efforts have been made on this subject.

Hot Electron Thermionic Emission

If electrons reside in a potential well their spatial density distribution is given by

$$n \propto \exp(e\,\phi/kT_c), \tag{39}$$

where ϕ is the potential. Considering a buried-channel device structure, this means that with increasing electron temperature T_c, the electrons are able to overcome higher potential barriers and occupy a larger volume. If kT_c becomes of the order of the barrier height of the potential well a considerable number of electrons will spread to the surface and consequently a transition of buried-channel device to surface-channel device will occur. Of course this is lethal for device performance and therefore the possibilities of high carrier temperatures have to be included in the device design. However, the thermionic emission of hot electrons from a potential well has not only destructive consequences. Consider for example layered heterojunction structures of GaAs-$Al_x\,Ga_{1-x}As$. The band gap of GaAs is smaller than the band gap of $Al_x\,Ga_{1-x}As$. Therefore most of the electrons reside in the GaAs layers at minimum potential energy. These electrons can be heated by a high electric field parallel to the layers. Again, if kT_c becomes of the order of the potential height, the electrons are thermionically emitted into the $Al_x\,Ga_{1-x}As$ and into neighboring cool GaAs layers. This process closely resembles the emission of electrons from a glow cathode. Since the typical width of these layers is 100Å the processes can be very fast and it should be possible to construct useful devices on the basis of hot electron thermionic emission. Note that the power needed to generate hot electrons is rather small (depending of course on the concentration n).

Disturbances of the Phonon Distribution

Disturbances of the phonon distribution have been included in device modeling only by estimating the lattice heating, i.e., by taking into account the higher phonon occupation numbers caused by a temperature rise. Recent experiments indicate, however, that large disturbances of specific phonon modes can be important in photo-pumped-optoelectronic devices. Shah (1978) observed strongly increased phonon occupation numbers of polar-optical phonons in bulk GaAs. N_q was obtained by measuring the ratio of the Stokes to the anit-Stokes Raman intensity. Even more pronounced disturbances of the phonon distribution have been observed by Holonyak, et al. (1979) in layered heterojunction structures. These authors observed laser operations of quantum well hetero-structures on a phonon (LO) sideband about 36 meV below the lowest confined particle transitions (Fig. 28).

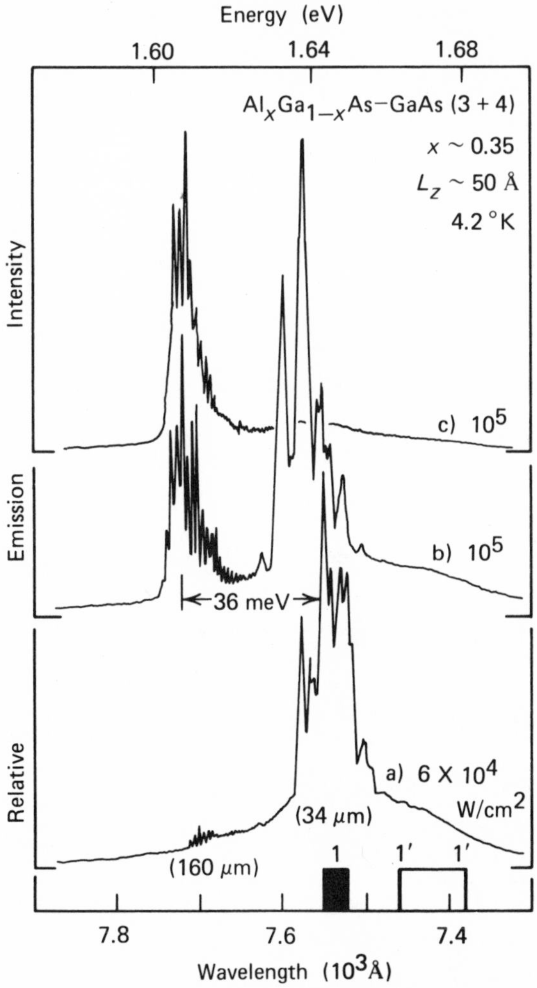

Fig. 28. Laser emission spectra (4.2K) of a photopumped (pulse-excited) multiple-quantum-well MO-CVD $Al_xGa_{1-x}As$-GaAs heterostructure.

The electron-phonon interaction is enhanced in layered structures over the interaction in the bulk, because the quantum well can "provide" the momentum $\hbar q_z$. The formulas for the scattering rates are given in the Appendix. This enhancement gives rise to even larger disturbances of N_q as compared to the bulk. Estimates show that the phonon emission rate is increased by a factor of at least $\pi(N_q + 1)/2$ for virtual processes, which gives rise to the observed phonon sideband laser operation [Kolbas, et al., (1979)].

APPENDIX

In this appendix formulas for the momentum relaxation times and, as far as explicitly possible, for the mobilities due to scattering by various mechanisms are given. First scattering in bulk semiconductors is discussed then the scattering times for special structures (at interfaces and in layered structures) are listed. The symbols are explained at the beginning of this review. Using the scattering times τ_{ac}, τ_{op} etc. given below, the formulas for the mobilities are obtained by assuming a Maxwellian-Boltzmann distribution, and:

$$\mu = \frac{e\langle\tau_{tot}\rangle}{m_\sigma^*} \qquad \text{where } \frac{1}{\tau_{tot}} = \frac{1}{\tau_{ac}} + \frac{1}{\tau_{op}} + \ldots$$

$$\langle\tau_{tot}\rangle = \int_0^\infty E^{3/2} e^{-E/kT_c} \tau_{tot} dE \Big/ \int_0^\infty E^{3/2} e^{-E/kT_c} dE$$

For two-dimensional systems the power 3/2 has to be replaced by 1.

MOMENTUM RELAXATION (BULK SEMICONDUCTORS)

Scattering by Impurities

1. Neutral-impurity scattering:

$$\frac{1}{\tau_m^N} \simeq \frac{20\,\varepsilon\varepsilon_o\,\hbar^3\,N_I}{m_d^{*2}\,e^2}$$

$$\mu = \frac{m_d^*\,e^3}{20\,\varepsilon\varepsilon_o\,\hbar^3\,N_I}\,\frac{m_d^*}{m_\sigma^*}$$

2. Ionized-impurity scattering:

$$\frac{1}{\tau_m^I} = E^{-3/2}\,\frac{N_I}{16\pi\sqrt{2m_d^*}}\left(\frac{e^2 Z}{\varepsilon\varepsilon_o}\right)^2\left[\ln(1+\beta^2) - \frac{\beta^2}{1+\beta^2}\right]$$

$$\beta = 2|\vec{k}|L_D\,,\quad L_D^2 = k\,T_c\,\varepsilon\varepsilon_o/(e^2 n)$$

An approximate value for the mobility is obtained by putting

$$|\vec{k}| \simeq \sqrt{3kT_c m_d^*}/\hbar$$

and replacing $E^{3/2}$ by $(kT_c)^{3/2}\Gamma(4)/\Gamma(5/2)$

3. Dislocation scattering:

See K. Seeger (1973) p. 230.

Scattering by Phonons

1. Acoustic phonons:

$$\frac{1}{\tau_{ac}} = \frac{m_d Z_A^2 kT_L}{\pi\hbar^3 v_\ell} |\vec{k}|$$

$$\mu = \frac{2\sqrt{2\pi}}{3} \frac{e\,\hbar^4\, v_\ell}{m_d^{3/2} m_\sigma kTZ_A^2 \sqrt{kT_c}}$$

For the many valley model Z_A has to be replaced by a combination of Ξ_u and Ξ_d.

2. Optical phonons:

$$\frac{1}{\tau_{opt}} = \frac{D^2\, m_d^{*3/2}}{\sqrt{2}\pi\,\hbar^3\, q\,\omega_o} [N_q(E + \hbar\omega_o)^{1/2} + (N_q + 1)\ (E - \hbar\omega_o)^{1/2}]$$

here it is assumed that $E \geq \hbar\omega_o$ in the emission term.

3. Polar Optical Phonons:

In this case a momentum relaxation time does not exist. The scattering rate $1/\tau_{po}$ is given by

$$\frac{1}{\tau_{op}} = \frac{2e\ \varepsilon_o}{(2m_d^*\ E)^{1/2}} [N_q \sinh^{-1}(E/\hbar\omega_o)^{1/2} + (N_q + 1)\sinh^{-1}\left(\frac{E-\hbar\omega_o}{h\omega_o}\right)^{1/2}]$$

The mobility cannot be obtained in a simple direct fashion from $1/\tau_{po}$ but an approximate value is obtained by averaging τ_{po} as usual. For a more thorough discussion see Conwell (1967) and the references therein.

4. Piezoelectric Scattering

$$\frac{1}{\tau_{pie}} = \frac{\sqrt{m_d}\, e^2 K^2\, k_B T_L}{2\sqrt{2\pi}\, \hbar^2\, \varepsilon\varepsilon_o \sqrt{E}}$$

$$\mu_{pie} = \frac{16\sqrt{2\pi}\, \hbar^2\, \varepsilon\varepsilon_o \sqrt{kT_c}}{3\sqrt{m_d}\, m_\sigma\, e\, K^2\, k\, T_L}$$

5. Alloy Scattering

See Littlejohn, et al., (1978).

B. Interfaces and Layered Structures

A description of momentum relaxation times at interfaces is complicated by the fact that electrons are confined to a layer of width L. We assume an electron wave function of the form:

$$\Psi = \sqrt{2/V}\, e^{i\vec{k}\cdot\vec{r}} \sin(\pi z/L)$$

where $\vec{k}$ and $\vec{r}$ are two-dimensional wavevectors. Clearly this functional form is only exact for square well potentials.

The phonons may, or may not be confined to the layers, depending e.g. on the elastic properties of the neighboring media. We distinguish therefore two cases:

a) Two-dimensional electrons-three-dimensional phonons:

1. Acoustic Phonons (Deformation Potential)

$$\frac{1}{\tau_{ac}} = \frac{3}{2\pi} \frac{kT_L m_d^* Z_A^2}{\hbar^3 v_\ell^2\, \rho L}$$

$$\mu_{ac} = \frac{2\pi\, \hbar^3 e v_\ell^2\, \rho}{3kT_L m_d^*\, m_\sigma^*\, Z_A^2}$$

2. Optical Phonons (Deformation Potential)

$$\frac{1}{\tau_{opt}} = \frac{3}{4} \frac{m_d^* D^2}{\hbar^2 L\rho\ \omega_o} [N_q + (N_q + 1) H (E - \hbar\omega_o)]$$

3. Polar Optical Phonons

(momentum relaxation time cannot be defined. $1/\tau_{po}$ represents scattering rate.)

$$\frac{1}{\tau_{po}} = \frac{e^2\ \omega_o}{4\ \pi\ \varepsilon_o} \left(\frac{1}{\varepsilon_\infty} - \frac{1}{\varepsilon_{st}}\right) \frac{|\vec{k}|}{E} \cdot$$

$$\cdot [N_q I_1' + (N_q + 1) H(E - \hbar\omega_o) I_2']$$

Here

$$I_1' = \int_{(1+\hbar\omega_o/E)^{1/2}-1}^{(1+\hbar\omega_o/E)^{1/2}+1} \frac{dx}{\sqrt{4x^2 - (x^2 + \hbar\omega_o/E)^2}}$$

$$I_2' = \int_{1-(1-\hbar\omega_o/E)^{1/2}}^{(1+(1-\hbar\omega_o/E)^{1/2}} \frac{dx}{\sqrt{4x^2 - (x^2 - \hbar\omega_o/E)^2}}$$

MOMENTUM RELAXATION (SURFACE PHONONS)

Acoustic Rayleigh Waves (Surfaces)

For a treatment of surfon-scattering see Hess, *et al.*, (1977) and references therein.

Remote Polar Optical Phonons

$$\frac{1}{\tau_{po}} = \frac{e^2\omega_o}{2\pi\ \varepsilon_o} \left(\frac{1}{\varepsilon_\infty + \varepsilon_n} - \frac{1}{\varepsilon_s + \varepsilon_n}\right) \cdot \frac{|\vec{k}|}{E}$$

$$\cdot\ [N_q I_1' + (N_q + 1)\ H(E - \hbar\omega_o) I_2']$$

This equation represents scattering of electrons in a nonpolar medium with dielectric constant ε_n adjacent to a polar semiconductor. The scattering agent is the polar interface mode. We have assumed $q_z L << 1$. The list of scattering mechanisms is, of course, by no means complete but the main scattering mechanisms should be covered by the above formulas.

ENERGY LOSS

To calculate the carrier temperature one has to know the energy loss $(dE/dt)_c$ to the lattice. Except for very low temperatures the energy loss is due to scattering by optical phonons. It can be calculated by multiplying the above given scattering rates $1/\tau_{opt}$ by $\hbar\omega_o$ (the phonon energy) and substracting the emission term (proportional to $N_q + 1$) from the absorption term (proportional to N_q) i.e.

$$\left(\frac{dE}{dt}\right)_c = \hbar\omega \left(\frac{1}{\tau_{abs}} - \frac{1}{\tau_{em}} \right)$$

To obtain the average energy loss $<dE/dt>$ one performs the following integration:

$$<\frac{dE}{dt}> = \int_0^\infty \left(\frac{dE}{dt}\right)_c E^{1/2} e^{-E/kT_c}\, dE \Big/ \int_0^\infty E^{1/2} e^{-E/kT_c}\, dE$$

Some values for the average energy loss can be found in Seeger (1973). The average rates of loss to interface phonons can be found in the literature for special cases. Note, however, that in many of these publications the mode of the phonons (if three-dimensional or interface modes) has not been clearly distinguished and the formulas are sometimes rather meaningless mixtures. A discussion is given in the references of the review by Hess (1978).

ACKNOWLEDGEMENT

It is a pleasure to thank Professors B. G. Streetman, G. E. Stillman and N. Holonyak, Jr., for their continual interest and support.

SYMBOLS

B_z	z-component of magnetic field $\vec{B}$
$c_{n,p}$	capture rate for electrons and holes
D	optical deformation potential constant
$D(T_c)$	diffusion constant
$<dE/dt>$	average energy loss of carriers to phonons
E	energy of the charge carriers
e	elementary charge
$<E>$	average energy of the charge carriers
E_G	band gap energy
ε_o	polar optical coupling field ($\approx$ 6000 V/cm for GaAs)
$e_{n,p}$	emission rate for electrons and holes
$\vec{F}$	electric field
F_i	components of electric field $\vec{F}$
f	energy distribution of charge carriers
f_o	spherical symmetrical part of f
f_1	drift term of f
$H(E-h\omega_o)$	Heaviside step function
$I_{n,p}$	electron and hole current
j	current density
k	Boltzmann constant
$\vec{k},\vec{k}'$	wave vector of charge carriers
$\vec{k}_\alpha$	component of $\vec{k}$
K	piezoelectric coupling constant
m^*	effective mass
m_o	free electron mass
m_d^*	density of states effective mass
m_σ^*	conductivity effective mass
$m_\alpha, \alpha=x,y,z$	component of the diagonal effective mass tensor in the coordinate system of the ellipsoidal axes.
$M, M_{\vec{k},\vec{k}'}$	matrix element for a transition from $\vec{k}$ to $\vec{k}'$
n	carrier concentration
N_I	density of impurities

N_q	phonon occupation number
$\vec{q}$	phonon wave vector
$q = \lvert\vec{q}\rvert$	
T_L	lattice temperature
T_c	carrier temperature
$\vec{v}$	velocity of charge carriers
v_ℓ	velocity of sound (longitudinal mode)
v_d	drift velocity
$X=E/kT_L$	
$X_o=h\omega_o/kT_L$	
x,y,z	space components
Z	charging state of impurity (Z = 1,2 . . .)
$\alpha_{n,p}$	ionization coefficient for electrons and holes
$\beta_n, \beta_{ijk\ell}$	expansion coefficients for the current density as a function of the electric field
δ_{xyz}	total antisymmetric tensor
$\varepsilon, \varepsilon_n$	semiconductor dielectric constant
ε_o	free space dielectric constant
θ	angle between $\vec{k}$, $\vec{k}'$
κ	thermal conductivity of free carriers
λ_{opt}	mean free path of electron when scattered by optical phonons
μ	mobility of charge carriers
$\hat{\mu}$	mobility tensor
μ_o	zero field mobility
Ξ_u, Ξ_d	components of deformation potential tensor
ρ	mass density
σ	conductivity
σ_{ij}	components of conductivity tensor
$1/\tau$	scattering rate
τ_{abs}, τ_{em}	relaxation times for absorption and emission of phonons
τ_ε	energy relaxation time
τ^{i-j}	intervalley scattering time
$1/\tau_m$	momentum relaxation rate

$<\tau_m>$ average momentum relaxation time

τ_{ph} phonon relaxation time

ω frequency of phonons

ω_o frequency of optical phonons

BIBLIOGRAPHY

Asche, M. and Borzeskowski, J. V., 1970, Phys. Stat. Sol. 37:433.
Baraff, G. A., 1962, Phys. Rev. 128:2507.
Butcher, P. N., and Fawcett, W., 1966, Phys. Letters 21:489.
Conwell, E. M., 1967, "High Field Transport in Semiconductors," Academic Press, New York.
Dumke, W. P., 1967, Phys. Rev. 167:783.
Fang, F., and Fowler, A. B., 1970, Phys. Rev. 41:1825
Ferry, D. K., 1976, Phys. Rev. B14:5364.
Ferry, D. K., 1978, Sol.-State Electr. 21:115.
Goetzberger, A., McDonald, B., Haitz, R. H., and Scarlett, R. M., 1963, J. Appl. Phys. 34:1591.
Gunn, J. B., 1963, Sol. State Commun. 1:88.
Heinrich, H., and Jantsch, W., 1976, Phys. Letters 57A:485.
Heinrich, H., and Kriechbaum, M., 1970, J. Phys. Chem. Sol. 31:927.
Heinrich, H., Jantsch, W., and Rozenbergs, J., 1975, Sol. State Commun. 17:1145.
Hess, K., 1975, Sol. State Commun. 17:157.
Hess, K., 1978, Sol.-State Electr. 21:123.
Hess, K., Neugebauer, T., Englert, T., Landwehr, G., and Dorda, G., 1977, Phys. Rev. B16:3652.
Hess, K., and Sah, C. T., 1974, Phys. Rev. B10:3375.
Hess, K., and Sah, C. T., 1975, Sol.-State Electr. 18:667.
Hess, K., and Sah, C. T., 1978, IEEE Trans. Electr. Dev. ED-25:1399.
Hess, K., and Vogl, P., 1979, Sol. State Commun., in press.
Hilsum, C., 1962, Proc. IRE 50:185.
Holonyak, N., Jr., Kolbas, R. M., Laidig, W. D., Dupuis, R. D., and Dapkus, P. D., Appl. Phys. Letters 34:501.
Jacoboni, C., Canali, C., Ottaviani, G., and Albiergi-Quaranta, A. A., 1977, Sol.-State Electr. 20:77.
Kolbas, R. M., Holonyak, N., Jr., Vojak, B. A., Hess, K., Altarelli, M., Dupuis, R. D., and Dapkus, P. D., 1979, Sol.-State Commun., in press.
Lax, M., 1960, Phys. Rev. 119:1502.
Littlejohn, M. A., Hauser, J. R., Glisson, T., Ferry, D. K., and Harrison, J. W., 1978, Sol.-State Electr. 21:107.
McCumber, D. E., andChynoweth, A. G., 1966, IEEE Trans. Electr. Dev. ED-13:4.
Ning, T., 1978, Sol.-State Electr. 21:273.
Ridley, B. K., and Watkins, T. B., 1961, Proc. Phys. Soc. 78:293.
Ruch, J. G., and Kino, G. S., 1967, Appl. Phys. Letters 10:40.

Ryder, E. J., 1953, Phys. Rev. 90:766.
Sah, C. T., 1976, Sol.-State Electr. 19:975.
Sasaki, W., Shibuya, M., and Mizuguchi, K., 1958, J. Phys. Soc. Jpn. 13:465.
Seeger, K. H., 1973, "Semiconductor Physics," Springer-Wien, New York.
Shah, J., 1978, Sol.-State Electr. 21:43.
Shockley, W., 1961, Sol.-State Electr. 2:36.
Sze, S. M., 1969, "Physics of Semiconductors and Semiconductor Devices," Wiley-Interscience, New York.
Wolff, P. A., 1954, Phys. Rev. 95:1415.

ELECTRONIC STRUCTURE OF SEMICONDUCTORS

C. Calandra

Istituto di Fisica
Università di Modena
Modena, Italy

The subject of the electronic structure of solids is quite extensive and cannot be exhausted in few lectures. Advances in this field have been enormous in the last few years, both in the development of new theoretical approaches and in the application of sophisticated experimental techniques. My aim in this chapter is to give an introduction to the topic of the electronic structure of crystalline semiconductors, mainly emphasizing important concepts, rather than technical details. To this end I have done a rather personal selection of topics, having in mind the illustration of the basic concepts and problems, whose appreciation is required to understand the behavior of the valence electrons in semiconductors. The exposition falls into two parts, the first being devoted to the one electron description of the electronic structure and the second going into the problems of the many body effects in pure and extrinsic materials. The aim of the first part is to provide a general background about the semiconducting phase of a crystal, i.e. the main features of the chemical bond, the mechanism responsible in its formation and the quantities of interest in studying transport properties. The level of the second part is a little higher: many body theory is considerably more complicated than the independent particle description. To illustrate the results concerning the inclusion of self-energy effects, an outline of the Green's function method is required. Modifications in the band structure arise when an allowance of electron-electron repulsion is made, and are very difficult to evaluate. The problems connected with this aspect of the theory are discussed in the text. Finally, it is convenient to mention that the subject is covered in many books and articles, to which the reader is referred for a deeper understanding.

THE SINGLE PARTICLE DESCRIPTION: BONDS AND BANDS

Bond Alternation in One-Dimensional Solids

Although strictly one-dimensional systems, where the valence electrons are forced to move along linear chains of atoms, are seldom encountered in practice, it is convenient to start our description of the electronic structure of semiconductors from one-dimensional problems. We do this not only because the results for these systems provide a simple and useful framework in which to discuss some general aspects of semiconductor physics, but also because they apply to a number of compounds, mainly organic materials, some of which show interesting transport properties. An example is provided by long conjugated carbon chains as polyenes, which are characterized by a large number of unsaturated (double) bonds:

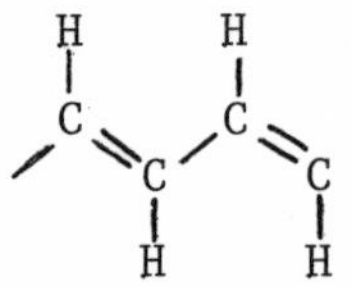

From the standpoint of the electronic structure this means that three of the four valence electrons of each carbon atom are in hybridized sp^2 orbitals and form saturated bonds (σ bonds) with the neighbouring C and H atoms; the fourth electron (π electron) has a $2p_z$ symmetry (we take the z-axis orthogonal to the plane of the molecule) and gives rise to a band of states, well-separated in energy from the σ-band. The one-dimensional character of these systems is due to the fact that π electrons are mobile over the whole chain and any interchain interaction is rather weak, so that a description of their behaviour in terms of localized one-dimensional band states is appropriate.

The most important feature of these systems is that the energy ΔE of the first optical transition does not vanish as the number of C atoms in the chain is increased, but remains finite, being approximately 2.2 eV for chains of several atoms. This indicates that the system stabilizes in a semiconducting rather than in a metallic state. According to the traditional chemical description this semiconducting state is achieved in a configuration of alternatively long (single) and short (double) bonds. This is simply understood in terms of the Huckel theory [see, e.g., Salem (1966)]. We write the one electron wave function in a tight binding form

$$\Psi_k(x) = c_1\Phi_1(k,x) + c_2\Phi_2(k,x), \tag{1}$$

where

$$\Phi_\alpha(k,x) = N^{-\frac{1}{2}} \sum_n \exp(ikna)\chi_\alpha(x-na) \qquad (2)$$

are Bloch combinations of atomic $2p_z$ wavefunctions centered on the first (second) atom of the unit cell. The coefficient c_1 and c_2 are to be determined through the conditions

$$(\chi_\alpha H \Psi) - \varepsilon_k(\chi_\alpha \Psi) = 0 . \qquad (3)$$

Calling α_1 and α_2 the Coulomb integrals for the carbon atoms in the unit cell, β_1 and β_2 the hopping integrals on the long and short bonds respectively and assuming vanishing overlap integrals, we arrive at the following set of equations

$$\begin{aligned} (\alpha_1 - \varepsilon_k)c_1 + \{\beta_1 + \beta_2\exp(-ika)\}c_2 &= 0, \\ (\alpha_2 - \varepsilon_k)c_2 + \{\beta_1 + \beta_2\exp(\ ika)\}c_1 &= 0, \end{aligned} \qquad (4)$$

from which the two band solution can be derived

$$\varepsilon_{\pm k} = \frac{\alpha_1 + \alpha_2}{2} \pm \{(\frac{\alpha_2 - \alpha_1}{2})^2 + \beta_2^2 + \beta_1^2 + 2\beta_2\beta_1\cos(ka)\}^{\frac{1}{2}}$$

Here k varies inside the Brillouin zone and the signs refer to the conduction (+) and valence (-) bands respectively. With one electron per site the valence band is filled and the conduction band is empty, the system being thus in a semiconducting state. The smallest gap occurs at the zone boundary ($ka = \pi$) and is

$$E_g = 2\{(\frac{\alpha_1 - \alpha_2}{2})^2 + (\beta_2 - \beta_1)^2\}^{\frac{1}{2}} \qquad (5)$$

For the special case $\alpha_1 = \alpha_2$ (homoatomic chain) we get

$$E_g = 2(\beta_2 - \beta_1) \qquad (6)$$

In spite of the particular model we chose, some of the results obtained are quite general. For this reason we shall spend a few words to comment on them.

First we notice that the presence of the gap in the electronic spectrum stems from the joint effect of the difference in the Coulomb integrals α_1 and α_2 and of the inter-atomic hopping integrals β_1 and β_2. Calling the first term ionic gap E_i and the second one covalent gap E_h, we can write

$$E_g = (E_i^2 + E_h^2)^{\frac{1}{2}}$$

a formula first given by Phillips (1970) in his theory of the covalent bond. To explain the above definitions we note that $\alpha_1 = \alpha_2$ implies that the two atoms in the unit cell behave in the same way and there is inversion symmetry with respect to an origin taken halfway between atoms. A difference in the Coulomb integrals corresponds to the introduction of some ionicity in the bond, since any inversion symmetry is cancelled. The difference in the hopping integrals is a measure of the covalent character of the bond. For a homoatomic chain and $\beta_1 = \beta_2$, the gap vanishes and we have a metallic phase. In this sense the ratio between the hopping integrals is a measure of the localization of the bond. Fairly localized states are present for strong bond alternation and correspond to a pronounced covalent character of the bond between the atoms in the unit cell.

To stress the generality of the above results we can notice that bond alternation is only a particular aspect of a more general kind of instability, which tends to favor the semiconducting rather than the metallic state, known as a Peierls instability [Peierls (1955)]. In a number of recently discovered materials, such as mixed valence planar transition metal compounds or the organic salts of tetracyanoquinodimethane, it is a Peierls distortion of the lattice, which originates the insulating phase [Belinsky (1976), Glaser (1974)]. A similar mechanism seems to be responsible of the reconstruction of surface atoms in the low index surfaces of semiconductors [Tosatti (1975)] and of a presence of a large optical absorption within the gap, due to single particle transitions between filled and empty surface states.

To illustrate the nature of this instability, it is best to begin by considering a system of free electrons in one dimension, whose hamiltonian is given by:

$$H = \Sigma_k \ \varepsilon_k \ a_k^+ a_k \tag{7}$$

where a_k^+ and a_k are creation and annihilation operators for electron states of energy ε_k and wave vector k. The one electron wavefunctions are simply plane waves, leading to a uniform distribution of the valence charge in the solid. It is easy to show that the electronic system described by (7) is unstable to any arbitrary small force of wave vector $2k_f, k_f$ being the Fermi momentum. Suppose in fact we apply a longitudinal external potential $\phi(x)$ to the electronic system: this will induce charge density fluctuations in the electron distribution. Assuming a sinusoidal dependence of the perturbation with a wavevector q, the system hamiltonian becomes

$$H = \Sigma_k \varepsilon_k a_k^+ a_k + \phi_q \rho_{-q} + \phi_{-q}\rho_q, \tag{8}$$

where ρ_q is the Fourier component of the induced charge density with momentum q. We can now express ρ_q in terms of the external potential by using linear response theory and the self-consistent field approximation. According to this theory

$$\rho_q = \chi(q)\phi_q, \tag{9}$$

where $\chi(q)$ is the density response function given by [Harrison (1967)].

$$\chi(q) = \Sigma_k \frac{f_k - f_{k+q}}{\varepsilon_k - \varepsilon_{k+q}} . \tag{10}$$

Here f_k are the equilibrium occupation numbers for the Fermi statistics. The main point is that $\chi(2k_f)$ diverges. To show this, we notice that for $|k| \sim k_f$, we can expand single particle energies linearly in k as

$$\varepsilon_k \sim \varepsilon_f + (|k| - k_f)\frac{k_f}{m}\hbar^2 . \tag{11}$$

Under such a condition, we have that

$$\varepsilon_{k-2k_f} - \varepsilon_f = -(\varepsilon_k - \varepsilon_f), \tag{12}$$

close to the Fermi energy. Assuming that the linear approximation holds in a range $-\Delta \leq \varepsilon \leq \Delta$, we can write $(2k_f)$ as:

$$\chi(2k_f) = -\frac{L\,m}{\pi\hbar^2 k_f}\int_0^\Delta \frac{d\varepsilon}{\varepsilon}\tanh(\varepsilon/2k_B T) + I \tag{13}$$

where I indicates the contribution to the response function from outside the range where the linear approximation holds and we replaced the sum over k with an integration. Neglecting I, which is small at low temperature, we get

$$\chi(2k_f) = -\frac{L\,m}{\pi\hbar^2 k_f}\ln(k_B T/1.14\Delta) \tag{14}$$

which diverges logarithmically at T = 0. It follows that any arbitrary small potential of wavevector $2k_f$ produces a finite distortion of the electron density at sufficiently low temperatures. Thus the electronic system is unstable and charge density waves tend to form with wave vector $2k_f$. The external perturbation giving rise to this charge distortion can be caused by the electron-phonon interaction. A lattice vibration of periodicity π/k_f and very low energy (soft-mode) can produce the required electronic instability. Due to the Coulomb interaction, ions will follow the electrons and a bond alternation or some other atomic distortion will occur, and favors the semiconducting phase.

The new electron eigenvalues will then be given by the secular equation

$$\begin{vmatrix} \varepsilon_k - E & \phi_q \\ \phi_{-q} & \varepsilon_{k-q} - E \end{vmatrix} = 0 \tag{15}$$

leading to the usual two band solution. The electron energy levels are split for wave vectors $k = \pm q/2$ and the potential ϕ_q creates a gap of the size of approximately $2\phi_q$ in the electronic spectrum. For $q = 2k_f$ the energy gaps occur at the Fermi energy, as expected.

Because of the Peierls instability one expects one-dimensional systems to prefer the semiconducting rather than the metallic phase at T = 0. For two- and three-dimensional systems, the occurrence of the charge density waves is less probable, since $\chi(q)$ shows weak singularities at $2k_f$. In systems as KCP or TTF-TCNQ, which show quasi one-dimensional behaviour, the Peierls distortion has been experimentally observed by diffuse X-ray scattering [Conies, et al., (1973)].

Bond Properties of Covalent Solids

Turning to three-dimensional systems, as the elemental semiconductors of group IV or III-V compounds, we can ask whether a bond picture, as the one sketched in the previous section, work in these cases. The problem has been investigated by several authors, notably Hall (1952), Leman (1962), and Weaire and Thorpe (1971). Recently Harrison (1979) has shown that a LCAO description of the energy bands can provide a useful tool to understand many important properties of semiconducting materials, as cohesive energies, effective charges, surface states, etc. In the following we shall restrict our attention to cubic semiconductors of group IV, as Si and Ge. The crystal structure in these cases is composed by two f.c.c. sublattices, with every atom tetrahedrally surrounded by atoms belonging to a different sublattice. The positions $\underline{R}_j$ of the atoms in a given sublattice are connected to those of the atoms of the other sublattice by $\underline{R}_j = \underline{R}_j + \tau$, where $\tau \equiv a/4\ (111)$. There are two atoms per unit cell separated by $\underline{\tau}$. The coordinates of the four neighbours of the atom at $\underline{R}_j$ are given by $\underline{R}'_j + \underline{a}_n$ (n = 0,1,2,3), where $\underline{a}_n$ are the following vectors:

$$\underline{a}_o \equiv (0,0,0),\ \underline{a}_1 \equiv a(110)/2,\ \underline{a}_2 \equiv a(101)/2,\ \underline{a}_3 \equiv a(011)/2.$$

The simplest possible description of the electronic structure of these systems is provided by the so called linear combination of hybrids. To illustrate it we notice that each atom in the unit cell has four atomic states and four electrons, so that a linear combination of atomic orbitals is expected to give rise to eight bands, half of which are filled in the ground state. In tetrahedral solids, rather than combining s and p atomic orbitals separately, it is convenient to combine sp^3 hybrids, which are orthogonal each other and lean in the direction of the nearest neighbor atoms [Coulson (1961)]. The hybrids and the directions they lean are

$$\begin{aligned} \chi_1 &= \tfrac{1}{2}(s+p_y+p_x+p_z) \text{ with } (111) \text{ orientation} \\ \chi_2 &= \tfrac{1}{2}(s-p_x-p_y+p_z) \quad " \quad (\bar{1}\bar{1}1) \quad " \\ \chi_3 &= \tfrac{1}{2}(s-p_x+p_y-p_z) \quad " \quad (\bar{1}1\bar{1}) \quad " \\ \chi_4 &= \tfrac{1}{2}(s+p_x-p_y-p_z) \quad " \quad (1\bar{1}\bar{1}) \quad " \end{aligned} \tag{16}$$

It is easy to show that these states are orthonormal and the expectation value of the single atom hamiltonian with respect to hybrids is

$$\varepsilon_h = \frac{E_s + 3E_p}{4}, \tag{17}$$

where E_s and E_p are the energies of s and p levels respectively. Because the hybrids are not eigenstates of H, there are matrix elements between them, given as

$$(\chi_\alpha H \chi_\beta) = (E_s - E_p)/4 = -V_1 \tag{18}$$

which is often the <u>promotion energy</u>. In the crystal, hybrids on different atoms are mixed through hopping integrals, which are responsible for broadening the atomic levels into energy bands. To show this we write the one electron Bloch function as

$$\Psi_{\underline{k}}(\underline{r}) = (2N)^{-\frac{1}{2}} \Sigma_j \exp(i\underline{k}\cdot\underline{R}_j)\ \Sigma_\alpha\{C_\alpha\chi_\alpha(\underline{r}-\underline{R}_j) + C'_\alpha\chi_\alpha(\underline{r}-\underline{R}'_j)\}. \tag{19}$$

Using conditions (4), together with the assumption of neglecting overlap integrals between atomic wavefunctions on different sites, and retaining only hopping integrals between hybrids on neighbor atoms and pointing at each other

$$\int \chi_\alpha(\underline{r}) V(\underline{r}) \chi_\beta(\underline{r}-\tau-a_\beta) = -V_2 \tag{20}$$

one arrives at the following secular equation

$$\begin{vmatrix} \varepsilon+V_1 & V_1 & V_1 & V_1 & V_2\gamma_o & 0 & 0 & 0 \\ V_1 & \varepsilon+V_1 & V_1 & V_1 & 0 & V_2\gamma_1 & 0 & 0 \\ V_1 & V_1 & \varepsilon+V_1 & V_1 & 0 & 0 & V_2\gamma_2 & 0 \\ V_1 & V_1 & V_1 & \varepsilon+V_1 & 0 & 0 & 0 & V_2\gamma_3 \\ V_2\gamma_o^* & 0 & 0 & 0 & \varepsilon+V_1 & V_1 & V_1 & V_1 \\ 0 & V_2\gamma_1^* & 0 & 0 & V_1 & \varepsilon+V_1 & V_1 & V_1 \\ 0 & 0 & V_2\gamma_2^* & 0 & V_1 & V_1 & \varepsilon+V_1 & V_1 \\ 0 & 0 & 0 & V_2\gamma_3^* & V_1 & V_1 & V_1 & \varepsilon+V_1 \end{vmatrix} = 0, \tag{21}$$

where $\gamma_n = \exp(i\underline{k} \cdot \underline{a}_n)$ and the energies are referred to E_p. The solution of this secular equation can be obtained analytically. For every $\underline{k}$ we get eight energies, which represent the eight bands arising from the atomic orbitals of the two atoms in the unit cell. Among them two are flat, showing no dispersion along the Brillouin zone, and doubly degenerate. Their energy is

$$\varepsilon_1 = E_1 - E_p = V_2 \qquad \varepsilon_2 = E_2 - E_p = -V_2 \tag{22}$$

and they are pure p bands. The other four bands arise from sp-hybridization and have energies given by

$$\begin{aligned} \varepsilon_3 &= -2V_1 + (4V_1^2 + V_2^2 + 4V_1V_2f)^{\frac{1}{2}} \\ \varepsilon_4 &= -2V_1 - (4V_1^2 + V_2^2 + 4V_1V_2f)^{\frac{1}{2}} \\ \varepsilon_5 &= -2V_1 + (4V_1^2 + V_2^2 - 4V_1V_2f)^{\frac{1}{2}} \\ \varepsilon_6 &= -2V_1 - (4V_1^2 - V_2^2 - 4V_1V_2f)^{\frac{1}{2}} \end{aligned} \tag{23}$$

where f is given by

$$f = \tfrac{1}{2}[1+\cos(k_x a/2)\cos(k_y a/2) + \cos(k_x a/2)\cos(k_y a/2) + \cos(k_y a/2)\cos(k_z a/2)].$$

These are simple bands and can accomodate two electrons per unit cell. The resulting band structure is rather different according to whether V_2 is lower than $2V_1$ or *vice-versa*. In the first case, ($V_2 < 2V_1$), the valence band bottom is $\varepsilon_4(\underline{k} = 0) = -4V_1 - V_2$ and the two lowest bands are given by:

$$-4V_1 - V_2 \le \varepsilon_4 \le -2V_1 - (4V_1^2 + V_2^2)^{\frac{1}{2}},$$
$$-2V_1 - (4V_1^2 + V_2^2)^{\frac{1}{2}} \le \varepsilon_6 \le -4V_1 + V_2, \tag{24}$$

while the other two sp-bands vary in the range

$$-V_2 \le \varepsilon_5 \le -2V_1 + (4V_1^2 + V_2^2)^{\frac{1}{2}},$$
$$-2V_1 + (4V_1^2 + V_2^2)^{\frac{1}{2}} \le \varepsilon_6 \le V_2, \tag{25}$$

These two bands are degenerate with the flat p-bands at the centre of the Brillouin zone. It follows that in this case there are two groups of bands separated by a gap $E_g = 2(V_1 - V_2)$. The first group can accomodate two electrons per spin component. The other four electrons are distributed in the second group of bands, which turns out to be a half-filled continuum of bands. The resulting band structure has a metallic character, the lowest bands having a dominant s-contribution, while those belonging to the second group have a dominant p-character.

A significantly different situation occurs when $2V_1 < V_2$. In this case, the bands are split into two groups of four bands separated by a gap. All the electrons are accomodated in the bands of the first group, which gives rise to a typical semiconducting phase, with the valence band varying in the range

$$-V_2 - 4V_1 \le \varepsilon_4 \le -2V_1 - (4V_1^2 + V_2^2)^{\frac{1}{2}} \le \varepsilon_6 \le \varepsilon_2 \le -V_2 \tag{26}$$

and the conduction band

$$-V_2 + 4V_1 \le \varepsilon_5 \le -2V_1 + (4V_1^2 + V_2^2)^{\frac{1}{2}} \le \varepsilon_3 \le \varepsilon_1 \le V_2. \tag{26'}$$

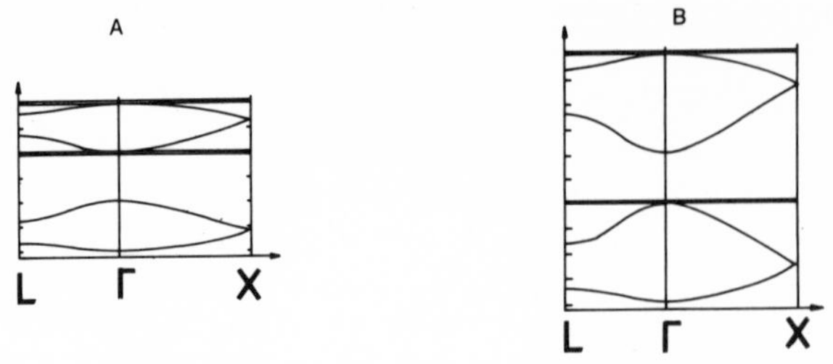

Fig. 1. Band structure of a diamond structure semiconductor in the bond model: A) $|V_2/V_1| = 1$; B) $|V_2/V_1| = 3$.

The fact that only for $V_2 \gtrsim 2V_1$ do we get a semiconducting phase for the diamond structure is essentially due to the crossing of the s and p bands, which occurs for increasing V_2 thus making the promotion energy smaller and smaller. For large V_2 in fact, the top of the valence band turns out to be given by the flat p band of energy $-V_2$, while the bottom of the conduction band has a predominant s-character. It is this crossing, called hybridization crossing, which is responsible of the existence of a gap between empty and filled band states. Typical band structures obtained for various ratios of V_2/V_1 are given in Fig. 1.

This is a particular case of a rather general kind of metal-insulator transition due to band structure effects. It occurs when some parameter, such as the atomic volume or the composition, changes in such a way that two bands overlap, producing a full valence band and an empty conduction band with a gap between them. This is the mechanism which causes the metal-semiconductor transition in the divalent metals Mg,Ba,Yb under pressure. For example, Yb has a semiconducting behavior above 10 kbar with a resistivity of 10^{-3} Ω-cm at room temperature [Mott (1974)].

If we now ask whether the description of the electronic structure provided by this approach is appropriate for the understanding of transport and optical properties, we have to conclude that the LCAO model in this simple form is not a good one. The top of the valence band turns out to be flat, giving infinite effective masses for the holes; moreover, the direct gap is not nearly constant over the Brillouin zone, as occurs in realistic band structure calculations of Si (see Fig. 2) but varies by a factor of two on passing from the centre to the border of the zone. Finally the conduction band turns out to be very narrow, compared with the free-electron-like dispersion of the energies of the lowest excited states in Si and Ge.

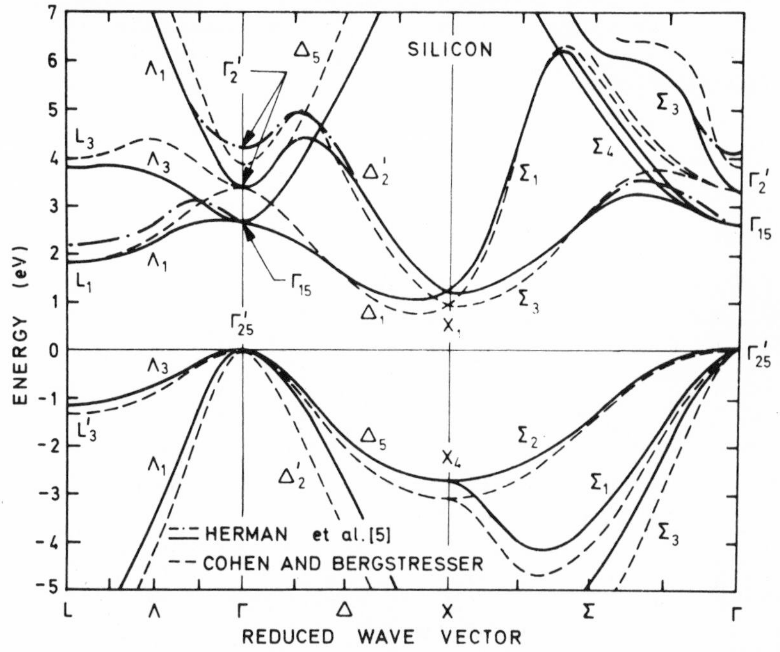

Fig. 2. Band structure of Silicon as obtained from different theoretical calculations (after A. Antoncik in "Electrons in Crystalline Solids" Vienna IAEA (1973)).

All these shortcomings can be eliminated by using a more realistic description of the electronic structure. If use is made of such a LCAO scheme, one has to include several hopping integrals before achieving reasonably accurate energies and effective masses [Reggiani and Calandra (1973)]. The point is that the valence charge distribution in Si and Ge crystals is considerably more uniform than that obtained from the simple bond picture. Although significant piling up of charge along the nearest-neighbour direction is present, the bonding charge is not so localized. This explains why theories based on a free-electron description of the chemical bond are appropriate to describe the electronic structure of these materials [Cohen and Heine (1970), Walter and Cohen (1971), Bertoni, et al., (1973)]. In treating perturbations to the electronic structure caused by extended defects, however, as surfaces or dislocations, the chemical approach turns out to be rather appropriate. Electronic surface states show a behavior rather similar to that of hybridized dangling-bond orbitals. The same is true for states localized on a dislocation lines [Bortolani, et al., (1973)]. Changes in hybridization can account for the modifications induced on the electronic structure by surface reconstruction or by atomic displacements along the dislocation line, while a description of these effects in terms of nearly free-electron states, although correct, is considerably more complicated.

Bands and Effective Masses

As we mentioned previously, a number of methods to calculate bulk band structures in semiconductors have been proposed in the last few years, which have led to reliable and accurate descriptions of the electronic structure of these materials. All these methods aim at the solution of the single particle equation

$$\{-\hbar^2 \frac{\nabla^2}{2m} + V_c(\underline{r})\; \psi_{\underline{k}n}(\underline{r}) = E_{\underline{k}n}\; \psi_{\underline{k}n}(\underline{r}), \tag{27}$$

where $V_c(\underline{r})$ is the crystal potential, i.e. the potential felt by an electron travelling in the crystal. In some cases this equation is replaced by a secular problem, where pseudopotentials and pseudo-wavefunctions are involved, the energies of the bands being the same [Harrison (1979)]. From the standpoint of the many-body hamiltonian the use of a single particle model to describe properties of the valence electrons is a rather drastic assumption. We will discuss this point, together with the prescriptions commonly used to evaluate the crystal potential, in the next section.

An analysis of the various theoretical approaches used in the calculation of the Bloch functions is beyond the purpose of these lectures. On the other hand, for the purpose of understanding transport properties a detailed analysis of the whole energy band structure is not necessary. When discussing transport phenomena, one is most interested in the top of the valence band and in the bottom of the conduction band only. A convenient and conceptually simple method to study the shape of an energy band round its edge in detail is provided by the so-called $\underline{k} \cdot \underline{p}$ method; it makes possible to estimate the behavior of the energy bands in the proximity of a point $\underline{k}_o$ of the Brilluoin zone, using a small number of parameters, which can be obtained from the experiments.

To illustrate this approach, suppose we know the energy bands and wavefunctions at some particular point $\underline{k}_o$. Then we look for these quantities at a neighbouring point $\underline{k}$. The $\underline{k} \cdot \underline{p}$ method allows us to relate energies and wavefunctions at these two points. To this end we introduce the set of functions

$$\chi_j(\underline{k},\underline{r}) = \exp\{i(\underline{k}-\underline{k}_o) \cdot \underline{r}\}\psi_{\underline{k}_o j}(\underline{r}) \tag{28}$$

where $\psi_{\underline{k}_0 j}$ is the wavefunction for the band j at $\underline{k}_0$. It can be shown that the set formed by the χ's is a complete and orthonormal set. Moreover, since $\psi_{\underline{k}_0 j}$ is a Bloch function, we have also

$$\chi_j(\underline{k},\underline{r}) = \exp(i\underline{k} \cdot \underline{r}) u_{\underline{k}_0}(\underline{r}), \tag{29}$$

where $u_{\underline{k}_0 j}$ is the strictly periodic part of $\psi_{\underline{k}_0 j}$. Equation (29) shows that χ_j satisfies the Bloch theorem for a state of wavevector $\underline{k}$. The unknown Bloch function $\psi_{n\underline{k}}$ giving the electron wavefunction for the n-th band at $\underline{k}$ can be expanded as

$$\psi_{n\underline{k}}(\underline{r}) = \Sigma_j\, c_{nj}(\underline{k})\, \chi_j(\underline{k},\underline{r}). \tag{30}$$

Inserting (30) into (27), we get the following equation for $u_{n\underline{k}}$

$$\Sigma_j c_{nj}(\underline{k})\{E_{\underline{k}_0 j} + \frac{\hbar}{m}(\underline{k}-\underline{k}_0) \cdot \nabla + \frac{\hbar^2}{2m}(\underline{k}^2-\underline{k}_0^2)\}u_{j\underline{k}_0} = E_{n\underline{k}}\Sigma_j c_{nj}(\underline{k})u_{j\underline{k}_0} \tag{31}$$

By multiplying by $u^*_{i\underline{k}_0}$ and integrating over the unit cell of the crystal, we obtain

$$\Sigma_j\{(E_{j\underline{k}_0} - E_{n\underline{k}} + \frac{\hbar^2}{2m}(\underline{k}^2-\underline{k}_0^2))\delta_{ji} + \frac{\hbar}{m}(\underline{k}-\underline{k}_0)\cdot\underline{p}_{ij}\}c_{nj} = 0, \tag{32}$$

where

$$\underline{p}_{ij} = \frac{(2\pi^3)}{\Omega}\int u_{j\underline{k}_0}(\underline{r})\ \underline{p}u_{j\underline{k}_0}(\underline{r})$$

are the momentum matrix elements. Equation (32) represents a set of simultaneous linear and homogeneous equations (one equation for every value of the band index i) which has nontrivial solutions only if the determinant of the coefficients vanishes

$$|H_{ji} - E_{\underline{k}}\delta_{ji}| = 0, \tag{33}$$

with

$$H_{ji} = \{E_{j\underline{k}_0} + \hbar^2(\underline{k}^2-\underline{k}_0^2)/2m\}\delta_{ji} + \hbar/m(\underline{k}-\underline{k}_0) \cdot \underline{p}_{ij}. \tag{33'}$$

Up to now no approximation has been introduced in the theory; if all the bands are included in the expansion, this is exact procedure, in principle, to get energies and wavefunctions throughout the Brillouin zone. To do this we must be able to evaluate the momentum matrix elements $\underline{p}_{ij}$. Theoretical information do not usually allow such a precise determination of $\underline{p}_{ij}$. However, if there is enough experimental information to fix the values of the matrix elements, this procedure can be used to generate the energy band structure.

An alternative approach consists in using a perturbation expansion, taking

$$V_k = \hbar(\underline{k}-\underline{k}_o) \cdot \underline{p}/m + \hbar^2(\underline{k}-\underline{k}_o)^2/2m$$

as a small perturbation and expressing u_{nk} in terms of $u_{j\underline{k}_o}$ by using either degenerate or nondegenerate perturbation theory, according to the degeneracy of the unperturbed level $E_{j\underline{k}_o}$. As an example, we calculate the energy bands at $\underline{k}$ assuming a nondegenerate level at $\underline{k}_o$. By retaining the lowest terms in a Rayleigh-Schrodinger perturbation expansion, we get

$$E_n(\underline{k}) = E_n(\underline{k}_o) + \hbar(\underline{k}-\underline{k}_o) \cdot \underline{p}_{nn}/m + \hbar^2(k^2-k_o^2)/2m$$

$$+ \frac{\hbar^2}{2m} \Sigma_j' \{((\underline{k}-\underline{k}_o) \cdot \underline{p}_{nk})(\underline{k}-\underline{k}_o) \cdot \underline{p}_{jn}/(E_{n\underline{k}_o} - E_{j\underline{k}_o})\}, \qquad (34)$$

and for the wave function

$$u_{n\underline{k}}(\underline{r}) = u_{n\underline{k}_o}(\underline{r}) + (\hbar/m)(\underline{k}-\underline{k}_o) \cdot \Sigma_j' \{\underline{p}_{jn} u_{j\underline{k}_o}(\underline{r})/(E_{n\underline{k}_o} - E_{j\underline{k}_o})\} \qquad (34')$$

Eq. (34) is commonly used to get the effective masses in case where the band n has an extremum at $\underline{k}_o$, so that the linear term vanishes. Calling s_α and s_β the components of the wavevector $(\underline{k}-\underline{k}_o)$ with respect to some fixed Cartesian axis and differentiating (34) twice, we get

$$\frac{m}{\hbar^2}\partial^2 E_n/\partial s_\alpha \partial s_\beta = \delta_{\alpha\beta} + 1/m \, \Sigma_j' (p^\alpha_{nj} p^\beta_{jn} + p^\beta_{nj} p^\alpha_{jn})/(E_{n\underline{k}_o} - E_{jk_o}). \qquad (35)$$

Defining a reciprocal effective mass tensor for the band n as

$$m/m_n^*|_{\alpha\beta} = \frac{m}{\hbar^2} \partial^2 E_n/\partial k_\alpha \partial k_\beta \qquad (36)$$

we get for the energy band the following dispersion

$$E_{n\underline{k}} = E_{n\underline{k}_o} + \tfrac{1}{2}\hbar^2 \, \Sigma_{\alpha\beta}(m/m_n)_{\alpha\beta}(k_\alpha - k_{o_\alpha})(k_\beta - k_{o_\beta})/m$$

As with any other tensor, the reciprocal effective mass tensor can always be diagonalized by a proper choice of coordinate axes. If $\underline{k}_o$ is a general point in the zone, the diagonalization depends on the details of the crystal potential. However, if the extremum occurs along a symmetry line or in a symmetry point of the Brillouin zone, the axes may be partially determined by symmetry. For example if $\underline{k}_o$ lies along the principal axes of a cubic crystal, the symmetry axis must also be a principal axis. In the case of (100) and (111) axis the surfaces of constant energy in the effective mass approximation must be ellipsoids of revolution, with a longitudinal effective mass referring to displacements along the symmetry axis and a transverse mass referring to perpendicular displacements. For $\underline{k}_o = 0$, the surface of constant energy in a cubic crystal must be spherical.

The case of the conduction band of Si is particularly interesting. The minimum of the conduction band lies along the (001) direction and symmetry considerations show that surfaces of constant energy in the neighbourhood of each of the conduction band minimum are prolate spheroids (ellipsoids) of revolution. There are six such ellipsoids oriented along the six equivalent (001) directions. If the x-axis is chosen parallel to the crystal axis the constant energy surfaces are given by

$$E_{n\underline{k}} + E_{n\underline{k}_o} + \hbar^2/2m_\ell(k_x - k_{ox})^2 + \hbar^2\{(k_y - k_{oy})^2 + (k_z - k_{oz})^2\}/2m_t$$

Here m_ℓ and m_t are the longitudinal and transverse effective mass respectively. For Si $m_\ell = 0.98m_0$ and $m_t = .19m_0$. Clearly the conduction band of Si (and in some other semiconductors as well) consists of several nondegenerate states, giving rise to the so called many-valley structure.

MANY-BODY EFFECTS ON THE ELECTRONIC STRUCTURE OF SEMICONDUCTORS

The theory sketched up to now relies heavily on the single particle description of the electron gas. At first view it is surprising that many problems of semiconductor physics can be handled with an independent particle model; the electron-electron interaction is certainly important in optical and transport properties or in studying the effects of external perturbations. The only possible answer to the question of why single particle theory enjoys such a large degree of success is that many features of the many-body problem can be included in such a single particle description.

The aim of the present section is to clarify this point and to show the limitations of the independent electron model in the description of the electronic structure of semiconductors.

Green's Function and Self-energy

The fundamental tool to investigate the properties of a many-body system is the Green's function method. A full description of it is beyond the scope of the present discussion and can be found in many important books and articles [See, e.g., Kodanoff and Baym (1962), Nozierres (1964), Hedin and Lundquist (1969)]. We are more interested in the results of the theory rather than in the detailed derivations. Let us start with the single particle Green's function, whose definition at T = 0 is

$$G(\underline{x}t;\ \underline{x}'t') = -i < T\{\psi(\underline{x}t)\psi^{+}(\underline{x}'t')\}>, \tag{36}$$

where $\underline{x}$ stands for both space and spin coordinates $\underline{x} \equiv (\underline{r},\sigma)$, $\psi(\underline{x}t)$ and $\psi^{+}(\underline{x}t)$ are electron field operators in the Heisenberg representation. The brackets stand for averaging with respect to the exact ground state, which we assume not to be degenerate, and T arranges the operator in such a way that time increases from right to left. A minus sign is supplied; if the order of the operators established by T is different from that presented in (36), then

$$\begin{aligned} T\{\psi(\underline{x}t)\psi^{+}(\underline{x}'t')\} &= \psi(\underline{x}t)\psi^{+}(\underline{x}'t') \quad && t' < t, \\ &= \psi^{+}(\underline{x}'t')\psi(\underline{x}t) \quad && t' > t. \end{aligned} \tag{37}$$

In the first case G describes the propagation of an extra-particle added to the system at $\underline{r}'t'$ and removed subsequently at $\underline{r}t$; specifically G gives the probability amplitude for finding an electron at $\underline{r}t$ with spin component σ once one has been added at $\underline{r}'t'$ with spin component σ'. For $t < t'$ the particle is first removed and then added: G describes the propagation of a hole. The relevance of the definition of the Green's function to the study of the electronic properties can be understood by noticing that the way in which we study the electronic properties of a system, for example in photoemission, optical or energy loss spectroscopy, is by removing electrons or creating holes and looking at the decay of the excited states.

As is the case of the Green's function of an ordinary differential equation, the one particle Green's function obeys a differential equation known as the equation of motion. This is easily obtained starting from the equation of motion of field operators in the Heisenberg representation, and it is given by

$$\{i \frac{\partial}{\partial t} + \frac{\nabla^2}{2m} + V_i\} \, G(\underline{x}t;\underline{x}'t') = \delta(\underline{x}-\underline{x}') \, \delta(t-t')$$

$$- i\int v(\underline{x},\underline{x}')K(\underline{x}''t;\underline{x}t;\underline{x}''t_+;\underline{x}'t')d\underline{x}'', \tag{38}$$

where V_i is the potential (or pseudopotential) felt by an electron due to all the ions in the crystal; $v(\underline{x},\underline{x}') = e^2/|\underline{r}-\underline{r}'|\delta_{\sigma\sigma'}$, is the Coulomb interaction between the electrons and $K(\underline{x}_1t_1;\underline{x}_2t_2;\underline{x}_3t_3;\underline{x}_4t_4)$ is the two particle Green's function, defined by

$$K(\underline{x}_1t_1;\underline{x}_2t_2;\underline{x}_3t_3;\underline{x}_4t_4) = - \langle T\{\psi(\underline{x}_1t_1)\psi(\underline{x}_2t_2)\psi^+(\underline{x}_3t_3)\psi^+(x_4t_4)\}\rangle,$$

gives information about states of N+2 particles. It can be interpreted as the propagator of a pair of excitations (electron-electron, electron-hole, hole-hole) in the same way as G gives indications on single particle excitations of the many-body system. As for G, we could derive a dynamical equation for K and this would involve the Green's function for a higher number of particles. Equation (38) could be taken as the starting point for generating an infinite chain of equations, introducing successively more complicated correlations. Such a chain could be determined using some convenient assumption about decoupling, but the procedure is rather cumbersome and we will not pursue it further. Rather, it is useful to decompose K in the following way:

$$K(\underline{x}_1t_1;\underline{x}_2t_2;\underline{x}_3t_3;\underline{x}_4t_4) = G(\underline{x}_1t_1;\underline{x}_3t_3)G(\underline{x}_2t_2;\underline{x}_4t_4)$$

$$-G(\underline{x}_1t_1;\underline{x}_4t_4)G(\underline{x}_2t_2;\underline{x}_3t_3) + \delta K(\underline{x}_1t_1;\underline{x}_2t_2;\underline{x}_3t_3;\underline{x}_4t_4). \tag{38'}$$

To understand the physical meaning of this decomposition, we note that if two indistinguishable particles propagate independently, K is the antisymmetrized product of two Green's functions and δK vanishes. Therefore δK, represents the interaction between the two particles. This can be written as

$$\delta K(\underline{x}_1t_1;\underline{x}_2t_2;\underline{x}_3t_3;\underline{x}_4t_4) = \int\int\int\int d\underline{x}_1'dt_1'd\underline{x}_2'dt_2'd\underline{x}_3'dt_3'd\underline{x}_4'dt_4' \times$$

$$\times G(\underline{x}_1t_1;\underline{x}_1't_1')G(\underline{x}_2t_2;\underline{x}_2't_2')\gamma(\underline{x}_1't_1';\underline{x}_2't_2';\underline{x}_3't_3';\underline{x}_4't_4')G(\underline{x}_3't_3';\underline{x}_3t_3)\times$$

$$\times G(\underline{x}_4't_4';\underline{x}_4t_4), \tag{38''}$$

We thus separate the interaction process into three steps: i) propagation from $\underline{x}_1t_1,\underline{x}_2t_2$ to $\underline{x}_1't_1',\underline{x}_2't_2'$; ii) interaction extending from $\underline{x}_1't_1',\underline{x}_2't_2'$ to $\underline{x}_3't_3',\underline{x}_4't_4'$; iii) propagation from $\underline{x}_3't_3',\underline{x}_4't_4'$ to $\underline{x}_3t_3,\underline{x}_4t_4$. The function describes the interaction between two elementary excitations. Inserting (38') into the dynamical equation, we get

$$\{i \frac{\partial}{\partial t} + \frac{\nabla^2}{2m} + V_i\}G(\underline{x}t;\underline{x}'t') = \delta(\underline{x}-\underline{x}')\delta(t-t') - i\int d\underline{x}''v(\underline{x},\underline{x}'') \times$$

$$\times\ G(\underline{x}''t;\underline{x}''t_+)G(\underline{x}t;\underline{x}'t') + i\int d\underline{x}''v(\underline{x},\underline{x}'')G(\underline{x}''t;\underline{x}'t')G(\underline{x}t;\underline{x}''t_+)$$

$$-i\int v(\underline{x},\underline{x}'')\delta K(\underline{x}''t;\underline{x}t;\underline{x}''t^+;\underline{x}'t')d\underline{x}''. \qquad (39)$$

By introducing the definition of <u>effective potential</u> $V(\underline{x}t)$, through the equation

$$V(\underline{x}t) = i\int d\underline{x}''v(\underline{x},\underline{x}'')G(\underline{x}''t;\underline{x}''t_+) = \int d\underline{x}''v(\underline{x},\underline{x}'')\langle\psi^+(\underline{x}''t)\psi(\underline{x}''t)\rangle, \qquad (40)$$

we can write (39) in the form

$$\{i \frac{\partial}{\partial t} + \frac{\nabla^2}{2m} + V_i - V(\underline{x},t)\}G(\underline{x}t;\underline{x}'t')$$

$$= \delta(\underline{x}-\underline{x}')\delta(t-t') + \int\Sigma(\underline{x}t;\underline{x}''t'')G(\underline{x}''t'';\underline{x}'t')d\underline{x}''dt'', \qquad (41)$$

where the operator Σ is called the <u>self-energy operator</u>. Its explicit expression can be easily written down from (38'-38").

In the following we assume the dependence upon time to be

$$G(\underline{x}t,\underline{x}'t') = G(\underline{x},\underline{x}';t-t'),$$

$$\Sigma(\underline{x}t;\underline{x}'t') = \Sigma(\underline{x},\underline{x}';t-t'),$$

so that we can write the following equation for the Fourier component of G

$$\{\omega + \frac{\nabla^2}{2m} + V_i - V(\underline{x})\}G(\underline{x},\underline{x}',\omega) = \delta(\underline{x}-\underline{x}')$$

$$+ \int\Sigma(\underline{x},\underline{x}'',\omega)G(\underline{x}'',\underline{x}',\omega)d\underline{x}''. \qquad (42)$$

If we define an operator

$$L(\underline{x},\underline{x}',\omega) = \frac{-\nabla^2}{2m} - V_i + V(\underline{x}) + \Sigma(\underline{x},\underline{x}',\omega),$$

(42) can be written as

$$\{\omega - L(\underline{x},\underline{x}',\omega)\}\ G(\underline{x},\underline{x}',\omega) = \delta(\underline{x}-\underline{x}').$$

In this form, (42) looks like the equation for the resolvent of an ordinary differential equation. The trouble here is that Σ is a complex, non-local and energy-dependent operator so that L is not self-adjoint. The main consequence of this fact is that the eigenvalues of L, which correspond with the poles of the single particle Green's function, are not real [Morse and Feshbach (1953)]; since single particle excitations have not an infinite lifetime, their energy has an imaginary part, which corresponds to the damping of the excited states due to the electron-electron interaction.

G is susceptible to a biorthogonal expansion, known as the spectral representation. To achieve it, we introduce a complete set of energy eigenstates for the N+1 and N-1 particle systems, which we call $|N+1,s\rangle$ $|N-1,s\rangle$. Because we are in a semiconductor, we have that

$$\varepsilon_s^e = E_{N+1,s} - E_N > \mu + E_g/2,$$

$$\varepsilon_s^h = E_{N-1,s} - E_N > \mu - E_g/2. \tag{43}$$

Here μ is the chemical potential, E_N is the energy of the ground state of the N particle system $|N\rangle$ and E_g is an energy, which in the single particle picture corresponds to the energy gap between valence and conduction bands. Defining

$$f_s^e = \langle N|\psi(\underline{x})|N+1,s\rangle \qquad f_s^h = \langle N-1,s|\psi(x)|N\rangle, \tag{44}$$

it can be easily shown that

$$G(\underline{x},\underline{x}'\omega) = \Sigma_s f_s^e(\underline{x}) f_s^{e*}(\underline{x}')/(\omega-\varepsilon_s^e)\ \theta(\omega-\mu-E_g/2) +$$

$$\Sigma_s f_s^h(\underline{x}) f_s^{h*}(\underline{x}')/(\omega-\varepsilon_s^h)\ \theta(\omega-\mu+E_g/2) \tag{45}$$

from which eigenvalue equations for ε_s and $f_s(\underline{x})$ are obtained

$$\{\varepsilon_s^e + \frac{\nabla^2}{2m} + V_i - V^e(\underline{x})\}\ f_s^e(\underline{x}) - \int \Sigma^e(\underline{x},\underline{x}',\varepsilon) f_s^e(\underline{x}') = 0 \tag{46}$$

$$\{\varepsilon_s^h + \frac{\nabla^2}{2m} + V_i - V^h(\underline{x})\}\ f_s^h(\underline{x}) - \int \Sigma^h(\underline{x},\underline{x}'\varepsilon) f_s^h(\underline{x}') = 0 \tag{46'}$$

These are the equations for the eigenvalues and eigenfunctions of the electrons and holes respectively [Pratt (1960)]. They are to be solved, if we want to know the single particles excitations of the many-body system.

Independent particle models rely upon approximation to the self-energy term in this equation. The most common are: A) Hartree approximation: self-energy effects are simply neglected. The two particle Green's function in the dynamical equation is approximated by K(1234) = G(13)G(24). The quasi-particles eigenvalues are given by the equation

$$\{\varepsilon_s + \frac{\nabla^2}{2m} + V_i - V(\underline{x})\} f_s(\underline{x}) = 0$$

and turn out to be real. Elementary excitations have an infinite lifetime. B) Hartree-Fock approximation: this is achieved by writing the two particle Green's function as in eq. (38') with $\delta K = 0$. Elementary excitations are given by the solution of the non-local equation

$$\{\varepsilon_s + \frac{\nabla^2}{2m} + V_i - V(\underline{x})\} f_s(\underline{x}) - i\int v(\underline{x},\underline{x}')\langle\psi^+(\underline{x})\psi(\underline{x}')\rangle f_x(\underline{x}')dx'=0,$$

and the eigenvalues are real. C) Density-functional approach: this approach derives from the theory of the inhomogenous electron gas [Hohenberg and Kohn (1964), Kohn and Sham (1965)], and it is based on a theorem establishing that the ground-state energy of any system of interacting electrons is uniquely determined if the charge density is specified. More precisely the ground state energy is a functional of the charge density $E = E\{\rho\}$. This suggests that one can replace the complex, non-local and energy-dependent self-energy operator with a local potential $V_{xc}(\underline{x})$, to be found as the functional derivative of the exchange and correlation contribution to the ground-state energy. However the functional $E\{\rho\}$ is generally unknown. For those systems where the valence-charge density is slowly varying, one expects V_{xc} to be well-approximated by expression derived from the electron gas theory, as a Kohn-Sham-Gaspar potential

$$V_{xc}(\underline{r}) = -4(3\rho/8\pi)^{1/3}, \tag{47}$$

or the Wigner formula

$$V_{xc}(\underline{r}) = -\rho^{1/3}\{0.984 + \frac{0.944 + 8.77\rho^{1/3}}{(1 + 12.57\rho^{1/3})}\}. \tag{47'}$$

Expressions of this kind are currently used in many theoretical calculations of the energy-band structure of metals and semiconductors. Elementary excitations are obtained by solving the equation

$$\{\varepsilon_s + \frac{\nabla^2}{2m} + V_i - V(\underline{x}) - V_{xc}(\underline{x})\} f_x(\underline{x}) = 0. \tag{47''}$$

With respect to the Hartree-Fock approximation, this latter approach has the advantage of eliminating some undesirable properties of the Hartree-Fock eigenvalues, such as the vanishing of the density-of-states at the Fermi surface in nearly-free electron materials. As a matter of fact, the Hartree-Fock approximation overestimates the effects of the electron-electron interaction, which turn out to be weaker when a more correct theory is used. The density-functional approach has provided a remarkably successful description of the electronic structure of many materials [Chelikowsky and Cohen (1974)].

Although the relation between the density-functional method and the full many-body theory has been clarified by several works, notably Lang (1973), effective local potentials as those in equations (47-47') do not have a simple connection with the many-body problem. Since they have been derived from an electron gas theory, they are expected to work in solids, where the charge density does not vary too drastically. In practice their range of validity is considerably larger and they work even in systems, as semiconductors and transition metals, where the valence-charge density undergoes significant variations.

Self-energy Effects in Band Structure Calculations

The reason why local exchange and correlation potentials can mimic quite well the non-local self-energy is still unclear. Some work aimed at elucidating this aspect has been done by Inkson (1973). We can go into the details of this work a little, since it can also be useful in understanding gap renormalization effects in heavily doped materials.

An important result of the many-body theory is the fact that it is possible to express the self-energy formally as a series expansion in powers of the dynamically screened interaction [Hedin and Lundquist (1969), Bassani et al. (1962)].

$$W(12) = \int v(13)\varepsilon^{-1}(32)d(3), \tag{48}$$

Here we have used an abbreviated notation

$$(1) = (\underline{x}_1 t_1), \ v(12) = v(|\underline{r}_1 - \underline{r}_2|)\delta(t_1 - t_2)\delta_{\sigma_1\sigma_2}.$$

The inverse dielectric function ε^{-1} measures the screening of the Coulomb interaction in the system. To calculate it we need to know the irreducible polarization propagator P

$$\varepsilon(12) = \delta(12) - \int P(32)v(13)d(3), \tag{48'}$$

which in turn implies the knowledge of the Green's function and of the dressed electron-electron interaction. By using these definitions it is possible to arrive at the following set of equations, which define the self-energy in terms of the dynamically screened interaction [Hedin and Lundquist (1969)]

$$\Sigma(12) = i\int W(1^{+}2)G(14)\Gamma(42;3)d(3)d(4) \tag{49a}$$

$$W(12) = v(12) + \int W(13)P(34)v(42)d(3)d(4) \tag{49b}$$

$$P(12) = -i\int G(23)G(42)\Gamma(34;1)d(3)d(4) \tag{49c}$$

$$\Gamma(12;3)=\delta(12)\delta(13)+\int\delta\Sigma(12)/\delta G(45)\times$$

$$\times G(46)G(75)\Gamma(67;3)d(4)d(5)d(6)d(7). \tag{49d}$$

The quantity Γ is called the vertex function and involves the functional derivative of the self-energy with respect to the Green's function. One of the advantages of expressing the many-body problem through this set of equations is that they can be used to generate expressions for Σ through an iterative procedure. One can for example start with the Hartree approximation ($\Sigma = 0$) and insert it into (49d), leading to a zero-order expression of the vertex function

$$\Gamma^{0}(12;3) = \delta(12)\delta(13). \tag{50a}$$

This can then be inserted into the first three equations leading to

$$\Sigma^{1}(12) = i\, W(1^{+}2)G(12) \tag{50b}$$

$$W^{1}(12) = v(12) + \int W(13)P(34)v(42)d(3)d(4) \tag{50c}$$

$$P^{0}(12) = -iG(21)G(12). \tag{50d}$$

From these equations one can deduce a value for $\delta\Sigma/\delta G$ and start with a new iteration. The expressions obtained in this way become increasingly complicated since the integral equation for Γ does not possess an explicit solution and no attempt to go beyond the first approximation in Σ has been made up to now. Equation (50b) is the first term in the expansion of the self-energy in powers of the screened interaction. This is the main difference with respect to the Hartree-Fock approximation, which is the first term in an expansion in powers of the bare Coulomb interaction. Σ^{1} is definitely a better approximation to the full many-body problem than Σ^{HF} in those systems where the correlation effects are of much the same size as exchange effects. Having this approximation at our disposal, we can try to estimate the effects of the non-locality of the self-energy operator on band structure calculations. Inkson (1973) argues that if these effects turn out to be small, even for materials with appreciable variations of the charge density, then the validity of local approximations, such as those which enter equation (47"), is proved.

Non-locality effects are conveniently evaluated by using the Fourier transform

$$\Sigma^1(\underline{x},\underline{x}',\omega) = \frac{i}{2\pi}\int d\omega' G(\underline{x},\underline{x}',\omega-\omega')W(\underline{x},\underline{x}',\omega')\exp(-i\delta\omega'), \qquad (51)$$

with δ a positive infinitesimal. This is the way Σ enter equations (46). To evaluate expression (51), we start by assuming that single particle energies $\phi(\underline{k})$ are known, and are obtained by solving (46) with some local approximation to the exchange and correlation energy. Valence-band energies and eigenstates correspond to the previously defined energies ε_s^h and wavefunctions f_s^h for the holes; conduction-band states to those of the electrons. The Green's function for this problem is given simply by

$$G(\underline{x},\underline{x}',\omega) = \Sigma_{\underline{k}} \frac{\phi_k^*(\underline{x}')\phi_k(\underline{x})}{\omega-\varepsilon_{\underline{k}}+i\delta\mathrm{sgn}(\omega-\mu)} . \qquad (52)$$

(Notice that, since we label band states with the lattice wave-vector $\underline{k}$ only, we are working in an extended zone scheme.)

The other quantity we need to evaluate Σ^1 is the dynamically-screened interaction, whose spatial-Fourier component can be written in terms of the longitudinal-dielectric function as

$$W(\underline{q},\omega) = \frac{4\pi e^2}{q^2\varepsilon(\underline{q},\omega)} , \qquad (53)$$

if local field effects are neglected. The most convenient way to calculate $\varepsilon(q,\omega)$ is to use the self-consistent field approximation [Ehrenreich and Cohen (1959)], which allows us to calculate the polarization propagator using the single particle energies and wave-functions as

$$\varepsilon(\underline{q},\omega) = 1 + \frac{4\pi e^2}{q^2\Omega}\Sigma_{\underline{k}} \frac{f(\varepsilon_{k+q})-f(\varepsilon_k)}{\hbar\omega-\varepsilon_{\underline{k}}-\varepsilon_{\underline{k+q}}} |\Omega_c\int u_{\underline{k+q}}(\underline{r})u_{\underline{k}}(\underline{r})dv|^2, \qquad (54)$$

where Ω_c and Ω are the unit cell and crystal volumes respectively, u_k is the periodic part of the wavefunction $\phi(\underline{k})$, and the integration is over the unit-cell volume. The computation of (54) however is a formidable task if a realistic band structure is used. To avoid this problem one can adopt some model dielectric function, which preserves the main features of the semiconductor dielectric function, but has a simple analytical form. This is the way followed by Inkson (1973), who has used the following expression

$$\varepsilon(q,\omega) = 1 + (\varepsilon_o-1)/\{1+q^2(\varepsilon_o-1)/q_{TF}^2 - \omega^2(\varepsilon_o-1)/\omega_p^2\}, \qquad (55)$$

where q_{TF} is the Thomas-Fermi momentum, ω_p is the plasma frequency, and ε_o is the static dielectric function. The inverse dielectric function is then given by

$$\varepsilon^{-1}(q,\omega) = \{\omega_R(q)^2 - \omega_p^2 - \omega^2\}/\{\omega_R(q)^2-\omega^2\}, \qquad (56)$$

where

$$\omega_R(q)^2 = \omega_p^2\varepsilon_o/\{\varepsilon_o-1\}\{1 + \frac{q^2}{q_{TF}^2}(\varepsilon_o-1)/\varepsilon_o\}.$$

This shows that the dynamically screened interaction has poles at $\omega = \pm\omega_R$, in correspondence with the so-called reduced-plasma frequency. Together with the poles of the Green's function, these give the contributions to the integral in (51). Using these approximations Inkson and Bennett (1978,1979) have calculated self-energy effects in the electronic spectrum of Si and ZnS, assuming only first-order corrections to be important. The self-energy expectation value, giving the shift of the one-electron energies, is thus given by

$$\Delta\varepsilon^1(\underline{k}) = \int\phi_k(\underline{x}) \int\Sigma(\underline{x},\underline{x}',\varepsilon)\phi_k(\underline{x}')d\underline{x}d\underline{x}'. \qquad (57)$$

Table I shows the results of such non-local calculations and compares them with those of a local calculation. It is seen that in both Si and ZnS appreciable differences appear between the outcomes of the non-local and local theory, indicating that the non-locality of Σ is not negligible. However in both cases the non-local effects act to increase discrepancies with experimental data. The self-energy contribution to the band gap turns out to be significant and tends to enhance the gap with respect to local calculations, but the resulting description of the band structure is worse than that obtained with a purely local exchange and correlation potential. The only possible explanation of these discrepancies is that, although non-locality is significant, its effects are almost completely cancelled in the band calculations from other sources of non-locality such as core-valence exchange, pseudopotential etc. It is for this reason, and not for an intrinsic weakness of the non-locality of the valence electrons, that approximations as those given in eqs. (47-47')

Table I. Comparison between band structures of Si and ZnS from local (A) and non-local (B) theory. Energies in eV.

	Silicon		
	Exp	(A)	(B)
$L_{3'}-L_1$	3.40	3.40	3.40
$\Gamma_{25'}-\Gamma_{15}$	3.45	3.45	3.45
$\Gamma_{25'}-\Gamma_{2'}$	4.21±.02	4.20	4.20
L_3	3.9±.1	4.00	3.98
L_1	2.04±.06	2.15	2.02
X_4	-2.9	-2.98	-3.40
Γ_1	7.6	8.24	9.54
$\Gamma_{12'}$	8.3±.1	7.85	8.93
E_g	1.15	1.11	1.21
	Zinc-Sulphide		
$\Gamma_{15}-\Gamma_1$	3.73	3.73	3.73
$L_3 - L_1$	5.75	5.75	5.75
$\Delta_5-\Delta_1$	7.35	8.03	8.40
$\Delta_3-\Delta_3$	9.10	9.10	9.12

work and provide a reliable description of the electronic structure of semiconductors, even if large valence-charge-density fluctuations are present.

Gap Renormalization in Extrinsic Semiconductors

We turn now to a problem, which arises in the study of the electronic properties of heavily doped semiconductors, i.e. extrinsic semiconductors, where the doping is so high that a description in terms of individual impurity states is inadequate. Rather than studying the one electron states in the presence of a disordered array of impurity centres, we will focus our attention on the many-body interactions, which arise in the presence of a high number of free carriers in a semiconductor with a high donor concentration. For this purpose, we use a jellium model, which replaces the array of donor centres with a uniform positive background charge. The many-body interactions are expected to have important effects on the band structure, causing band-gap shrinking and modifications in the carrier effective masses. The theoretical approach to be used in this context is the same as in the previous section. The interaction between the electrons comes into the equation of the quasi-particle states at an energy ε_k through the self-energy $\Sigma(\underline{x},\underline{x}',\varepsilon_k)$ and we have to see how this quantity changes when electrons are added to the conduction band. Since Σ gives a significant contribution to the band gap, modifications in the electron-electron interactions are expected to cause corresponding changes in the gap, which might be observed through some experimental technique, as optical absorption.

The analysis can be convenient carried out using (51). However, in this particular problem, only contributions arising from the poles of the Green's function are to be retained if we want to evaluate modifications in the band gap. Terms arising from the poles of the inverse dielectric function do not influence the band gap, since they give the same energy shift to both the conduction and valence bands. For our purposes we can then write the self-energy as

$$\Sigma^1_{ex}(\underline{x},\underline{x}',\omega) = -\Sigma_{\underline{k}occ.}\,\phi_{\underline{k}}(\underline{x})\,\phi^*_{\underline{k}}(\underline{x}')\,W(\underline{x},\underline{x}',\varepsilon_{\underline{k}}-\omega), \tag{58}$$

which we call "screened dynamical exchange". Notice that the sum is over occupied states only.

Assuming a direct-gap semiconductor with minima at the centre of the Brillouin zone, we can calculate the first-order energy shift in the lowest conduction band state ε_o through the formula

$$\Delta E_c = \int \phi^*_{co}(\underline{x})\,\Delta\Sigma^1_{ex}(\underline{x},\underline{x}',\varepsilon_{co})\,\phi_{co}(\underline{x}\ )\,d\underline{x}d\underline{x}', \tag{59}$$

and for the valence band

$$E_v = \int \phi^*_{vo}(\underline{x})\,\Delta\Sigma^1_{ex}(\underline{x},\underline{x}',\varepsilon_{vo})\,\phi_{vo}(\underline{x}')\,d\underline{x}d\underline{x}'. \tag{59'}$$

Here the quantity $\Delta\Sigma_{ex}$ represents the change in the self-energy due to the added free carriers. This change is caused by two different mechanisms. First, the screened interaction is changed, because the dielectric function is altered by the free carrier gas. Second, the sum over the occupied states extends to states in the conduction band as well, since states here are occupied. We can calculate these two corrections to the band gap separately.

As far as the changes in the screening are concerned, one can approximate the dielectric function when electrons are present in the conduction band by

$$\varepsilon(q,\omega) = \varepsilon_s(q\omega) + \{\varepsilon_g(q,\omega) - 1\}, \tag{60}$$

where ε_s is the intrinsic semiconductor dielectric function and ε_g is the dielectric function of a free electron gas with the appropriate effective masses. The second term accounts for the change in the irreducible polarization propagator due to free carriers. It follows from (60) that the change in the screened interaction is given by

$$\Delta W(q,\omega) = \frac{4\pi e^2}{q^2}\{\varepsilon_s^{-1}(q,\omega) - 1/\{\varepsilon_s(q,\omega) + \varepsilon_g(q,\omega) - 1\}\}, \tag{61}$$

When (61) is inserted into (58), this yields the change in the self-energy due to the modifications of the screening properties of the electron gas. The corresponding shift of the valence band edge is then

$$\Delta E_v^1 = \frac{-\Omega}{(2\pi)^6}\iiiint \phi_{vo}^*(\underline{x})\,\phi_{v\underline{k}}(\underline{x})\,\Delta W(q,\varepsilon_v(\underline{k})$$

$$- \varepsilon_v(o))e^{i\underline{q}\cdot(\underline{x}-\underline{x}')}\,\phi_{vo}(\underline{x}')\,\phi_{v\underline{k}}^*(\underline{x}')\,d\underline{x}d\underline{x}'d\underline{q}d\underline{k}, \tag{62}$$

where we have replaced the sum over the occupied states with an integral over $\underline{k}$ for the valence band. ΔE_c^1 is obtained by changing ϕ_{vo} to ϕ_{co} and $\omega_v(o)$ to $\omega_c(o)$. In the case of the conduction-band shift, the free electron dielectric function is evaluated at energies greater than or equal to the semiconductor band gap. At these frequencies $\varepsilon_g(\omega)\sim 1$ with good approximation for the electron densities characteristic of large doping level. It follows that ΔE_c^1 vanishes. Therefore the change in the band gap caused by the modifications in W is given by

$$\Delta E_g^1 = -\Delta E_v^1 \tag{63}$$

which is always negative, leading to a gap shrinkage.

As far as the modification in Σ^1_{ex} due to occupied conduction band states is concerned, it is given by

$$\Delta E_c^2 = -\frac{\Omega}{(2\pi)^6} \iiiint \phi^*_{co}(\underline{x})\phi_{c\underline{k}}(\underline{x}')W(q,\varepsilon_c(\underline{k}) - \varepsilon_c(o)) \times$$

$$\times e^{i\underline{q}\cdot(\underline{x}-\underline{x}')}\phi_{co}(\underline{x})\phi^*_{c\underline{k}}(\underline{x}')d\underline{x}d\underline{x}'d\underline{q}d\underline{k}. \qquad (64)$$

A similar expression holds for the valence band shift with the replacement of ϕ_{co} by ϕ_{vo}. In this case, however, since newly occupied states are confined at small $\underline{q}$, the exponential in the equation can be dropped and, because of the orthogonality between valence and conduction states, the integral turns out to be very small. It follows that the valence edge shift can be neglected and the change in the gap is almost entirely due to that of the conduction band. We can therefore conclude that many-body interaction in extrinsic materials lead to a band-gap renormalization given by

$$\Delta E_g = \Delta E_c^2 - \Delta E_v^1 \qquad (65)$$

Inkson (1976) has evaluated this correction using a Thomas-Fermi approximation to the dielectric screening. The results are given by

$$\Delta E_g^{TF} = -\frac{2e^2}{\pi\varepsilon_o} k_f\{\pi\kappa/(2k_f) + 1 - \frac{\kappa}{k_f}\tan^{-1}(k_f/\kappa)\}, \qquad (66)$$

where κ is the Thomas-Fermi wavenumber. ΔE_g turns out to always be a negative quantity, the main contribution arising from changes in the screened interaction at low carrier concentrations. At very high doping, ΔE_c^2 is rather important.

The theory can be extended rather easily to indirect gap semiconductors and more correct expressions for the free carrier dielectric function, such as the Lindhard dielectric function for the free electron gas, can be used [Abrams, et al., (1978)]. Although corrections to the Thomas-Fermi results are obtained from these improvements, the qualitative results do not change very much. Theoretical carrier concentration dependences of the band gap narrowing in GaAs at T = 0 are presented in Figs. 3-4.

A direct comparison of this theory with experiments is difficult; the presence of impurity bands and the disorder of the true array of impurity centres can give significant contributions to the gap shrinkage. However, most of the theoretical results obtained up to now seem to suggest that this is the basic mechanism for band-gap renormalization in extrinsic semiconductors.

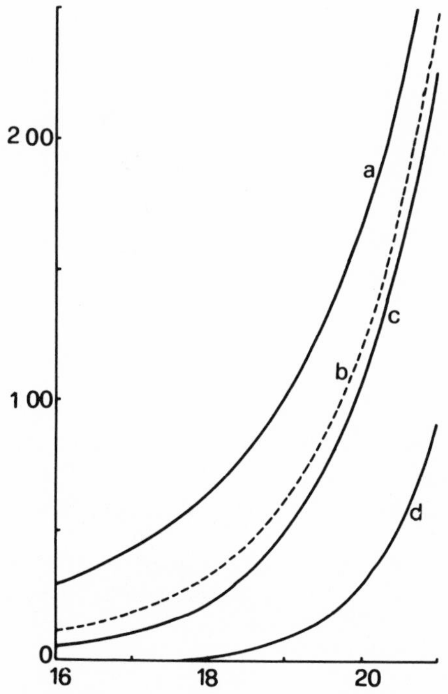

Fig. 3. Theoretical dependence of the gap narrowing (in meV) upon Log of carrier concentration for n-type GaAs at T = OK. Dashed line gives the results of a calculation with the Lindhard dielectric function. Solid lines: a) results of Thomas-Fermi calculation; b) unscreened exchange contribution; c) ΔE_c^2 conduction band shift.

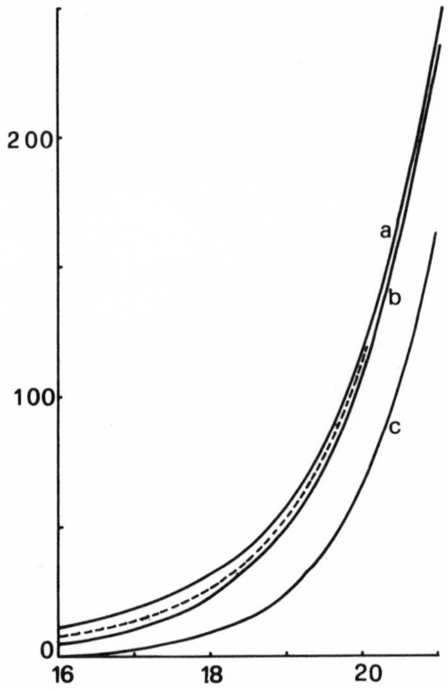

Fig. 4. Theoretical dependence of the gap narrowing (in meV) upon Log of carrier concentration for p-type GaAs at T = OK. a) results for the Thomas-Fermi model; b) with the Linhard dielectric function; c) unscreened exchange contribution; d) ΔE_v^2.

BIBLIOGRAPHY

Abrams, R. A., Rees, G. J., and Wilson, B. L. H., 1978, Adv. Phys. 27:799.

Bassani, F., Robinson, J., Goodman, B., and Schrieffer, J. R., 1962, Phys. Rev. 127:1969.

Bennet, H., and Inkson, J. C., 1979, J. Phys. C 12:283.

Berlinsky, A. J., 1976, Contemp. Phys. 17:331.

Bertoni, C. M., Bortolani, V., Calandra, C. and Nizzoli, F., 1973, J. Phys. C 6:3612.

Bortolani, V., Calandra, C., and Kelly, M. J., 1973, J. Phys. C 6:L349.

Chelikowsky, J. R., and Cohen, M. L., 1974, Phys. Rev. B10:5095.

Cohen, M. L., and Heine, V., 1970, Sol. State Phys. 24:37.

Conies, R., Lambert, M., Launois, H., and Zeller, H. R., 1973, Phys. Rev. B8:952.

Coulson, C. A., 1961, "Valence", Oxford Univ. Press, London.
Ehrenreich, H., and Cohen, M. H., 1959, Phys. Rev. 115:786.
Glaser, W., 1974, in "Festkorperprobleme" Vol. XIV, Ed. by H. J. Quiessen, Pergamon-Vieweg, Berlin.
Hall, G. G., 1952, Phil. Mag. 43:338.
Harrison, W. A., 1967, "Solid State Theory", McGraw-Hill, New York.
Harrison, W. A., 1979, "The Physics of the Chemical Bond", Freeman, San Francisco.
Hedin, L., 1965, Phys. Rev. 139A:796.
Hedin, L., and Lundquist, S., 1969, Solid State Physics 23:1
Hohenberg, P., and Khon, W., 1964, Phys. Rev. 136:864.
Inkson, J. C., 1973, J. Phys. C. 6:L181.
Inkson, J. C., 1976, J. Phys. C. 9:1177.
Inkson, J. C., and Bennet, H., 1978, J. Phys. C. 11:2017.
Kadanoff, L. P., and Baym, G., 1962, "Quantum Statistical Mechanics", W. A. Benjamin, New York.
Kohn, W., and Sham, L. J., 1965, Phys. Rev. 140:A1133.
Lang, N., 1973, Solid State Phys. 28:192.
Leman, G., 1962, Ann. de Phys. (Paris) 7:505.
Morse, P. M., and Feshbach, H., 1953, "Methods of Theoretical Physics," McGraw-Hill, New York.
Mott, N. F., 1974, "Metal-Insulator Transitions," Taylor and Francis, Ltd., London.
Nozierres, P., 1964, "Theory of Interacting Fermi Systems," W. A. Benjamin, New York.
Peierls, R., 1955, "Quantum Theory of Solids," Clarendon, Oxford.
Phillips, J. C., 1970, Rev. Mod. Phys. 42:317.
Pratt, G., 1960, Phys. Rev. 118:462.
Reggiani, L., and Calandra, C., 1973, Phys. Letters 43A:339.
Salem, L., 1966, "Molecular Orbital Theory of Conjugated Systems," W. A. Benjamin, New York.
Tosatti, E., 1975, "Festkorperprobleme," Vol. XV, Ed. by Q. J. Quiessen, Pergamon-Vieweg, Berlin.
Walter, J. P., and Cohen, M. L., 1971, Phys. Rev. B4:1877.
Weaire, D., and Thorpe, M. F., 1971, Phys. Rev. B4:2058.

THE ELECTRON-PHONON INTERACTION IN SEMICONDUCTORS

P. Vogl

Institut für Theoretische Physik
Universität Graz
A-8010 Graz, Austria

INTRODUCTION

The interaction between electrons and lattice vibrations is one of the fundamental interaction processes in solids. In this lecture, we wish to examine this interaction in semiconductors. Although the methods we are going to develop are general enough to be applicable to all classes of semiconductors and insulators, the analysis is most transparent and probably also most useful if we confine our concern to the technologically important class of semiconductors, i.e. group IV, III-V, and II-VI materials. With the exception of some of the II-VI's, these can be grown with high purity, can easily be doped and have high mobility — all the properties desirable for device applications.

From the point of view of electron transport, the interaction of electrons and phonons provides the dominant scattering mechanism in semiconductors except at the lowest temperatures. Specifically, in high electric fields it gives the dominating electron energy- and momentum-loss mechanism which limits the electron mobility (see Conwell, 1967, for a general review). From the point of view of optics, the electron-phonon (el-ph) interaction determines the inelastic light scattering cross-sections (Raman and Brillouin scattering) and the free carrier absorption (Richter, 1976, Cardona, 1969, 1977).

The elementary process is the scattering of an electron or hole in a Bloch state with wave vector $\vec{k}$ into a state $\vec{k}'$, absorbing or emitting one or more phonons. Up to room temperature, one-phonon processes are the most probable ones, where $\vec{k}' = \vec{k} \pm \vec{q}$, $\vec{q}$ denoting the phonon wave vector. The potential causing the scattering depends on the material - whether non-polar or polar - and also on the particular phonon mode. In semiconductors, the various el-ph scattering processes are mostly described in terms of classical concepts involving phenomenological parameters which are fit to experiment (Conwell, 1967, Cardona, 1969, 1977, Richter, 1976, Sham and Ziman, 1963). The simplicity of these concepts makes them appealing and they are in fact sufficient to explain most of the transport and optical data for pure, bulk samples.

On the other hand, as the general interest in semiconductor physics shifts increasingly to imperfect structures, small samples, interfaces, very impure materials, etc., discrepancies between new data and theoretical models based on these simple concepts developed for the pure, periodic solid have to be expected. We then need a systematic means of locating the physical origin of these discrepancies; it is desirable to know which el-ph interactions depend sensitively on perturbations and imperfections, which concepts have to be modified and to what extent.

In this lecture, we will deal with these fundamental aspects of the el-ph interaction. We shall first review in some detail the standard el-ph scattering theory for optical and acoustic phonons in semiconductors with emphasis on the physical concepts involved: The deformation potential concept, the polar optical Fröhlich interaction and the piezoacoustic coupling. Then we present a microscopic, unified treatment of the el-ph interaction in semiconductors which (i) allows us to understand the standard concepts in a rigorous, quantum mechanical framework, (ii) exhibits the approximations in the phenomenological treatments and their validity, (iii) provides a practical scheme of computing the el-ph scattering rates, without phenomenological or adjustable parameters, from the band structure of the perfect crystals, and (iv) gives a systematic scheme to include screening, external fields, surfaces, etc. The latter sections are devoted to examples and applications of the general methods developed and a discussion of generalizations for nonperfect crystals.

Central to the tractability of the problem of the interaction of electrons with phonons in semiconductors — and in any material — is the adiabatic approximation which we put in the beginning of this lecture. We shall use atomic units ($\hbar = e = m = 1$) and also take the crystal volume equal to unity.

THE ADIABATIC APPROXIMATION

The high mobility of the standard semiconductors indicates that the el-ph interaction can be treated in perturbation theory. The basis of this simplification is the adiabatic approximation (Born and Huang, 1954, Maradudin, 1974, Haug, 1972). Consider a perfect, intrinsic semiconductor and write the total crystal Hamiltonian as

$$H = T(\vec{R}) + U(\vec{R}) + H_{el}(\vec{r}) + V(\vec{r},\vec{R}), \tag{1}$$

where H_{el} includes the kinetic energy of the electrons and the Coulomb interaction between them, $T(\vec{R}) + U(\vec{R})$ is the kinetic plus potential energy of the ions and $V(\vec{r},\vec{R})$ is the electron-ion interaction. By the term "ions" we mean the nuclei together with the core electrons which we consider as rigidly attached to the nuclei. Accordingly, by "electrons" we mean only the valence electrons of the semiconductor. The electronic and ionic coordinates are $\vec{r}$ and $\vec{R}$, respectively.

The adiabatic approximation achieves a separation of the electronic and ionic motion to such an extent that their coupling can be treated in perturbation theory. The term adiabatic means that the change of some external parameter does not induce transitions between eigenstates of a system. In particular, for a system of electrons and ions, the electrons can be expected to respond adiabatically when the ionic configuration is changed. The reason lies in the very different masses of electrons and ions; the ions move slowly relative to the electrons so that the electrons see in every instant a frozen configuration of ions. This means, that for the electronic motion, we can neglect the kinetic energy of the nuclei. For a non-degenerate system, the adiabatic approximation consists in the following ansatz for the total wave function $\phi(\vec{r},\vec{R})$:

$$\phi_{nv}(\vec{r},\vec{R}) = \psi_n(\vec{r};\vec{R})\ \chi_{nv}(\vec{R}), \tag{2}$$

where $\psi_n(\vec{r};\vec{R})$ is the electronic part and χ_{nv} is the vibrational wave function, determined by the properties of only the n-th electronic state. The electronic wave function $\psi_n(\vec{r};\vec{R})$ satisfies

$$\{U(\vec{R}) + H_{el}(\vec{r}) + V(\vec{r},\vec{R})\}\ \psi_n(\vec{r};\vec{R}) = E_n(\vec{R})\ \psi_n(\vec{r};\vec{R}). \tag{3}$$

The electronic problem is solved with the ionic coordinates fixed at $\vec{R}$, i.e. in (3) the ions are considered as infinitely heavy particles in fixed positions $\vec{R}$. The equation for the ionic motion can be found variationally (Longuet-Higgins, 1948) and is, up to a small term which we neglect,

$$\{T(\vec{R}) + E_n(\vec{R})\}\ \chi_{nv}(\vec{R}) = \omega_{nv}\ \chi_{nv}(\vec{R}). \tag{4}$$

The effective potential energy for the ionic motion is given by the electronic eigenvalue $E_n(\vec{R})$, which incorporates the direct interaction $U(\vec{R})$ between ions. Notice that the ionic motion depends on the electronic state n and that the equilibrium positions of the ions have therefore to be determined selfconsistently. This plays a role for electronic recombination processes near deep impurities (luminescence) (Markham, 1959).

For perfect materials the equilibrium positions are known in the ground state of the electronic system and a further approximation can be made in (2). In the so-called static approximation, the wave function $\psi_n(\vec{r};\vec{R})$ is assumed to be independent of $\vec{R}$, i.e. the electronic part is solved with the ions in their equilibrium position $\vec{R}_o$ in (2):

$$\phi_{nv}(\vec{r},\vec{R}) = \psi_n(\vec{r};\vec{R}_o)\, \chi_{nv}(\vec{R}) = \psi_n(\vec{r})\, \chi_{nv}(\vec{R}). \tag{5}$$

When (2) is inserted into the Schrödinger equation for H, given by (1), and (3) and (4) are used, extra terms appear which mix the different adiabatic states $\phi_{nv}(\vec{r},\vec{R})$ and cause electronic transitions during the nuclear motion. These terms are the el-ph interaction, given by

$$<\phi_{nv}|\delta V(\vec{r})|\phi_{n'v'}> ,$$

$$\delta V(\vec{r}) = V(\vec{r},\vec{R}) - V(\vec{r},\vec{R}_o). \tag{6}$$

In the adiabatic approximation, this interaction is treated in perturbation theory. For the calculation of the el-ph scattering rate, the ionic displacements are considered as a "frozen-in" perturbation creating an electrostatic potential felt by the electrons.

A large amount of literature exists on the validity of this useful approximation (Sham and Ziman, 1963, Born and Huang, 1954, Maradudin, 1974, Haug, 1972). Its justification is particularly difficult, since in most cases the (properly defined) dimensionless coupling coefficient between electrons and phonons is not small but of the order of 1. Nevertheless, the adiabatic approximation can be established for insulators and semiconductors as well as for metals if the self-consistent electron-ion potential is of short range (Combes, 1976, Fetter and Walecka, 1971). A simple illustrative model system was solved by Moshinsky and Kittel (1968). The proofs generally rest on the observation that two systems with comparable momentum ($\vec{q}_{ph} \sim \vec{k}_{el}$) but very different energy scales ($\omega_{ph} << E_{el}$) cannot interact efficiently with one another even if the coupling coefficients are not small. For a long-range electron-ion potential, as it occurs for longitudinal optical modes in polar crystals, the adiabatic approximation holds only if the el-ph coupling is truly small so that a perturbative approach is justified (see all lectures in Devreese, 1972). A definition of the adiabatic approximation from the modern viewpoint has been given by Sham (1974).

THE DEFORMATION POTENTIAL INTERACTION

We shall now review the standard el-ph couplings in semiconductors from the phenomenological standpoint and particularly concentrate on the physical concepts involved (see, e.g., Conwell, 1967, Sham and Ziman, 1963, Haug, 1972, Herring and Vogt, 1956). Common to all phenomenological el-ph coupling terms is that they are based on a continuum approximation for the crystal. Clearly, this can be expected to be valid only for spatially slowly varying perturbations, i.e. long-wavelength phonons.

The deformation potential concept is based on the observation that the lattice distortion created by a long-wavelength acoustic phonon can be viewed as produced by a homogeneous strain of magnitude equal to the local strain set up by the acoustic mode. In other words, one assumes that the interaction of electrons with a long-wavelength acoustic mode is equivalent to the effect of a locally homogeneous strain. This conjecture, which has been called the "deformation potential theorem" (Sham and Ziman, 1963), is of great practical importance. It implies that the coupling between electrons and long-wavelength acoustic phonons can be obtained either from transport measurements or from uniaxial stress data. Indeed, the latter method gives more direct and more accurate information on the el-ph interaction than transport data. For scattering at band edges, the equivalence of the effect of a long acoustic wave and static strain could indeed be proven by Sham and Ziman (1963). The proof is very lengthy and we shall not repeat it here. Let us demonstrate instead how this equivalence is used. Consider the shift of a non-degenerate band edge in a semiconductor with a cubic lattice upon applying stress. For a spherical constant energy surface, this can be written as

$$\delta E_{edge}(\vec{r}) = E_1(S_{xx} + S_{yy} + S_{zz}) = E_1\, \Delta(\vec{r}), \tag{7}$$

where the S_{ii} are the diagonal components of the strain tensor which sum up to the volume dilatation $\Delta(\vec{r}) = \delta V/V$ and E_1 is the shift of the band edge per unit dilatation. E_1 is the simplest case of a deformation potential for acoustic mode scattering. $\Delta(\vec{r})$ acts as an effective potential for the electrons near the band edge. Notice that we are considering the macroscopic strain which can be associated with an acoustic wave, but only the volume dilatation enters in (7) because of symmetry. Because of this, E_1 is given by the shift of the band extremum under pure hydrostatic pressure:

$$\delta E_{edge}(\vec{r}) = \frac{\partial E_{edge}}{\partial V}\, \delta V = E_1 \frac{\delta V}{V}, \tag{8}$$

which gives

$$E_1 = \frac{a}{3} \frac{\partial E_{edge}}{\partial a} \tag{9}$$

with a being the lattice constant. In standard semiconductors, E_1 is of the order 10 eV.

We can write the displacement field associated with a single long wavelength acoustic mode of wave vector $\vec{q}$ as

$$\vec{u}(\vec{r}) = \vec{u}\, e^{i\vec{q}\cdot\vec{r}}, \tag{10}$$

with $\vec{u}$ representing the phonon amplitude and polarization. In cubic crystals, acoustic modes separate in the long-wavelength limit into one longitudinal (LA) and two transverse (TA) modes, where $\vec{u}//\vec{q}$ and $\vec{u} \perp \vec{q}$, respectively. We can obtain the scattering matrix element for electrons near the band edge by using the relation

$$\Delta(\vec{r}) = \vec{\nabla} \cdot \vec{u}(\vec{r}) = i\vec{q}\cdot\vec{u}\, e^{i\vec{q}\cdot\vec{r}}, \tag{11}$$

and inserting this into (7). The electron states are approximated by plane waves. First we see from (11) that $\Delta(\vec{r}) = 0$ for TA waves where $\vec{q}\cdot\vec{u} = 0$, i.e. only LA waves scatter carriers near the band edge. The matrix element for scattering from an electron state $\vec{k}$ to $\vec{k} \pm \vec{q}$ with creation or anihilation of a phonon $\vec{q}$ follows to be

$$M^2_{kq}(LA) = |\langle\vec{k} \pm \vec{q}|\delta E_{edge}(\vec{r})|\vec{k}\rangle|^2 = E_1^2 q^2 \langle u\rangle^2, \tag{12}$$

$$\langle u\rangle^2 = \frac{1}{2NM\omega_{LA}(\vec{q})} (N_{\vec{q}} + \frac{1}{2} \pm \frac{1}{2}). \tag{13}$$

The factor $\langle u\rangle^2$ is the phonon part of the matrix element, which is obtained from a normal mode expansion of $\vec{u}$. M (not to be confused with M_{kq}) denotes the mass of the unit cell, N is the number of unit cells in the crystal, and $\omega_{LA}(\vec{q})$ is the phonon frequency. $N_{\vec{q}}$ is the phonon occupation and the plus sign in (13) is to be used for emission, the minus sign for absorption. Crucial for transport applications is the wave-vector dependence of the matrix element (12). Because $\omega_{LA}(\vec{q}) = c_L \cdot q$, the matrix element (12) for acoustic phonon scattering goes like $M_{kq} \sim q^{\frac{1}{2}}$.

One can generalize the above treatment to ellipsoidal energy surfaces or degenerate band edges. For a valley i along <100> or <111> directions in a cubic crystal one has, for example, (Conwell, 1967, Herring and Vogt, 1956),

$$\delta E^{(i)}_{edge} = \Xi_d \Delta(\vec{r}) + \Xi_u \hat{k}^{(i)} \cdot S \cdot \hat{k}^{(i)}, \tag{14}$$

where $\hat{k}^{(i)}$ is a unit vector pointing from the center of the BZ to the edge of valley i and S is the strain tensor. Symmetry gives two independent deformation potentials in this case, Ξ_d and Ξ_u which come from the dilatational and uniaxial shear part of the deformation, respectively. In this case the scattering matrix element for TA as well as LA modes is nonzero.

We also note that in a uniaxial stress experiment, at least two band edges are involved. Only the changes in the optical band gaps, i.e. combined shifts of the conduction and valence band edges, can be measured. By applying stress in various crystal directions or by measuring the stress-induced line splitting of shallow impurity levels, however, the deformation potentials of conduction and valence band edges can be obtained separately.

Intrinsic to the whole deformation potential concept is the assumption that the potential produced by the acoustic phonon wave is of short range. On the basis of this assumption the form of the matrix element (12) can be derived without using strain arguments (Haug, 1972). We can generally write

$$M_{kq} = \langle\vec{k} \pm \vec{q}|\delta V(\vec{r})|\vec{k}\rangle, \tag{15}$$

where $\delta V(\vec{r})$ is the electric potential induced by the small (acoustic wave-like) displacements $\vec{u}(\vec{r})$. We can ssume $\delta V(\vec{r})$ to be linear in $\vec{u}$. Higher order terms give rise to more-phonon processes. The Bloch functions $\vec{k}$ and $\vec{k} \pm \vec{q}$ contain a periodic part which we absorb in $\delta V(\vec{r})$. One can then write

$$M_{kq} = \int d^3r\ e^{\pm i\vec{q}\cdot\vec{r}}\ \delta V(\vec{r}). \tag{16}$$

We are interested in the long-wavelength limit of this matrix element. If $\delta V(\vec{r})$ is of short-range, the exponential in the integral in (16) can be expanded in powers of $\vec{q}$. The lowest order term of this expansion corresponds to a rigid shift of the whole crystal ($\vec{q} = 0$ acoustic phonon) and vanishes. This gives

$$M_{kq} = i\vec{q} \cdot [\int d^3r\ \vec{r}\ \partial\vec{V}(r)] \cdot \langle\vec{u}\rangle, \tag{17}$$

where we have written $\delta V(\vec{r}) = \partial\vec{V}(\vec{r})\cdot\vec{u}$. The form of (17) is exactly the same as of (12); the expression in the brackets of (17) can be interpreted as the deformation potential. We see that the result (12) implies an expansion of the full electron-phonon matrix element in powers of $\vec{q}$ and retaining only the lowest order non-vanishing term. It also implies, as already mentioned, the assumption of a short-range electron-ion potential, which is really far from being evident in view of the Coulomb nature of the ionic potentials in the crystal. We shall investigate this systematically in a later section.

Another point which should be made concerns screening. If the screening of the potential produced by phonons plays a significant role due to the presence of many carriers, the equivalence of phonons and strain breaks down. A phonon produces long wavelength density waves and corresponding potential variations to which the free carriers respond, while a homogeneously strained crystal produces no macroscopic potential which redistributes the free carriers.

Experimental and theoretical values for acoustic deformation potentials in a variety of semiconductors are given e.g. in Cardona (1969), Tsay et al. (1974), Cerdeira and Cardona (1972), Savaria and Brust (1969), Camphausen et al. (1971), Müller et al. (1979), Mathei et al. (1979), Nash and Holm-Kennedy (1978), Jorgenson (1978), Yee and Myers (1976), and Tekippe et al. (1972).

NONPOLAR-OPTICAL PHONON SCATTERING

For this scattering also, the term deformation potential is used. This is somewhat misleading since the displacement associated with a long-wavelength optical mode is not a macroscopic deformation of the crystal; it is a relative displacement of the sublattices. However, common to both acoustic and nonpolar optical deformation potentials is the assumption of a short-range el-ph interaction.

We consider the scattering of electrons near a non-degenerate band edge in a cubic, binary semiconductor by long-wavelength optical phonons. The "deformation potential" D is defined by (Harrison, 1956)

$$|\langle\vec{k} \pm \vec{q}|H_{op}|\vec{k}\rangle|^2 = D^2 \frac{1}{2NM\omega_o(\vec{q})} (N_{\vec{q}} + \frac{1}{2} \pm \frac{1}{2}). \qquad (18)$$

D is the shift of the band edge per unit relative displacement of the two sublattices. Experimentally D - and its generalizations for complex bands - can be obtained either from transport measurements (Conwell, 1967) or, more accurately, from Raman scattering (Cardona, 1969, 1977). For Ge, e.g., one finds approximately $D = 5\times10^8$ eV/cm. Equation (18) can again be obtained from an expansion of the general el-ph matrix element (15) in powers of $\vec{q}$. For optic phonons, the lowest order term $\mathcal{O}(q^0)$ — which was zero for modes — represents the leading term:

$$M_{kq} = \int d^3r \; u^*_{\vec{k}}(\vec{r}) \;\; \delta V(\vec{r}) u_{\vec{k}}(\vec{r}), \tag{19}$$

where $u_{\vec{k}}(\vec{r})$ denotes the periodic part of the Bloch function. Based on the symmetry of the Bloch functions and the potential $\delta V(\vec{r})$ one can work out selection rules for this scattering. In nonpolar cyrstals, for example, the transverse and longitudinal optical (TO and LO) phonons are degenerate for $\vec{q} \to 0$ and the normal modes transform like a vector, i.e. they are p-like. The matrix element (19) then vanishes for s-like band minima since the product representation s x p does not contain s. This remains valid for the III-V materials with their s-like conduction bands. As a consequence, the optical deformation potential interaction (18) vanishes for scattering near the conduction band edge in the III-V's and in Si, but not in Ge. The interaction matrix element (18) is also used for intervalley scattering where the phonon vector $\vec{q} \approx \vec{k}_{valley}$ is not small but varies little. In this case $u^*_{\vec{k}}(\vec{r})$ and $u_{\vec{k}}(\vec{r})$ in (19) belong to different minima.

Detailed listings of selection rules for optical phonon scattering are given by Birman (1966), Streitwolf (1970), and Bir and Pikus (1974). Experimental data and calculations for optical deformation potentials can be found in Conwell (1967), Cardona (1969, 1977), Richter (1976), Tsay et al. (1974), Cerdeira and Cardona (1972), Savaria and Brust (1969), Camphausen et al. (1971), Müller et al. (1979), Mathieu et al. (1979), Nash and Holm-Kennedy (1978), and Jorgenson (1978).

POLAR-OPTICAL INTERACTION

Electrons in polar cyrstals interact with longitudinal optical (LO) phonons through the electric field of the polarization wave. This is a long-range Coulomb interaction and is different from the deformation potential interaction, which was considered as a short-range interaction. Transverse optical phonons do not produce macroscopic electric fields and their interaction with electrons in polar crystals is conceptually the same as in nonpolar crystals.

The interaction matrix element for LO-phonon scattering in a heteropolar crystal is derived phenomenologically taking into account the long-range macroscopic electrostatic field associated with the polarization in a LO phonon mode and considering long waves where a continuum approximation can be made. This so-called Fröhlich coupling is the basis of the large polaron theory (Devreese, 1972, Haug, 1972). It was originally developed for ionic crystals where the crystal potential can be approximated by a sum of positive and negative point charge potentials. One can, however, derive the polar optical el-ph coupling without this assumption. Again we restrict ourselves for simplicity to cubic semiconductors with two ions per unit cell.

Consider a long-wavelength optical displacement, where the ions move relative to each other inside each cell. For a TO-phonon mode, the ionic motion induces a dipole moment per ion pair of

$$\vec{P}_T = \frac{1}{\Omega_o} e_T^* \vec{u}_{rel}(\vec{r}), \tag{20}$$

where $\vec{u}_{rel}(\vec{r})$ is the relative displacement of the two ion lattices and Ω_o the volume of the unit cell. Equation (20) defines an effective ionic charge e_T^* which is called the transverse or Born dynamical effective charge. When the ions vibrate longitudinally, a macroscopic electric field develops in the deformation. As stated above, we consider long-wavelength phonons. To be specific, we mean phonons with wavelengths which are large compared to unit cell dimensions (so that continuous displacement fields and macroscopic electrostatics can be used), but wavelengths which are still small compared to sample dimensions, so that no macroscopic polarization is induced on the surface of the crystal which would give rise to boundary effects.

The electric field in the LO mode satisfies the field equation

$$\vec{\nabla} \cdot \vec{D} = \vec{\nabla} \cdot (\vec{E} + 4\pi\vec{P}_L) = 0, \tag{21}$$

where $\vec{E}$ and $\vec{D}$ are the macroscopic electric field and the electric displacement, respectively. This leads to $\vec{E} = -4\pi\vec{P}_L$ because of the longitudinal character of the fields. $\vec{P}_L$ is the polarization field set up by the LO wave-like displacements, and can be written

$$\vec{P}_L = \frac{1}{\Omega_o} e_T^* \vec{u}_{rel}(\vec{r}) + \frac{1}{4\pi} (\varepsilon_\infty - 1)\vec{E}. \tag{22}$$

In (22), ε_∞ is the optical dielectric constant of the crystal. The difference of $\vec{P}_L$ and $\vec{P}_T$ comes from the electronic polarization of the electrons induced by the macroscopic electric field set up by the LO phonon. If an external field $\vec{E}_{ext}$ is applied additionally, $\vec{E}$ in (22) represents the total macroscopic electric field and $\vec{u}_{rel}$ and $\vec{E}$ can be considered as independent variables. Usually we take $\vec{E}_{ext} = 0$. Then $\vec{E} = -4\pi\vec{P}_L$ and one gets

$$\vec{P}_L = \frac{1}{\Omega_o} \frac{e_T^*}{\varepsilon_\infty} \vec{u}_{rel}(\vec{r}) = \frac{1}{\Omega_o} e_L^* \vec{u}_{rel}(\vec{r}), \tag{23}$$

which defines the longitudinal or Callen dynamical effective charge. It is possible to express e_T^* or e_L^* in terms of experimentally known quantities.

The difference between the long-wavelength transverse and longitudinal optical lattice vibrations originates in the additional restoring force for LO modes from the electric field $\vec{E} = -4\pi\vec{P}_L$ which counteracts the polarization $\vec{P}_L$:

$$-\omega_{LO}^2 M \vec{u}_{rel} = -\omega_{TO}^2 M \vec{u}_{rel} + e_T^* \vec{E}, \tag{24}$$

and

$$\omega_{LO}^2 = \omega_{TO}^2 + \frac{4\pi e_T^{*2}}{M\Omega_o \varepsilon_\infty} \tag{25}$$

M denotes the reduced mass of the two ions in the unit cell. Using (25) and the Lydanne-Sachs-Teller relation

$$\frac{\omega_{LO}^2}{\omega_{TO}^2} = \frac{\varepsilon_o}{\varepsilon_\infty}, \tag{26}$$

the effective ionic charge e_L^* can be expressed in terms of the static and optical dielectric constants ε_o and ε_∞:

$$e_L^{*2} = \frac{M\Omega_o}{4\pi} \omega_{LO}^2 \left(\frac{1}{\varepsilon_\infty} - \frac{1}{\varepsilon_o}\right). \tag{27}$$

The electron-LO-phonon interaction is obtained from the electric potential $\delta V(\vec{r})$ produced by the LO phonons. $\delta V(\vec{r})$ is the potential of an electron in the medium with the polarization $\vec{P}_L(\vec{r})$. From $\vec{E} = -\vec{\nabla}\delta V(\vec{r})$ one gets

$$\delta V(\vec{r}) = \int d^3r' \vec{P}_L(\vec{r}') \cdot \frac{(\vec{r} - \vec{r}')}{|\vec{r} - \vec{r}'|^3}. \tag{28}$$

We now expand the displacements in normal modes and write

$$\vec{u}_{rel}(\vec{r}) = \vec{u}_{rel}\, e^{i\vec{q}\cdot\vec{r}}. \tag{29}$$

With the help of the relation

$$\int d^3r'\, e^{i\vec{q}\cdot\vec{r}'} \vec{\nabla} \frac{1}{|\vec{r} - \vec{r}'|} = \frac{4\pi i\vec{q}}{q^2} e^{i\vec{q}\cdot\vec{r}}, \tag{30}$$

one gets the el-ph potential

$$\delta V(\vec{r}) = -\frac{4\pi i}{\Omega_o}\frac{e_T^* q}{\varepsilon_\infty q^2}\cdot\vec{u}_{rel}\, e^{i\vec{q}\cdot\vec{r}}, \tag{31}$$

which is just the Fourier transform of a macroscopic dipole potential. The electron-phonon matrix element is

$$M_{kq} = \langle\vec{k}+\vec{q}|\delta V(\vec{r})|\vec{k}\rangle. \tag{32}$$

The electronic matrix element involves the periodic parts of the Bloch functions $u_{\vec{k}}(\vec{r})$ as

$$\langle\vec{k}+\vec{q}|e^{i\vec{q}\cdot\vec{r}}|\vec{k}\rangle = \int d^3r\, u^*_{\vec{k}+\vec{q}}(\vec{r})\, u_{\vec{k}}(\vec{r}) \xrightarrow[q\to 0]{} 1, \tag{33}$$

which gives the normalization integral for $q\to 0$. For long-wavelength LO phonons, one finally arrives at

$$|M_{kq}|^2 = \frac{4\pi}{q^2}\left(\frac{\omega_{LO}}{2}\right)\left(\frac{1}{\varepsilon_\infty}-\frac{1}{\varepsilon_o}\right)\left(N_{\vec{q}}+\frac{1}{2}\pm\frac{1}{2}\right), \tag{34}$$

where a normal mode expansion for $\vec{u}_{rel}$ analogous to (13) has been used.

Crucial for applications is again the wave-vector dependence of $M_{kq} \sim q^{-1}$. We should like to emphasize that the result in (31) can be interpreted as the dipole potential produced by point charges with an effective charge e_T^*/ε_∞ but we obtained the result without assuming rigid point charges. In the strongly covalent III-V semiconductors, it is impossible to assign the electrons in the perfect crystal in a sensible way to anions and cations. However, a dynamical effective charge - defined as the induced dipole moment per unit displacement - always has a well-defined meaning. From the observed values for ω_{TO} and ω_{LO} one can deduce $|e_T^*| \cong 2$ for the III-V and II-VI semiconductors.

PIEZOELECTRIC INTERACTION

An internal electric field is accompanied not only with LO phonons, but in polar crystals also appears with acoustic phonons. This occurs because the zincblence crystals - and all crystals without an inversion center - are piezoelectric. In an acoustic mode the two ions in the unit cell move in the same direction but there is a small net change in the distance between them which gives rise to a macroscopic polarization $\vec{P}$ (Mahan, 1972). This polarization is proportional to the macroscopic strain. The piezoelectric constants $e_{\lambda,\mu\nu}$ are defined by

$$P_\lambda = e_{\lambda,\mu\nu} S_{\mu\nu} + \frac{1}{4\pi} (\varepsilon^o_{\lambda\mu} - \delta_{\lambda\mu}) E_\mu , \qquad (35)$$

where $S_{\mu\nu} = (\partial u_\mu/\partial x_\nu + \partial u_\nu/\partial x_\mu)/2$ is the strain field, u_μ is the amplitude of the ionic displacement in the μ-direction, and $\varepsilon^o_{\lambda\mu}$ is the static dielectric tensor. Summation over repeated Greek indices is implied throughout this paper. $\vec{E}$ is the macroscopic electric field in the crystal.

The splitting of the polarization into a contribution proportional to the strain and a part proportional to the macroscopic electric field as in Eq. (35) [or in an optical displacement part and $\vec{E}$ as in (22)] is crucial for polar materials. Let us discuss this point in more detail. A macroscopic strain induces an electric field which shows up as piezoelectric charge on the surface of the crystal. This surface polarization alters the total field inside the crystal and depends on the geometry of the surface. It is clearly desirable to separate the volume effects from these surface effects. One therefore attempts to separate the short-range fields from the long-range fields induced by a perturbation - in this case by the strain. This is achieved by (35). The electric field in (35) can be composed of the field induced by the strain, $\vec{E}_{strain}$, and an externally applied field $\vec{E}_{ext}$. By varying this external field, one can adjust the total field $\vec{E} = \vec{E}_{ext} + \vec{E}_{strain}$ independently of the strain and S and $\vec{E}$ consequently can be considered as independent variables. If we choose particularly $\vec{E}_{ext} = - \vec{E}_{strain}$, no macroscopic electric field is present in the strained crystal. The polarization $\vec{P} = \vec{P}|_{\vec{E}=0}$ is independent of the boundary of the crystal in this case; all long-range fields induced by the ionic displacements are compensated and no surface depolarization fields occur. Therefore $e_{\lambda,\mu\nu}$ is manifestly a bulk quantity. Thus, while $\vec{P}$ as well as $\vec{E}$ in (35) generally depends on the geometry of the sample, the coefficients in (35) are bulk quantities, i.e. determined by the local electronic structure of the crystal.

Consider now a long-wavelength acoustic phonon wave in the crystal. As long as the wavelength is small compared to the size of the crystal, no (macroscopic) charges are induced on the surface and the problems just discussed for a homogeneously strained crystal do not occur. However, it is still useful to separate the long-range and the short-range parts of the polarization in order to be able to express the el-ph coupling in terms of the measured piezoelectric constants $e_{\lambda,\mu\nu}$.

For an acoustic wave of wave-vector $\vec{q}$, the strain field is

$$S_{\mu\nu}(\vec{q}) = \frac{i}{2}(q_\mu u_\nu + q_\nu u_\mu). \tag{36}$$

The el-ph interaction is given by the macroscopic electric potential set up by this strain field. Using $\vec{\nabla}\cdot\vec{D} = \vec{\nabla}\cdot(\vec{E} + 4\pi\vec{P}) = 0$, one has

$$\vec{q}\cdot\vec{D}(\vec{q}) = 0 = 4\pi\vec{q}\cdot e\cdot S(\vec{q}) + \vec{q}\cdot\varepsilon^o\cdot\vec{E}(\vec{q}). \tag{37}$$

The electric field is an electrostatic field and therefore longitudinal. Note that we use the adiabatic approximation where the phonons can be viewed as frozen-in perturbations. We write, with $\hat{q} = \vec{q}/q$,

$$\vec{E}(\vec{q}) = \hat{q}E(\vec{q}) = -i\vec{q}\delta V(\vec{q}), \tag{38}$$

with $\delta V(\vec{q})$ being the induced el-ph potential. From (37) and (36), we find

$$\delta V(\vec{q}) = 4\pi \frac{q_\lambda e_{\lambda,\mu\nu} q_\mu}{\vec{q}\cdot\varepsilon^o\cdot\vec{q}} u_\nu. \tag{39}$$

In cubic materials, the denominator can be written as $q^2\varepsilon_o$. Expressing $\vec{u}$ in terms of the normalized phonon polarization vector $\vec{\xi}(j)$ - where j labels the acoustic mode - and a phonon amplitude, and forming the el-ph matrix element, one finally gets

$$M_{kq}(j) = \frac{4\pi q_\lambda e_{\lambda,\mu\nu} q_\mu}{q^2\varepsilon_o}\xi_\nu(j)\left(\frac{1}{2MN\omega_j(\vec{q})}\right)^{\frac{1}{2}}\left(N_{\vec{q}} + \frac{1}{2} \pm \frac{1}{2}\right)^{\frac{1}{2}}. \tag{40}$$

In the zincblende structure, the only non-zero, independent piezoelectric constant is $e_{x,yz}$. This gives for j = TA or LA

$$M_{kq}(j) = \frac{8\pi e_{x,yz}}{q^2\varepsilon_o}\left(\frac{1}{2MN\omega_j(\vec{q})}\right)^{\frac{1}{2}}(q_x q_y \xi_z + \text{perm.})$$

$$\times\left(N_{\vec{q}} + \frac{1}{2} \pm \frac{1}{2}\right)^{\frac{1}{2}}. \tag{41}$$

With a denoting the lattice constant, $a^2 e_{x,yz}$ has the dimension of charge and ranges from (Martin, 1972) -0.1 to +0.1 for the III-V's and II-VI's. In contrast to the polar-optical interaction, longitudinal as well as transverse (acoustic) modes give rise to this electrostatic coupling. Notice that the electric field (38) is longitudinal, but this does not require or imply $\vec{u}//\vec{q}$ in the piezoelectric case. The wave-vector dependence is in this case $M_{kq} \sim q^{-\frac{1}{2}}$.

MICROSCOPIC APPROACH

We shall now investigate the entire el-ph interaction from a fundamental or first-principles point of view. Two questions will be dealt with particularly: To what extent can the phenomenological concepts be understood from a microscopic approach, which will be investigated in this section and how can the various coupling terms be calculated, which will be discussed in the following section. The only essential approximation in our investigation (Vogl, 1976, 1978) is the adiabatic approximation. Since it makes the discussion much clearer, we shall additionally adopt a self-consistent field or one-electron approximation. We also ignore relativistic effects.

We consider a perfect semiconductor as made up of ions which include the core electrons and a corresponding number of valence electrons. The ions are treated as point charges Z_α with equilibrium positions $\vec{x}_L$, $L \equiv (\ell\alpha)$ denoting the lattice cell ℓ and α the ions in the cell. The one-electron Hamiltonian of the perfect crystal is given by

$$[-\frac{1}{2}\nabla^2 + V(\vec{r})]|n\vec{k}\rangle = E_{n\vec{k}}|n\vec{k}\rangle, \tag{42}$$

$$V(\vec{r}) = \sum_{\ell\alpha} V(\vec{r} - \vec{x}_L, \alpha). \tag{43}$$

$V(\vec{r})$ is the self-consistent periodic lattice potential, which we describe by a pseudopotential, $|n\vec{k}\rangle$ are the Bloch states, and the band structure is given by the eigenvalues $E_{n\vec{k}}$ of (42). Several pseudopotential models are commonly used for $V(\vec{r})$ and its Fourier coefficients are tabulated for practically all materials (see e.g. Cohen and Heine, 1970). $V(\vec{r} - \vec{x}_L, \alpha)$ consists of the bare ion pseudopotential $v(\vec{r}-\vec{x}_L, \alpha)$ plus the self-consistent Hartree potential of the other valence electrons. For definiteness, we take for the bare el-ion potential

$$v(\vec{r} - \vec{x}_L, \alpha) = \frac{-Z_\alpha}{|\vec{r} - \vec{x}_L|} \qquad \text{outside the ion core,}$$

$$= 0 \qquad \text{inside the ion core} \tag{44}$$

For the III-V's, $Z_\alpha = 3, 5$ for the group-III and group-V atoms, respectively. We now consider ionic displacements $\vec{u}_L$ which cause a change $[\partial v(\vec{r} - \vec{x}_L, \alpha)/\partial \vec{x}_L] \cdot \vec{u}_L$ in the bare el-ion potential. This in turn will redistribute the valence electrons and set up a change in the total self-consistent potential. Since the displacements are small, this induced total potential will also be linear in $\vec{u}_L$ and can be expressed in terms of the dielectric function $\varepsilon(\vec{r},\vec{r}')$ of the crystal (Sham and Ziman, 1963, Sham, 1974, Pines and Nozières, 1966):

$$\delta V(\vec{r}) = \int d^3r' \; \varepsilon^{-1}(\vec{r},\vec{r}')\,\delta v(\vec{r}'), \tag{45}$$

$$\delta v(\vec{r}) = \sum_L \frac{\partial v(\vec{r} - \vec{x}_L, \alpha)}{\partial \vec{x}_L} \cdot \vec{u}_L. \tag{46}$$

In homogeneous materials, like metals, the dielectric function depends only on $|\vec{r} - \vec{r}'|$ and one obtains the "screened" potential simply by dividing the applied perturbation by ε. In general, however, and particularly in semiconductors, (45) gives the total potential set up by a small perturbation, in this case by $\delta v(\vec{r})$ from (46).

We already make use of the adiabatic approximation by considering the displacements $\vec{u}_L$ as "frozen-in" - perturbations of the perfect crystal and, correspondingly, the potential (46) as a static applied or bare potential. $\delta V(\vec{r})$ is the total potential felt by an extra electron or hole in the crystal, e.g. by a conduction electron, if we are interested in the scattering of conduction electrons by phonons. The dielectric function $\varepsilon(\vec{r},\vec{r}')$ and its inverse can most easily be calculated in Fourier space. Since it is a property of the perfect crystal, the translational invariance of the crystal gives

$$\varepsilon(\vec{r} + \vec{x}_L, \vec{r}' + \vec{x}_L) = \varepsilon(\vec{r},\vec{r}'). \tag{47}$$

The Fourier transform consequently has the form $\varepsilon(\vec{q} + \vec{G}, \vec{q} + \vec{G}')$, where $\vec{q}$ is a wave vector in the first Brillouin zone and $\vec{G}$ and $\vec{G}'$ are reciprocal lattice vectors. In the Hartree approximation, one obtains the Lindhard-type expression (Sham, 1974, Pines and Nozières, 1966)

$$\varepsilon(\vec{q} + \vec{G}, \vec{q} + \vec{G}') = \delta_{GG'} - v(\vec{q} + \vec{G})\chi^P(\vec{q} + \vec{G}, \vec{q} + \vec{G}'), \tag{48}$$

$$\chi^p(\vec{q}+\vec{G}, \vec{q}+\vec{G}') = \sum_{nn'\vec{k}} \frac{f_{n'\vec{k}+\vec{q}} - f_{n\vec{k}}}{E_{n'\vec{k}+\vec{q}} - E_{n\vec{k}}}$$

$$\times \langle n\vec{k}|e^{-i(\vec{q}+\vec{G})\vec{r}}|n'\vec{k}+\vec{q}\rangle\langle n'\vec{k}+\vec{q}|e^{i(\vec{q}+\vec{G}')\vec{r}}|n\vec{k}\rangle. \tag{49}$$

In (49), $|n\vec{k}\rangle$ and $E_{n\vec{k}}$ are the Bloch solutions of (42); $f_{n\vec{k}} = 1$ if n is one of the valence bands and $f_{n\vec{k}} = 0$ otherwise. The Fourier transform of the Coulomb potential in (48) is given by

$$v(\vec{q}) = \frac{4\pi}{q^2}, \quad q \neq 0$$

$$= 0 \quad , \quad q = 0. \tag{50}$$

The average of the Coulomb potential is the (infinite) self-energy of the electrons which is exactly compensated by the positive ions. By setting $v(\vec{q} = 0) = 0$ we take into account the neutrality of the whole system.

The dielectric function ε in (48) can be viewed as a matrix, the rows and columns being labeled by $\vec{G}$ and $\vec{G}'$. This matrix has to be inverted to obtain $\varepsilon^{-1}(\vec{q}+\vec{G}, \vec{q}+\vec{G}')$ which is needed in the el-ph potential (45). We now expand the displacements $\vec{u}_L$ in normal modes and consider a single phonon characterized by the wave-vector $\vec{q}$, branch j, frequency $\omega_j(\vec{q})$, polarization vector $\vec{\xi}(\alpha|\vec{q}j)$ and amplitudes $a_{\vec{q}j}$, $a^\dagger_{\vec{q}j}$:

$$\vec{u}_L = \vec{u}_\alpha e^{i\vec{q}\cdot\vec{x}_L}, \tag{51}$$

$$\vec{u}_\alpha = (2NM_\alpha\omega_j(\vec{q}))^{-\frac{1}{2}} \vec{\xi}(\alpha|\vec{q}j)\ (a_{\vec{q}j} + a^\dagger_{-\vec{q}j}). \tag{52}$$

Here M_α is the mass of the ions of type α. In the electron-phonon matrix element, only the matrix element of $\vec{u}_\alpha$ with respect to the phonon part of the wave function enters [see (5), (6)], which we can write as

$$\langle N_{\vec{q}} \pm 1|\vec{u}_\alpha|N_{\vec{q}}\rangle = (2NM_\alpha\omega_j(\vec{q}))^{-\frac{1}{2}} \vec{\xi}(\alpha|\vec{q}j)(N_{\vec{q}}+\tfrac{1}{2}\pm\tfrac{1}{2})^{\frac{1}{2}}. \tag{53}$$

$N_{\vec{q}}$ is the phonon occupation number. In the following way, we shall denote this matrix element simply by $\vec{u}_\alpha$.

After all these preparations, we can write down the total el-ph scattering matrix element (Sham and Ziman, 1963, Sham, 1974, Vogl, 1976):

$$M_{n\vec{k},n'\vec{k}+\vec{q}}(qj) = \langle n'\vec{k}+\vec{q}|\delta V(\vec{r})|n\vec{k}\rangle, \tag{54}$$

$$\delta V(\vec{r}) = \sum_{G} e^{i(\vec{q}+\vec{G})\cdot\vec{r}}\, \delta V(\vec{q}+\vec{G}), \tag{55}$$

$$\delta V(\vec{q}+\vec{G}) = \sum_{\vec{G}'} \varepsilon^{-1}(\vec{q}+\vec{G}, \vec{q}+\vec{G}')\,\delta v(\vec{q}+\vec{G}'), \tag{56}$$

$$\delta v(\vec{q}+\vec{G}) = \frac{-i}{\Omega_o} \sum_{\alpha} v(\vec{q}+\vec{G},\alpha)(\vec{q}+\vec{G})\cdot\vec{u}_\alpha\, e^{-i\vec{G}\cdot\vec{x}_\alpha}. \tag{57}$$

Equation (54) is a completely general definition of the el-ph scattering amplitude. Apart from the neglect of many-body effects, which can also be included and are discussed by Vogl (1976), the only approximation is the adiabatic approximation because of which only static potentials and the static dielectric function enter. The use of the linear response of the electrons to the displacements is also rigorous as long as we consider only one-phonon processes. Many-phonon processes are coupled with terms of quadratic and higher order in the $\vec{u}_L$'s and then higher-order electronic response functions enter the scattering matrix (Vogl, 1976).

In principle, we could stop here. Equations (54) - (57) give the el-ph scattering matrix element for any mode in any crystal. The bare pseudopotential (57) is given in (44) and the dielectric matrix can be calculated from (48) and (49). Actually, the expression for the el-ph interaction as given in (54) is neither transparent nor practical - the inversion of the full dielectric matrix $\varepsilon(\vec{q}+\vec{G}, \vec{q}+\vec{G}')$ is far from being simple.

We shall now analyze (54) - (57) in order to exhibit the connection of this general el-ph matrix element with the various phenomenological matrix elements. Since we want to make contact with the previous sections, we shall study particularly the long-wavelength form of M_{kq}.

Due to the Coulombic interactions the el-ph interaction $\delta V(\vec{r})$ in (54) contains long-range interaction terms and short-range contributions. We shall separate these terms from one another. This can be done by Fourier analyzing the potential $\delta V(\vec{r})$. The Fourier expansion of $\delta V(\vec{r})$ can be written

$$\delta V(\vec{r}) = e^{i\vec{q}\cdot\vec{r}}\delta V(\vec{q}) + \sum_{\vec{G}\neq 0} e^{i(\vec{q}+\vec{G})\cdot\vec{r}}\delta V(\vec{q}+\vec{G}). \tag{58}$$

For long wavelengths, $\exp(i\vec{q}\cdot\vec{r})$ is practically constant within a unit cell and the first term on the r.h.s. of (58) therefore represents the macroscopic part of the potential $\delta V(\vec{r})$, giving rise to a macroscopic, long-range electric field in the crystal. On the other hand, $\exp(i\vec{q}\cdot\vec{r}+i\vec{G}\cdot\vec{r})$ with $\vec{G}\neq 0$ varies rapidly within each unit cell; its mean value over a length comparable with the cell dimension vanishes. However, the second term on the r.h.s. of (58) does not yet represent the actual short-range part of the potential; it behaves singularly for $\vec{q}\to 0$ (which corresponds to $\vec{r}\to\infty$):

$$\vec{G}\neq 0: \quad \delta V(\vec{q}+\vec{G}) \xrightarrow[\vec{q}\to 0]{} \infty. \tag{59}$$

This strange behaviour can easily be understood. We write (56) symbolically

$$\delta V = \varepsilon^{-1}\delta v. \tag{56'}$$

The dielectric function ε from (48) can similarly be written

$$\varepsilon = 1 - v\chi^{p}. \tag{48'}$$

The inverse of ε can formally be obtained as a series,

$$\varepsilon^{-1} = 1 + v\chi^{p} + v\chi^{p}v\chi^{p} + \ldots, \tag{60}$$

which contains repeatedly the Coulomb potential $v(\vec{q}+\vec{G})$ as well as the $\vec{G}=0$ component $v(\vec{q}) = 4\pi e^2/q^2$ which gives rise to the singularity of (59). χ^{p} itself is a smooth function of the wavevector due to the finite energy denominators of the order of E_{gap} in (49).

In order to separate completely the long-range and the short-range terms in the el-ph interaction potential $\delta V(\vec{r})$, one has therefore to extract all terms in $\delta V(\vec{q}+\vec{G})$, $\vec{G}\neq 0$, which are proportional to $v(\vec{q})$. This is done in the Appendix. The result is

$$\delta V(\vec{r}) = e^{i\vec{q}\cdot\vec{r}}\,\delta V(\vec{q}) + \sum_{\vec{G}\neq 0} e^{i(\vec{q}+\vec{G})\cdot\vec{r}}\,\delta V^{sr}(\vec{q}+\vec{G})$$

$$+\, e^{i\vec{q}\cdot\vec{r}}\Big(\sum_{\vec{G}\neq 0} e^{i\vec{G}\cdot\vec{r}}\,\frac{\varepsilon^{-1}(\vec{q}+\vec{G},\vec{q})}{\varepsilon^{-1}(\vec{q},\vec{q})}\Big)\,\delta V(\vec{q}). \tag{61}$$

The last term on the r.h.s. of (61) can be viewed as a correction to the first term, being proportional to the long-range part of the total el-ph potential $\delta V(\vec{q})$ · $\delta V^{sr}(\vec{q}+\vec{G})$ is the actual short-range part of the el-ph potential. It can be expressed in terms of the dielectric matrix ε and is given explicitly in the Appendix. For the following, only two properties of $\delta V^{sr}(\vec{q}+\vec{G})$ are of importance: First, it behaves regularly for $\vec{q}\rightarrow 0$, i.e.

$$\delta V^{sr}(\vec{q}+\vec{G}) \xrightarrow[\vec{q}\rightarrow\vec{0}]{} \delta V^{sr}(\vec{G}) + q \cdot \delta V^{sr'}(\vec{G}) + \ldots . \tag{62}$$

This implies that $\delta V^{sr}(\vec{r})$ is truly of short-range and drops off rapidly for large $\vec{r}$. Second, within a rigid-ion approximation, it is given by

$$\delta V^{sr}(\vec{q}+\vec{G}) = (1-\delta_{\vec{G}0})\int d^3r\, e^{-i(\vec{q}+\vec{G})\cdot\vec{r}} \sum_L \frac{\partial V(\vec{r}-\vec{x}_L,\alpha)}{\partial\vec{r}} \cdot \vec{u}_L$$

$$= (1-\delta_{\vec{G}0}) \frac{-i}{\Omega_o} \sum_\alpha V(\vec{q}+\vec{G},\alpha)(\vec{q}+\vec{G}) \cdot \vec{u}_\alpha\, e^{-i\vec{G}\cdot\vec{x}_\alpha}, \tag{63}$$

where $V(\vec{r},\alpha)$ and its Fourier transform $V(\vec{q}+\vec{G},\alpha)$ is the self-consistent pseudopotential of the perfect crystal for the ions of sublattice α which was defined in (43). The crucial point is that with (63) we can obtain $\delta V^{sr}(\vec{q}+\vec{G})$ directly from the tabulated pseudopotentials of the perfect crystals.

For the following, it is convenient to extract the displacements explicitly from the el-ph potential and write expression (56) for $\delta V(\vec{q}+\vec{G})$ in the form

$$\delta V(\vec{q}+\vec{G}) = \frac{1}{\Omega_o} \sum_\alpha \partial\vec{V}(\vec{q}+\vec{G},\alpha) \cdot \vec{u}_\alpha . \tag{64}$$

Then, $\partial\vec{V}(\vec{q}+\vec{G},\alpha)$ is the change of the el-ph potential per unit displacement of the sublattice α in the lattice wave with wave vector $\vec{q}$. Similarly, we extract the $\vec{u}_\alpha$'s from $\delta V^{sr}(\vec{q}+\vec{G})$ and introduce $\partial\vec{V}^{sr}$ by

$$\delta V^{sr}(\vec{q}+\vec{G}) = \frac{1}{\Omega_o} \sum_\alpha \partial\vec{V}^{sr}(\vec{q}+\vec{G},\alpha) \cdot \vec{u}_\alpha . \tag{65}$$

Having separated the long-range and the short-range parts of the el-ph interaction, we now investigate the long-range part of the el-ph interaction $\delta V(\vec{q})$, which is given by (56) and reads

$$\delta V(\vec{q}) = \sum_G \varepsilon^{-1}(\vec{q},\vec{q}+\vec{G})\ \delta v(\vec{q}+\vec{G}). \tag{66}$$

It is again advantageous to extract from ε^{-1} all terms proportional to the long-range part of the Coulomb potential $v(\vec{q})$ as indicated in (60). This is also done in the Appendix. Then, (66) can be written in the form

$$\delta V(\vec{q}) = -\frac{4\pi i}{\Omega_o} \frac{q}{q^2 \varepsilon(\vec{q})} \sum_\alpha \vec{Z}(\vec{q}|\alpha) \cdot \vec{u}_\alpha, \qquad (67)$$

$$\vec{Z}(\vec{q}|\alpha) = \hat{q} \frac{v(\vec{q},\alpha)}{v(\vec{q})} + i \sum_{\vec{G}\neq 0} q^{-1} \chi^P(\vec{q},\vec{q}+\vec{G})\, \partial\vec{V}^{sr}(\vec{q}+\vec{G},\alpha). \qquad (68)$$

Here $\varepsilon(\vec{q}) = 1/\varepsilon^{-1}(\vec{q},\vec{q})$ and $\hat{q} = \vec{q}/q$. Equation (67) is an exact relation but it offers a simple physical interpretation. It represents the macroscopic dipole potential induced by the lattice wave of wave vector $\vec{q}$ in a dielectric medium with the macroscopic dielectric function $\varepsilon(\vec{q})$ and with ions characterized by an effective charge per unit displacement $\vec{Z}(\vec{q}|\alpha)$. Equation (67) resembles the phenomenological Fröhlich interaction (31), but it involves no approximations (except the adiabatic theorem). This analogy can be carried still further. The longitudinal macroscopic polarization $\hat{q}\cdot\vec{P}(\vec{q})$ set up by the displacement is

$$\hat{q} \cdot \vec{P}(\vec{q}) = -\frac{1}{4\pi} E(\vec{q}) = \frac{1}{4\pi} iq\, \delta V(\vec{q}), \qquad (69)$$

which is, with (67), given by

$$\hat{q} \cdot \vec{P}(\vec{q}) = \frac{1}{\Omega_o \varepsilon(\vec{q})} \sum_\alpha \vec{Z}(\vec{q}|\alpha) \cdot \vec{u}_\alpha. \qquad (70)$$

With this, the long-range part of the el-ph interaction can be written as

$$\delta V(\vec{q}) = -4\pi i \frac{\vec{q} \cdot \vec{P}(\vec{q})}{q^2}. \qquad (71)$$

The $\vec{q}$-dependence of the polarization can be understood as follows: A spatially varying polarization can always be expanded to yield a dipole density, quadrupole density, etc. Take Poisson's equation: $\vec{\nabla} \cdot \vec{P} = -\delta\rho$,

$$i\vec{q} \cdot \vec{P}(\vec{q}) = -\int d^3r\, e^{-i\vec{q}\cdot\vec{r}}\, \delta\rho(\vec{r}), \qquad (72)$$

and expand both sides in powers of q:

$$i\vec{q}\cdot(\vec{P}^{(o)}+iq\,\vec{P}^{(1)}+\ldots)=\int d^3r[i\vec{q}\cdot\vec{r}+\frac{1}{2}(\vec{q}\cdot\vec{r})^2+\ldots]\delta\rho(\vec{r}). \quad (73)$$

The zeroth order term on the r.h.s. vanishes due to charge neutrality. One sees that an expansion of $\vec{P}(\vec{q})$ in powers of q gives the dipole, quadrupole, etc., polarization.

Let us now collect the results obtained so far and apply them to the el-ph scattering matrix element M_{kq} of (54). We shall first discuss intra-band scattering, where the Bloch functions belong to the same band. We have

$$M_{kq}^{intra}=\langle n\vec{k}+\vec{q}|e^{i\vec{q}\cdot\vec{r}}|n\vec{k}\rangle\ (-4\pi i)\ \frac{\vec{q}\cdot\vec{P}(\vec{q})}{q^2}$$

$$+\langle n\vec{k}+\vec{q}|\sum_{\vec{G}\neq 0}e^{i(\vec{q}+\vec{G})\cdot\vec{r}}\ \delta V^{sr}(\vec{q}+\vec{G})\,|n\vec{k}\rangle$$

$$+\langle n\vec{k}+\vec{q}|e^{i\vec{q}\cdot\vec{r}}\sum_{\vec{G}\neq 0}e^{i\vec{G}\cdot\vec{r}}\frac{\varepsilon^{-1}(\vec{q}+\vec{G},\vec{q})}{\varepsilon^{-1}(\vec{q},\vec{q})}\ \delta V(\vec{q})\,|n\vec{k}\rangle. \quad (74)$$

In order to compare with the phenomenological theory, we study M_{kq} in the long-wavelength limit. We first consider the third term on the r.h.s. of (74). As mentioned above, it can be considered as a correction to the first term. From the long-wavelength behaviour of ε^{-1} one can show that this third term is always of higher order in q than the first term (Vogl, 1976). For cubic materials in particular, it is smaller by a factor $(\vec{q}/\vec{G})^2$ than the first term and we shall therefore neglect it. We then write

$$M_{kq}^{intra}=M_{kq}^{\ell r}+M_{kq}^{sr}\ . \quad (75)$$

For the long-range part, we have

$$M_{kq}^{\ell r}=\langle n\vec{k}+\vec{q}|e^{i\vec{q}\cdot\vec{r}}|n\vec{k}\rangle(-4\pi i)\ \cdot\ \frac{\vec{q}\cdot\vec{P}(\vec{q})}{q^2}\ , \quad (76)$$

with the macroscopic polarization $\vec{P}(\vec{q})$ given by (70),

$$\vec{q}\cdot\vec{P}(\vec{q})=\frac{1}{\Omega_o\varepsilon(\vec{q})}\sum_{\alpha}\vec{Z}(\vec{q}|\alpha)\cdot\vec{u}_{\alpha}. \quad (77)$$

Let us now expand all wave-vector dependent quantities in (76) in powers of $\vec{q}$ and consider the leading terms.

The Bloch matrix element tends to the normalization integral for $\vec{q}\to 0$ and one has (Antoncik and Landsberg, 1963)

$$\langle n\vec{k}+\vec{q}|e^{i\vec{q}\cdot\vec{r}}|n\vec{k}\rangle \underset{q\to 0}{=} 1 + O(q^2). \tag{78}$$

The leading terms of M^{lr}_{kq} are then

$$M^{lr}_{kq} = -4\pi i\,\frac{\vec{q}\cdot\vec{P}^{(0)}}{q^2} + 4\pi\,\frac{q\vec{q}\cdot\vec{P}^{(1)}}{q^2}\,, \tag{79}$$

where the macroscopic polarization $\vec{P}(\vec{q})$ was expanded according to (73) and the first term in (79) is a macroscopic dipole potential, the second term a macroscopic quadrupole potential etc. In order to obtain these terms explicitly, we expand the q-dependent effective charge from (68),

$$Z_\lambda(\vec{q}|\alpha) = e^*_T(\alpha)_{\lambda\mu}\hat{q}_\mu + \frac{i}{2}\,Q(\alpha)_{\lambda\mu\nu}\hat{q}_\mu q_\nu + \ldots\,, \tag{80}$$

the macroscopic dielectric function $\varepsilon(\vec{q})$,

$$\varepsilon(\vec{q}) = \varepsilon_\infty + O(q^2), \tag{81}$$

where ε_∞ represents the optical dielectric constant, and the displacements $\vec{u}_\alpha$ which depend on the wave-vector through the phonon polarization vectors [see (52)],

$$\vec{u}_\alpha = \vec{u}^{(0)}_\alpha + iq\vec{u}^{(1)}_\alpha + \ldots\,. \tag{82}$$

In order to avoid too many indices, we write (80) in a simpler fashion as

$$\vec{Z}(\vec{q}|\alpha) = e^*_T(\alpha)\cdot\hat{q} + \frac{i}{2}\,Q(\alpha)\cdot\hat{q}\cdot\vec{q} + \ldots\,, \tag{80'}$$

keeping in mind that $e^*_T(\alpha)$ and $Q(\alpha)$ are second and third rank tensors, respectively. The leading term of M^{lr}_{kq} then becomes

$$M_{kq}^{lr} = -\frac{4\pi i}{\Omega_o}\frac{\vec{q}}{q^2\varepsilon_\infty}\cdot\sum_\alpha e_T^*(\alpha)\cdot\vec{u}_\alpha^{(0)} + \ldots . \tag{83}$$

For a cubic crystal with two sublattices $\alpha = 1,2$ we define $e_T^* = (e_T^*(1) - e_T^*(2))/2$ and this expression then becomes, for a long optical wave,

$$M_{kq}^{lr} = -\frac{4\pi i}{\Omega_o}\frac{e_T^*\vec{q}}{\varepsilon_\infty q^2}\cdot\vec{u}_{rel}, \tag{84}$$

in exact agreement with the phenomenological Fröhlich polar-optical interaction. This gives us a rigorous expression for the transverse and longitudinal dynamical effective charge $e_T^*(\alpha)$ and $e_L^*(\alpha) = e_T^*(\alpha)/\varepsilon$ of the ions of sublattice α. Mathematically, $e_T^*(\alpha)$ is

$$e_T^*(\alpha)_{\lambda\mu} = -Z_\alpha\hat{q}_\lambda\delta_{\lambda\mu} + \sum_{\vec{G}\neq 0}\left(\frac{\partial\chi^p(\vec{q},\vec{q}+\vec{G})}{\partial q_\lambda}\right)_{\vec{q}=0} i\partial V_\mu^{sr}(\vec{G},\alpha), \tag{85}$$

which follows from the long-wavelength form of $\vec{Z}(\vec{q}|\alpha)$, (68), which is also discussed in the Appendix. Physically, $e_T^*(\alpha)$ represents the dipole polarization for a rigid unit displacement of the sublattice α and for fixed macroscopic electric field; (85) is the microscopic expression for the Born charge. We note that ε_∞ can also be expressed in terms of the band structure of the perfect crystal by expression (48) for ε. With (83) and (84) we have a microscopic justification of the Fröhlich Hamiltonian. It is the correct el-ph interaction for LO modes in the limit of long wavelengths and dominates for $\vec{q}\to 0$. It is independent of a point charge model and includes correctly the polarizability of the valence electrons.

For TO modes in a cubic crystal, $\vec{q}\cdot\vec{u}_{rel} = 0$ and the interaction (84) is zero. What gives this interaction term for acoustic modes? For long acoustic modes, the zeroth-order displacements $\vec{u}_\alpha^{(0)}$ are independent of the sublattice α, i.e. $\vec{u}_\alpha^{(0)} = \vec{u}^{(0)}$, corresponding to a rigid shift of the unit cell. A rigid displacement of the whole crystal does not induce a dipole moment, as can be seen from (73):

$$\vec{P}^{(0)} = \int d^3r\,\vec{r}\,\frac{\partial\rho(r)}{\partial\vec{r}}\cdot\vec{u} = -\left(\int d^3r\,\rho(\vec{r})\right)\vec{u} = 0. \tag{86}$$

From (77) this gives the sum rule

$$\sum_\alpha e_T^*(\alpha) = 0, \tag{87}$$

which guarantees that the dipole type el-ph interaction (83) vanishes for acoustic modes. The sum rule (87) is called the acoustic sum rule and was first derived by Sham (1969) and Pick et al. (1970) and again by Sham (1974) within a microscopic theory of lattice dynamics. In these papers, microscopic expressions for the effective charge were also obtained for the first time.

Consider nonpolar crystals like Si and Ge. From symmetry, in this case $e_T^*(1) = e_T^*(2)$ for the two sublattices. The acoustic sum rule gives additionally $e_T^*(1) + e_T^*(2) = 0$, which gives $e_T^*(1) = e_T^*(2) = 0$ in nonpolar cubic crystals.

To summarize, the leading term of the el-ph interaction in semiconductors is

$$M_{kq}^{lr} = -4\pi i \frac{\vec{q}\cdot\vec{P}^{(0)}}{q^2} + \ldots . \tag{88}$$

For LO modes in polar crystals, this is the Fröhlich interaction. This interaction term vanishes for acoustic modes. For cubic crystals, the interaction (88) also vanishes for TO modes in polar semiconductors and it vanishes for all modes in nonpolar materials.

The next order term in the el-ph matrix element is

$$M_{kq}^{lr} = 4\pi \frac{q\vec{q}\cdot\vec{P}^{(1)}}{q^2} + \ldots . \tag{89}$$

By the expansions (80) - (82), this has the form

$$M_{kq}^{lr} = \frac{4\pi}{\Omega_0} \frac{q\vec{q}}{q^2\varepsilon_\infty} \cdot \sum_\alpha \left(\frac{1}{2}\hat{q}\cdot Q(\alpha)\cdot\vec{u}_\alpha^{(0)} + e_T^*(\alpha)\cdot\vec{u}_\alpha^{(1)}\right) + \ldots \tag{90}$$

and is a quadrupole-type el-ph interaction of the order q^0. The quadrupole moments $Q(\alpha)$ can be obtained straightforwardly in terms of the Hartree polarizability χ^p and the short-range potential ∂V^{sr} in analogy to $e_T^*(\alpha)$ in (85).

For optical modes in zincblende or diamond type crystals, $\vec{u}_\alpha^{(1)} = 0$ for the symmetry directions and the interaction (90) is a pure multipole (quadrupole) contribution to the optical el-ph interaction which adds to the short-range deformation potential interaction discussed below. Defining $Q = (Q(1) - Q(2))/2$, we get for (90)

$$M_{kq}^{lr}(\text{optic}) = \frac{4\pi}{\Omega_0\varepsilon_\infty}\frac{1}{2}\hat{q}_\lambda\hat{q}_\mu Q_{\lambda\mu\nu}\, u_{rel,\nu} + \ldots . \tag{91}$$

This is the leading term of the el-ph interaction in nonpolar crystals, where the effective charge $e_T^* = 0$. From a phenomenological point of view, these multipole couplings were discussed by Bir and Pikus (1974) and Lawaetz (1969).

For acoustic modes in polar semiconductors, the quadrupole type interaction (89) is the piezoelectric interaction. This can be shown as follows. For acoustic waves, $\vec{u}_\alpha^{(o)} = \vec{u}^{(o)}$ is a rigid shift of the unit cell and $iq_\nu u_\mu^{(o)}$ is the (unsymmetrized) macroscopic strain. $\vec{u}_\alpha^{(1)}$ represents the so-called internal strain which describes the relative displacement of the sublattices in an acoustic mode. From the harmonic force constants and the equations of motion for the displacements, one can derive a relation between the internal strain, the macroscopic strain and the macroscopic polarization in the crystal associated with the acoustic mode (Maradudin, 1974, Vogl, 1976). Moreover, the piezoelectric constants can be expressed in terms of the quadrupole moments $Q(\alpha)$ and dynamical effective charges $e_T^*(\alpha)$ (Maradudin, 1974, Martin, 1972, Vogl, 1976):

$$e_T^*(\alpha) \cdot \vec{u}_\alpha^{(1)} = e_T^*(\alpha) \cdot \Gamma(\alpha) \cdot \hat{q} \cdot \vec{u}^{(0)} + (\varepsilon_\infty - \varepsilon_o) \cdot \vec{P}^{(1)}, \tag{92}$$

$$e_{\lambda,\mu\nu} = \frac{1}{\Omega_o} \sum_\alpha Q_{\lambda\mu\nu}(\alpha) + \sum_\alpha e^*_{T\lambda\lambda'}(\alpha)\Gamma_{\lambda'\mu\nu}(\alpha). \tag{93}$$

The third rank tensor $\Gamma(\alpha)$ is the internal strain tensor determined by the interatomic force constants and ε_o is the static dielectric tensor. With these relations, the coupling term (89) becomes

$$M_{kq}^{lr}(\text{acoustic}) = 4\pi \frac{q_\lambda q_\mu e_{\lambda,\mu\nu}}{q\cdot\varepsilon_o\cdot q} u_\nu + \ldots, \tag{94}$$

in exact agreement with the phenomenological piezoacoustic coupling term (39). In nonpolar crystals, $e_{\lambda,\mu\nu} = 0$ from symmetry. Notice that this implies from (93) that $\Sigma_\alpha Q(\alpha) = 0$ but not $Q(\alpha) = 0$ itself which is actually non-zero in nonpolar semiconductors.

To summarize these findings, the el-ph coupling of LO modes in polar crystals is dominated by the Fröhlich interaction. For TO modes or any optical modes in nonpolar crystals, the quadrupole term (91) is the leading long-range el-ph interaction. For acoustic phonons, (94) is the leading long-range interaction and represents the phenomenological piezoelectric interaction, which vanishes in nonpolar crystals. For all modes there are higher order long-range terms proportional to q and higher order, which represent octopole and higher multipole interaction terms.

Let us now turn to the short-range part of the el-ph interaction, (74),

$$M_{kq}^{sr} = \langle n\vec{k} + \vec{q}| \sum_{\vec{G}\neq 0} e^{i(\vec{q}+\vec{G})\cdot\vec{r}} \delta V^{sr}(\vec{q}+\vec{G})\,|nk\rangle. \qquad (95)$$

This represents the short-range el-ph interaction and can be identified with the acoustic or optical deformation potential interaction. For $\vec{q}\to 0$, (95) becomes

$$M_{kq}^{sr} = \langle n\vec{k}| \sum_{\vec{G}\neq 0} e^{i\vec{G}\cdot\vec{r}} \delta V^{sr}(\vec{G})\,|n\vec{k}\rangle + i\vec{q}\cdot\sum_{\alpha} \underset{\approx}{\Xi}_{\alpha}\cdot\vec{u}_{\alpha} + \ldots \qquad (96)$$

The first term in (96) is the expectation value of a periodic potential. For acoustic phonons, this is simply the el-ion potential induced by a rigid shift of the whole crystal,

$$\delta V(\vec{r}) = \frac{\partial V(\vec{r})}{\partial\vec{r}}\cdot\vec{u}. \qquad (97)$$

This potential does not induce any scattering. This follows from the translational invariance of the Bloch functions. They are eigenfunctions of $H = -(1/2)\nabla^2 + V(\vec{r})$ and therefore

$$\langle n\vec{k}\,|\frac{\partial V(\vec{r})}{\partial\vec{r}}\,|n\vec{k}\rangle = \langle n\vec{k}|[\vec{\nabla},H]|n\vec{k}\rangle = 0. \qquad (98)$$

Thus, for acoustic modes the $\vec{q}$-independent term in (96) vanishes which expresses the fact that the el-ph interaction vanishes for a uniform translation of the crystal. We note that in deriving this result we implicitly used the relation $\delta V^{sr}(\vec{G}) = \delta V(\vec{G})$ which is shown in the Appendix. For acoustic modes, the short-range part of the el-ph interaction can therefore be written

$$M_{kq}^{sr}(\text{acoustic}) = (\sum_{\alpha} \underset{\approx}{\Xi}_{\alpha})_{\lambda\mu} S_{\lambda\mu} \qquad (99)$$

where $S_{\lambda\mu} = i(q_{\lambda}u_{\mu} + q_{\mu}u_{\lambda})/2$ is the strain tensor. We identify $\underset{\approx}{\Xi}$ with the acoustic deformation potential. It is microscopically well defined; in a rigid-ion approximation it is easily obtained from (63) in terms of the pseudopotential of the perfect crystal. For optical modes, the q-independent term in (96) corresponds to a rigid shift of the sublattices relative to one another and is therefore non-zero. It represents the short-range part of the optical deformation potential for intraband scattering.

We can now summarize the results of this section. We studied the el-ph interaction in semiconductors from a microscopic standpoint and started with the most general definition of the el-ph scattering matrix element in terms of the inverse dielectric function of the crystal. By carefully separating the long-range interactions and the short-range interactions and using only general analytical properties of the dielectric function, we find the el-ph matrix element for intra-band scattering in the long-wavelength limit to be

$$M_{kq}(\text{optic}) = -4\pi i \frac{\vec{q}\cdot\vec{P}^{(0)}}{q^2} + \left(\vec{D}_{opt}(\vec{k}) + \frac{4\pi}{\Omega_o \varepsilon_\infty}\frac{1}{2}\,\hat{q}\cdot Q\cdot\hat{q}\right)\cdot\vec{u}^{(0)}_{rel} + O(q), \tag{100}$$

$$M_{kq}(\text{acoustic}) = \frac{4\pi}{q^2\varepsilon_0}\, q_\lambda e_{\lambda,\mu\nu} q_\mu u_\nu + \left(\underset{\approx}{\Xi} + \hat{q}\cdot O\cdot\hat{q}\right)_{\mu\nu} q_\mu u_\nu + O(q^2). \tag{101}$$

The dominating term for optical phonons agrees with the phenomenological Fröhlich interaction. The second term on the r.h.s. of (100) is the optical deformation potential which contains a short-range term as well as an electrical multipole contribution. The latter is not taken into account in the standard phenomenological deformation potential theory. However, Lawaetz (1969) has shown how to implement multipole interactions in the usual scattering theory. Equation (101) is the acoustic scattering matrix element, which is dominated by the piezoelectric interaction. The second term is the acoustic deformation potential, again consisting of short-range and multipole terms.

For inter-band scattering, the analytical form of the short-range part of the el-ph matrix element remains unchanged. However, from (76) and the properties of the Bloch functions (Antoncik and Landsberg, 1963), the long-range part of the el-ph matrix element is now dominated by the "interband" Fröhlich interaction,

$$-\,4\pi i\langle n'\vec{k}|\vec{q}\cdot\vec{\nabla}_{\vec{k}}|n\vec{k}\rangle \frac{\vec{q}\cdot\vec{P}^{(0)}}{q^2}, \tag{102}$$

other contributions being already $O(q)$.

APPLICATIONS

In this section we show how the formalism developed in the previous section can be used for practical calculations and discuss examples and applications.

For a long time the dielectric formalism was thought to be impractical for actual calculations (Sham, 1974). This comes from the numerical difficulties with inverting the dielectric matrix. It is particularly difficult to get convergence for the off-diagonal elements $\varepsilon^{-1}(\vec{q},\vec{q}+\vec{G})$ by a straightforward numerical inversion because all these elements turn out to be coupled by translational invariance (Sham, 1969, Pick et al., 1970). Several schemes for inverting ε have been proposed (Bertoni et al., 1974, Bilz et al., 1974, Hanke, 1973) and only recently convergent results were reported for semiconductors (Louie and Cohen, 1978, Baldereschi and Tosatti, 1978, Shindo and Nara, 1977). We demonstrate here how one can sidestep the trouble of inverting ε and calculate the el-ph matrix element M_{kq}, given by (54), in a simple but self-consistent way (Vogl, 1978).

The basis of this simplification is the observation that the short-range part of the el-ph potential $\delta V^{sr}(\vec{q}+\vec{G})$ is in a good approximation given by the rigid-ion form (63). In principle, the calculation of $\delta V^{sr}(\vec{q}+\vec{G})$ involves the calculation of the inverse of the dielectric function (see Appendix). However, by making use of (63), δV^{sr} can be directly obtained from the tabulated perfect crystal pseudopotentials. The short-range part as well as the long-range part of the el-ph interaction can then be obtained from a calculation of only the Bloch states of the semiconductor, (42), without requiring an inversion of ε. This can be seen from (95) and (76) when the expression (68) for $\vec{Z}(\vec{q}|\alpha)$ is used: the long-range part M_{kq}^{lr} contains $\delta V^{sr}(\vec{q}+\vec{G})$ and the polarizability χ^{p} which is proportional to ε [see (48)], but does not involve ε^{-1} any more. The short-range part also contains only $\delta V^{sr}(\vec{q}+\vec{G})$ and the Bloch functions.

We shall justify the rigid-ion approximation for δV^{sr} by showing that it obeys the general invariance requirements and gives good agreement with experiment when used in a calculation of el-ph couplings. First, as can be deduced from (63), the rigid-ion approximation naturally fulfills the translational invariance condition (97) and therefore guarantees the el-ph interaction to vanish for a rigid shift of the whole crystal. Second, with this approximation, the transverse dynamical effective charge $e_T^*(\alpha)$, as given in (85), becomes

$$e_T^*(\alpha) = -Z_\alpha - 4 \sum_{\vec{G}\neq 0} \sum_{nn'\vec{k}} \frac{f n\vec{k}(1 - f n\vec{k})}{(E_{n\vec{k}} - E_{n'\vec{k}})^2} \langle n\vec{k}|p_x|n\vec{k}\rangle$$

$$\times \langle n'\vec{k}|\exp(i\vec{G}\cdot\vec{r})|n\vec{k}\rangle\, V(\vec{G},\alpha) G_x \exp(-i\vec{G}\cdot\vec{x}_\alpha), \qquad (103)$$

where the Hartree polarizability (49) was used. In (103), we have to observe a self-consistency requirement: The same semiconductor-pseudopotential has to be used for $V(\vec{G},\alpha)$ as for the calculation of the Bloch functions $|n\vec{k}\rangle$ and $E_{n\vec{k}}$. More generally, for $\vec{Z}(\vec{q}|\alpha)$ the same pseudopotential has to be used in $\delta V^{sr}(\vec{q}+\vec{G})$ as for the Bloch functions which enter the polarizability $\chi^p(\vec{q},\vec{q}+\vec{G})$. With this it can be shown that the acoustic sum rule (87), which is a consequence of translational invariance and charge neutrality, is fulfilled (Vogl, 1978).

How well does the rigid-ion approximation for δV^{sr} work in practice? Equation (103) was used to calculate transverse dynamical effective charges together with the macroscopic dielectric constant and, thereby, the Fröhlich coupling constant, for 10 semiconductors by Vogl (1978). Good agreement for III-V and II-VI compounds was obtained. Recently, we applied this to IV-VI materials which have exceptionally large e_T^* and also obtained good agreement with experiment; e.g. we find for SnTe $e_T^*(\text{theor}) = 7.3$, where $e_T^*(\text{exp}) = 8.1$ (Kawamura, 1978). Also, the experimental pressure dependence of e_T^* can be reasonably well accounted for by (103) (Trommer et al., 1979). Pseudopotential calculations of deformation potentials in semiconductors are, to our knowledge, exclusively based on the rigid ion approximation for the el-ph potential. In all calculations, onl the short-range part of the deformation potentials were included; th r.h.s. of (63) was always implicitly assumed to represent the total induced el-ph potential. Deformation potentials have been calculate for many semiconductors and give generally good agreement with experiment.

The higher order multipole moments Q and O in (100) and (101) have not yet been calculated in the pseudopotential framework presented here. Early shell-model calculations (Pindor, 1972) predicte these long-range contributions to the deformation potentials to be of the same order as the standard (short-range) deformation potentials. However, stress experiments revealed that the quadrupole moments Q, which determine the long-range part of the optical deformation potential, are at least one order of magnitude smaller than predicted by the shell-model calculation (Onton et al., 1967, Vogl et al., 1976). Two tight-binding calculations by Pindor (1972) and Vogl et al. (1976) also gave small multipole coupling terms in agreement with this finding. An attempt to determine these multipol moments from transport data (Gram and Jorgenson, 1973) cannot be considered as conclusive since many other small effects were neglected (see, e.g., Ngai, 1974). In view of the stress data and the anisotropy of the multipole coupling terms (i.e. their $\hat{q}$-dependence) we conclude that the multipole contributions to the deformation potentials in pure semiconductors are negligible for transport applications.

THE ELECTRON-PHONON INTERACTION IN NONPERFECT SEMICONDUCTORS

Screening in Highly Impure Semiconductors

Screening is important when the screening length of the free carriers becomes comparable to or smaller than the phonon wavelength, i.e. when the wave-vectors obey the condition

$$q_{sc} \gtrsim q_{phonon}. \tag{104}$$

Here,

$$\begin{aligned} q_{sc} &= \left(\frac{4\pi n}{\varepsilon_\infty k_B T}\right)^{1/2} && \text{non-degenerate carriers} \\ &= \left(\frac{4m^*}{\varepsilon_\infty}\right)^{1/2} \left(\frac{3n}{\pi}\right)^{1/6} && \text{degenerate carriers} \end{aligned} \tag{105}$$

and q_{sc} is the Debye or Thomas-Fermi screening wave vector for the non-degenerate and degenerate case, respectively, and n denotes the carrier concentration. For the standard semiconductors, the condition (104) is fulfilled typically for carrier concentrations $n = 10^{16}$ to $10^{17} cm^{-3}$.

Let us investigate how the el-ph interaction is modified in the presence of free carriers. The el-ph scattering amplitude M_{kq} as given in (54) - (57) is completely general and is valid for any (periodic) solid. When free carriers are present, their contribution to the dielectric function has to be taken into account. One can approximately write (Bogusławski and Mycielski, 1977)

$$\varepsilon(\vec{q}+\vec{G},\vec{q}+\vec{G}') = \varepsilon_{val}(\vec{q}+\vec{G},\vec{q}+\vec{G}') + (1-\delta_{\vec{G}O})v(\vec{q})\chi^{p}_{\text{free carriers}}(\vec{q},\vec{q}+\vec{G}') \tag{106}$$

where ε_{val} is the dielectric function of the pure semiconductor, unmodified by the free carriers, and χ^p is the intra-band polarizability of the free carriers in the conduction (or valence) band. The approximation (106) is based on the fact that $\chi^p_{\text{free carriers}}$

remains finite for $\vec{q} \to 0$ so that the dominating free carrier contribution to ε appears in $\varepsilon(\vec{q},\vec{q}+\vec{G})$ and goes like $v(\vec{q}) \sim q^{-2}$. The short range part of the el-ph interaction $\delta V^{sr}(\vec{q}+\vec{G})$ involves only matrix elements of ε with $\vec{G}$ and $\vec{G}' \neq 0$ (Appendix). Consequently the short-range part of the el-ph matrix element is not modified by free carriers. However, the macroscopic potential $\delta V(\vec{q})$ is screened: We had

$$\delta V(\vec{q}) = - \frac{4\pi i}{\Omega_o} \frac{q}{q^2 \varepsilon(\vec{q})} \sum_\alpha \vec{Z}(\vec{q}|\alpha) \cdot \vec{u}_\alpha, \tag{67'}$$

where now, for small $\vec{q}$,

$$\varepsilon(\vec{q}) = \varepsilon_\infty (1 + \frac{q_{sc}^2}{q^2}). \tag{107}$$

As a consequence, the polar-optical Fröhlich interaction and the piezoelectric interaction are screened by multiplying ε_∞ in the denominator of (31) by $(1 + q_{sc}^2/q^2)$ and ε_o in the denominator of (41) by $(1 + q_{sc}^2 \varepsilon_\infty / q^2 \varepsilon_o)$. A complication arises from the fact that, once screening is important, short-range and long-range terms become of the same order in q. From (107), it follows that, for $\vec{q} \to 0$, the macroscopic el-ph potential behaves like

$$\delta V(\vec{q}) \xrightarrow[q \to 0]{} q. \tag{108}$$

This is of the same order as the short-range part of the acoustic deformation potential. We also note that $\vec{Z}(q|\alpha)$ involves $\chi^P(\vec{q},\vec{q}+\vec{G})$ and is therefore also modified by free carriers, although the analytical form of (80) is still valid. As a result, the el-ph matrix element M_{kq}, which was given in (100) and (101) for the pure semiconductor, becomes

$$M_{kq}(\text{optic, impure}) = \vec{D}_{opt}(\vec{k}) \cdot \vec{u}^{(0)}_{rel} + O(q), \tag{109}$$

$$M_{kq}(\text{acoustic, impure}) = (\approx + 0')_{\mu\nu} q_\mu u_\nu + O(q^2), \tag{110}$$

where 0' is the octopole contribution modified by the free carriers. The important point is that the el-ph interaction for optical phonon in impure semiconductors is dominated by the unscreened optical deformation potential of the pure semiconductor. This was also found by Lawaetz (1978). The term $O(q)$ in (109) contains the screened Fröhlich interaction as well as contributions from the short-range part of the el-ph interaction which are of the same order in q but have - to our knowledge - not yet been considered. The matrix element for acoustic phonon scattering is given in (110) and is seen to be dominated by the acoustic deformation potential interaction. We mentioned before that the octopole terms are small for pure materials; this is not necessarily true for the free carrier contribution to O' in (110). Indeed, Bogusławski and Mycielski (1977) have shown that the long-range terms in (110) have the net effect of screening the acoustic deformation potential $\underset{\approx}{\Xi}$ of the pure material by the dielectric function (107). We refer to these authors for a further discussion of these subtleties. The effects of frequency dependent screening on the long-range el-ph interactions are discussed by Mahan (1972).

Surfaces, Interfaces

Near surfaces or interfaces the long-range el-ph interactions are modified because of the electrostatic boundary conditions for the macroscopic fields at the surface. The short-range el-ph interactions are determined by the short-range forces and are therefore unaffected by the presence of a surface except very close to it. As a result of the microscopic theory, the interaction of electrons with long-wavelength polar optical phonons can be derived by considering the ions plus valence electrons as point charges with a transverse effective charge e_T^* and using macroscopic electrostatics. e_T^* itself is determined by the short-range forces in the crystal and is therefore unaltered up to very close to the surface.

Electrons near a surface (interface) are scattered by the bulk phonons as well as by the surface (interface) modes. Long wavelength LO bulk phonons produce an el-ph potential near a free surface which is given by (Licari and Evrard, 1977, Mahan, 1974)

$$\delta V(\vec{r}_{//},z) = \frac{c_B}{(q_{//}^2 + q_z^2)^{\frac{1}{2}}} e^{i\vec{q}_{//}\cdot\vec{r}_{//}} \sin(q_z z)\,\theta(z), \tag{111}$$

$$c_B^2 = 4\pi\omega_{LO}\left(\frac{1}{\varepsilon_\infty} - \frac{1}{\varepsilon_0}\right). \tag{112}$$

In (111), we decomposed the phonon wave vector $\vec{q} = (\vec{q}_{//}, q_z)$ and the electron position operator $\vec{r} = (\vec{r}_{//}, z)$ in components parallel and perpendicular to the surface. The bulk modes produce zero potential outside the crystal which is expressed by the θ-function. The surface modes corresponding to the bulk LO phonons, on the other hand, produce a potential

$$\delta V(\vec{r}_{//},z) = \frac{c_s}{\sqrt{q}}\, e^{i\vec{q}_{//}\cdot\vec{r}_{//}} e^{-q_{//}|z|}, \tag{113}$$

$$c_s^2 = 2\pi\omega_s\left(\frac{1}{\varepsilon_\infty + 1} - \frac{1}{\varepsilon_o + 1}\right), \tag{114}$$

$$\omega_s^2 = \omega_{TO}^2 \frac{\varepsilon_o + 1}{\varepsilon_\infty + 1} . \tag{115}$$

Here ω_s is the frequency of the surface mode. An interesting phenomenon occurs at an interface of a polar and a non-polar material (Hess and Vogl, 1979). The inversion-layer electrons in Si near a Si-SiO_2 interface, for example, are coupled to the polar-optical phonons of the insulator. The polarization field set up by the LO-phonons of SiO_2 across the interface is an effective scattering mechanism for the energy loss of the carriers in Si near the interface. The interaction potential of the inversion layer carriers in Si with the polar-optical interface modes is given by (113), with C_S^2 replaced by (Hess and Vogl, 1979)

$$C_I^2 = 2\pi\omega_I\left(\frac{1}{\varepsilon_\infty + \varepsilon_\infty^{Si}} - \frac{1}{\varepsilon_o + \varepsilon_\infty^{Si}}\right), \tag{116}$$

$$\omega_I^2 = \omega_{TO}^2 \frac{\varepsilon_o + \varepsilon_\infty^{Si}}{\varepsilon_\infty + \varepsilon_\infty^{Si}}, \tag{117}$$

where ε_o, ε_∞ and ω_{TO}^2 are the SiO_2-parameters, given by Ferry (1978). A further discussion of el-ph interactions near surfaces and interfaces is given in the above references and by Maradudin and Sham (1977).

Amorphous Structures

Covalent semiconductors tend also to keep the short-range order in the amorphous phase because of the strong bonding (Weaire and Thorpe, 1971). Consequently, the short-range el-ph interactions which probe only the local structure of the solid, are not strongly affected and the deformation potential concept is still applicable. An interesting problem concerns the long-range interactions - can the Fröhlich concept still be used, does the splitting between ω_{LO} and ω_{TO} persist in the amorphous phase? To discuss this, it is instructive to write (25) in the form

$$\omega_{LO}^2 = \omega_{TO}^2 + \omega_p^2, \tag{118}$$

where ω_p^2 is the plasma frequency of the ions with an effective charge $e_T^*/\sqrt{\varepsilon_\infty}$. Consider a general system of N one-particle excitations with energy ω_T which are coupled by a long-range Coulomb interaction. Then it is known that this interaction leaves N-1 excitations essentially unchanged but leads to one collective mode with frequency $\omega_L^2 = \omega_T^2 + \omega_p^2$ (Thouless, 1961). In our case, the ω_{TO}-mode is the optical mode determined by the short-range force constants alone and can be viewed as the "single-particle" mode of the system, the ω_{LO}-mode represents the collective mode. From this argument the splitting of ω_{LO} and ω_{TO} appears not to require long-range order. Short-range order is required so that ω_{TO} is not broadened into many different quasi-local modes and so that the Coulomb interaction can actually "align" the vibrations in the (approximate) unit cells over a distance large compared to the cell dimensions. Raman experiments in amorphous SiO_2, GeO_2 (Galeener and Lucovsky, 1976) and in solid versus liquid CF_4 (Gilbert and Drifford, 1977) indicate that the splitting of LO and TO-modes and therefore also the Fröhlich coupling mechanism indeed persists in the noncrystalline phases.

ACKNOWLEDGEMENT

Work supported in part by Fonds zur Förderung der wissenschaftlichen Forschung in Österreich (Project No. S22).

APPENDIX

In this Appendix, we summarize the linear response theory and carry out explicitly the separation of long-range and short-range el-ph interactions used in (61) and (67). Let us first review the linear response or dielectric formalism (Sham, 1974, Pines and Nozieres, 1966, Adler, 1962, Wiser, 1966): A small perturbation $\delta v(r)$ induces an electron charge density

$$\delta n(\vec{r}) = \int d^3r' \ \chi(\vec{r},\vec{r}')\delta v(\vec{r}'), \tag{A.1}$$

which we write in a short-hand notation

$$\delta n = \chi \ \delta v. \tag{A.1'}$$

The total induced potential consists of the applied potential δv and the Hartree potential due to δn:

$$\delta V = \delta v + v\chi\delta v, \tag{A.2}$$

where $v(\vec{r}) = 1/r$ is the Coulomb interaction. Equation (A.2) can be written

$$\delta V = \varepsilon^{-1}\, \delta v, \tag{A.3}$$

$$\varepsilon^{-1} = 1 + v\chi. \tag{A.4}$$

We can also define a polarizability χ^P by

$$\delta n = \chi^P \delta V. \tag{A.5}$$

With the above equations, one finds the relations

$$\chi = \chi^P\, \varepsilon^{-1}, \tag{A.6}$$

$$\varepsilon = 1 - v\chi^P. \tag{A.7}$$

Equations (A.6) and (A.4) show that ε^{-1} and χ fulfill the following equations:

$$\varepsilon^{-1} = 1 + v\chi^P\varepsilon^{-1} = 1 + v\chi^P + v\chi^P v\chi^P + \ldots, \tag{A.8}$$

$$\chi = \chi^P\varepsilon^{-1} = \chi^P + \chi^P v\chi^P + \ldots. \tag{A.9}$$

As discussed in the main text, the Fourier transform of χ or χ^P has the form $\chi^P(\vec{q}+\vec{G},\ \vec{q}+\vec{G}')$ with $\vec{q} \in$ 1.BZ and $\vec{G},\vec{G}'$ reciporcal lattice vectors. In a Hartree approximation, χ^P is given by (49). Let us study χ^P in the long-wavelength limit. The energy denominators in (49) remain finite for $q \to 0$. The plane wave matrix elements are

$$\langle n\vec{k}|\exp[-i(\vec{q}+\vec{G})\cdot r]|n'\vec{k}+\vec{q}\rangle \underset{\vec{q}\to 0}{\longrightarrow} \langle n\vec{k}|\exp(-i\vec{G}\cdot\vec{r})|n'\vec{k}\rangle + \ldots, \tag{A.10}$$

$$\langle n\vec{k}|\exp(-i\vec{q}\cdot\vec{r})|n'\vec{k}+\vec{q}\rangle \underset{\vec{q}\to 0}{\longrightarrow} \langle n\vec{k}|n'\vec{k}\rangle + \vec{q}\cdot\langle n\vec{k}|\vec{\nabla}_{\vec{k}}|n'\vec{k}\rangle + \ldots \tag{A.11}$$

Because $n \neq n'$ in χ^P, $\langle n\vec{k}|n'\vec{k}\rangle = 0$. This gives the following behaviour of χ^P:

$$\chi^P(\vec{q},\vec{q}) \underset{\vec{q}\to 0}{\longrightarrow} q^2, \quad \chi^P(\vec{q},\vec{q}+\vec{G}) \underset{\vec{q}\to 0}{\longrightarrow} q,$$

$$\chi^P(\vec{q}+\vec{G},\ \vec{q}+\vec{G}') \underset{\vec{q}\to 0}{\longrightarrow} \text{const.} \tag{A.12}$$

One can show that (A.12) is generally valid independent of the Hartree approximation (Sham, 1974, 1969, Pick et al., 1970). According to (A.9), χ contains repeatedly the Coulomb interaction, and one can extract from χ all parts containing the Coulomb interaction $v(\vec{q})$. One defines a polarizability p by

$$\chi(\vec{q}+\vec{G}, \vec{q}+\vec{G}') = p(\vec{q}+\vec{G}, \vec{q}+\vec{G}') + p(\vec{q}+\vec{G}, \vec{q})v(\vec{q})\chi(\vec{q},\vec{q}+\vec{G}') \quad (A.13)$$

which we symbolically write as

$$\chi_{GG'} = p_{GG'} + p_{GO}v_O\chi_{OG'}. \quad (A.13')$$

From the definition of χ^p it then follows that

$$p_{GG'} = \chi^p_{GG'} + \sum_{G''\neq 0} \chi^p_{GG''}v_{G''}p_{G''G'}. \quad (A.14)$$

This equation implies that the long-wavelength behaviour of χ^p, (A.12), is also valid for p.

We are now ready to derive the results (61) and (67), following Vogl (1978). From (A.13) one obtains

$$\chi_{GO} = \varepsilon^{-1}_{00}\, p_{GO}, \quad (A.15)$$

$$\chi_{GG'} = p_{GG'} + \frac{\chi_{GO}v_O\chi_{OG'}}{\varepsilon^{-1}_{00}}. \quad (A.16)$$

With (A.4), one can now easily verify that the following relation holds:

$$\varepsilon^{-1}_{GG'} = (1-\delta_{GO})(\delta_{GG'} + v_G p_{GG'})(1-\delta_{G'O}) + \frac{\varepsilon^{-1}_{G0}\,\varepsilon^{-1}_{0G'}}{\varepsilon^{-1}_{00}}. \quad (A.17)$$

Equation (A.17) is verified by considering separately the cases G,G' = 0 and ≠ 0. Defining

$$\tilde{\varepsilon}^{-1}_{GG'} = (1-\delta_{GO})(\delta_{GG'} + v_G p_{GG'})(1-\delta_{G'O}), \quad (A.18)$$

we can write (A.17)

$$\varepsilon^{-1}_{GG'} = \tilde{\varepsilon}^{-1}_{GG'} + \frac{\varepsilon^{-1}_{G0}\,\varepsilon^{-1}_{0G'}}{\varepsilon^{-1}_{\infty}} . \qquad (A.19)$$

With this, (A.14) reads for $G \neq 0$

$$p_{0G} = \sum_{G'} \chi^{p}_{0G'}\,\tilde{\varepsilon}^{-1}_{G'G} . \qquad (A.20)$$

Equations (A.19) and (A.20) are the results needed to establish (61) and (67). We define

$$\delta V^{sr}_{G} = \sum_{G'} \tilde{\varepsilon}^{-1}_{GG'}\,\delta v_{G'} , \qquad (A.21)$$

and obtain from (A.3) and (A.19)

$$\delta V_{G} = \delta V^{sr}_{G} + \frac{\varepsilon^{-1}_{G0}}{\varepsilon^{-1}_{00}}\,\delta V_{0} , \qquad (A.22)$$

which is (61). In addition, (A.20) and (A.15) give

$$\delta V_{0} = \sum_{G} \varepsilon^{-1}_{0G}\,\delta v_{G} = \varepsilon^{-1}_{00}\delta v_{0} + v_{0} \sum_{G\neq 0} \chi_{0G}\delta v_{G}$$

$$= \varepsilon^{-1}_{00}\,\delta v_{0} + \varepsilon^{-1}_{00}\,v_{0} \sum_{G,G'\neq 0} \chi^{p}_{0G'}\,\tilde{\varepsilon}^{-1}_{G'G}\,\delta v_{G}$$

$$= \varepsilon^{-1}_{00}\delta v_{0} + \varepsilon^{-1}_{00}\,v_{0} \sum_{G} \chi^{p}_{0G}\,\delta V^{sr}_{G} , \qquad (A.23)$$

which establishes (67), when $1/\varepsilon(\vec{q}) = \varepsilon^{-1}(\vec{q},\vec{q})$ is used. We finally show that

$$\vec{q} = 0: \quad \delta V(\vec{G}) = \delta V^{sr}(\vec{G}) \qquad (A.24)$$

This follows from the fact that the mean value of the Fourier transform $v(\vec{q} = 0) = 0$ in (50). Because of this, (A.13) gives $\chi_{GG'} = p_{GG'}$ for $\vec{q} = 0$ and also $\delta V(\vec{q} = 0) = 0$. This proves (A.24).

REFERENCES

Adler, S. L., 1962, Phys. Rev. 126:413.
Antoncik, E., and Landsberg, T., 1963, Proc. Phys. Soc. (London) 82:337.
Baldereschi, A., and Tosatti, E., 1978, Phys. Rev. B 17:3174.
Bertoni, C. M., Bortolani, V., and Calandra, C., 1974, Phys. Rev. B 9:1710.
Bilz, H., Gliss, B., and Hanke, W., in "Dynamical Properties of Solids," Ed. by G. K. Horton and A. A. Maradudin, North Holland, Amsterdam, Vol. I.
Bir, G. L., and Pikus, G. E., 1974, "Symmetry and Strain Induced Effects in Semiconductors," Wiley, New York, p. 337.
Birman, J. L., 1966, Phys. Rev. 145:620.
Bogusławski, P., and Mycielski, J., 1977, J. Phys. C 10:2413.
Born, M. and Huang, K., 1954, "Dynamical Theory of Crystal Lattices," Oxford Univ. Press, Oxford.
Camphausen, D. L., Connell, G. A. N., and Paul, W., 1971, Phys. Rev. Letters 26:184.
Cardona, M., 1969, "Modulation Spectroscopy," Academic Press, New York.
Cardona, M., Ed., 1977, "Light Scattering in Solids," Springer-Verlag Berlin.
Cerdeira, F., and Cardona, M., 1972, Phys. Rev. B 5:1440.
Cohen, M. L., and Heine, V., 1970, in "Solid State Physics," Ed. by F. Seitz and D. Turnbull, Academic Press, New York, 24:38.
Combes, J. M., 1976, Acta Phys. Austriaca, Suppl., 17:139.
Conwell, E. M., 1967, "High Field Transport in Semiconductors," Academic Press, New York.
Devreese, J. T., Ed., 1972, "Polarons in Ionic Crystals and Polar Semiconductors," North Holland, Amsterdam.
Ferry, D. K., 1978, in "Proc. Int. Conf. Phys. of SiO_2 and Its Interfaces," Ed. by S. Pantelides, Plenum, New York.
Fetter, A. L., and Walecka, J. D., 1971, "Quantum Theory of Many Particle Systems," McGraw-Hill, New York.
Galeener, F. L., and Lucovsky, G., 1976, Phys. Rev. Letters 37:1474.
Gilbert, M., and Drifford, M., 1977, J. Chem. Phys. 66:3205.
Gram, N. O. and Jorgenson, M. H., 1973, Phys. Rev. B 8:3902.
Hanke, W., 1973, Phys. Rev. B 8:4585, 8:4591.
Harrison, W. A., 1956, Phys. Rev. 104:1281.
Haug, A., 1972, "Theoretical Solid State Physics," Pergamon, New York.
Herring, C., and Vogt, E., 1956, Phys. Rev. 101:944.
Hess, K., and Vogl, P., 1979, Sol. State Commun. 30:807.
Jorgenson, M. H., 1978, Phys. Rev. B 18:5657.
Kawamura, H., 1978, in "Proc. 3rd Intern. Conf. on Narrow-Gap Semiconductors," Warsaw, p. 7.
Lawaetz, P., 1969, Phys. Rev. 183:730.
Lawaetz, P., 1978, thesis, Techn. Univ. Denmark, unpublished.
Licari, J. J., and Evrard, R., 1977, Phys. Rev. B 15:2254.

Longuet-Higgins, H. C., 1948, Proc. Phys. Soc. 60:270.
Louie, S. G., and Cohen, M. L., 1978, Phys. Rev. B 17:3174.
Mahan, G. D., 1972, in "Polarons in Ionic Crystals and Polar Semiconductors," Ed. by J. T. Devreese, North-Holland, Amsterdam, p. 553.
Mahan, G. D., 1974, in "Elementary Excitations in Solids, Molecules, and Atoms," Ed. by J. T. Devreese, A. B. Kunz, and T. C. Collins, Plenum, New York, p. 93B.
Maradudin, A. A., 1974, in "Dynamical Properties of Solids," Ed. by G. K. Horton and A. A. Maradudin, North-Holland, Amsterdam, 1:1
Maradudin, A. A., and Sham, L. J., 1977, in "Proc. Int. Conf. on Lattice Dynamics," Ed. by M. Balkanski, Flammarion, Paris, p. 29
Markham, J. J., 1959, Rev. Mod. Phys. 31:956.
Martin, R. M., 1972, Phys. Rev. B 5:1607.
Mathieu, H., Merle, P., Ameziane, E. L., Archilla, B., and Camassel, J., 1979, Phys. Rev. B 19:2209.
Moshinsky, M., and Kittel, C., 1968, Nat. Ac. Sci. Proc. 60:1110.
Müller, H., Trommer, R., Cardona, M., and Vogl, P., 1979, Phys. Rev. B, in press.
Nash, J. G., and Holm-Kennedy, J. W., 1977, Phys. Rev. B 15:3994.
Ngai, K. L., 1974, in "Proc. 12th Intern. Conf. Phys. Semiconductors Teubner, Stuttgart, p. 489.
Onton, A., Fisher, P., and Ramdas, A. K., 1967, Phys. Rev. Letters 19:781.
Pick, R. M., Cohen, M. H., and Martin, R. M., 1970, Phys. Rev. B 1:910.
Pines, D., and Noziéres, P., "The Theory of Quantum Liquids," Benjamin, New York.
Pindor, A. J., 1972, J. Phys. C 5:2357.
Richter, W., in "Springer Tracts in Modern Physics," Ed. by G. Höhler, Springer-Verlag, New York, 78:121.
Savaria, L. R., and Brust, D., 1969, Phys. Rev. 178:1240.
Sham, L. J., 1969, Phys. Rev. 188:1431.
Sham, L. J., 1974, in "Dynamical Properties of Solids," Ed. by G. K. Horton and A. A. Maradudin, North Holland, Amsterdam, 1:301.
Sham, L. J., and Ziman, J. M., 1963, in "Solid State Physics," Ed. by F. Seitz and D. Turnbull, Academic, New York, 15:223.
Shindo, K., and Nara, H., 1977, J. Phys. Soc. Jpn. 43:899.
Streitwolf, H. W., 1970, Phys. Stat. Sol. (b) 37:K47.
Tekippe, V. J., Chandrasekhar, H. R., Fisher, P., and Ramdas, A. K., 1972, Phys. Rev. B 6:2348.
Thouless, D. J., "The Quantum Mechanics of Many-Body Systems," Academic, New York.
Trommer, R., Müller, H., Cardona, M., and Vogl, P., 1979, Phys. Rev. B, to be published.
Tsay, Y. F., Mitra, S. S., and Bendow, B., 1974, Phys. Rev. B 10:1476.
Vogl, P., 1976, Phys. Rev. B 13:694. The notation in this and the following paper differ somewhat from the present work.

Vogl, P., 1978, J. Phys. C 11:251.
Vogl. P., Kocevar, P., and Baumann, K., 1976, in "Proc. 13th Intern. Conf. on Phys. of Semiconductors," Marves, Rome, p. 251.
Weaire, D., and Thorpe, M. F., 1971, Phys. Rev. B 4:2508.
Wiser, N., 1963, Phys. Rev. 129:62.
Yu, J. H., and Myers, G., 1976, Phys. Stat. Sol. (b) 77:K81.

SEMI-CLASSICAL BOLTZMANN TRANSPORT THEORY IN SEMICONDUCTORS

D. K. Ferry

Colorado State University
Fort Collins, Colorado 80523

INTRODUCTION

The over-riding theoretical problem in hot electron behavior remains that of trying to understand the manner in which the distribution function of the electrons is modified by the presence of the electric field. This is true whether we are dealing with a bulk material or the current response in a device. It is also a formidable experimental problem [see, e.g., Bauer (1974)]. In general, the Boltzmann transport equation can be expressed in its most general form as

$$\frac{\partial f}{\partial t} + \vec{v}\cdot\frac{\partial f}{\partial \vec{r}} + e\vec{F}\cdot\frac{\partial f}{\partial \vec{p}} = \int d\vec{p}'[f(\vec{p}')W(\vec{p}',\vec{p}) - f(\vec{p})W(\vec{p},\vec{p}')] \tag{1}$$

where $f(\vec{r},\vec{p},t)$ is the carrier distribution function. Generally for the steady-state response, the first two terms on the left (1) are ignored and the distribution is a function of the carrier pseudo-momentum $\vec{p} = \hbar\vec{k}$ and the energy. It is important to note that the Boltzmann transport equation (BTE) assumes that the collisions are instantaneous in both space and time, and that the field and scattering are different perturbations. In spatially varying problems, the addition of the position vector $\vec{r}$ brings about an effective lowering of the symmetry of the problem and complicates the solution of the Boltzmann equation (or an equivalent formulation). One can usefully classify the various phenomena or solutions upon the level of this symmetry. In the preceding sections, where hot carrier behavior was not considered, a single dimension, the electron energy E, was sufficient. If an axis of rotational symmetry exists for the hot electron problem, perhaps along the electric field direction, then two variables, $\vec{p}$ (or E) and θ, are all that is required. In many cases the problem is more complicated, however But, in some circumstances an analytical form can be assumed for the distribution function and if this form depends only upon a small number of parameters, then the zero-dimensional case results. In this latter case, simple equations for the evaluation of these parameters can be found, usually from moments of the BTE. In other cases, no approximations can be made, and detailed numerical techniques must be used. In modern transport theory, approximation techniques based upon Legendre expansions for $f(\vec{p},t)$ no longer find much usage. Modern computers allow relatively rapid solution of detailed equations, and so the two approaches found above receive extensive usage. Which of the two methods is to be preferred depends upon the rate of inter-carrier energy exchange, discussed in these proceedings by Hearn.

DISPLACED MAXWELLIAN

Fröhlich (1947) first pointed out that the isotropic part of the carrier distribution function is Maxwellian provided that the carrier concentration exceeds a certain critical concentration; i.e. -- provided that the rate of inter-carrier energy exchange was sufficiently large. Under conditions for which the anisotropic term can be taken as small, Fröhlich and Paranjape (1956) have pointed out that a displaced Maxwellian distribution

$$f(E) = A\exp[-(E - \vec{v}_d\cdot\vec{p})/k_BT_e], \tag{2}$$

containing the electron temperature T_e and drift velocity $\vec{v}_d$ as parameters, could be utilized. This is then the hot electron distribution in the approximation for which the critical carrier concentration is exceeded, and the parameters are then determined from balance equations, which in turn are obtained from the Boltzmann equation. Following the approach of Price (1977), (1) is multiplied by any function $\phi(\vec{p})$ and integrated over $\vec{p}$. For electrons within a single valley, this gives

$$\frac{\partial}{\partial t}\langle\phi\rangle + \frac{\partial}{\partial\vec{r}}\cdot\langle\phi\vec{v}\rangle = \langle e\vec{F}\cdot\frac{\partial}{\partial\vec{p}}\phi\rangle + \langle\int d\vec{p}'[\phi(\vec{p}') - \phi(\vec{p})]W(\vec{p},\vec{p}')\rangle. \tag{3}$$

The last term on the right arises from an interchange of variables involved in the double integration. The terms on the left-hand side vanish for the homogeneous steady-state. For ϕ equal to $\vec{p}$, the firs term on the right is just the force $\vec{F}$, and (3) is the momentum balance equation. For ϕ equal to the energy E, the first term on the right is $\vec{v}_d \cdot \vec{F}$, and (3) is the energy balance equation. In particular, the factor within the angle brackets in the last term on the right serves to define the average rate of energy loss to the lattice by collisions

$$\left\langle \frac{dE}{dt} \right\rangle_{coll.} = \int d\vec{p}\,'[E(\vec{p}\,') - E(\vec{p})]W(\vec{p},\vec{p}\,'), \tag{4}$$

and the average rate of momentum loss due to collisions

$$\left.\frac{d\vec{p}}{dt}\right|_{coll.} = \int d\vec{p}\,'[\vec{p}\,' - \vec{p}]W(\vec{p},\vec{p}\,'). \tag{5}$$

As an example, we consider a material such as gallium arsenide, where the scattering in the central valley is dominated by the polar-optical phonon. In this case, the energy balance equation becomes

$$ev_dF = A_1[\exp(y - x) - 1]x^{-\frac{1}{2}}\exp(x/2)K_1(x/2), \tag{6}$$

where $A_1 = eF_0(2\hbar\omega_0/\pi m^*)^{\frac{1}{2}}/[\exp(y) - 1]$, $x = \hbar\omega_0/k_BT_e$, $y = \hbar\omega_0/k_BT_0$, and F_0 is an effective electric field describing the coupling between the electrons and the phonons. Similarly, the momentum balance equation becomes

$$eF = (A_1m^*v_d/3\hbar\omega_0)\{[\exp(y - x) + 1]K_1(x/2) \tag{7}$$

$$+ [\exp(y - x) - 1]K_0(x/2)\}x^{3/2}\exp(x/2) + 3(m^*/2\pi k_BT_e)^{\frac{1}{2}}v_c1_{ac},$$

where the last term on the right takes into account the momentum loss to the elastic scattering by acoustic modes. Now, (6) and (7) can be solved simultaneously to yield T_e for a given electric field F, and then this result used in (7) to find v_d, and hence μ. It should be pointed out that this is an exceedingly simplified model, as transport in GaAs is much more complicated and factors such as intervalley transfer and band non-parabolicity should be considered. However, the results given here are useful as an illustrative example of the application of the displaced Maxwellian technique.

The accuracy of the balance equations obtained from (3) ranges, in its applications to solving transport problems, from very good to exceedingly poor, the latter in cases in which the assumptions made above are just not valid. Perhaps the most easily violated condition is the critical carrier concentration required. In the cases in which the balance equations are good, one can use them to infer energy and momentum relaxation times as

$$e\tau_E = d\langle E\rangle/d(\vec{v}_d\cdot\vec{F}), \tag{8}$$

$$e\tau_p = d\langle\vec{p}\rangle/d\vec{F}. \tag{9}$$

It should be pointed out that although these definitions appear in the balance equations, their validity goes beyond the displaced Maxwellian. Equations (4) and (5) can be averaged over any distribution f(E) to define effective energy and momentum relaxation times but the connection of this to the electric field as in (8) and (9) must be used carefully. Since these times are based upon the balance equations, the validity of their definitions are inherently tied into the results, and although any distribution function can be used, it must be done judiciously. One finds from the above considerations of the relaxation times that the energy relaxation time τ_E is considerably longer than the momentum relaxation time τ_p, so that the drift velocity responds to a change in $\vec{F}$ faster than the electron temperature responds. This can lead to overshoot effects in the velocity, in which $\vec{v}_d$ rises to a value corresponding to an electron temperature at the starting time just before the change in $\vec{F}$, then changes further as the temperature and distribution relax to the new temperature. This time-dependent behavior and the steady-state a.c. cases can be treated by using the time-dependent form of (3), using for example a large steady d.c. $\vec{F}_0$ and a small sinusoidal a.c. field $\vec{F}_1$ superimposed upon it. In this case one finds that the a.c. mobility includes terms like $(1 + i\omega\tau_p)$ and $(1 + i\omega\tau_E)$, reflecting the two time scales in the problem [see, e.g., Das and Ferry (1976)]. More complicated, spatial variations of the displaced Maxwellian have been considered by Bosch and Thim (1974), and the complicated time and space variations found in solid state devices can be treated by these techniques.

NUMERICAL TECHNIQUES

The truncation of the expansion for the distribution function to the first two terms, as is done in the displaced Maxwellian, is in general not a valid approach. It usually is justified only in the rare cases that $eF\ell$ is small compared to the energy range over which f_0 varies appreciably, where ℓ is a composite mean free path. The rapid variation of f_0 in the region about the optical phonon energy limits this truncation to small values of the field, or to cases of very rapid energy exchange via carrier-carrier scattering. The problem is to derive the distribution function for free electrons in a semiconductor from a knowledge of the various scattering processes and the applied fields. Although it is possible to justify various numerical solutions of the electron transport problem without specific reference to the integro-differential form of the Boltzmann equation, it is more illustrative to take this as a starting point, and thus clarify the basic properties of the numerical methods. For a spatially uniform electron system in an external field F, the time dependent Boltzmann equation is given from (1) as

$$\left[\frac{\partial}{\partial t} + e\vec{F}\cdot\frac{\partial}{\partial\vec{p}} + \lambda(\vec{k})\right]f(\vec{k},t) = \int W(\vec{k}',\vec{k})f(\vec{k}',t)d\vec{k}', \tag{10}$$

where $\vec{p} = \hbar\vec{k}$ and $\lambda(\vec{k})$ is the total out-scattering rate ($\lambda = 1/\tau$) and represents the second term on the right-hand-side of (1). The term on the right-hand-side of (10) represents just the in-scattering contributions. However, the definitions in $\lambda(\vec{k})$ are relatively incomplete since part, or all, of the $\lambda(\vec{k})f(\vec{k},t)$ term could be absorbed into the term on the right. The precise definition of $\lambda(\vec{k})$, and consequently of $W(\vec{k},\vec{k}')$ will usually depend upon the particular calculation to be undertaken, but the formal theory is independent of these considerations. We will therefore proceed as defined above.

The inverse of the differential operator of (10) is just an integral operator. Budd (1966) pointed out that the integration is a generalization of the Chambers (1952) path integral, and the result can be written as

$$f(\vec{k},t) = \int_0^\infty ds \int d\vec{k}' f(\vec{k}',t-s)W(\vec{k}',\vec{k}-e\vec{F}s/\hbar)\exp\{-\int_0^\infty \lambda(\vec{k}-e\vec{F}y/\hbar)\,dy\}. \tag{11}$$

The transformation of the Boltzmann equation into this form, when supplemented by numerical techniques on a grid of points in $\vec{k}$-space, as may be performed on modern digital computers, provides an extremely powerful technique to solve the Boltzmann equation. It was demonstrated, quite elegantly, by Rees (1968,1972) that an iterative approach can be utilized to great advantage. Moreover, he also pointed out that the iterative solution was a surrogate for the time evolution of $f(\vec{k},t)$, a point we will discuss further below.

An important technical innovation introduced by Rees (1968) is the concept of self-scattering, a fictitious scattering process, which does not alter the physics but allows a great simplification of the mathematical detail. To each side of (11), we add a term of the form

$$\Gamma(\vec{k})\delta(\vec{k}-\vec{k}'). \tag{12}$$

In particular, the simplification arises if we define this term as

$$\Gamma(\vec{k}) = \Gamma - \lambda(\vec{k}), \qquad (13)$$

with Γ sufficiently large that $\Gamma(\vec{k}) > 0$ for all k, although adequate results can often still be obtained if this condition is relaxed. Then, (11) becomes

$$f(\vec{k},t) = \int_0^\infty ds \int d\vec{k}' f(\vec{k}',t-s) W^*(\vec{k}',\vec{k}-e\vec{F}s/\hbar) \exp(-\Gamma s), \qquad (14)$$

where

$$W^*(\vec{k}',\vec{k}) = W(\vec{k}',\vec{k}) + \Gamma(\vec{k})\delta(\vec{k}-\vec{k}'). \qquad (15)$$

The iteration presented by Rees (1968,1972) consists of two distinct parts or steps. The first part of each iteration is represented in the time domain by the evaluation of an intermediate function $g_n(\vec{k},t)$ from the n-th iterate $f_n(\vec{k},t)$ according to

$$g_n(\vec{k},t) = \int d\vec{k}' f_n(\vec{k}',t) W^*(\vec{k}',\vec{k}). \qquad (16)$$

The second part of the iteration generates the (n+1)-st iterate $f_{n+1}(\vec{k},t)$ as the causal solution of (1), recognizing that $g_n(\vec{k},t)$ can be the right-hand side and $f(\vec{k})$ is replaced by $f_{n+1}(\vec{k},t)$. This causal solution is then

$$f_{n+1}(\vec{k},t) = \int_0^\infty ds\ g(\vec{k}-e\vec{F}s/\hbar,t-s) \exp(-\Gamma s). \qquad (17)$$

Physically, this last integral represents integration along the trajectory (the path integral) and the exponential factor is just the probability that no scattering has occurred during the traverse of the path. The appeal to the stability of the steady-state gives the result that the final distribution function arises from

$$f(\vec{k}) = \lim_{n\to\infty} f_{n+1}(\vec{k},t), \qquad (18)$$

and since the scattering factors and path variables shape $f(\vec{k})$, the initial guess for $f_1(\vec{k})$ is not critical. There is the further useful result that with $\Gamma(\vec{k})$ defined as in (19),

$$\lim_{n\to\infty} f_n(\vec{k}) = f(\vec{k},n/\Gamma). \qquad (19)$$

Rees (1969), in arriving at this result, has shown that each iteration is equivalent to a time step of $1/\Gamma$. However, whatever value of Γ is chosen, the resulting steady-state $f(\vec{k})$ is the same. The time development capability, however, allows a description of the approach to equilibrium from any given initial function to be ascertained, and by varying the field between iterations, the time dependent response can be found.

It can be readily observed that the iterative approach represents a chain of integrations, which represent alternate applications of a path-integral and a scattering integral. The entire chain operates on an initial trial function. This chain suggests an alternative approach to the solution of (10), the Monte Carlo evaluation of the integrals. In this latter case, a two-step iteration is also followed. The first step is a path traversal, terminated at a time t selected on a random value of the function $\exp(-\Gamma t)$. The second step involves scattering from the state resulting at the end of this traverse to a new state. The new state is governed by the type of scattering process used and this latter quantity is randomly selected from those present. That is, a typical electron is considered at $t = 0$ to be accelerated to t_1, where $t_1 = \ln(R_1)/\Gamma$, and R_1 is a random number. At t_1, the electron has been accelerated to a state $(\vec{k}_1, E_1)$. At this point, the relative probability of the i-th scattering event is $\lambda_i(E_1)/\Gamma, (\lambda_1 + \lambda_2 + \ldots + \lambda_n = \Gamma)$, including self-scattering. The particular scattering event is selected by a second random number R_2 as the k-th process when $\lambda_1 + \ldots + \lambda_{k-1} < \Gamma R_2 < \lambda_1 + \ldots + \lambda_{k-1} + \lambda_k$. The properties of this scattering event are then used to determine the final state $(\vec{k}_2, E_2)$, which is then used as the new $t = 0$ state and the process repeated. If the states $(\vec{k}_1, E_1)$ are tabulated in a $\vec{k}$-space grid, then their distribution becomes a representation of $f(\vec{k})$. An estimator for the physical variable ϕ, such as the velocity, is generated as $\Sigma\phi(\vec{k}_1)/(\text{no. of scatterings})$, where the sum runs over all of the scatterings. The validity of such an average lies in the ergodicity of the physical process, providing that a sufficiently large number of iterations has been used. The basic Monte Carlo technique was first put forward by Kurosawa (1966), but its full capabilities were not evident until the introduction of self-scattering by Boardman, Fawcett, and Rees (1968). The application of Monte Carlo to time-dependent phenomena is restricted to cases where $\omega \ll \Gamma$, due to the need to establish equilibrium among the states.

A velocity-field curve for electrons as found from the two methods of calculation would show no variation in the steady-state. Such a curve calculated from a Monte Carlo technique by Littlejohn, Hauser, and Glisson (1975) for GaN for example, is the same as that calculated by iterative calculation by Ferry (1975). The difference between these methods and the displaced Maxwellian lies in the anisotropy introduced by the polar-optical phonon scattering. For low carrier concentrations, this anisotropy, due to the small angle of scattering in this situation, leads to streaming of the carriers. The streaming effect is generated by a spike in $\vec{k}$-space directed along the electric field. To adequately model this by a Legendre expansion would require an exceedingly large number of terms to be retained. At high densities, however, carrier-carrier scattering will cause the spike to be dissipated. Which of the techniques is the correct one depends upon the degree to which the carrier-carrier interaction is significant in the development of the distribution function.

The explicit representation of the actual distribution function, such as occurs in the iterative solution and to a lessor extent in the Monte Carlo approach, has natural advantages. Effects that are non-linear in f(E), such as carrier-carrier scattering and degeneracy of f(E), can readily be incorporated into the calculations. The details of the scattering are fully tied up in the terms $W^*(\vec{k},\vec{k}')$ and $\Gamma(\vec{k})$, so no conceptual difficulty arises in incorporating non-phonon scattering events such as impact ionization, optical carrier generation, or even cyclotron resonance type transitions.

There is some concern however, over the use of Monte Carlo techniques for transient calculations. Ferry and Barker (1979) have recently shown that Monte Carlo results for transient response and overshoot velocity tend to indicate longer transient response times than found for either displaced Maxwellian or iterative calculations. One possible cause is the lack of a direct calculation of $f(\vec{k})$ in the Monte Carlo technique and the large number of trials required for the $\vec{k}$-space tabulation to converge to a true $f(\vec{k})$.

BIBLIOGRAPHY

Bauer, G., 1974, in "Springer Tracts in Modern Physics", No. 74, Ed. by G. Höhler, Springer-Verlag, Heidelberg.

Boardman, A. D., Fawcett, W. and Rees, H. D., 1968, Sol. State Communications 6:305.

Bosch, R. and Thim, H., 1974, IEEE Trans. Electr. Dev. ED-21:16.

Budd, H. F., 1966, Proc. Intern. Conf. on Phys. Semiconductors, Kyoto, in J. Phys. Soc. Jpn. 21 (suppl.):420.

Chambers, R. G., 1952, Proc. Phys. Soc. (London) A65:458.

Das, P. and Ferry, D. K., 1976, Sol.-State Electr. 19:851.

Ferry, D. K. and Barker, J. R., 1979, Sol.-State Electr., in press.

Ferry, D. K. 1975, unpublished.

Fröhlich, H., 1947, Proc. Roy. Soc. A188:521.

Fröhlich, H. and Paranjape, B. V., 1956, Proc. Phys. Soc. (London) B69:21.

Kurosawa, T., 1966, Proc. Intern. Conf. on Phys. Semiconductors, Kyoto, in J. Phys. Soc. Jpn. 21 (suppl.):424.

Littlejohn, M. A., Hauser, J. R. and Glisson, T., Appl. Phys. Letters 26:625.
Price, P., 1977, Sol.-State Electr. 21:9.
Rees, H. D., 1968, Sol. State Commun. 26A:416.
Rees, H. D., 1969, J. Phys. Chem. Solids 30:643.
Rees, H. D., 1972, J. Phys. C 5:64.

QUANTUM TRANSPORT THEORY

J. R. Barker

Physics Department
Warwick University
Coventry CV4 7AL, UK

CONCEPTS

Boltzmann transport theory (BTT) is an ideal theory. It has the twin virtues of conceptual and mathematical simplicity. It also works far better than one could reasonably expect from its origin as a graft from the classical theory of dilute gases. Quantum transport theory (QTT) (Kohn and Luttinger, 1957, 1958; Kubo, 1957; Dresden, 1961; Chester, 1963; Kubo, 1966; Luttinger, 1968) enjoys no such status: it is neither conceptually nor mathematically simple; it is very hard to make it work; and it often reduces, after considerable labour, to the Boltzmann picture (Peierls, 1974; Cohen and Thirrig, 1973). But even if there were no manifestly quantum transport phenomena (and we might single out transport in quantizing magnetic fields, but equally: hopping conduction, impurity conduction, polaron transport, high-frequency transport, quasi-1 D transport, the Kondo effect, size limited transport, and of course superconductivity) (Barker, 1978, 1979), we would still require QTT as an *ab initio* theory to explain how the phenomenological BTT picture and its related concepts actually arise from the underlying framework of reversible quantum statistical mechanics. QTT is thus concerned with: (1) an explanatory and supportive theory for the Boltzmann picture, where that exists; (2) setting confidence limits for the application of BTT; (3) developing the necessary novel concepts and transport kinetics for genuine quantum transport phenomena (the latter may be loosely defined as those effects which depend explicitly on the quantum mechanical nature of the electron, and/or those processes for which the simple relaxive local Boltzmann description fails.

With the benefit of hindsight, let us approach the deeper perspective offered by QTT by first examining the prejudices which underlie BTT, leaving aside more obvious aspects such as the use of effective mass theory.

Assumptions in BTT

BTT models the conduction electrons as an approximately independent-particle dilute gas in which the electronic states are nearly stationary and free-electron-like with a well defined momentum $\vec{K}$ (Peierls, 1974). Non-stationarity arises from the assumption that the perfect crystal periodicity is violated by imperfections, impurities and phonons. The latter are assumed to cause weak, infrequent scattering of the electrons amongst the states $\{\vec{K}\}$. The imposition of an applied electric field is supposed to solely accelerate carriers through the momentum states without otherwise distorting the states or interfering with the scattering processes. It then seems reasonable to describe the carriers by a classical distribution function $f_\sigma(\vec{K},\vec{R};t)$ (σ is a spin index) over a vaguely defined phase space $\{\vec{K},\vec{R}\}$ (in which $\vec{R}$, the spatial location of the carrier is not too closely defined because of fears of the uncertainty principle [Jaynes, 1967]). The next assumptions concern how f_σ is determined by the fields and the scattering processes.

Usually one deals with relatively macroscopic systems so that the intuitive concepts of τ_d: transit-time through the semiconductor; τ: the mean-free-time between collisions; and τ_c the atomic duration of a collision, are assumed to satisfy the inequality

$$\tau_c << \tau << \tau_d \tag{1}$$

Transport processes are thus conventionally viewed on a coarse-grained time scale $\tau >> \tau_c$ such that many completed independent collisions are supposed to occur in the passage of a carrier through the system. Moreover, each collision event is treated as an irreversible process which is completed: (a) locally in space; (b) locally in time (instantaneously); (c) independently of any driving fields and other scattering processes; and (d) with low frequency: weak dissipativity--no self-energy effects. Given these assumptions it then seems "obvious" how to write down the equation of motion for f_σ as a local (in time and space variables) irreversible equation:

$$\frac{\partial f}{\partial t} + \left(\frac{\partial f}{\partial t}\right)_{\text{diffusion}} + \left(\frac{\partial f}{\partial t}\right)_{\text{fields}} = - \left(\frac{\partial f}{\partial t}\right)_{\text{collisions}} \tag{2}$$

Here, the LHS is time-reversible, but the overall equation is irreversible through the gain-loss structure of the collision integral which depends on time only through its functional dependence on f.

Features that are neglected in BTT provide warning signs for failure of the semiclassical approach. They may be summarized under the headings: (1) non-locality of scattering processes: each collision event is actually extended in space and time. If the spatial or temporal variations (described by wavevector $\vec{q}$ and frequency ω) of the applied driving forces approach the microscopic scale, then collisions will only be partially completed. Normally BTT assumes $\omega\tau < 1$ and $qL < 1$ where L is the mean free path: many collisions are completed in one cycle of the applied forces. If however, $\omega\tau_c \gtrsim 1$, $qv\tau_c \gtrsim 1$ (where τ_c is estimated by $\hbar/\varepsilon$, ε is a characteristic carrier energy; $v\tau_c$ is the de Broglie wavelength) appreciable quantum effects can arise and the irreversible character of completed collisions will be lost. Inter-band effects may also occur if ω, $\hbar^2q^2/m^* \gtrsim \varepsilon_b$ where ε_b is the vertical energy separation to the band controlling m^*. Very high frequencies correspond to a quantal behaviour more reminiscent of optical response (Ron, 1964; Price, 1966; Fujita, 1966). The elementary treatment of collisions must also be reconsidered if the mean free time becomes comparable to τ_c: multiple scattering involving at least two scatterers is then possible. (2) Strong driving forces: once the extended nature of a collision is recognized it becomes obvious that applied fields can transfer energy and momentum to the carrier during the collision: an interference or intra-collisional field effect (Barker, 1973, 1978, 1979; Thornber, 1979; Ferry and Barker, 1979a). The effect will be very large if $eEv\tau_c \sim \varepsilon$. The reverse effect may also occur: scattering can forestall the instantaneous accelerative effect of the driving field (this effect also occurs for low fields). In general the driving and scattering terms in BTT cannot be independent. (3) Strong scattering: this exacerbates the previous problems and weakens the assumption that the electronic states are of long lifetime and are free-electron like. Polaron and co-operative effects are typical consequences. (4) Dense systems: many-body effects appear and a single carrier description fails. (5) Small systems: size quantization or surface limited transport effects become important (Ferry and Barker, 1979b; Barker, 1979a). Ultimately the conditions of expression (1) break down. (6) Non-classical influence of driving fields: sufficiently strong electric or magnetic fields lead to Stark or Landau quantization of the electronic states (Barker, 1979b).

The above concepts are intuitive but may be given precision meaning within QTT, to which we now turn.

Generalized Distribution Functions

QTT is generally based on the Liouville-von Neumann equation for the statistical density matrix $\rho(t)$:

$$i \frac{\partial}{\partial t} \rho(t) = \hat{H}_F \rho(t) \tag{3}$$

$$\hat{H}_F \rho \equiv [H_F, \rho] \equiv H_F \rho - \rho H_F \tag{4}$$

where $H_F = H + F$, and the Hamiltonian H describes the full system in the absence of the coupling F to the externally applied driving forces. The usual boundary condition is that $\rho = \rho_0(H)$ for $t < 0$, where ρ_0 is a thermal equilibrium solution (e.g. the grand canonical density matrix). The driving perturbation is initiated at $t = 0$. The above starting point needs modification for the description of small systems embedded in an interactive environment (Barker, 1979a; Barker and Ferry, 1979b). We use units for which $h = 1$.

All observable properties, e.g. the current or charge densities, labelled generically by J_i may be evaluated as quantum statistical expectation values determined by $\rho(t)$

$$\langle J_i(t) \rangle = TR[J_i \rho(t)] \equiv \sum_{\lambda,\lambda} \langle \lambda | J_i | \lambda' \rangle \quad \langle \lambda' | \rho | \lambda \rangle$$

$$= \sum_{\lambda,\lambda'} J_i^{\lambda\lambda'} \rho_{\lambda'\lambda} \tag{5}$$

Here $\{|\lambda\rangle\}$ is any complete set of states. Usually, the Hamiltonian H_F is partitioned into "free" carrier, "free" scatterer, carrier-scatterer interaction, and driving force components as

$$H_F = H_e + H_s + V + F \equiv H_o + V + F \quad . \tag{6}$$

This partioning is not unique. H_e might describe small polaron states by incorporating part of the electron-phonon interaction, or H_e might include the coupling to a magnetic field and describe Landau states. The basis states $|\lambda\rangle$ are then chosen to diagonalise $H_o = H_e + H_s$, usually via $|\lambda\rangle \equiv |e\rangle \times |s\rangle$ where $\{|e\rangle\}$, $\{|s\rangle\}$ diagonalise H_e and H_s respectively. The choice of representation $\{|\lambda\rangle\}$ decides the character and interpretation of the subsequent transport theory.

If the current density operator $\vec{J}$ depends only on electronic variables and commutes with H_e (which is true for extended state, homogeneous transport in zero magnetic fields) the observable response is

$$\langle \vec{J}(t) \rangle = \sum_e \vec{J}^{ee} f(e) \tag{7}$$

where

$$f(e) = \langle e|\sum_s \langle s|\rho|s\rangle|e\rangle \equiv TR_s\langle e|\rho|e\rangle \equiv \langle\langle e|\rho|e\rangle\rangle_s \tag{8}$$

defines a real, time dependent generalized electron distribution function over the free carrier states $\{|e\rangle\}$. A transport equation for f(e) may then be constructable from the Liouville equation. It may turn out to have Boltzmann-like form although the quantum nature of the states $|e\rangle$ will be reflected in the detailed forms for the collision rates. However, for inhomogeneous transport and for transport in quantizing magnetic fields, for example, $\vec{J}$ is not necessarily diagonal and we must consider the off-diagonal matrix elements of the electron density matrix $f = \langle\rho\rangle_s$. Various methods exist for expressing f(e) in terms of f(e,e') (Barker, 1979b) but the subsequent transport theory will not have a Boltzmann-like form. In general, a closed equation of motion for f(t) can only be obtained for the special case of independent carrier transport in a stationary scattering system (i.e., where the scatterers remain in thermal equilibrium at all times). This situation has been extensively studied for homogeneous non-linear transport.

Wigner (1932) has shown that QTT can get quite close to the classical concept of a phase-space distribution function. The Wigner one-electron distribution function for example, is defined for free carrier states by

$$f_\sigma(\vec{K},\vec{R};t) \equiv \int d^3r e^{-i\vec{K}\cdot\vec{r}} TR\{\rho(t)\psi_\sigma^+(\vec{R}-\tfrac{1}{2}\vec{r})\psi_\sigma(\vec{R}+\tfrac{1}{2}\vec{r})$$

$$\equiv \int d^3r e^{-i\vec{K}\cdot\vec{r}} f_\sigma(\vec{r},\vec{R};t), \tag{9}$$

where $\psi_\sigma^+(\vec{r})$, $\psi_\sigma(\vec{r})$ are the second quantized creation and annihilation operators for a carrier of spin σ at location $\vec{r}$. Here ρ is second quantized, and the trace is a many-body trace. Similarly, a phonon Wigner distribution may be defined by

$$N_\alpha(\vec{K},\vec{R};t) \equiv \sum_k e^{i\vec{k}\cdot\vec{R}} \langle b_\alpha^+(\vec{K}-\tfrac{1}{2}\vec{k})b_\alpha(\vec{K}+\tfrac{1}{2}\vec{k})\rangle, \tag{10}$$

where $\langle\ldots\rangle \equiv TR(\ldots)$, b_α^+, $b_\alpha(\vec{k})$ are creation or annihilation operators for a phonon of type α and momentum $\vec{k}$. The Wigner construction utilizes a partial, generalized Fourier transform over the full (off-diagonal) density matrix, and is easily generalized to other basis states (e.g. Bloch states or Landau states). One may easily prove that f_σ, N_α are real-valued generalized distributions which give the correct statistical expectation values, e.g. the carrier density and current density in an inhomogeneous system are given by

$$\langle n(\vec{R},t)\rangle = e \sum_{\sigma} \int \frac{d^3K}{(2\pi)^3} f_{\sigma}(\vec{K},\vec{R};t), \tag{11}$$

$$\langle \vec{J}(\vec{R},t)\rangle = \sum_{\sigma} \int \frac{d^3K}{(2\pi)^3} e\vec{v}(K) f_{\sigma}(\vec{K},\vec{R};t), \tag{12}$$

where $\vec{v}(\vec{K}) \equiv \nabla_K \varepsilon(\vec{K})$ is the (c-number) group-velocity of the electron momentum state $|\vec{K}\rangle$. Homogeneous systems are translationally invariant ($f_{\sigma}(\vec{r},\vec{R})$ is independent of $\vec{R}$) which implies

$$f(\vec{K},\vec{k};t) \equiv \int e^{-i\vec{k}\cdot\vec{R}} f(\vec{K},\vec{R})\, d^3R$$

is independent of $\vec{k}$ and the equivalent electron density matrix is diagonal in momentum space.

There are some difficulties with interpreting Wigner distributions as probability densities: they are not necessarily positive definite (usually a sign of strong quantum interference effects). However, the Wigner method does allow one to construct a properly gauge invariant transport theory (Fujita, 1966; Stinchcombe, 1974). For independent carriers, f_{σ} reduces to $\langle \vec{R} - \tfrac{1}{2}\vec{r}|\langle\rho\rangle_s|\vec{R} + \tfrac{1}{2}\vec{r}\rangle$ where ρ is now a functional of the one-electron Hamiltonian. QTT has been extensively developed for this case, which we now discuss.

STRUCTURE OF HIGH FIELD QUANTUM KINETIC THEORY

Let us now sketch the general features of QTT (Barker, 1973) using the Wigner distribution for the simple case of: (1) uniform constant electric field E; (2) non-degenerate homogeneous systems; (3) stationary phonon and impurity distributions; and (4) $H_e = p^2/2m^*$ a simple one-band spin-free model. Alternative methodologies are listed in Appendix 1. (As before we use units in which h = 1.)

The Master Equation in the Super-Operator Picture

The electron density matrix f(t) is determined from the full density matrix by $f(t) = TR_S[\rho(t)]$. From assumption (3) we factorize approximately the initial thermal-equilibrium density matrix as

$$\rho(t=0) \approx f_o(H_e + V)\,\Omega_s(H_s),$$

where f_o will be taken as the Maxwellian equilibrium form, and Ω_s describes the equilibrium distribution of scatterers. Let us now Laplace transform the Liouville equation (3) and re-arrange the terms using the definition of f(t) to find

$$f(s) \equiv \int_0^\infty dt e^{-st} f(t) = TR_s\{(\hat{H}_F - is)^{-1} f_o(H_e + V)\Omega_s\}(-i)$$

$$\equiv -i\langle(\hat{H}_F - is)^{-1} f_o\rangle_s, \tag{13}$$

where the super-operator $\hat{R}(s) \equiv [\hat{H}_F - is]^{-1}$ is the resolvent (Green's function) of the Liouville equation and we recall the super-operator notation $\hat{H}_F A \equiv H_F A - AH_F$. The complete transport kinetics are embodied in $\hat{R}(s)$. Expression (13) cannot be immediately factorized as $\langle\hat{R}\rangle_s\langle f_o\rangle_s$ because $f_o(H_e + V)$ contains scatterer variables. But with the aid of projection-operator calculus we can easily obtain a kinetic equation for f(s) involving only averages over Ω_s.

Let us introduce projection super-operators $\hat{P}$ and $\hat{Q}$ via:

$$\hat{P} \equiv TR_s(. \; . \; . \; \Omega_s) \equiv \langle. \; . \; .\rangle_s; \; \hat{Q} = 1 - \hat{P}, \tag{14}$$

where we may confirm

$$\hat{P}^2 = \hat{P}; \; \hat{Q}^2 = \hat{Q}; \; \hat{Q}\hat{P} = \hat{P}\hat{Q} = 0. \tag{15}$$

Evidently the electron density matrix is expressible as

$$f(s) = -i \; \hat{P} \; \hat{R}(s) \; [\hat{P}f_o + \hat{Q}f_o]. \tag{16}$$

The super-resolvent may be formally expressed in terms of an important identity (derived in Appendix 2):

$$\hat{R} \equiv [1 - (\hat{Q}\hat{H}_F\hat{Q} - is)^{-1}\hat{Q}\hat{H}_F\hat{P}][\hat{P}\hat{H}_F\hat{P} - \hat{C}_F - is]^{-1}[1 - \hat{P}\hat{H}_F\hat{Q}(\hat{Q}\hat{H}_F\hat{Q} - is^{-1}]$$
$$+ (\hat{Q}\hat{H}_F\hat{Q} - is)Q^{-1} \tag{17}$$

where $\hat{C}_F$ is the collision super-operator

$$\hat{C}_F \equiv \hat{P}\hat{H}_F\hat{Q}(\hat{Q}\hat{H}_F\hat{Q} - is)^{-1}\hat{Q}\hat{H}_F\hat{P}, \tag{18}$$

which is essentially the "self-energy" for the Green-function-like quantity $\langle\hat{R}\rangle_s$. Using expressions (13) - (17) we find

$$if(s) = (\hat{P}\hat{H}_F\hat{P} - \hat{C}_F - is)^{-1}\{\langle f_o\rangle_s - \hat{P}\hat{H}_F\hat{Q}(\hat{Q}\hat{H}_F\hat{Q} - is)^{-1}\hat{Q}f_o\},$$

or, re-arranging to isolate the transform of $\partial f/\partial t$,

$$sf(s) - \langle f_o\rangle_s + i\hat{P}\hat{H}_F\hat{P}f(s) = i\hat{C}_F(s)f(s) + M_F(s), \tag{19}$$

$$M_F(s) \equiv -\langle \hat{H}_F \hat{Q} (\hat{Q} \hat{H}_F \hat{Q} - is)^{-1} \hat{Q} f_o \rangle_s . \tag{20}$$

We may now transform back to the time-domain to obtain the general master equation

$$\frac{\partial f(t)}{\partial t} + i[H_e, f(t)] + i[F, f(t)] = i \int_0^t d\tau \hat{C}_F(\tau) f(t-\tau) + M_F(t). \tag{21}$$

Here we have used $\hat{P}\hat{H}_F\hat{P} \equiv (\hat{H}_e + \hat{F})\hat{P}$ and anticipated $\langle V \rangle_s$ = constant. The LHS of (21) describes the collision-free diffusion and acceleration of the carriers. For homogeneous systems, as remarked earlier, f is a function of momentum only and if $H_e = p^2/2m^*$, the term $[H_e, f]$ vanishes (for inhomogeneous transport it gives rise to a term $v(\vec{K}).\partial f/\partial \vec{R}$; $[H_e, f]$ is also non-vanishing if H_e includes coupling to a quantizing magnetic field). With our previous model assumption the coupling to the electric field is simply $F \equiv -e\vec{E}.\vec{r}$ and in the momentum representation $i[F,f]$ reduces to $e\vec{E}.\partial f/\partial \vec{K}$. The RHS is proportional to the scattering interaction and describes collision effects via $\hat{C}_F$ and memory effects via M_F. At this stage, (21) is still time-reversible. An irreversible equation may be obtained if we seek the asymptotic steady-response for $t \to \infty$ provided $\hat{C}_F(s \to o^+)$ exists. Then asymptotically we may use the theorem $A(t \to \infty) \sim sA(s)$; $s \to o^+$; to obtain

$$\partial f/\partial t + i[F,f] = i\hat{C}_F(s \to o^+) f(t) + M_F(t), \tag{22}$$

where the collision operator $\hat{C}_F$ will automatically generate a gain-loss structure linear in f. The asymptotic limit is somewhat delicate and we return to it after examining some special limiting cases.

The Boltzmann Limit

BTT may be exactly recovered under the conditions (Barker, 1973): (1) weak, infrequent scattering; (2) point collisions; (3) translational invariance of the scattering system; and (4) asymptotic time scale $t \gg \tau_c$ (actually related to (2)). For simplicity we concentrate on phonon scattering and choose

$$H_s = \sum_q \hbar\omega_q (b_q^+ b_q + \tfrac{1}{2}); \quad V = \sum_q \{C(q) e^{i\vec{q}\cdot\vec{r}} b_q + C^*(q) e^{-i\vec{q}\cdot\vec{r}} b_q^+\}, \tag{23}$$

$$\langle b_q^+ b_{q'} \rangle_s = N_q \delta_{qq'}, \tag{24}$$

where N_q is the Bose-Einstein thermal distribution [and we have used condition (3)]. Condition (1) is modeled by setting $M_F \approx 0$ (since it is proportional to EV^2), and modeling C_F by the lowest order non-vanishing contribution. Since $\hat{P}(H_0 + F)\hat{Q} = \hat{Q}(H_0 + F)\hat{P} = 0$ we find

$$\hat{C}_F(o+) \sim \langle \hat{V}(\hat{H}_0 + \hat{F} - io^+)^{-1}\hat{V}\rangle_s . \tag{25}$$

Condition (2) is modelled by ignoring the V dependence of $\{\hat{Q}\hat{H}_F - is\}^{-1}$ and by setting $\hat{F} \simeq 0$ in expression (25), i.e. no self-energy or intra-collisional field effects. The resulting expression for $\hat{C}_0(o^+)$ is then field independent, and from condition (3) couples to just the diagonal momentum matrix elements of f. Thus in the momentum representation we find

$$\frac{\partial f(\vec{K},t)}{\partial t} + e\vec{E}\cdot\nabla_k f(\vec{k},\vec{t}) = i\sum_{K'} C_o(\vec{K},\vec{K};\vec{K}',\vec{K}')f(\vec{K}',t), \tag{26}$$

where $C_o(KK,K'K')$ are the semi-diagonal matrix elements of $\hat{C}_o(o^+)$. The latter contain linear combinations of terms like

$$\sum_q |C(q)|^2 \langle b_q^+\{\varepsilon_K - \varepsilon_{K-q} - \omega_q - io^+\}^{-1} b_q\rangle_s \delta_{K',K-q} . \tag{27}$$

Since $\varepsilon_K - \varepsilon_{K-q}$ is a continuous function of K,q for a sufficiently macroscopic system, (27) can be expanded as

$$\sum_q |C(q)|^2 N_q \delta_{K',K-q} \left\{\frac{P}{\varepsilon_K - \varepsilon_{K-q} - \omega_q} + i\pi\delta(\varepsilon_K - \varepsilon_{K-q} - \omega_q)\right\}, \tag{28}$$

where we use $(x - io^+)^{-1} \to P/x + i\pi\delta(x)$. The principal part terms $P/\varepsilon-\varepsilon'$ actually cancel out of the complete collision integral $\hat{C}_o f$ leaving just those terms containing energy conservation factors $\delta(\varepsilon-\varepsilon')$. The RHS of (26) is then just the BTT expression for $(\partial f/\partial t)_{e-p}$

$$\left(\frac{\partial f}{\partial t}\right)_{ep} = \sum_{K'} R(\vec{K},\vec{K}')f(\vec{K}') - R(\vec{K}',\vec{K})f(\vec{K}), \tag{29}$$

where the $R(\vec{K},\vec{K}')$ are the usual second-order perturbation theory scattering rates. Equation (26) is thus just the Boltzmann equation.

We have arrived at an irreversible transport equation. The irreversibility arises from a number of conditions: (1) using the causal (retarded) solution of the Liouville equation thus forcing the transforms into the upper half complex energy plane $s > o$; (2) the long-time limit $t \to \infty$, corresponding to completed collisions and the use of a coarse grained time scale; (3) the continuous spectrum for ε_K [which picks out the imaginary part of $\hat{C}_o(K,K;K',K')$. This latter condition assumes a large system such that the actual discrete momentum states have very small separation $\Delta\varepsilon_K$. The neglect of this spacing is equivalent to ignoring time scales in excess of the Poincare recurrence periods which are of the order $1/\Delta\varepsilon_K$. If the spectrum of H_e were not to be discrete even in the infinite volume limit (as in the case of atomic bound states) the term $\mathrm{Im}\,\hat{C}_o \to 0$ and the transport equation would be completely time-reversible even in the asymptotic limit.

It is instructive to view the appearance of irreversibility in the time domain. The collision integral then contains the structure

$$\sum_{K'} |C(\vec{K}-\vec{K}')| \int_0^t d\tau e^{i(\varepsilon_K - \varepsilon_{K'} - \hbar\omega_q)\tau} f(K', t-\tau). \tag{30}$$

We only get back to the structure of (27) if we follow f on a time scale $t > \tau_c$ where τ_c is the effective collision duration. The latter may now be given more precision: it is roughly the time for at least a few oscillations of the propagation kernel $\exp[i\tau(\varepsilon_K - \varepsilon_{K'} - \hbar\omega]$. Thus if we have a short-range potential, $|C(q)|$ = const., we estimate max $\tau_c \sim 1/\varepsilon_K$. If the potential has range a [e.g. the gaussian structure $|C|^2 \sim \exp(-q^2a^2)$] we have alternatively max $\tau_c \sim 2/\varepsilon_K - \varepsilon_{K'} \sim a/v_K$, where v_K is the mean speed. Provided f varies slowly on the scale τ_c (a variation controlled by the electric field in general) we may isolate the fast exponential in (30) to obtain the asymptotic result of (27). These considerations need modification if the influence of the field on the collision is considered and for fast variation of f(t). The BTT result fails seriously if τ_c becomes large (i.e. low velocity or long de Broglie wavelength states) and strong quantum structure can emerge. This happens for example in band-tail transport where a localization edge may occur.

Self-Energy Effects

The BTT point-collision approximation describes real scattering between sharp unperturbed momentum states $\vec{K}$. But the very existence of scattering, and the presence of many scatterers, ensures that this assumption is inconsistent. We do not have an isolated scattering problem. Each momentum state will actually be displaced $\varepsilon_K \to \varepsilon_K + \Delta(K)$ and have a finite lifetime (broadened) $\tau_\Gamma \sim 1/\Gamma(\varepsilon_K \to \varepsilon_K + \Delta - i\Gamma)$ by virtual collision processes. Thus within each scattering event a carrier will propagate in a perturbed state which is controlled by virtual scattering on the entire scattering medium. These self-energy effects are often small, but for strong coupling may lead to significant level shifts and broadening of the otherwise sharp energy conservation in the collision process. Indeed a whole range of collective electron-scatterer states may arise to destroy the elementary BTT picture.

In the detailed full expansion of $\hat{C}_F$, the self-energy effects show up to lowest order in the generalized Born Approximation

$$\hat{C}_F \sim \langle \hat{V}\{\hat{H}_e + \hat{H}_s + \hat{F} - \hat{C}_F - io^+\}^{-1}\hat{V}\rangle_s . \tag{31}$$

If the field dependence of $\hat{C}_F$ is again neglected, the energy conservation factors of BTT generalize to the smeared-out structure

$$\delta(\varepsilon_K - \varepsilon_{K-q} - \omega_q) \to \frac{1}{\pi} \frac{\tilde{\Gamma}}{(\varepsilon_K - \varepsilon_{K-q} - \tilde{\Delta})^2 + \tilde{\Gamma}^2} , \tag{32}$$

where $\tilde{\Delta}$ and $\tilde{\Gamma}$ are related to the real and imaginary parts of $\hat{C}_0(K,K';K'',K''')$, respectively. One may demonstrate that if we further model $\hat{C}$ within the propagation kernel by the lowest-order expression (25), then $\tilde{\Delta}$, $\tilde{\Gamma}$ relate to the real and imaginary parts of the true carrier self-energy $\Sigma = \Delta - i\Gamma$ defined in this approximation by

$$\Sigma(\varepsilon + io^+, K) = \sum_q |C(q)|^2 N_q \{\varepsilon - \varepsilon_{K-q} + io^+\}^{-1} \tag{33}$$

and illustrated for high-temperature (quasi-elastic) electron-phonon scattering. Then

$$\tilde{\Delta} = \Delta(\vec{K}, \varepsilon_{K-q}) - \Delta(\vec{K}-\vec{q}, \varepsilon_K); \ \tilde{\Gamma} = \Gamma(\vec{K}, \varepsilon_{K-q}) + \Gamma(\vec{K}-\vec{q}, \varepsilon_K). \tag{34}$$

The collision integral is therefore generally a convolution over the renormalized joint spectral density functions for the initial and final momentum states. Self-energy effects are clearly important if Δ/ε, $\Gamma/\varepsilon \sim 1$. They are crucial to an understanding of quantum magnetotransport. If self-energy effects do arise, the memory term M_F of the full master equation cannot be neglected.

Multiple-Scattering Effects

If the electron-scatterer interaction is expressed as a sum over different species of scatterer $V = \sum_i V_i$ (where i refers, for example, to ionized impurities, acoustic phonons, optical phonons, etc.), we may use infinite-order renormalized perturbation theory to show that $\hat{C}_F$ has the structure

$$\hat{C}_F = \sum_i \hat{C}_F(i) + \sum_{i \neq j} \hat{C}_F(i,j) + \sum_{i \neq j \neq k} \hat{C}_F(i,j,k) + \ldots \qquad (35)$$

where, for example,

$$\hat{C}_F(i,j) \sim \langle \hat{V}_i \{\hat{H}_o - \hat{C}_F - io^+\}^{-1} \hat{V}_j \{\hat{H}_o - \hat{C}_F - io^+\}^{-1} \hat{V}_i \{\hat{H}_o - \hat{C}_F - io^+\}^{-1} \hat{V}_j \rangle_s + \ldots$$

$\hat{C}_F(i)$ contains all multiple-scattering associated with a single species but involves an internal propagation structure which implicitly depends on $\hat{C}_F$, i.e. propagation during each multiple-scattering process takes place in the average field of all other scatterers. $\hat{C}_F(i,j)$ describes interference-scattering between two species and so on. Multiple-scattering becomes non-negligible when the carrier wavelength can encompass more than one scattering centre of the same or different species, or when the scattering potentials are very strong or of long range. If either self-energy effects or mixed multiple-scattering are important, the resulting collision integrals are no longer linear in the separate scattering processes. Some forms of multiple-scattering can be described in terms of self-consistent T-matrices (Scott and Moore, 1972). Mixed scattering is discussed in more detail by Lodder and van Zuylen (1970) and Barker (1973, 1979b).

Memory-Interference Terms

The master equation (21) contains a memory term $M_F(t)$ which is one consequence of the interference between the electric field and the scattering processes, and is a functional of the initial equilibrium state $f_o(H_e + V)$. It represents the correction to the otherwise instantaneous accelerative effect of the field due to the field having to break up the correlations in the electron states induced by scattering processes. Indeed M_F may be interpreted as a renormalization of the driving-force term in the kinetic equation (Barker, 1979b; Scott and Moore, 1972), because its non-vanishing part is proportional to $\vec{E}$. Under suitable conditions it may be cleanly combined with the normal driving term by transforming to a quasi-particle representation. To see this we may exploit the stationarity of the equilibrium state to exactly rewrite $M_F(s)$ as

$$M_F(s) = \left(\frac{-i}{s}\right) \langle \hat{V} \{\hat{Q}\hat{H}_F - is\} \hat{Q}\hat{F}\{f_o - \langle f_o \rangle_s\} \rangle_s, \qquad (36)$$

Further simplification is possible by introducing $\vec{X}_V$ as the solution of the operator equation

$$\hat{F}f_o = \vec{E} \cdot [V, \vec{X}_V] + \hat{F}\langle f_o \rangle_s, \qquad (37)$$

We then find

$$M_F(s) = \frac{i}{s}\,\hat{C}_F'\,\vec{E}\cdot\vec{X}_v;\quad M_F(t\to\infty) \sim i\hat{C}_F'(s\to o^+)\vec{E}\cdot\vec{X}_v, \tag{38}$$

$$\hat{C}_F' \equiv \hat{P}\hat{V}\{\hat{Q}\hat{H}_F - is\}^{-1}\hat{Q}\hat{V}. \tag{39}$$

$\hat{M}_F(s)$ is evidently at least of order $E\lambda^2$ where λ is the coupling parameter $V \equiv \lambda V$. Whereas for Boltzmann transport the collision integral contributes $f(K) \sim 1/\lambda^2$, the memory-term contributes terms of order λ^0 and λ^2 to the response, and is therefore apparently negligible for weak coupling. The memory-term can never be neglected for high-frequency time-dependent response (Barker, 1979b; Price, 1966); it then contributes terms of the same order as the relaxive high-frequency Boltzmann theory, i.e. $\sim \lambda^2$. Studies of memory effects have been mainly confined to linear transport problems, although such results should be valid for high fields provided the intracollisional field effect is not important. The memory-effect is conjectured to be crucial in the transport physics of ultra-small devices for which $\tau_c \sim \tau_d$ (Barker and Ferry, 1979). We mention in passing that M_F controls the entire driving effect of the electric field for certain problems, e.g. very high-frequency response, hopping-conduction and quantum transverse magnetoresistance in the quantum limit.

The Intra-Collisional Field Effect

The influence of an electric field within a collision event is analogous to a self-energy effect (Barker, 1973, 1978; Thornber, 1978; Ferry and Barker, 1979a): it generates a level shift Δ_E and level broadening γ_E which destroy the sharp energy-conservation of BTT rates. For a uniform static field $\vec{E}$ the coupling $F = -e\vec{E}\cdot\vec{r}$ makes $\hat{F}$ a differential super-operator in the momentum representation

$$\langle\vec{\underset{\sim}{K}}|\hat{F}A|\underset{\sim}{K}'\rangle \equiv -ie\vec{E}\cdot(\partial/\partial\vec{K} + \partial/\partial\vec{K}')\langle\vec{K}|A|\vec{K}'\rangle, \tag{40}$$

which may be compared with the algebraic linear form for $\hat{H}_e$

$$\langle\vec{K}|\hat{H}_e A|\vec{K}'\rangle \equiv [\varepsilon_K - \varepsilon_{K'}]\langle\vec{K}|A|\vec{K}'\rangle. \tag{41}$$

The modification to the BTT collision integral is most easily seen for weak coupling approximations to $\hat{C}_F$ given by (25). The evaluation of terms like

$$G(K,K') = \{\varepsilon_K - \varepsilon_{K'} - \omega_q - ie\vec{E}\cdot(\frac{\partial}{\partial K} + \frac{\partial}{\partial K'}) - is\}^{-1}A(K,K'), \tag{42}$$

which appear in $\hat{C}_F f$ is best handled by solving the equivalent linear equation $A = \langle R\rangle^{-1} G$ using path-variable techniques. The end result, which for generality we quote for time-dependent fields and in the time-domain is, for phonon scattering,

$$[\partial/\partial t + e\vec{E}\cdot\partial/\partial\vec{K}]f(\vec{K},t) = \int_0^t d\tau \sum_{K'} \{S(\vec{\tilde{K}},\vec{\tilde{K}}';t,\tau)f(\vec{\tilde{K}}',\tau) - S(\vec{\tilde{K}}',\vec{\tilde{K}};t,\tau)f(\vec{\tilde{K}},\tau)\} \tag{43}$$

$$S = 2\underset{\sim}{Re}\sum_q e^{-(t-\tau)/\tau_\Gamma}(N_q + \tfrac{1}{2} + \tfrac{1}{2}\eta)|C(q)|^2 \delta_{\vec{\tilde{K}},\vec{\tilde{K}}' + \eta q} \times \exp[-i\int_t^\tau d\tau' \beta(K,K',\tau')] \tag{44}$$

$$\beta \equiv \varepsilon_{\vec{\tilde{K}}(\tau')} - \varepsilon_{\vec{\tilde{K}}(\tau')} + \eta\omega_q. \tag{45}$$

Here $\eta = +1$ is for phonon emission events, $\eta = -1$ describes phonon absorption. $\tau_\Gamma \sim 1/\Gamma$ is inserted to show qualitatively how self-energy broadening terms enter from ordinary collision-broadening. There are two points to notice: the collision integral is field-dependent through the retarded momenta $\tilde{K},\tilde{K}'$ evaluated along the collision-free trajectory

$$\vec{\tilde{K}}(\tau) \equiv \vec{K} - \int_\tau^t d\tau' e\vec{E}(\tau'),$$
$$\vec{\tilde{K}}'(\tau) \equiv \vec{K}' - \int_\tau^t d\tau' e\vec{E}(\tau'), \tag{46}$$

and the entire form is retarded in momentum space as well as time.

If we now seek the steady-state response by evaluating the RHS of (43) for $\tau_\Gamma \to \infty$, in the case $t \to \infty$, and for a constant electric field, we recover essentially the usual BTT collision integral but the energy-conserving δ-functions are replaced by field-dependent joint spectral density functions given approximately by

$$\delta(\varepsilon_K - \varepsilon_{K'} + \omega_q) \to \frac{1}{\pi}\frac{\gamma_E}{(\varepsilon_{K-eE\tau_\gamma} - \varepsilon_{K'-eK\tau_\gamma} + \omega_q)^2 + \gamma_E^2} \equiv \delta_E(\beta) \tag{47}$$

where the field dependent level shifts and widths are given by

$$\gamma_E \approx |e\vec{E}\cdot(\vec{K}-\vec{K}')/2m^*\pi|^{\frac{1}{2}} \equiv 1/\tau_\gamma, \tag{48}$$

$$\Delta_E \simeq -e\vec{E}\cdot(\vec{K}-\vec{K}')\tau_\gamma/m^*. \tag{49}$$

The intra-collisional field effect is only significant if γ_E, Δ_E exceed the natural line-widths and shifts $\tilde{\Gamma}, \tilde{\Delta}$ discussed above, i.e. if $eEv\tau \gg \Gamma$ (easily achieved in weak-coupling systems). However, the effect on collision processes is only important if γ_E or Δ_E are a significant fraction of ε_K or ω_q, i.e. if $eEv\tau_c \sim eEv\tau_\varepsilon$ where τ_ε is the energy-relaxation time. The whole effect is velocity dependent and thus favours low-effective mass systems (a not surprising result since energy is gained from the field at a rate eEv).

What are the consequences of this effect? The level shift Δ_E may sufficiently upset the normal energy balance that the threshold for certain collision processes, e.g. optical phonon emission, may be substantially lowered. At sufficiently high fields the damping γ_E may wipe out the collision events altogether. If the Bloch representation is properly used the field-effect automatically describes Stark quantization effects. At relatively low fields ($>5kV.cm^{-1}$ for GaAs) the transient response becomes significantly retarded and the fully retarded collision integral must be used (it leads to a general quickening of the transient response).

We have so far discussed single parabolic band transport. The equivalent form for $\delta_E(\beta)$ for non-equivalent phonon-assisted intervalley scattering (between different mass valleys) is readily obtained from (44). The energy deficit $\varepsilon_{K-eE\tau} - \varepsilon_{K'-eE\tau}$ now contains a term quadratic in E which gives rise to a joint-spectral function of the form sketched in Fig. 1. The shift Δ_0 is about 36meV at $100kVcm^{-1}$ for the GaAs (Γ - L) transition, and predicts a substantial virtual lowering of the intervalley threshold.

The intra-collisional field effect also occurs in quantum magneto-transport where it has been known for some time in a different guise. It also occurs in the screening of the interactions. A restricted version of the field effect has been discussed recently by Thornber (1978) in the context of strong coupling polaron theory which exploits Feynman path-integral methodology (see Appendix 1).

Fast Temporal Response

Let us now briefly consider time-dependent homogeneous electric fields, using expressions (43) - (45). Suppose there exists a high-frequency component $e^{-i\omega t}f(\vec{K},\omega)$ of $f(\vec{K},t)$ where $\omega \sim \tau_c$. The asymptotic evaluation of $\int_0^t d\tau C_F(\tau) f(t-\tau)$ will contain

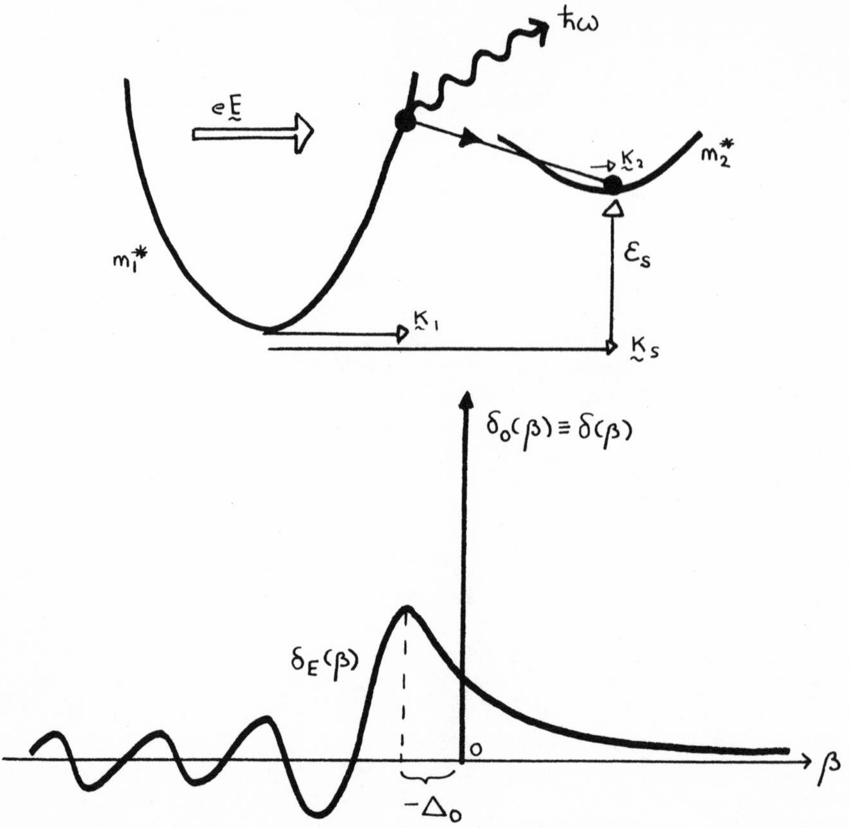

Fig. 1. Intra-collisional Field Effect for Phonon Assisted Intervalley Scattering.

$$\delta_E(\beta) = \frac{1}{\pi} \int_0^\infty d\tau \cos[\tau(\frac{[\underset{\sim}{K}_1 - e\underset{\sim}{E}\tau]^2}{2m_1*} - \varepsilon_s - \frac{[\underset{\sim}{K}_2 - e\underset{\sim}{E}\tau]^2}{2m_2*} - \omega)]$$

$$\beta \equiv \varepsilon_{K_1} - \varepsilon_{K_2} - \omega - \varepsilon_s$$

$$\Delta_0 \propto |E|^{2/3} (\frac{1}{m_1*} - \frac{1}{m_2*})^{1/3}$$

$$m_2* > m_1*$$

$$\int_0^t d\tau \hat{C}_F(\tau) e^{i\omega\tau} f(\vec{K},\omega) e^{-i\omega t} \rightarrow \hat{C}_F(-i\omega + o^+) f(\vec{K},\omega) e^{-i\omega t}$$

This form cannot be found in the BTT picture because the Boltzmann collision integral is evaluated from $\hat{C}_F(o^+)$ not $\hat{C}_F(-i\omega + o^+)$. The QTT thus shows that the collision-kernel is generally frequency-dependent. The energy conservation factors δ_E then generalize to terms involving $\delta_E(\omega)$ and $iP_E(\omega)$ where, neglecting the intra-collisional field effect, $\delta_E(\omega) \sim \delta(\varepsilon - \varepsilon' + \omega_q + \omega)$; $P_E(\omega) \sim P/(\varepsilon - \varepsilon' + \omega_q + \omega)$. The resultant collision-kernel evidently vanishes at very high frequencies: $\omega >> \varepsilon_K$, ω_q. The collisional response is then controlled by the memory term $M_F(t)$. The full response is then more typical of high-frequency optical response. BTT is thus not applicable at high frequencies even for linear fields.

MANY-BODY FORMULATION AND THE SCREENING PROBLEM

Let us conclude with a brief look at the almost unexplored general quantum transport problem: many-carrier, high-field, inhomogeneous transport in the completely coupled hot carrier-hot phonon-impurity system. The overall transport kinetics will be the same as for the independent carrier homogeneous systems discussed previously. But, three new features must be considered: (1) the phonon kinetics, (2) effects of spatial inhomogeneity, and (3) co-operative effects. Of these, I want to emphasize the self-consistent non-equilibrium dynamic carrier shielding of the various scattering processes (Barker, 1979: in the press).

The screening problem has been largely handled phenomenologically in the hot electron literature, usually by basing the scattering rates on a Debye, pre-shielded Hamiltonian with at most an effective electron temperature appearing in the Debye length. Actually the screening process is a dynamic effect and pre-screened initial density matrices and Hamiltonians cannot be rigorously defined even for linear transport (Martin, 1967). Instead one should self-consistently evaluate the non-equilibrium screening within the transport problem. Without such an approach it is difficult to allow for phonon softening effects and other renormalization processes which affect the electronic excitations. In the following we sketch one possible method for handling screening but we do not further discuss the renormalization of the electron and scatterer states for lack of space.

Many-body QTT is conveniently analyzed in the Heisenberg picture so that the Wigner density matrices for one-electron, two-electron, one-phonon and mixed electron-phonon forms, etc., are evaluated as averages of appropriate annihiliation and creation operators over the initial equilibrium density matrix $\rho(H)$ (which is constant in this representation). Given a coupling to the self-consistent local electric potential $\phi(\vec{r},t)$ (determined from Poisson's equation and initiated at $t = 0$), we can build up the dynamical evolution of the full system from the time dependence of the elementary field operators $\psi_\sigma^+(\vec{r},t)$, ψ_σ, b_q^+, b_q which is unfolded by the second quantized Hamiltonian $H_1 = H + e\sum_\sigma \int d^3r\psi_\sigma^+(r)\phi(r,t)\phi_\sigma(r)$. For H we take an elementary model

$$H = \sum_\sigma \int d^3r\psi_\sigma^+(\vec{r})\{-\frac{1}{2m^*}\frac{\partial^2}{\partial r^2} + \frac{1}{2}\sum_{\sigma'}\int d^3r'\psi_{\sigma'}^+(\vec{r}')\phi_c(\vec{r}-\vec{r}')\psi_{\sigma'}(\vec{r}')$$

$$+ \sum_{R_i}\sum_q C_I(q)e^{i\vec{q}.(\vec{r}-\vec{R}_i)} + \sum_{\alpha,q}[C_\alpha(\vec{q})e^{i\vec{q}.\vec{r}}b_{q\alpha}^+ + C_\alpha^*(q)e^{-i\vec{q}\cdot\vec{r}}b_{q\alpha}]\}\psi_\sigma(\vec{r}$$

$$+ \sum_{\alpha,q}\omega_q^\alpha(b_{q\alpha}^+ b_{q\alpha} + \frac{1}{2}), \tag{50}$$

where $\phi_c = e^2/\varepsilon r$ is the bare Coulomb potential statically screened by the bound electrons, impurity and equilibrium lattice configuration. C, C_I describe respectively the electron-phonon and electron-impurity coupling.

Coupled kinetic equations for the one-electron and one-phonon Wigner distributions $f_\sigma(\vec{K},\vec{R};t), N_\alpha(\vec{K},\vec{R};t)$ (see eqns. 9,10) may be obtained from the averaged Heisenberg equations for $\psi^+\psi$ and b^+b,

$$\frac{\partial}{\partial t}\langle A\rangle = (-\frac{1}{i})\langle[H_F,A]\rangle \tag{51}$$

(where $A \equiv \psi^+\psi$ or b^+b; $\langle\ldots\rangle \equiv TR\{\rho_o(H)\ldots\}$). The presence of field operator products in H_F leads to the appearance of higher order density matrices such as the two carrier distribution $f_{\sigma\sigma'}(\vec{r},\vec{R};\vec{r}',\vec{R}')$ and the mixed electron-phonon density matrix $D_\sigma(\vec{r},\vec{R},\vec{q};t) = \langle b_q\psi_\sigma^+(R-\tfrac{1}{2}r)\psi_\sigma(R+\tfrac{1}{2}r)\rangle$, etc. $\partial f\sigma/\partial t$ and $\partial N\alpha/\partial t$ are therefore functionals of $f_{\sigma\sigma'}$, D_σ and so on. Indeed, the full system dynamics is governed by a hierarchy of coupled Heisenberg equations for the many-body Wigner distributions. This chain can only be rigorously decoupled for independent carriers and fixed scatterer distribution. More generally one must truncate the hierarchy. The level of truncation (which may include an infinite-order resummation) decides which many-body and/or correlation effects are to be included in the description. Thus if we ignore self-energy effects and multiple scattering (e.g. 2-phonon processes, impurity cluster scattering, phonon-assisted carrier-carrier scattering, etc.), but include the recoil of the scattering system and the dynamic shielding of the lowest order scattering rates (1-phonon, independent impurity,

2-carrier) it is sufficient to terminate the hierarchy at the 3-particle level. The procedure is faciliated by introducing functions which isolate the many-electron and phonon correlations. For example, the two-carrier density matrix $f_{\sigma\sigma'}$ defines the two-carrier irreducible correlation function $g_{\sigma\sigma'}$ as

$$f_{\sigma\sigma'}(\vec{r},\vec{R};\vec{r}',\vec{R}';t) = \gamma_2\{f_\sigma(\vec{r},\vec{R};t)f_{\sigma'}(\vec{r}',\vec{R}';t) + g_{\sigma\sigma'}(\vec{r},\vec{R};\vec{r}',\vec{R}',t)\},$$

where (52)

$$\gamma_n \equiv \prod_{j=2}^{n} [1 - \sum_{k=1}^{j-1} \delta_{\sigma j\sigma k} P_{jk}]$$

is the anti-symmetrization operator (P_{jk} permutes the variables $\vec{r}_j,\vec{r}_k$) where $\vec{r}_1 = \vec{R} - \frac{1}{2}\vec{r};\vec{r}_1' = \vec{R} + \frac{1}{2}\vec{r};\ \vec{r}_2 = \vec{R}' - \frac{1}{2}\vec{r}';\ \vec{r}_3 = \vec{R}' + \frac{1}{2}\vec{r}'$). $g_{\sigma\sigma'}$ is evidently a measure of the deviation from independent motion for two carriers, and is roughly proportional to the ratio between the average potential energy of the renormalized interaction and the average kinetic energy per carrier, and subsequently becomes negligible when the inter-carrier spatial separation exceeds the range of the self-consistently screened potential. We shall therefore assume that for normal transport $g_{\sigma\sigma'}$ is a first order quantity compared to f_σ.

If we now retain just terms of first and zeroth-order in the hierarchy, we may close the lower members of the chain so that the electron and phonon distributions satisfy

$$[\partial/\partial t + \vec{K}/m^* \cdot \partial/\partial\vec{R} + e\vec{E}(\vec{R},t) \cdot \partial/\partial\vec{K}]f_\sigma(\vec{K},\vec{R};t) = (\partial f/\partial t)_{ee}$$

$$+ (\partial f/\partial t)_{e-ph} + (\partial f/\partial t)_{e-imp}, \tag{53}$$

$$[\partial/\partial t + \vec{V}_\alpha(Q) \cdot \partial/\partial\vec{R}]N_\alpha(\vec{Q},\vec{R},t) = (\partial N/\partial t)_{ph-ph} + (\partial N/\partial t)_{e-ph}, \tag{54}$$

where the generalized scattering rates are functionals of f_σ, N_α, $q_{\sigma\sigma'}$, D_σ, D_σ^I and where $V_\alpha(\vec{Q})$ is the phonon (type α) group velocity. The correlation functions $g_{\sigma\sigma'}$, D_σ, D_σ^I satisfy equations of motion depending on the same five density matrices. The latter equations may be formally solved in terms of f_σ and N_α to give a closed pair of coupled equations for f_σ and N_α alone. To illustrate, let us quote the electron-phonon result:

$$(\frac{\partial f}{\partial t})_{e-ph} \equiv -2\underset{\sim}{\mathrm{Im}} \sum_{q,\alpha} e^{-i\vec{q}.\vec{R}} C^{\alpha}(q)\{D^{\alpha}_{\sigma}(\vec{K}+\vec{q}/2;\vec{R};\vec{q},t)$$

$$- D^{\alpha}_{\sigma}(\vec{K}-\vec{q}/2;\ \vec{R};\ \vec{q},t)\}, \tag{55}$$

where D^{α}_{σ} is found to follow the equation of motion

$$[\partial/\partial t + i\omega^{\alpha}_{q} + \frac{\vec{K}}{m^*}\cdot \partial/\partial\vec{R} + e\vec{E}(\vec{R},t)\cdot \partial/\partial\vec{K}]D^{\alpha}_{\sigma}(\vec{K},\vec{R};\ \vec{q};\ t)$$

$$= S^{\sigma}_{e-p} + S^{\sigma}_{e-e-p} \equiv S^{\sigma}, \tag{56}$$

$$S^{\sigma}_{e-p} \equiv i^{-1}\sum_{q'} C^*(q')e^{-i\vec{q}'.\vec{R}}\{N(\vec{q}',\vec{q},t)[f_{\sigma}(\vec{K}-\vec{q}'/2,\vec{R},t) \tag{57}$$

$$- f_{\sigma}(\vec{K}+\vec{q}'/2,\vec{R},t)] + \delta_{qq'}f_{\sigma}(\vec{K}-\vec{q}'/2,\vec{R},t)[1-f_{\sigma}(\vec{K}+\vec{q}'/2,\vec{R},t)]\}$$

$$S^{\sigma}_{e-e-p} = i^{-1}\sum_{q'}\phi_c(q')e^{-i\vec{q}'.\vec{R}}[f_{\sigma}(\vec{K}+\vec{q}'/2,\vec{R},t)$$

$$- f_{\sigma}(\vec{K}-\vec{q}'/2,\vec{R},t)]X(\vec{q},\vec{q}'), \tag{58}$$

$$X(q,q') \equiv \sum_{K'\sigma'}\int d^3R' D^{\alpha}_{\sigma'}(\vec{K}',\vec{R}';\vec{q};t)e^{-i\vec{q}'.\vec{R}'}. \tag{59}$$

Here S^{σ}_{e-p} alone generates the unscreened electron-phonon scattering rate, while the addition of the Coulombic self-consistent term S^{σ}_{e-e-p} gives rise to the fully dynamically screened field-dependent scattering rate. The other correlation functions and scattering rates have similar forms. It should be noted that in deriving (53) and (56), we use a driving term $e\vec{E}.\partial/\partial\vec{K}$ which is actually a local-homogeneity approximation to an exact form which derives from the Wigner transform of $(-1/i)e[\phi(\vec{R}+\frac{1}{2}\vec{r},t) - \phi(\vec{R}-\frac{1}{2}\vec{r},t)]f_{\sigma}(\vec{r},\vec{R};t)$.

Closed form solutions for the correlation functions D_{σ}, $g_{\sigma\sigma'}$ etc. may be easily obtained from the respective equations of motion by using a path-variable (retarded Green function) method. Thus for (56), we introduce a temporal parameter τ such that

$$\frac{d}{d\tau} \equiv \frac{dt}{d\tau}\frac{\partial}{\partial t} + \frac{d\vec{R}}{d\tau}\cdot\frac{\partial}{\partial\vec{R}} + \frac{d\vec{K}}{d\tau}\frac{\partial}{\partial\vec{K}}, \tag{60}$$

$$\frac{dt}{d\tau} = 1;\ \frac{d\vec{R}}{d\tau} = \frac{\vec{K}(\tau)}{m^*};\ \frac{d\vec{K}}{d\tau} = e\vec{E}(\tau). \tag{61}$$

Equations (61) define the collision-free carrier trajectory in the field $\vec{E}(\vec{R}(\tau),\tau)$, and $\vec{K}(t) \equiv \vec{K}; \vec{R}(t) \equiv \vec{R}$, $t(t) \equiv t$. Equation (56) then has the retarded solution

$$D^{\alpha}_{\sigma}(\vec{K},\vec{R},\vec{q},t) = D^{\alpha}_{\sigma}[\vec{K} - \int_0^t d\tau e\vec{E}(\tau), \vec{R} - \int_0^t d\tau \vec{K}(\tau)/m^*, q, 0]e^{-i\omega_q t}$$

$$+ \int_0^t d\tau e^{-i\omega_q \tau} S_{\sigma}[\vec{K}(t-\tau), \vec{R}(t-\tau); \vec{q}; t-\tau], \tag{62}$$

which is a retarded functional of f_σ and N_α along the free-carrier orbit. The leading term in (62) is the memory-interference term (as before it includes the field-dependent memory of the initial state and vanishes for long times provided E,f and N are slowly varying on the atomic scale τ_c). If we now insert the non-transient part of (62) into (55) we get the most general form of the one-phonon dynamically screened electron-phonon collision rate $(\partial f/\partial t)_{e-ph}$. We shall just consider one limiting case. Under the conditions: (a) negligible intra-collisional field effect; (b) local homogeneity f_σ, N_α, E vary slowly in space and time across the collision range, and (c) asymptotic limit $t \gg \tau_c$, the non-memory part of (55) reduces to the usual Boltzmann-Bloch collision integral with the exceptions that: (i) f_σ, N_α are both non-equilibrium and evaluated locally at $(\underset{\sim}{R},t)$; (ii) the electron-phonon potential $|C(q)|^2$ is dynamically screened by a generalized dielectric function $\varepsilon(\vec{\underset{\sim}{Q}},\omega)$ of Lindhard-like form

$$|C(q)|^2 \rightarrow \frac{|C(Q)|^2}{|\varepsilon(\vec{Q}, -\omega_Q)|^2} \times \underset{\sim}{\mathrm{Re}}\,\varepsilon(\vec{Q}, -\omega_Q) \tag{63}$$

$$\varepsilon(\vec{Q},\omega) \equiv 1 - \phi_c(\underset{\sim}{Q}) \sum_{K,\sigma} \int \frac{d^3R}{\Omega} \left\{ \frac{f_\sigma(\vec{K}+\tfrac{1}{2}\vec{Q},\vec{R},t) - f_\sigma(\vec{K}-\tfrac{1}{2}\vec{Q},\vec{R},t)}{\varepsilon_{K+Q/2} - \varepsilon_{K-Q/2} + \omega + io^+} \right\}, \tag{64}$$

(iii) an additional Landau-damped electron-phonon scattering rate $(\partial f/\partial t)'_{e-ph}$ which is proportional to $\underset{\sim}{\mathrm{Im}}\,\varepsilon(\vec{\underset{\sim}{Q}},\omega)$ occurs. Similar forms have been obtained for a weakly-coupled homogeneous electron-phonon gas in the absence of an electric field (Ron, 1963, 1964). Conditions (a-c) also generate an inhomogeneous screened variant of the phonon Boltzmann transport equation, and forms for $(\partial f/\partial t)_{e-e}$, $(\partial f/\partial t)_{e-imp}$ which are again similar to BTT forms but are dynamically screened by $\varepsilon(\vec{\underset{\sim}{Q}},\omega)$. The dielectric function appears naturally in this approach and depends on the non-equilibrium field dependent distribution f_σ. Generalized Thomas-Fermi and Debye screening and Friedel oscillations are all special limits of (64). We must stress that if the intra-collisional field effect is important, the exact expressions for $\varepsilon_E(Q,\omega)$ contain an additional field-dependence of the form $\frac{1}{2}\vec{E}(\vec{R},t) \cdot \vec{Q}\tau\gamma/m^*$ in the energy denominator of (64). Generally (64) is replaced by a retarded path integral as are the energy conservation factors in $(\partial f/\partial t)_{e-ph}$ etc.

The main consequence of the above many-body QTT may be summarized as: (1) explicitly and implicitly field-dependent, self-consistent dynamic screening of the scattering processes is automatically introduced (this makes the electron kinetic equation non-linear in f even for non-degenerate statistics); (2) Pauli degeneracy factors are included; (3) the collision rates are field-dependent, retarded (in space, time and momentum) functions of f_σ, N_α; (4) coupled hot carrier and hot phonon transport equations are generated.

As an interesting application, we note that the above theory may be deployed to give quantitative confirmation of Herbert's (1973) old conjecture that the gallium arsenide non-equilibrium intervalley ($\Gamma - X$) electron-phonon scattering rate might de-screen in fields exceeding 10V/cm. This occurs because a sufficiently strong electric field will separate two different mass states connected by an inter-valley phonon faster than the plasma period and the screening is too slow to respond in the subsequent short scattering time.

APPENDIX 1: ALTERNATIVE METHODOLOGIES

Feynman Path Integral Method

Especially useful for strong electron phonon coupling, see Thornber (1978a, 1978b), Thornber and Feynman (1970).

Green Function Methods

Useful for rigorous treatment of collision broadening effects but so far have proved intractable outside linear transport problems see Fujita (1966), Scott and Moore (1972); Dewel (1969), Luttinger (1969), Gould (1967), Gerhardt and Hajdu (1971), Langer (1960, 1961, 1962), Elliott, Leath, and Krumhansl (1974).

Parameterized Density Matrices

An approach similar to the displaced Maxwellian models of classical hot electron transport based on constructing a non-equilibrium density matrix parameterized by the local constants of the motion. See Barker and Fer--y (1979), Zubarev and Kalishnikov (1970), and Kalishnikov (1970).

Memory Functionals

An approach based on expressing the current response functions as generalized scalar products in the super space spanned by the super resolvent. The detailed theory is analogous to Langevin Brownian motion theory. The linear theory is discussed by Mori (1965) and Kubo (1974). A non-linear adaptation is given by Barker and Ferry (1979). An important critique of this method which contains many pitfalls for the unwary is given by Kubo (1974).

APPENDIX 2: RESOLVENT IDENTITY

$\hat{R}(z)$ may be formally expanded in two ways (Dyson-like equations):

$$\hat{R}(z) = (\hat{P}\hat{H}\hat{P} + \hat{P}\hat{H}\hat{Q} + \hat{Q}\hat{H}\hat{P} + \hat{Q}\hat{H}\hat{Q} - z)^{-1} \equiv (\hat{H} - z)^{-1} \tag{A2.1}$$

$$= (\hat{P}\hat{H}\hat{P} - z)^{-1} [1 - (\hat{P}\hat{H}\hat{Q} + \hat{Q}\hat{H}\hat{P} + \hat{Q}\hat{H}\hat{Q})\hat{R}(z)] \tag{A2.2}$$

$$= (\hat{Q}\hat{H}\hat{Q} - z)^{-1} [1 - (\hat{Q}\hat{H}\hat{P} + \hat{P}\hat{H}\hat{Q} + \hat{P}\hat{H}\hat{P})\hat{R}(z)] \tag{A2.3}$$

We may now project out two separate equations for PR and QR, using (A2.2) and (A2.3) respectively:

$$\hat{P}\hat{R} = (\hat{P}\hat{H}\hat{P} - z)^{-1} [1 - (\hat{P}\hat{H}\hat{Q})\ \hat{Q}\hat{R}] \tag{A2.4}$$

$$\hat{Q}\hat{R} = (\hat{Q}\hat{H}\hat{Q} - z)^{-1} [1 - (\hat{Q}\hat{H}\hat{P})\ \hat{P}\hat{R}] \tag{A2.5}$$

Next solve (A2.5) for $\hat{Q}\hat{R}$ in terms of $\hat{P}\hat{R}$ and insert into (A2.4), to get a closed equation for $\hat{P}\hat{R}$. Add the two equations for $\hat{P}\hat{R} + \hat{Q}\hat{R} \equiv \hat{R}$; we get

$$\hat{R} \equiv [1 - (\hat{Q}\hat{H}\hat{Q}) - z)^{-1}\hat{Q}\hat{H}\hat{P}][\hat{P}\hat{H}\hat{P} - \hat{C} - \hat{z}]^{-1}[1 - \hat{P}\hat{H}\hat{Q}(\hat{Q}\hat{H}\hat{Q} - z)^{-1}] + (\hat{Q}\hat{H}\hat{Q} - z)^{-1}\hat{Q}$$

where

$$\hat{C} \equiv \hat{P}\hat{H}\hat{Q}(\hat{Q}\hat{H}\hat{Q} - z)^{-1}\hat{Q}\hat{H}\hat{P}$$

BIBLIOGRAPHY

Barker, J. R., 1973, J. Phys. C 6:2663.
Barker, J. R., 1978a, Sol.-State Electronics 21:197.
Barker, J. R., 1978b, Sol.-State Electronics 21:267.
Barker, J. R., 1979a, in "Opportunities for Microstructures Science, Engineering, and Technology," Ed. by J. Ballantyne, Cornell Univ. Press, Ithaca, NY, 116.
Barker, J. R., 1979b, in "Handbook of Semiconductors," Ed. by W. Paul, North-Holland, Amsterdam, in press.
Barker, J. R., and Ferry, D. K., 1979, Sol.-State Electr., in press.
Chester, G. V., 1963, Rept. Prog. Phys. 26:411.
Cohen, E., and Thirrig, W., Eds., 1973, "The Boltzmann Equation," Acta Phys. Austr., Suppl. X.
Dewel, G., 1969, Physica 44:120; 44:473.
Dresden, M., 1961, Rev. Mod. Phys. 33:265.
Elliott, R. J., Leath, P. L., and Krumhansl, J. A., 1974, Rev. Mod. Phys. 46:465.
Ferry, D. K., and Barker, J. R., 1979a, Sol. State Commun. 30:361.
Ferry, D. K., and Barker, J. R., 1979b, Sol.-State Electr., in press
Fujita, S., 1966, "Non-equilibrium Statistical Mechanics," Saunders, San Francisco.
Gerhardt, R., and Hajdu, J., 1971, Z. Physik, 245:176.
Gould, H. A., 1967, "Lectures in Theoretical Physics IXC: Kinetic Theory," Gordon and Breach, New York.
Herbert, D. C., 1973, J. Phys. C 6:2788.
Jaynes, E. T., 1967, in "Proceedings of the Delaware Seminar on Foundations of Physics," Ed. by M. Bunge, Springer-Verlag, New York.
Kalashnikov, V. P., 1970, Physica 48:93.
Kohn, W., and Luttinger, J. M., 1957, Phys. Rev. 108:590.
Kohn, W., and Luttinger, J. M., 1958, Phys. Rev. 109:1892.
Kubo, R., 1957, J. Phys. Soc. Jpn. 12:570.
Kubo, R., 1966, Rept. Prog. Phys. 29:225.
Kubo, R., 1974, in "Lecture Notes in Physics 31: Transport Phenomena Springer-Verlag, Berlin.
Langer, J. S., 1960, Phys. Rev. 120:714.
Langer, J. S., 1961, Phys. Rev. 124:1003.
Langer, J. S., 1962, Phys. Rev. 127:5.
Lodder, A., and van Zuylen, H., 1970, Physica 50:524.
Luttinger, J. M., 1968, "Mathematical Methods in Solid State and Superfluid Physics," Oliver and Boyd, New York.
Martin, P. C., 1967, Phys. Rev. 161:143.
Mori, H., 1965, Prog. Theor. Phys. 33:424.
Peierls', R., 1974, in "Lecture Notes in Physics 31: Transport Phenomena," Springer-Verlag, Berlin.
Price, P. J., 1966, IBM J. Res. Develop. 10:395.
Ron, A., 1963, J. Math. Phys. 4:1182.
Ron, A., 1964, Nuovo Cimento 34:1511.

Scott, C. H., and Moore, E. J., 1972, Physica 62:312.
Stinchcombe, R. B., 1974, in "Lecture Notes in Physics 31: Transport Phenomena," Springer-Verlag, Berlin.
Thornber, K. K., 1978a, Sol.-State Electr. 21:259.
Thornber, K. K., 1978b, in "Path Integrals in Quantum, Statistical, and Solid-State Physics", Ed. by Devreese, J. T., and de Sitter, J., Plenum Press, New York.
Thornber, K. K., and Feynman, R. P., 1970, Phys. Rev. B 1:4900.
Wigner, E. P., 1932, Phys. Rev. 40:749.
Zubarev, D. N., and Kalashnikov, V. P., 1970, Physica 46:550.

CARRIER-CARRIER INTERACTIONS AND SCREENING

C. J. Hearn

Department of Physical Oceanography
Marine Science Laboratories, U.C.N.W.
Menai Bridge, Gwynedd, U.K.

INTRODUCTION

We shall be concerned with the role of intercarrier scattering in hot carrier systems. The intercarrier scattering within a band acts to drive the carriers towards a displaced Maxwellian distribution. In so doing it alters both the energy distribution and the angular distribution in momentum space. Hence we recognise two facets of the scattering, namely momentum transfer and energy transfer between the carriers. In low-field transport the momentum transfer influences the mobility in an indirect manner by altering the angular distribution and this changes the momentum loss to the lattice by the other scattering mechanisms. This effect is important at high carrier densities and its presence should be signalled by a dependence of mobility upon density but in practice it is necessary to allow for accompanying changes in other factors such as impurity scattering. If there are several species of carrier, momentum transfer by inter-band scattering will occur and this has a direct effect upon mobility because of the difference in effective mass of the interacting carriers. For example, in electron-hole scattering, momentum is usually lost from the electrons to the holes in a manner similar to the momentum loss to stationary ionized impurities. This topic of intercarrier scattering in low-field transport had been extensively explored by the early 1960's.

The energy transfer between carriers is relevant to hot carrier problems and tends to become significant at lower densities than momentum transfer, because acoustic phonon scattering is very nearly elastic and is therefore a far more efficient process for momentum loss than for energy loss, and impurity scattering is entirely elastic. This lecture reviews the calculation of the critical concentration above which intercarrier scattering can be expected to influence the energy distribution of the carriers. Since we are dealing with an essentially Coulombic interaction the energy transfe rate will be greatest at low carrier energies and it is here that th lattice scattering is weakest. Thus the critical concentration becomes lowest for carriers which are only moderately heated in low temperature samples. This corresponds to a typical situation for photoexcited carriers and we can therefore expect intercarrier scattering to have a significant effect on some part of the photocarrier distribution. For the usual high-field experiments conducte at higher temperatures the critical concentration tends to be above the concentrations that are normally employed and intercarrier scattering is therefore of only marginal importance.

THE INTERCARRIER INTERACTION

The unperturbed states which describe the system are the independent particle states of the Hartree-Fock approximation. These are Slater determinants of Bloch states calculated for a self-consistent field. The interaction is a correlation effect and derives from the potential V surrounding a single electron situated in the combined media of electron gas, impurities and ion lattice. The calculation of this potential involves the dielectric function $\varepsilon(q,\omega)$ of the composite medium and this measures the polarization of the electrons and the lattice by the Coulomb field of a single electron. In a semiconductor the polarization of the lattice plays the major role of screening the magnitude of V whilst the electrons give V a finite screened range which is in contrast to the infinite range of the original Coulomb potential. It is well known that this infinite range potential leads to a divergence in the cross-section for small angle scattering. In a semiconductor the range is very large, as a consequence of the low density of carriers and impuritie and whilst no divergence occurs there remains a very large amount of small angle scattering which has little effect upon the carrier distribution function. It therefore follows that the actual value used for the screened range is not of critical importance and usually an order-of-magnitude estimate is sufficient. The total rate of intercarrier scattering per carrier is clearly proportional to the carrie density n for a given screening length. However, if the screening length is considered to vary with n we find the slightly paradoxical result that the scattering rate then becomes largely independent of n. In fact as n is reduced, and the screening length increases, so more of the scattering becomes small-angle and of little significanc Hence the momentum and energy loss rates, which are the quantities o real physical significance, remain essentially proportional to n.

Lattice polarization is represented by the low-frequency dielectric constant for small q in view of the long range of the interaction and the assumption of quasi-static conditions. This assumption is by far the most important in the application of intercarrier scattering in semiconductors and yet has received virtually no attention. The extent of screening provided by the lattice could well differ for the direct and exchange terms. The portions of the one-electron Hamiltonian of interest to us are firstly the Coulomb interaction between Bloch electrons

$$H_{ee} = \frac{e^2}{2\varepsilon_o} \sum_{\underline{k},\underline{p},\underline{q}} \frac{<\underline{k}+\underline{q}|\underline{k}+\underline{p}><\underline{k}-\underline{q}|\underline{k}-\underline{p}>}{(\underline{q}-\underline{p})^2} c^+_{\underline{k}-\underline{q}} c_{\underline{k}-\underline{p}} c^+_{\underline{k}+\underline{q}} c_{\underline{k}+\underline{p}} \quad (1)$$

where the overlap between the modulating functions of the Bloch waves is

$$<\underline{k}|\underline{k}'> = \int u^*_{\underline{k}}(\underline{r}) u_{\underline{k}'}(\underline{r}) d\underline{r} \quad (2)$$

and secondly the polar interaction with the L.O. phonons,

$$H_{ep} = \sum_{\underline{k},\underline{q}} A_{\underline{q}} <\underline{k}+\underline{q}|\underline{k}>\{a_{\underline{q}} + a^+_{-\underline{q}}\} c^+_{\underline{k}+\underline{q}} c_{\underline{k}} \quad (3)$$

If we consider exchange of a single virtual phonon in the phonon ground state

$$H_{ep} \rightarrow -\frac{1}{\hbar\omega} \sum_{\underline{k},\underline{k}',\underline{q}} |A_{\underline{q}}|^2 <\underline{k}+\underline{q}|\underline{k}><\underline{k}'-\underline{q}|\underline{k}'> c^+_{\underline{k}'-\underline{q}} c_{\underline{k}'} c^+_{\underline{k}+\underline{q}} c_{\underline{k}} \quad (4)$$

and denoting the static permittivity by μ,

$$|A_{\underline{q}}|^2 = e^2\hbar\omega\left(\frac{1}{\varepsilon_o} - \frac{1}{\mu}\right)/2q^2$$

hence we have the screening process

$$H_{ee} + H_{ep} \rightarrow (\varepsilon_o/\mu) H_{ee}$$

We now consider the screening of this effective potential by the electrons. Taking Fourier components

$$V_q = \frac{e^2}{\mu q^2 \chi(q,\omega)} \quad (5)$$

where χ is the dielectric function for the electrons. χ is related to the linear response function $X(q,\omega)$ which gives the change in electron density Δn_q due to the potential V_q,

$$\Delta n_q = X(q,\omega)V_q \tag{6}$$

by

$$\chi = 1 - (e^2X/\mu q^2) \tag{7}$$

For quasi-static, small q conditions the electron occupancy $f(\varepsilon,V_q)$ is taken as

$$f(\varepsilon,V_q) \sim f^0(\varepsilon + V_q) \tag{8}$$

where f^0 is the unperturbed occupancy so that

$$X \to X_0 = \Sigma \frac{\partial f^0}{\partial \varepsilon} \tag{9}$$

and in the Thomas-Fermi approximation this is used to derive V_q as

$$V_q = \frac{e^2}{\mu(q^2 + \lambda^2)} \tag{10}$$

where

$$\lambda^2 = - e^2X_0/\mu \tag{11}$$

and hence inverting (10), we obtain the screened Coulomb potential

$$V = \frac{e^2}{4\pi\mu r} \exp(-\lambda r) \tag{12}$$

where λ is the inverse screening length. This potential is used for intercarrier scattering and also for ionized impurity and polar phonon scattering. The complete form for X, within the self-consistent field approach, is

$$X(\underline{q},\omega) = \sum_{\underline{k}} \frac{f^0(\varepsilon_{\underline{k}+\underline{q}}) - f^0(\varepsilon_{\underline{k}})}{\varepsilon_{\underline{k}+\underline{q}} - \varepsilon_{\underline{k}} - \hbar\omega - ih\alpha} \tag{13}$$

(see for example Harrison 1970, p. 296) where α tends to zero to give a principal value to the sum. This leads to (9) in that limit.

It is obviously possible to revise (12) using (13) and this is of vital importance in high density systems such as metals. However, in semiconductors the screening by the electrons is of such little importance (i.e. X is so small) and we are so close to using the pure Coulomb interaction (heavily screened by the lattice) that there is no justification for the use of a more sophisticated potential. Indeed, the computation tasks posed by the simple screened Coulomb potential between carriers are very arduous and furthermore we shall employ much more serious approximations in regard to either the overlap of the Bloch states or the effects of exchange. The function (13) describes many properties of the electron gas and the zeros of χ give the eigen-frequencies representing single particle behaviour on a scale $r \lesssim \lambda^{-1}$ and collective plasma modes which incorporate the residual part of the Coulomb interaction. In semiconductors the density of such modes is obviously very small.

If we consider a number of carrier species, and assume that the bands are parabolic in the regions where the gradient of the distribution is significant, and include a set of impurity levels of density N_j of which N_j^+ are ionized, (11) and (9) give

$$\lambda^2 = \frac{e^2}{\mu}\left[\frac{1}{2}\sum_i n_i\langle 1/\varepsilon\rangle_i + \frac{1}{kT}\sum_j \frac{N_j^+(N_j - N_j^+)}{N_j}\right] \tag{14}$$

where we have assumed that the impurity levels are in thermal equilibrium. For a Maxwellian carrier distribution $\langle 1/\varepsilon\rangle = 2/3kT$ and we retrieve the usual Debye Formula. It is to be stressed that (14) is a consequence of (8) which means that the expression may not be well-founded for nonequilibrium carriers but is adequate for our order-of-magnitude requirements. For an extrinsic semiconductor the major doping level is always compensated to some degree and will therefore give the major contribution to screening at very low temperatures. At higher temperatures, as ionization occurs, the free carriers will control the screening. For photoexcited carriers these results are valid for weak illumination.

The matrix element of H_{ee} in (1), with the inverse screening length in the denominator and the use of the lattice permittivity, is after averaging over spin configurations

$$|M|^2 \equiv |\langle \underline{k}+\underline{q}, \underline{k}-\underline{q}|H_{ee}|\underline{k}+\underline{p}, \underline{k}-\underline{p}\rangle|^2$$

$$= \frac{e^4}{4\mu^2}\left[|M_+ - M_-|^2 + |M_+|^2 + |M_-|^2\right] \tag{15}$$

$$M_{\pm} = \frac{\langle \underline{k} \pm \underline{q} | \underline{k} + \underline{p} \rangle \langle \underline{k} \mp \underline{q} | \underline{k} - \underline{p} \rangle}{\lambda^2 + (\underline{p} \mp \underline{q})^2} \tag{16}$$

In using this expression the density of states in $\underline{k}$ - space must be assigned the usual value of $2V/(2\pi)^3$. The term $|M_+ - M_-|^2$ comes from collisions between parallel spin electrons and is lower than the Hartree form $|M_+|^2 + |M_-|^2$ by an exchange term which arises from the spatial separation of parallel spins, due to exclusion, reducing the collision probability. The exchange term is difficult to integrate so it is expedient to adopt the form

$$|M|^2 = \frac{ce^4}{4\mu^2} [|M_+|^2 + |M_-|^2] \tag{17}$$

where c is a constant satisfying $1 \leq c \leq 2$. The upper limit of c neglects exchange and in the lower limit parallel spin collisions are totally excluded as would occur for a zero-range interaction. The distinction between the denominators of the two terms in (17) disappears after some integration.

Intercarrier scattering can involve either intra- or inter-band, or inter-valley, transitions and so the wave-vector labels in (15), (16) formally include a band or valley index. The overlap functions arise in all scattering processes and for intra-band transitions they are usually assumed to be unity, corresponding to complete overlap. For intra-band transitions it can be shown that, more precisely,

$$| \langle \underline{k} + \underline{q} | \underline{k} \rangle |^2 \sim 1 - O(q^2)$$

where q^2 represents an energy which is significant relative to a characteristic energy such as the band gap. When considerable accuracy is required the overlap functions can be included in hot carrier problems involving intra-band (or inter-valley) scattering. However, for an inter-band transition the overlap vanishes with q^2 and hence is zero to a first approximation. If one considers intercarrier scattering between bands in which the initial and final states both involve the same two bands (e.g. electron-hole scattering) the zero overlap between bands results in either M_+ or M_- in (15) being zero. Thus $c = 2$ in (17) since there is no exchange (because the particles are distinguishable). If, however, the final state involves at least one change of band index from the initial state (e.g. impact ionization or Auger recombination) both M_+ and M_- are zero to a first approximation and a non-zero transition rate is only obtained when the more precise form of the overlap is included.

THE CRITICAL CONCENTRATIONS

Following our comments in the Introduction, we shall consider only the critical concentration for energy transfer n_c. The effect of dominant intercarrier scattering is to produce a Maxwellian distribution in energy with an effective temperature T differing from the lattice temperature T_0. Hence for high-field heating the energy input from the field is redistributed by intercarrier scattering and subsequently lost to the lattice. In such a case the electron-phonon interaction merely determines the value of T. However, in reality the condition $n >> n_c$ does not ensure a Maxwellian distribution but if $n \gtrsim n_c$ we know that intercarrier scattering is important. The reason for this non-Maxwellian behaviour lies in the dependence of the scattering rates on carrier energy so that the dominance of the total intercarrier scattering does not ensure that intercarrier scattering exceeds the phonon scattering everywhere in the distribution. Thus, for example, in InSb at 4.2°K $n_c = 10^{15}$ cm^{-3} for $T = 100$°K. Since $T < \theta$ (the optical phonon energy $h\omega/K$) the strong optical phonon scattering is restricted to the small fraction of carriers with $\varepsilon > \hbar\omega$. Hence the Maxwellian below $\hbar\omega$ will be truncated above $\hbar\omega$ because intercarrier scattering is much weaker there than the optical phonon emission process. However, at $T_0 = 300$°K and $T \sim 10^3$ °K the Maxwellian would extend through the optical phonon emission thresholds because the optical phonon scattering is more uniform over the distribution. For photoexcited carriers non-Maxwellian behaviour can result from the generation-recombination processes being faster than the intercarrier scattering, and this is studied in the next section.

Considering intra-band processes, the rate of loss of energy by a carrier of energy η_s kT to a Maxwellian of temperature T can be derived by perturbation theory from (17) and (16) as (Hearn, 1965)

$$W_s = c\, n\, \sigma\, g(\eta_s, \eta_\lambda) \tag{18}$$

where

$$\eta_\lambda \equiv \hbar^2\lambda^2/(2m^*kT)$$

$$\sigma \equiv e^4/[4\pi\mu^2(2m^*kT)^{\frac{1}{2}}]$$

The function g is shown in Fig. 1 together with the limiting form for $\eta_s >> 1$

$$g \to \frac{1}{2\sqrt{\eta_s}}\left\{\log\left(\frac{\eta_s+\eta_\lambda}{\eta_\lambda}\right) - \frac{\eta_s}{\eta_s+\eta_\lambda}\right\} \tag{19}$$

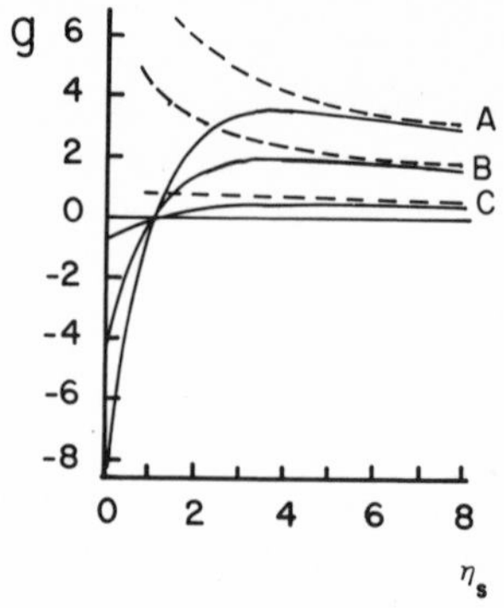

Fig. 1. The full curves show the variation of g with η_5:
Curve A, $\eta_\lambda = 10^{-7}$
Curve B, $\eta_\lambda = 10^{-4}$
Curve C, $\eta_\lambda = 10^{-1}$
The broken curves are the high energy asymptotes (19).

which makes W_s independent of T and corresponds to the loss rate to stationary electrons. Notice that g becomes negative for η_s below unity as a carrier with energy less than kT gains energy from the other carriers. Notice also how weakly g varies with η_λ. The critical concentration n_c is given by

$$n_c \sigma B(\eta_\lambda) = |S(T,T_0)|$$

where $S(T,T_0)$ is the rate of loss of energy from the carriers to the lattice and

$$B(\eta_\lambda) = \frac{4c}{\sqrt{\pi}} \int_R \exp(-\eta_s) |g(\eta_s,\eta_\lambda)| \sqrt{\eta_s}\, d\eta_s$$

where R is the range $g > 0$ or the range $g < 0$. For $\eta_\lambda \ll 1$ it is likely that $c(\eta_\lambda) \sim 2$ hence $B(10^{-1}) = 0.55$, $B(10^{-4}) = 2.99$, $B(10^{-7}) = 5.76$ so in practical terms $B \sim 1$ and for small η_λ varies logarithmically. Figure 2 shows $Bn_c(T)$ for InSb at 4.2 K. This clearly shows that photoexcited carriers would be strongly influenced by intercarrier scattering at this temperature since they usually involve only weak heating. Field-heating experiments under similar conditions do not seem to be consistent with Maxwellian behaviour probably because of effects at the optical phonon threshold.

The rate of energy transfer between two heated carrier systems, such as an electron and hole system or the electrons in two distinct

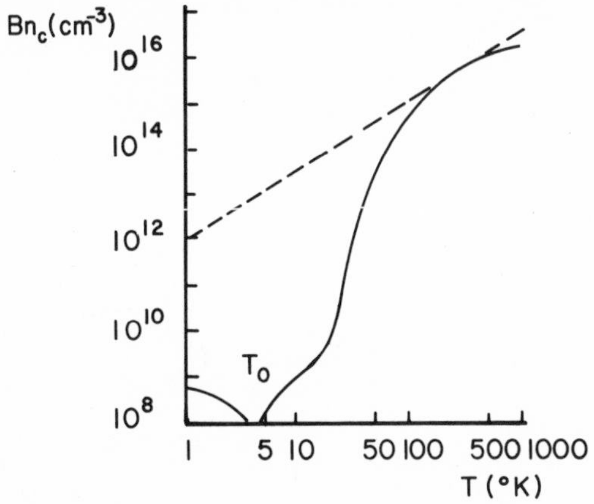

Fig. 2. The dependence of Bn_c on T for electrons in InSb with T_0 = 4.2 K. The broken line indicates the maximum electron concentration which is compatible with non-degeneracy.

valleys of the conduction band, can also be derived from (17), (16). In this case it is useful to adopt displaced Maxwellian distributions to allow the dependence of the transfer rate on drift velocity to be investigated. It is found (Hearn, 1974) that the rate of loss from system 1 to system 2 is, taking M_- as zero and c = 2,

$$R = \frac{e^4 n_1 n_2 \bar{m}^{3/2} \exp(-\zeta^2)}{4(2\pi^3)^{1/2} \mu^2 \,\bar{\varepsilon}^{1/2} m_1 m_2} \{P + Q\}$$

where $\bar{m}^{-1} = m_1^{-1} + m_2^{-1}$ is the reduced effective mass, thermal energies $\varepsilon_i = kT_i$ are used, $\bar{\varepsilon}$ is the mean thermal energy,

$$\bar{\varepsilon} = \bar{m}\left(\frac{\varepsilon_1}{m_1} + \frac{\varepsilon_2}{m_2}\right)$$

and ζ^2 is the drift energy relative to $\bar{\varepsilon}$,

$$\zeta^2 = \frac{\bar{m}}{2\bar{\varepsilon}}(\underline{v}_1 - \underline{v}_2)^2$$

P and Q are numbers which refer respectively to the thermal and mechanical contributions to R. P depends on the temperature difference and Q on the velocity difference and for the usual long range potential and $\zeta^2 \ll 1$,

$$P \sim C(\varepsilon_1 - \varepsilon_2)/\bar{\varepsilon}$$

$$Q \sim \frac{C\varepsilon_1\varepsilon_2}{3\bar{\varepsilon}^2}(\underline{v}_1 - \underline{v}_2)\cdot\left[\frac{m_1\underline{v}_1}{\varepsilon_1} + \frac{m_2\underline{v}_2}{\varepsilon_2}\right]$$

where C is a number of order unity having a similar logarithmic dependence on λ as B mentioned above.

Combining Q with P by defining an effective thermal difference $\Delta\varepsilon$, we can define the value of a parameter F,

$$F \equiv \frac{n_2}{\bar{\varepsilon}} \frac{\Delta\varepsilon}{\varepsilon_1 - kT_0} \left(\frac{\varepsilon_1}{\bar{\varepsilon}}\right)^{1/2} C,$$

such that the rate of energy loss to the lattice $S(T_1,T_0)$ by system 1 is equal to R. Here we have taken $T_1^{1/2}\, S(T_1,T_0)$ to be proportional to $(T_1 - T_0)$; for extreme heating F becomes a function of ε_1. For electrons in InSb at room temperature $F \sim 3 \times 10^{18}$ $cm^{-3}(eV)^{-1}$ for energy loss to holes. Thus for a concentration of 10^{16} cm^{-3} unheate holes the electron-hole energy loss is 10% of the loss to the lattic For $T_0 = 4.2$ K, $F \sim 2 \times 10^{13}$ $cm^{-3}(eV)^{-1}$ so that the electron-hole energy loss would exceed the lattice loss by an order of magnitude for a hole concentration of 10^{11} cm^{-3}. Thus electron-hole scatterin is apparently important for across-the-gap photoexcited hot carriers in the III-V compounds.

In GaAs just above the threshold field for the Gunn effect ther are comparable populations of the central and satellite valleys. Fo the central valley, $\varepsilon_1 \sim 0.1$ eV whilst for the satellite valleys, $\varepsilon_2 \sim 0.03$ eV and the mechanical term is found to contribute only 0.006 eV to their effective temperature difference. For a total electron concentration of 10^{17} cm^{-3} the satellite valleys are found to derive about equal quantities of energy from interelectronic and intervalley electron-phonon scattering, but this is an order of magnitude lower than the energy they receive directly from the field. Uniform electron densities of 10^{17} cm^{-3} are not used in Gunn devices but such densities do occur in the accumulation layers of domains bu are accompanied by a depopulation of the central valley. In Ge and Si at 77°K one finds significant intervalley interelectronic scatter ing for concentrations above 10^{15} - 10^{16} cm^{-3}.

DISTRIBUTION FUNCTIONS

Intercarrier scattering is likely to be far more significant for photoexcited carriers, which are only moderately heated in low temperature samples, than for the typical high-field experiments at higher temperatures. The concentration of photocarriers can readily be raised above the critical concentration but as we remarked above, this does not ensure a Maxwellian distribution because of the effects of the generation-recombination terms. Essentially we can consider that under monochromatic excitation all carriers enter the band with some initial energy $\bar{\varepsilon}$ and are then scattered within the band until they leave, by some recombination process, after a mean lifetime τ. The form of the steady-state distribution function depends upon the number of scattering events of different types which can occur during the carrier lifetime. If $n \gg n_c$ we know that these events are mostly intercarrier scattering so if τ is sufficiently long a Maxwellian distribution at some effective temperature T will occur and the actual value of T will be determined by the electron-phonon scattering. If however τ is sufficiently short very few intercarrier scattering events will occur during a carrier's lifetime and the distribution will resemble the narrow excitation pulse at $\bar{\varepsilon}$. The question to be raised here concerns the value of the critical lifetime beyond which the distribution is effectively Maxwellian. Clearly this critical lifetime τ_c must be inversely proportional to the population n since this controls the rate of scattering for a particular carrier. Thus we can alternatively introduce a second critical concentration n_c' (inversely proportional to τ) so that if $n \gg n_c'$ for a particular lifetime (and $n \gg n_c$) the distribution is Maxwellian. The product $n_c\tau$ is a critical excitation rate $G_c\tau^2$, so that if the rate of photoexcitation per unit volume greatly exceeds G_c we have Maxwellian photocarriers.

For the critical condition that we have mentioned the number of intercarrier scattering events is barely sufficient to establish a Maxwellian distribution. Thus if $n \gg n_c$, we know that the number of electron-phonon events is negligible during τ. Thus the effective temperature T is simply given by $3kT/2 = \bar{\varepsilon}$. The critical condition corresponds to the rate of intercarrier energy transfer being equal to the rate of energy output by the recombination process (which is equal here to the rate of energy input by photoexcitation). Hence referring back to (18),

$$\frac{3}{2}\, kT/\tau = \frac{3cne^4 D(\eta_\lambda)}{4\pi^{3/2}\mu^2(2m^*kT)^{1/2}} \tag{20}$$

$$D \equiv B\sqrt{\pi}/(3c)$$

where D is a logarithmic function of η_λ of order unity. If we incorporate the uncertainty of c within the lifetime we have a critical lifetime u^{-1} with

$$u \equiv \frac{ne^4}{\mu^2 m^{*1/2}(2\pi kT)^{3/2}} = \frac{\alpha n}{(\mu/\varepsilon_o)^2(m^*/m)^{1/2}T^{3/2}} \tag{21}$$

where $\alpha \sim 2.7 \times 10^{-6}\ m^3(\text{degK})^{3/2}s^{-1}$.

Hearn (1966) made a detailed calculation of the distribution of photoexcited carriers for a range of η_λ, and various values of τu, which confirms that the distribution is beginning to adopt a recognizable Maxwellian profile when $\tau u \gtrsim 1$. In deriving this distribution function the carriers were assumed to be almost Maxwellian so that the intercarrier scattering rate could be linearized which made the numerical task tractable within the computation facilities then available. The distribution becomes Maxwellian for $\varepsilon > \bar{\varepsilon}$ at much shorter lifetimes than for the region $\varepsilon < \bar{\varepsilon}$ which contains the bulk of the carriers. In Fig. 3, we show the variation of the value of the normalized distribution at $\varepsilon = \bar{\varepsilon}$ for various η_λ. Thus adopting u^{-1} as the critical lifetime we obtain the expression for n_c' and G_c,

$$\tau^2 G_c = n_c'\tau = \beta(\mu/\varepsilon_o)^2(m^*/m)^{1/2}T^{3/2}, \tag{22}$$

where $\beta \sim 4 \times 10^5\ m^{-3}\ (\text{degK})^{-3/2}s^{-1}$. If the carriers are excited by a radiation spectrum which consists of a pulse of width comparable to the mean excitation energy in the band, $\bar{\varepsilon}$ (or by a broad band spectrum), the effective temperature is again given by $3kT/2 = \bar{\varepsilon}$ but the value of n_c' is lower than (22) because less redistribution of the carriers is required to produce a Maxwellian profile. For InSb at 4.2 K and T = 10 K we find $n_c' \sim 10^{11}\ cm^{-3}$ when $\tau = 10^{-9}$ sec as compared with $n_c \sim 10^8\ cm^{-3}$ from Fig. 2.

Maksym (1979) has made some calculations of the steady-state distribution for photoexcited carriers in GaAs at 1.7 K with $\bar{\varepsilon}$ = 4meV and $\tau = 10^{-7}$ sec, which include both intercarrier and electron-phonon scattering. The most significant effect of intercarrier scattering is to be seen in the 'anti-Stokes' region $\varepsilon > \bar{\varepsilon}$. In the absence of intercarrier scattering the distribution decays exponentially above the excitation energy with the approximate shape of a Maxwellian at the lattice temperature T_o. Carriers have entered this region by acoustic phonon absorption from states at $\bar{\varepsilon}$, or just below $\bar{\varepsilon}$, and the Maxwellian decay reflects the Boltzmann factor in the equilibrium Planck distribution of the phonons. Intercarrier scattering tends to establish a Maxwellian above the excitation energy at an effective temperature T which is higher than T_o. Phonon absorption is a very weak process at low temperature and so intercarrier scattering will evidently dominate the anti-Stokes region when $n \sim n_c$. Maksym finds a figure of $n \sim 10^{11}\ cm^{-3}$ for intercarrier

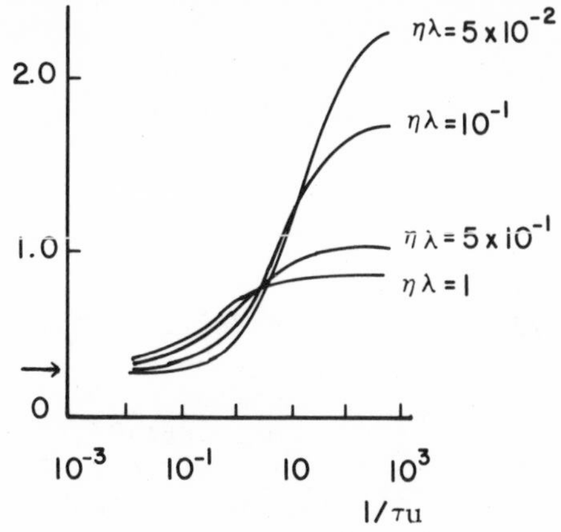

Fig. 3. The variation of the normalised photocarrier distribution function at the excitation energy as a function of $(1/\tau u)$ for various η_λ. The arrow indicates the value for a Maxwellian distribution.

scattering to be significant in this region which is of the same order as n_c. Unfortunately there are considerable computational difficulties in calculating distributions which include both electron-phonon and the full non-linear long-range intercarrier scattering. Maksym employed a Taylor series expansion of the intercarrier term and a mesh size for the numerical evaluation of the distribution well in excess of that required to correctly represent the nearly elastic acoustic phonon scattering.

The exact solution of the Boltzmann equation for carriers in a high-field has only become possible within the last decade because of the available computational facilities. Prior to this it was usual to calculate high-field transport parameters using either some approximate analytical procedure or a displaced Maxwellian distribution. The use of the Maxwellian was originally defended on the basis of densities well in excess of the critical concentration. However, in reality such densities are not employed and the Maxwellian was a calculational device which usually allowed a good assessment to be made of high-field transport parameters. If an exact solution is compared with the Maxwellian solution one is able to deduce the maximum effect of intercarrier scattering. Both solutions are often available as in the case of electrons in the multi-valley conduction band of GaAs. However revised electron-phonon scattering parameters and the inclusion of details of band-structure and overlap functions usually invalidate such comparisons and indeed place the role of intercarrier scattering in a more realistic context. This is expecially true for inter-valley transition rates in view of our earlier comments regarding the failure of a Maxwellian distribution above a threshold energy. However true intercarrier scattering processes are detectable at high densities and the variable path and Monte Carlo methods can assess their contribution to the transpor parameters.

REFERENCES

Harrison, W. A., 1970, "Solid State Theory", McGraw-Hill, New York.
Hearn, C. J., 1965, Proc. Phys. Soc., 86:881.
Hearn, C. J., 1966, Phys. Lett., 20:113
Hearn, C. J., 1966b, Proc. Phys. Soc., 88:407.
Hearn, C. J., 1974, Phys. Stat. Sol. (b), 64:527.
Maksym, P. A., 1979, Ph.D. Thesis, Univ. of Warwick, U.K., unpublished.

MULTIPHONON SCATTERING

P. Kocevar

University of Graz
Graz. Austria

ELECTRON-MULTIPHONON PROCESSES

Electron-phonon ("e-ph") interactions of second or higher order can for almost all cases of practical interest be considered as minor scattering-mechanisms for charge carriers in metals and semiconductors. It is quite obvious that any strong contribution of such processes would necessarily imply a failure of the adiabatic principle and therefore of the usually quite successful description of the dynamics of electrons and phonons in the frame-work of perturbation theory. On the other hand this perturbative approach in terms of free carriers and phonons in the initial and final states of a scattering event can only be justified for sufficiently weak e-ph-coupling. The well-known exception to this frequently encountered "weak-coupling" situation are strongly ionic substances, where each charge carrier continuously interacts with the polarization-field of the polar lattice modes; the corresponding quasi-particle, the polaron, can in the terminology of quantum field theory be considered as a bare electron, dressed by a cloud of virtual phonons. For not too strong e-ph-coupling, the lattice-distortion by the moving charge carrier can be reasonably well described by the usual e-ph-dynamics, and the average number of virtual phonons accompanying the carrier can be estimated, and has been found to reach values of 3 or more for the most interesting case of the alkali-halides. In this very restricted context of polaron theory, "multiphonon" processes have been known for a long time (Devreese, 1972; Kuper and Whitfield, 1963). On the other hand, as must be expected from our introductory remarks, instantaneous electron-N-phonon processes, with $N \geq 3$, occurring as well separated events, in the sense of ordinary perturbation theory, have never been isolated in any conventional conducting material even with the very selective experimental methods under which electron-2 phonon process have been studied recently (see below). The following discussion of higher-order e-ph-interactions will therefore necessarily be restricted to our present understanding of e-2ph-processes.

ELECTRON-TWO PHONON PROCESSES

For the sake of simplicity we shall restrict the basic consider tions of this section to model semiconductors with electrons in a standard conduction-band with the corresponding energy-dispersion $E(k) = \hbar^2k^2/2m$, where $\vec{k}$ and m denote the wavevector and effective mass of the carriers, respectively, and $k = |\vec{k}|$.

Since a very detailed treatment of various aspects of e-2ph-processes for optical phonons was presented earlier by Ngai and Ganguly (1973), the present section will focus on general facts, including some supplementary remarks on acoustic phonon processes, whereas the last sections will contain a critical description of the existing literature.

Within the weak coupling regime of interest for the present discussion, the smallness of second-order transition-amplitudes, as compared to the corresponding (allowed) first-order transitions, wil in general be partly compensated by the strongly relaxed selection-rules for instantaneous 2-phonon processes (in group-theoretical language: the decomposition of a product of the four representations for the initial and final electron-states and the two phonons will practically always contain the identical representation) and by the strongly increased volume of phase-space. This latter effect is mos prominent in second order Raman-scattering, but also is very importa for second order acoustic phonon processes. While conservation of crystal-momentum and energy restricts the final-states for electroni transitions to one energy-surface in $\vec{k}$-space for 1 ac. ph-absorption and -emission respectively, the additional degrees of freedom of the second phonon in a 2 ph-process relax these restrictions, so tha the final wavevectors can be in a much larger (2-dimensional) region of $\vec{k}$-space. Especially in the case of energy-dissipation of hot electrons, the energy of phonons participating in a 2 ph-emission can be much larger than in a 1ph-process: assuming a Debye-spectrum $\hbar\omega(q) = sq$, where ω, q and s denote the frequency, wavevector- and energy-conservation (neglecting Umklapp) for electron-transitions $\vec{k} \to \vec{k}'$ via 1ph- or 2ph-emission read

$$\text{1 ph:} \quad \vec{k} = \vec{k}' + \vec{q}, \tag{1}$$

$$\frac{\hbar^2}{2m} k^2 = \frac{\hbar^2}{2m}(\vec{k} - \vec{q})^2 + \hbar sq; \tag{2}$$

$$2\text{ ph:} \quad \vec{k} = \vec{k}' + \vec{q}_1 + \vec{q}_2, \tag{3}$$

$$\frac{\hbar^2}{2m} k^2 = \frac{\hbar^2}{2m}(\vec{k} - \vec{Q})^2 + \hbar s(q_1 + q_2)_1 \tag{4}$$

where $\vec{Q} = \vec{q}_1 + \vec{q}_2$. Equations (1) and (2) lead to a maximum energy loss $\hbar s q_{max} = 2\hbar s(k - ms/\hbar)$, which is much smaller than $E(k) = \hbar^2 k^2/2m$ for the vast majority of carriers (with velocity $\hbar k/m >> 2s$), whereas (3) and (4) imply $Q \leq 2(\vec{k}\vec{Q}/Q - ms/\hbar)$, so that exactly these electrons can lose all their momentum ($\vec{Q} = \vec{k}$) and energy in a 2ac.ph-transition (Aldredge and Blatt, 1967).

Turning from kinematics to the dynamics of 2-ph-processes, we have first to refine our terminology by distinguishing between direct and indirect second-order phonon processes. Quite generally the e-ph-interaction is obtained by expanding the (self-consistent) one-electron potential V(r) in the phonon-induced displacements $\vec{u}_\alpha$ of the ions from their equilibrium position $\vec{R}_\alpha$. In the rigid ion approximation the interaction Hamiltonian becomes

$$H_{e-ph} = \sum_\alpha \vec{u}_\alpha \cdot \vec{\nabla} V(\vec{r} - \vec{R}_\alpha) + \frac{1}{2} \sum_\alpha \vec{u}_\alpha \cdot [\vec{\nabla}\vec{\nabla} V(\vec{r} - \vec{R}_\alpha)] \cdot \vec{u}_\alpha + \dots \tag{5}$$

$$\equiv H^{(1)} + H^{(2)} + \text{terms of third or higher order in u.}$$

Accordingly, an e-2ph process can arise from either the bilinear term $H^{(2)}$ (the instantaneous e-2ph-process) or from an iterated 1ph-process of second order through an intermediate state, formally written as $H^{(1,1)} \equiv H^{(1)} (E-H)^{-1} H^{(1)}$, where H is the total Hamiltonian for the electron and the phonons. The corresponding transition rates will contain interference terms between matrix elements of $H^{(1,1)}$ and $H^{(2)}$. Complete cancellation in each order of perturbation theory has to result for acoustic modes in the limit of infinite wave-length due to translational invariance of the electron-ion system. The general belief in strong cancellations for quite general (nonpolar) lattice modes is also supported by the fact that all calculations of direct 2-ph-transitions have led to coupling constants several orders of magnitude larger than the experimentally determined values. According to Cardona (1974) and Weinstein and Cardona (1973), we can see an analogous case in the extreme tight-binding limit, because the energy spectrum of an isolated atom is invariant with respect to an adiabatic shift of its core, since the valence-electrons will instantaneously adapt themselves to the core-displacement. Nevertheless an understanding of the actual situation of strong bonding to neighbouring atoms in all the (predominantly covalent) materials of interest would require a much more detailed investigation. For the case of a general "cancellation-effect", we would expect that even in materials where the energetically most favourable $H^{(1,1)}$-transitions (e.g.

1ph-intraband) are forbidden by symmetry, $H^{(2)}$ will be strongly counterbalanced by $H^{(1,1)}$-contributions of all possible intermediate states from all bands (Devreese, 1974). The importance of these latter contributions, which are beyond even a qualitative theoretical treatment, has strongly impeded any quantitative calculation of second order e-ph couplings, because the determination of either $H^{(2)}$ or some leading contributions to $H^{(1,1)}$ cannot yield even a rough estimate of the actual e-2ph coupling-strength.

As a final important point on matters of terminology, we have to discuss a second type of interference concerning conventional 1ph-processes and resonant $H^{(1,1)}$-contributions. We follow Lawaetz, et al., (1974) and expand the total wave function of one electron plus the phonon-system $\Psi(t) = \sum_n a_n(t)\psi_n$ into the time-dependent eigenfunctions Ψ_n (with eigenvalues E_n) for the uncoupled e-ph-system. The Fourier-transform $G_n(E)$ of $a_n(t)$ is defined by

$$a_n(t) = -\frac{1}{2\pi i}\int_{-\infty}^{+\infty} dE\, G_n(E) e^{-iEt/\hbar}, \tag{6}$$

and has to be determined from the Schrödinger equation

$$(E - E_n)G_n(E) = \sum_m H^{(1)}_{nm}\, G_m(E) + \delta_{no}, \tag{7}$$

with $\delta_{no} = 1$ for $n = 0$, and $\delta_{no} = 0$ otherwise, where we restrict ourselves for the present purpose to 1ph-couplings (with matrix-elements $H^{(1)}_{nm}$, setting $H^{(1)}_{nn} = 0$) and start from $t = 0$ with an electron in state $n = 0$. The general solution of (7) can be written as

$$G_n(E) = U^{(1)}_{no}\frac{1}{E - E_n}G_o(E), \; n \neq 0, \tag{8}$$

with $U^{(1)}_{n_o}(E)$ to be determined from

$$U^{(1)}_{no}(E) = H^{(1)}_{nm} + \sum_{m\neq 0} H^{(1)}_{nm}\frac{1}{E - E_m}U^{(1)}_{mo}(E) \tag{9}$$

Next we have to solve (9) for all 1ph- and 2ph-processes to obtain

$$G_o(E) = [E - E_o + i\gamma^{(1)}_o(E)]^{-1}, \tag{10}$$

where the lifetime-broadening of the state 0,

$$\mathrm{Re}\gamma_o^{(1)}(E_o) = \mathrm{Re}\{i \sum_{n\neq 0} H_{on} \frac{1}{E - E_n} U_{no}(E)\}_{E=E_o}, \tag{11}$$

gives the decay-rate, via (6) and (10), through

$$|a_o(t)|^2 = e^{-2\mathrm{Re}\gamma_o^{(1)}(E_o)\cdot t/\hbar}. \tag{12}$$

Using n and n' as labels for states which differ from state 0 by one phonon, and similarly using ν for states differing from state 0 by two phonons, we solve (9) to obtain

$$\gamma_o^{(1)} = \bar{\gamma}_o^{(1)} + \bar{\bar{\gamma}}_o^{(1)}, \tag{13}$$

with

$$\bar{\gamma}_o^{(1)} = i \sum_n \frac{H_{on}^{(1)} H_{no}^{(1)}}{E - E_n + i\gamma_n^{(1)}} \tag{14}$$

and

$$\bar{\bar{\gamma}}_o^{(1)} = \sum_n \sum_{\nu\neq 0} \sum_{n'\neq n} \frac{H_{on}^{(1)} H_{n\nu}^{(1)}}{E - E_n + i\gamma_n^{(1)}} \frac{1}{E - E_\nu} \frac{H_{\nu n'}^{(1)} H_{n'o}}{E - E_{n'} + i\gamma_n^{(1)}}, \tag{15}$$

where

$$\gamma_n^{(1)}(E) = i \sum_{\nu\neq 0} H_n^{(1)} \frac{1}{E - E_\nu} H_{\nu n}^{(1)} \tag{16}$$

is the broadening of the intermediate state n through 1ph-processes.

The effect of a resonant intermediate state n [for which the real part of the denominator in (14) vanishes] is now easily seen to give

$$\mathrm{Re}\bar{\gamma}_o^{(1)} \to \pi \sum_n |H_{on}^{(1)}|^2 \delta(E - E_n) \tag{17}$$

in the limit $\gamma_n^{(1)} \to 0$. But this expression is identical to the "golden rule" for first order 1ph-transitions, so that transition-rates of resonant indirect 1ph-processes cannot be distinguished from ordinary 1ph-coupling-constant. This point has been overlooked in some earlier transport calculations, as we shall see in the following section. If (14) contains no resonant intermediate states, $\bar{\gamma}_0^{(1)}$ describes ordinary second order processes via intermediate states $n \neq 0$, as is the case for $\bar{\bar{\gamma}}_0^{(1)}$, since the two poles in (15) never coincide.

RESULTS

Since limited space does not permit a detailed analysis of the existing theoretical and experimental work on e-2ph processes, we can only give a chronological list of some of the most important references, stressing some points relevant to our discussion of the previous section.

Holstein (1959) shows explicitly the almost complete cancellation of $H^{(2)}$ and $H^{(1,1)}$ for the case of one long-wave (ultrasonic) and one thermal phonon. Herring (1962) cites unpublished calculations which show strong $H^{(2)}$ and $H^{(1,1)}$ cancellations for the genera case. Zeinstein and Cardona (1973) give a simple argument for the dominance of the core part of OPW-wavefunctions in $H^{(2)}$ [$H^{(2)}$ $\nabla^2 V(r) \sim \delta(r)$ for Coulomb-like ion potentials and matrix elements like $|\psi(o)^2|$], therefore the importance of the details of the wavefunctions near the core is considerable [which implies that higher order e-ph-couplings cannot be calculated with pseudo-wavefunctions and pseudo-potentials replacing ψ and V, as stressed by Lin-Chung (1973)]. These latter authors also experimentally derived 2-ph deformation potentials which are two orders of magnitude smaller than the $H^{(2)}$ OPW results, and generated the cancellations argument of the previous section. Ivey (1974) showed that no great reductio of the $H^{(2)}$ matrix elements for 2 TO-deformation potentials which supports the cancellation argument.

Ansel'm and Lang (1959) demonstrate the formal identity of resonant second-order and ordinary 1ph rates, derive convenient formulas for numerical calculations of $H_{nn}^{(2)}$, and estimate that 2ph processes become comparable to 1ph processes at temperatures of the order of $10^2 - 10^3$ K. Franzak and Bailyn (1962) show that $H^{(2)}$ contributes less than 4% to the conductivity of the alkali metals. Alldredge and Blatt (1967) calculate the resonant $H_{e-acph}^{(1,1)}$ process (a "ghost" process according to the last section), apply the Glauber formalism to multiphonon processes, and give a good description of t e-2acph kinematics. Sher and Thornber (1967) have calculated the $H_{e-LO}^{(1,1)}$ ghost process. Lin-Chung and Ngai (1972) evaluated the $H^{(2)}$ contribution to 2TO-ph coupling with a simple OPW calculation and found that the plane-wave portion of the OPW wave function turns out to yield 1% of that of the core portion. Levinson and Rashba (1972) developed a nonlinear generalization of the Born and Huang lattice dynamics to obtain the c-2LOph coupling. Baumann (1973) carried out a pseudopotential calculation of $H^{(1,1)}$ for TA-phonon pairs at the Brillouin zone boundary to explain structure found in the hot electron (Kahlert and Bauer, 1971) and magnetophonon spectra (Stradling, et al., 1970) in various III-V compounds, but no estimate was made of the $H^{(2)}$ contribution.

Thomas and Queisser (1968) found 2TA and (LO + TA) structure in the second-derivative conduction characteristics at 2K in tunnel junction spectroscopy. Ngai and Johnson (1972) found 2TO, (TO - LO), and 2LO pinning effects in cyclotron resonance measurements and derived estimates of the 2TO-ph deformation potential assuming a negligible $H^{(1,1)}$ contribution because of the 1TO intraband phonon being symmetry forbidden (not justified according to the previous section). Their result still depends on an unknown fraction of the phonon spectrum belonging to the critical points in the 2TO-ph density-of-states.

Ngai (1974) gave a general survey of experimental and theoretical results including an attempt to fit the temperature dependence of the mobility of n-Si with 2TA and 2TO deformation potentials. Vogl (1976) gave a unified microscopic treatment of all standard e-ph couplings including two-phonon processes. Devreese (1978) showed that experimental 2LO pinnings seemed to agree with theoretical results obtained using only the $H^{(1,1)}$ contribution.

SUMMARY

As has to be expected for materials where the adiabatic principle seems to work well, higher-order e-ph-processes should be small compared to first-order interactions for nonpolar lattice modes in the standard semiconducting materials. This has so far been experimentally verified, because 2 ph-processes have most clearly been identified only with spectroscopic methods or in certain resonant magneto-conductivity effects. As far as ordinary transport is concerned there is a possibility of noticeable contributions in n-Si, but any conclusive analysis of the electron mobility in this material is still lacking, as can be seen from independent attempts to explain the discrepancies between theory and experiment in terms of long-range e-ph-multipole interactions (Lawaetz, 1969; Gram and Jorgensen, 1973). Much larger contributions of 2-ph-processes have to be expected and have been found in high-field transport because of their increased importance as an energy-dissipation mechanism.

BIBLIOGRAPHY

Alldredge, G. P., and Blatt, F. J., 1967, Annals of Phys. 45:191.
Ansel'm, A. I., and Lang, I. G., 1959, Sov. Phys.-Sol. State 1:621.
Baumann, K., 1973, Acta. Phys. Austriaca 37:350.
Cardona, M., 1974, in "Elementary Excitations in Solids, Molecules, and Atoms," Devreese, J. T., Ed., Plenum, New York.

Devreese, J. T., Ed., 1972, "Fröhlich Polarons and the Electron-Phonon Interaction", North Holland, Amsterdam.
Devreese, J. T., Ed., 1974, "Elementary Excitations in Solids, Molecules, and Atoms," Plenum, New York.
Devreese, J. T., 1978, in "Proc. Intern. Conf. on Applications of High Magnetic Fields in Semiconductors," Oxford.
Franzak, E., and Bailyn, M., 1962, Phys. Rev. 126:2033.
Gram, N. O., and Jörgenson, M. H., 1973, Phys. Rev. B 8:3902.
Herring, C., 1962, in "Proc. Intern. Conf. Physics of Semiconductors Prague, 1960," Czech. Acad. Sci., Prague.
Holstein, T., 1959, Phys. Rev. 113:479.
Ivey, J. L., 1974, Phys. Rev. B 10:2480.
Kahlert, H., and Bauer, G., 1971, Phys. Stat. Sol. (b) 46:535.
Kuper, C. G., and Whitfield, G. D., Eds., 1963, "Polarons and Excitons," Oliver and Boyd, London.
Lawaetz, P., 1969, Phys. Rev. 183:730.
Lawaetz, P., Sell, D. D., and Kane. E. O., 1974, in "Proc. Intern. Conf. Physics of Semiconductors, Stuttgart", Teubner, Stuttgart
Levinson, I. B., and Rashba, E. I., 1972, Sov. Phys.-JETP 35:788.
Lin-Chung, P. J., 1973, Phys. Rev. B, 8:4043.
Lin-Chung, P. J., and Ngai, K. L., 1972, Phys. Rev. Letters 29:1610.
Ngai, K. L., 1974, in "Proc. Intern. Conf. Physics of Semiconductors Stuttgart," Teubner, Stuttgart.
Ngai, K. L., and Johnson, E. J., 1972, Phys. Rev. Letters 29:1607.
Sher, A., and Thornber, K. K., 1967, Appl. Phys. Letters 11:3.
Strading, R. A., Eaves, L., Hoult, R. A., Mears, A. L., and Wood, R. A., 1970, in "Proc. Intern. Conf. Physics of Semiconductors, Boston," Nat. Bur. Stand., Washington.
Thomas, P., and Queisser, H. J., 1968, Phys. Rev. 175:983.
Vogl, P., 1976, Phys. Rev. B 13:694.
Weinstein, B. A., and Cardona, M., 1973, Phys. Rev. B 8:2795.

EXPERIMENTAL STUDIES OF NONLINEAR TRANSPORT IN SEMICONDUCTORS

G. Bauer

Institut für Physik
Montanuniversität Leoben
A-8700 Leoben, Austria

INTRODUCTION

This paper is a survey of experimental methods which have been used in hot electron physics. It is restricted to hot electron phenomena in bulk materials caused by high d.c. or a.c. electric fields. The methods for obtaining transport parameters, electron temperatures and hot electron distribution functions are presented.

The essential phenomenon is the following: charge carriers (electrons or holes) in a semiconductor gain energy and momentum from an applied electric field. Their mean energy $\langle\varepsilon\rangle$ will increase above the value in thermal equilibrium with the lattice ε_L, if the applied electric field is high enough. This increase is caused by a displacement of the distribution of electron energies. In a stationary state, the energy and momentum gain will be balanced by relaxation processes towards thermal equilibrium. These relaxation processes result from the interaction of the electrons with the lattice by various scattering processes.

The first experimental investigations of the conductivity σ of semiconductors in high electric fields were carried out by Shockley (1951), Ryder (1953), and Shockley and Ryder (1951). These authors showed, that the mobility of carriers in Ge and Si decreases with increasing electric field at both room temperature and liquid nitrogen temperature. Thus, for the first time, deviations from the Ohmic behaviour were found which had been anticipated earlier by Landau and coworkers in the 1930's.

Already in these early experiments, the main problems in applying high electric fields to semiconductor samples were fully considered, and excluded or restricted:

(i) variation of carrier density due to injection at the contacts or due to ionization (either of impurities or electron-hole pair generation across the gap)

(ii) changing of the lattice temperature due to Joule heating.

Since 1951, a variety of techniques has been developed to meet these requirements. Extensive literature exists both on experimental and theoretical studies and several reviews have been published (Gunn, 1957, Paige, 1964, Schmidt-Tiedemann, 1962, Conwell, 1967, Nag, 1972, Asche and Sarbej, 1969, Dienys, 1971, Fawcett, 1973, Price, 1978, Bauer, 1974, Hilsum, 1978, and Seeger, 1973).

This paper is organized as follows: there are three main sections. In the first one, the determination of transport parameters in the warm ($\langle\varepsilon\rangle \gtrsim \varepsilon_L$) and in the hot ($\langle\varepsilon\rangle \gg \varepsilon_L$) electron regime will be discussed. Both d.c. and a.c. (microwave) techniques will be considered. Bridge methods adapted for pulsed electric fields, field modulation techniques, and microwave harmonic mixing methods for obtaining the warm electron coefficient $\beta(\sigma = \sigma_{ohmic}\ (1+\beta E^2))$ are also presented. Short pulse techniques are used for the determination of the hot electron conductivity, the Hall coefficient, and the magnetoresistance. Time resolved conductivity and Hall measurements are discussed for an understanding of impurity breakdown and avalanche processes. Microwave techniques and ultrashort pulse techniques are discussed especially in connection with effects resulting from equivalent and non-equivalent intervalley scattering (Gunn effect).

Apart from the determination of transport parameters, methods are discussed in the following two sections for the determination of electron temperatures in high electric fields and experimental techniques for obtaining direct information on the most basic quantity: the non-equilibrium carrier distribution function.

TRANSPORT PARAMETERS

Warm Electron Effects

In the warm electron range, the transport parameters are represented by an expansion of the conductivity in powers of the field intensity and, since E is comparatively small, one keeps just the first non-trivial term. For the mobility, one thus obtains

$$\mu = \mu_o\ (1+\beta\ E^2), \tag{1}$$

where μ_o represents the Ohmic mobility

$$\mu_o = e\tau_{mo}/m, \tag{2}$$

where m is the effective mass of the carriers and τ_{mo} is the zero-field momentum relaxation time. The quantity β is called the warm electron coefficient. If we denote the momentum relaxation time (describing the momentum exchange between the carriers and the lattice) with τ_m, the momentum balance for stationary conditions is given by

$$e\,\vec{E} = m\,\vec{v}_d/\tau_m, \tag{3}$$

where $\vec{v}_d$ denotes the drift velocity and $\tau_m = \tau_m(<\varepsilon>, T_L)$.

The energy exchange of the carriers with the lattice is described by an energy relaxation time, and for the stationary case

$$e\,\vec{v}_d \cdot \vec{E} = \frac{<\varepsilon> - \varepsilon_L}{\tau_\varepsilon} \tag{4}$$

For isotropic and parabolic bands, $\mu(\vec{v}_d = \mu\,\vec{E})$ can be expanded according to (Schmidt-Tiedemann, 1962):

$$\mu = \mu_o + \left(\frac{e}{m}\right) \frac{d\,\tau_{mo}}{d\,<\varepsilon>}\,\delta(<\varepsilon>), \tag{5}$$

where the increase in mean electron energy $\delta(<\varepsilon>)$ is given by the energy balance equation:

$$\delta(<\varepsilon>) = e(\vec{v}_d \cdot \vec{E})\tau_\varepsilon = e\mu_o\,E^2\,\tau_\varepsilon; \tag{6}$$

and with (1), β is given by

$$\beta = \frac{e^2}{m} \cdot \tau_\varepsilon \cdot \frac{d\,\tau_{mo}}{d\,<\varepsilon>}\,. \tag{7}$$

Depending on the scattering mechanisms which determine τ_{mo} and its dependence on $<\varepsilon>$, β can be either positive or negative.

In nonspheroidal and many-valley cubic semiconductors, and in all non-cubic semiconductors, the mobility will not only depend on the lattice temperature T_L and on $\vec{E}$, but also on the direction of $\vec{E}$ with respect to the crystal axis. The resulting current $\vec{j}$ will in general not be parallel to $\vec{E}$ ($\overleftrightarrow{\mu}$ being a tensor). In this case the energy balance is given by

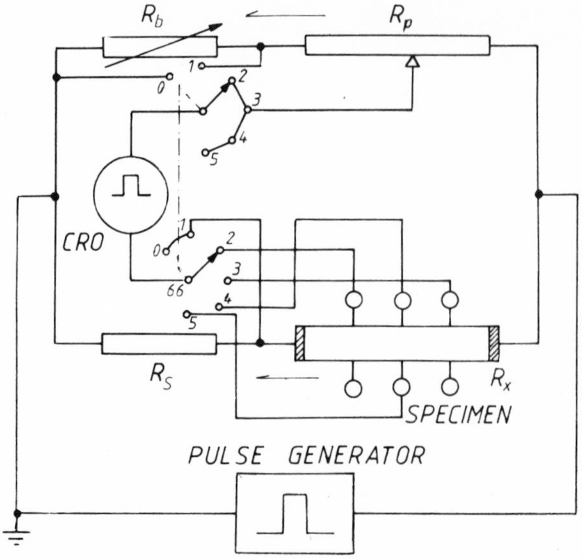

Fig. 1. Pulsed d.c. bridge for β- and $R_H(E)$ measurements (pulse-length: 1.5 . . . 3μs). The null indicator is a cathode ray oscilloscope; R_p is a potentiometer resistance with a dial scale R_b a precision variable resistance (after Miyazawa and Ikoma, 1967).

$$e \vec{E} \cdot \overleftrightarrow{\mu} \cdot \vec{E} = \frac{\langle \varepsilon \rangle - \varepsilon_L}{\tau_\varepsilon} \tag{8}$$

and $\overleftrightarrow{\sigma} = ne \vec{\mu}$ is given by (Schmidt-Tiedemann, 1963, Dienys et al., 1977)

$$\sigma_{ij}(\vec{E}) = \sigma_{ij}(0) \sigma_{ijkl} \vec{E}_k E_l. \tag{9}$$

The properties of the warm carriers are then determined by the fourt rank tensor σ_{ijkl}. The number of non-vanishing elements of σ_{ijkl} depends on the crystal symmetry.

D.C. methods. In practice β, values of the order of 10^{-4}. . . 10^{-8} $cm^2/V{\cdot}s$ have to be measured. To achieve this goal, a bridge circuitry is necessary. A possible experimental setup is shown in Fig. 1 due to Miyazawa and Ikoma (1967). To avoid Joule heating, voltage pulses of one to several microseconds time duration with a repetition rate of 10 - 50 Hz are used. As a null detector, an oscilloscope with a wide-band differential amplifier, of high common-mode rejection, is used. For shorter pulses, a coaxial version of this bridge is used. In Fig. 1, R_p represents a potentiometer resistance with a dial scale for a precise indication of the sliding contact.

Microwave methods. The deviations of the warm electron mobility μ from the Ohmic mobility μ_o can also be determined by using microwave techniques (Schmidt-Tiedemann, 1964). These techniques yield interesting and complementary information, i.e. on the relevant relaxation times if frequencies in the range 1 - 100 GHz are used (Morgan, 1959, Seeger, 1959, 1963a, Morgan and Kelly, 1965, Dienys and Pozhela, 1971). Usually, the samples are placed within a waveguide in a position where the electric field exhibits a maximum, and are subject to a superimposed d.c. field $\vec{E}_o$ oriented parallel to $\vec{E}_1$, so that

$$\vec{E} = \vec{E}_o + \vec{E}_1 \cos(\omega t). \tag{10}$$

In principle, two different techniques are then used:

(i) the microwave radiation heats the carriers, and the increase in the mean energy of the carriers is probed by measuring the change of the d.c. conductivity with and without $E_1 \cos \omega t$.

(ii) the d.c. field E_o is used to heat the carriers and the small signal microwaves are used to probe the conductivity. This is achieved by a measurement of the complex microwave conductivity determined from the microwave power absorbed and reflected by the semiconductor sample. From an amplitude and phase analysis, the microwave conductivity and microwave dielectric constant are obtained. This method was first used by Gibson, Granville and Paige (1961) (see also Schmidt-Tiedemann, 1964, 1963, Seeger, 1973, Morgan and Kelly, 1965).

Inserting E from (10) in the momentum and energy balance equations (6,7) Hess and Seeger (1969, 1970) have shown that the time average of the current density (averaged over a period of the microwave field) is given by

$$\langle j \rangle_{\|} = j_o + \frac{\sigma_o \, \beta \, E_o \, E_1^2}{2(1 + \omega^2 \tau_{mo}{}^2)} \left[1 + \frac{2 + \omega^2 \tau_{mo} \tau_\varepsilon}{1 + \omega^2 \tau_\varepsilon^2} \right], \tag{11}$$

where j_o is the current density for E_o. This is based on the assumption that in the warm electron range τ_ε is independent of energy, and thus τ_m is given by

$$\tau_m = \tau_{mo} [1 + \beta(\langle \varepsilon \rangle - \varepsilon_L)/e\mu_o \tau_\varepsilon]. \tag{12}$$

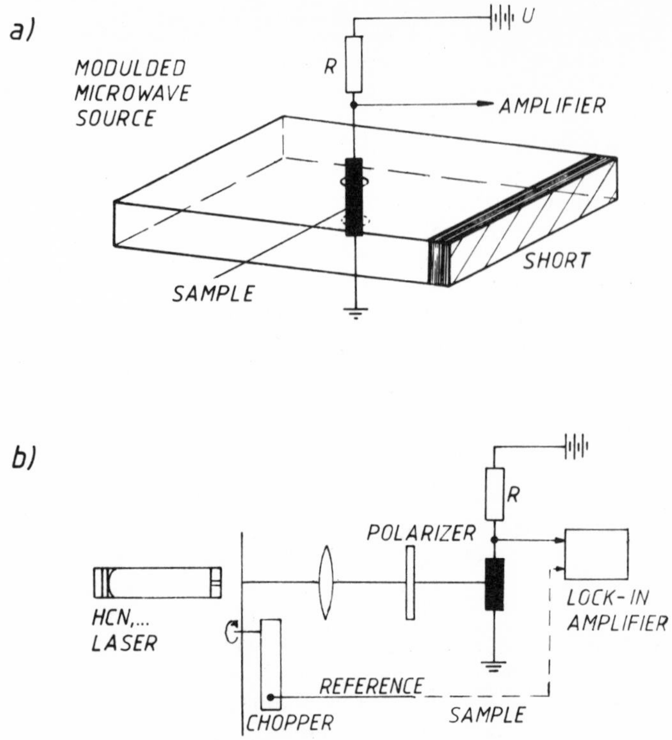

Fig. 2. a) Experimental set-up for determining β with a superposition of microwave and d.c. fields.
b) Arrangement for measuring warm carrier effects in the submillimeter wavelength range. The carriers are heated by radiation from an HCN-laser (890 GHz) and their response is measured by using a lock-in amplifier. (After Kuchar et al., 1972, Philipp et al., 1977).

It is important to note that the measured d.c. current $\langle j \rangle$ is determined by $\omega\tau_{mo}$, $\omega^2\tau_{mo}\tau_\varepsilon$ and $\omega\tau_\varepsilon$. If either $\omega\tau_\varepsilon$ or $\omega^2\tau_{mo}\tau_\varepsilon$ approach 1, remarkable influences on the measured $\langle j \rangle$ will occur, even if ω is still small compared to $1/\tau_m$.

For polar semiconductors, and temperatures where polar optical scattering is dominant, the relaxation times will be shorter than 10^{-12}s and therefore higher frequencies are needed to meet the criterion $\omega\tau \simeq 1$. Submillimeter sources, e.g. molecular gas lasers like HCN, DCN, H_2O lasers are suitable for this purpose and in Fig. 2 an experimental arrangement according to Kuchar et al. (1972) and Philipp et al. (1977) is shown (ν_{HCN} = 890 GHz). In this case measurements for the polarization $\vec{E}_1 \perp \vec{E}_o$ can also be made and the average current density is given by (Philipp et al., 1977)

$$\langle j \rangle_{\perp} = j_o + \frac{\sigma_o \beta E_o E_1^2}{2(1+\omega^2 \tau_{mo}{}^2)} \tag{13}$$

Thus, from a comparison of the d.c. conductivity change in parallel and perpendicular configurations, according to

$$\frac{\Delta\sigma_{||}}{\Delta\sigma_{\perp}} = 1 + \frac{2+\omega^2\tau_m\tau_\varepsilon}{1+\omega^2\tau_\varepsilon{}^2} , \tag{14}$$

rather precise information on τ_ε can be obtained.

The most accurate method for a determination of τ_ε using microwave techniques might be at present the harmonic mixing technique due to Schneider and Seeger (1966). For an applied field

$$E = E_1 \cos(\omega t + \rho) + E_2 \cos(2\omega t), \tag{15}$$

the current density has a d.c. component (Hess and Seeger, 1969, 1970)

$$\langle j \rangle = \frac{3}{4} \sigma_o \beta E_1^2 E_2 A \cos(2\rho + 2\psi) \tag{16}$$

where A and ψ are complicated functions of τ_m and τ_ε. For $\omega\tau_m \ll 1$, the phase shift ψ is given by

$$\tan 2\psi = 2\omega^3\tau_\varepsilon^3/(1+3\omega^2\tau_\varepsilon{}^2). \tag{17}$$

The experimental set-up used by Schneider and Seeger (1966) is shown in Fig. 3. Equations (16), (17) assume that E_1 and E_2 are parallel. This is accomplished by using a ridged waveguide, as shown in the insert of Fig. 3. In the experiments of this group, the fundamental (X-band) and second harmonic (K-band frequency) were fed into two arms of a single ridged-waveguide, double-Tee junction. In the other two arms, a short with the sample and a transition to X-band guide and a thermistor were placed.

For an increase in the sensitivity, especially for temperature dependent measurements, a reference sample of the same material at a constant temperature is used. This makes a direct determination of the change of the phase shift ψ with T possible.

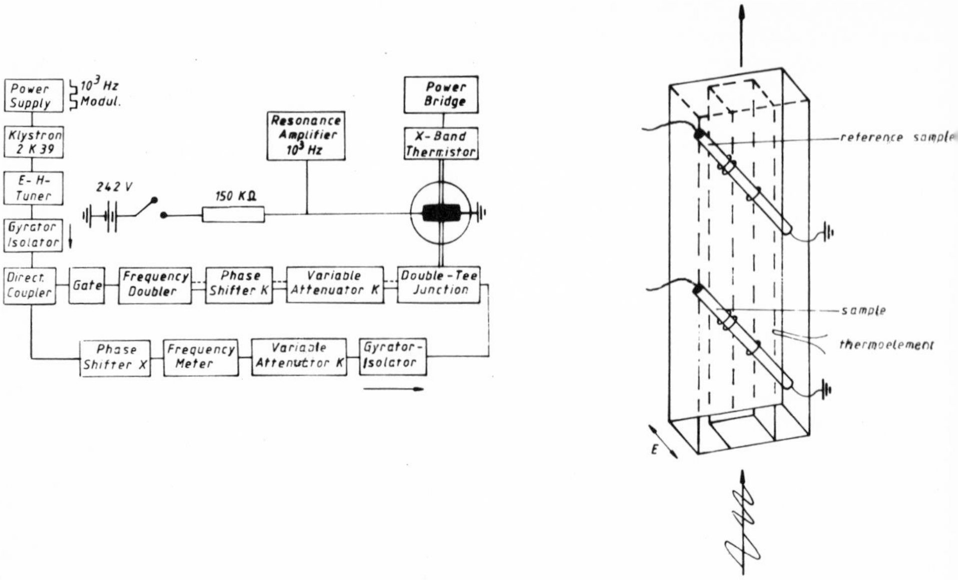

Fig. 3. Experimental arrangement for measuring warm carrier effects by the harmonic mixing method. Measurements are made in the X- and the doubled frequency (K-) band. The fields are super-imposed in parallel, using a ridged waveguide. The resulting d.c. current component from the nonlinearity of the sample is measured with a resonance amplifier. The set-up uses a reference sample for direct determination of the phaseshift, which yields τ_ε and its temperature dependence. (After Hess and Seeger, 1969, 1970, Schneider and Seeger, 1966).

As pointed out by Dienys and Kancleris (1975), the experiments with many-valley semiconductors have to be carried out in directions of E_1 and E_2 where carriers in all valleys are equivalently heated, in order to avoid problems arising from intervalley relaxation, characterized by a time τ_i (E_1, E_2 || <100> and <111> for Ge or Si, respectively). Using these methods, relaxation times τ_ε, as a function of temperature T between 77 and 200 K for n-Ge, n-Si, p-Ge, p-Si (Hess and Seeger, 1969, 1970) and n-GaAs (20-100 K) (Hess and Kahlert, 1970) were measured with τ_ε in the range of $1-6 \times 10^{-11}$s (GaAs: down to 2×10^{-11}s). Recently, the influence of electron-electron scattering on τ_ε in n- and p-Ge and n-Si has been established by this technique (Dienys and Kancleris, 1975).

Gibson et al. (1961) used the method of superimposing a small microwave field on a heating pulsed d.c. field to study warm carrier

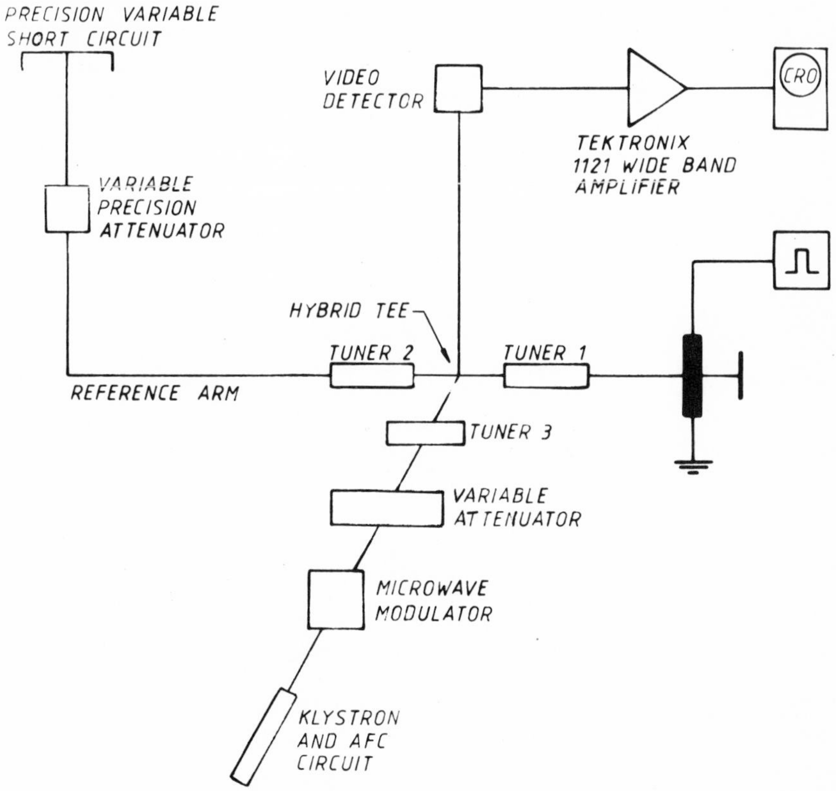

Fig. 4. Microwave bridge for measuring β by superimposing small microwave fields in parallel to a heating d.c. field. The microwave reflection coefficient and its change with $\vec{E}_o$ is determined. (After Holm-Kennedy and Champlin, 1972).

in Ge. The sample was situated in a slotted waveguide, which forms one arm of a microwave bridge (Fig. 4). From the microwave reflection coefficient (Holm-Kennedy and Champlin, 1972), the complex microwave conductivity and dielectric constant was determined by a numerical analysis. Holm-Kennedy and Champlin (1972) took into account the presence of odd-numbered modes in their analysis. Experiments performed with n-Si yielded information on the intervalley relaxation time (f-type relaxation) for $\vec{E}\,||\,\langle 100\rangle(\tau_i^{(100)} \simeq 1.2\times10^{-10}$s).

Warm electron effects in quantizing magnetic fields. High magnetic fields cause a quantization of the electron orbital motion in Landau levels. This produces a large density of states region near the bottom of each level and there is a high probability for scattering of carriers into these regions. In transport experiments, two different effects occur: the Shubnikov-de Haas (SdH) effect and the magnetophonon effect. Both are oscillatory magnetoresistance phenomena.

Oscillations in the SdH effect are connected with the passage of the bottom of Landau levels through the Fermi level. The magnetophonon resonances occur when the longitudinal optic phonon energy $\hbar\omega_{LO}$ corresponds to the separation between any two Landau levels of the carriers:

$$n \cdot \hbar\omega_c = \omega_{LO}, \; n = 1,2,3 \; . \; . \; . \; ,$$

where $\omega_c = eB/m$ (for parabolic bands).

Both the electric field dependence of the Shubnikov-de Haas effect (Bauer, 1974) and the magnetophonon effect (Harper et al., 1973, Hoult, 1975, Peterson, 1975, Parfenev et al., 1974) have been extensively studied in many semiconductors. These experimental investigations have proven to be a valuable tool for studying the effects of carrier heating in moderate electric fields.

One is faced with the problem of measuring the position of magnetoresistance extrema and of amplitudes under conditions where the oscillatory part ($\Delta\rho$) is smaller (sometimes by a factor of $10^{-4} - 10^{-5}$) than the nonoscillatory background magnetoresistance ρ_B. Several techniques have been developed for this purpose:

(i) time derivative techniques (Harper et al., 1973, Stradling and Wood, 1968),

(ii) third harmonic generation (Gulbrandsen et al., 1965, Shirakawa et al., 1973),

(iii) compensation techniques (Harper et al., 1973, Zawadzki et al., 1976, Racek and Bauer, 1976, Bauer et al., 1973),

(iv) magnetic field modulation technique (Kahlert and Seiler, 1977, Kasai et al., 1978).

Stradling and Wood (1968) have first used derivative techniques to display the second derivative of the magnetoresistance with respect to the magnetic field to accentuate the oscillatory part of the total signal. The differentiation was achieved by using two RC circuits to generate the second derivative with respect to time, while sweeping the magnetic field linearly in time.

This technique has been improved by Hoult (1975) and Bauer et al. (1973) who used active instead of passive differentiating networks in connection with active low-pass filters and additional

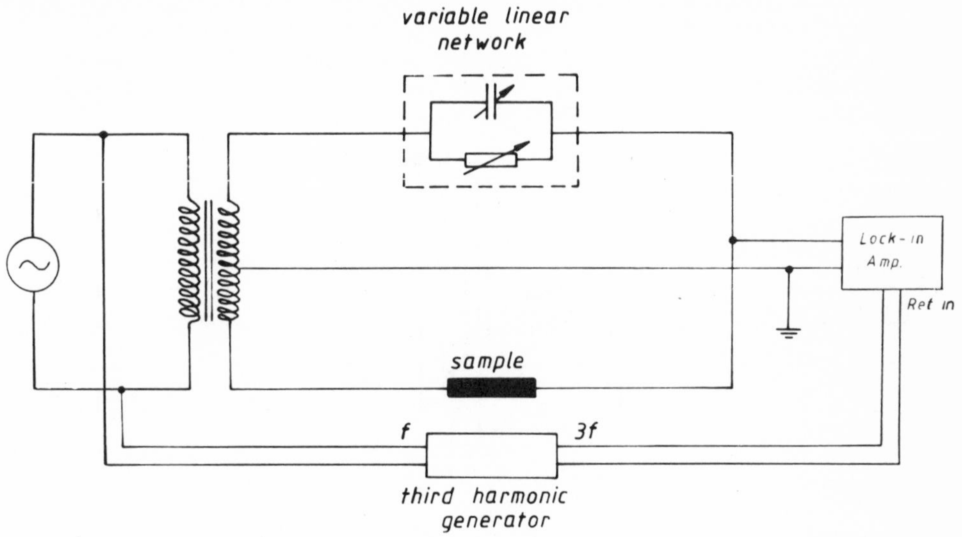

Fig. 5. Schematic diagram of a bridge circuit for measuring β with the aid of the third harmonic generated in the sample. The fundamental frequency is balanced out by the linear network. (After Shirakawa et al., 1973).

amplification of the signal in the desired frequency range. This method has also proved to be useful in connection with short pulsed electric fields (pulse duration: ≃ 100 ns) using sampling oscilloscopes. This technique has been improved by using a hyperbolic increase of magnetic field with time instead of a linear one and selection of the desired oscillations by a tuned amplification technique (Zawadzki et al., 1976, Racek and Bauer, 1976, Bauer et al., 1973).

The third harmonic generation technique is based on the fact that an applied field $E = E_1 \cos \omega t$ will result in a current which has components $\cos(3\omega t)$ if $\sigma = \sigma_0(1 - \beta E^2)$ (Gulbrandsen et al., 1965, Shirakawa et al., 1973).

$$j \simeq \sigma_0 E_1 \cos \omega t - \frac{1}{4} \sigma_0 E_1^3 \cos 3\omega t \equiv \sigma_0 (E_1 \cos \omega t - E_3 \cos 3\omega t), \quad (18)$$

$$\beta = \frac{4 E_3}{E_1^3} \quad (19)$$

A schematic diagram is shown in Fig. 5. The driving field oscillates at 2 kHz and is balanced out in a variable, linear network of the

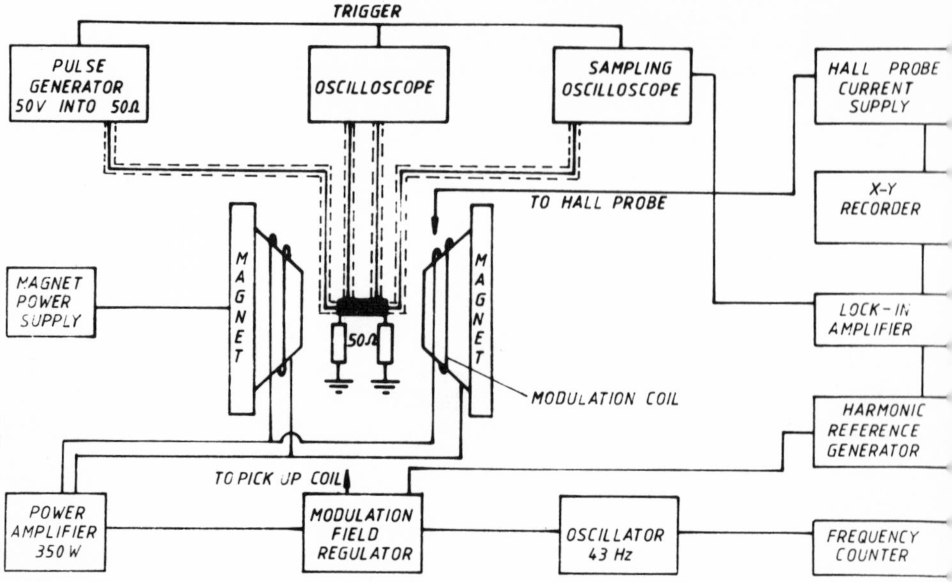

Fig. 6. Magnetic field modulation technique for pulsed measurements of SdH or magnetophononoscillations (pulse length: 100 ns) (After Kahlert and Seiler, 1977).

bridge, and the generated third harmonic (due to the nonlinear response of the sample) is detected by a lock-in amplifier. For low impedance materials a sensitivity of this bridge of the order of $\Delta\mu/\mu_o \simeq 10^{-7}$ could be achieved.

The magnetic field modulation technique, a well known tool in Ohmic SdH measurements, can also be used in pulsed measurements where the duration of the electric field pulse is in the submicrosecond region as has been demonstrated by Kahlert and Seiler (1977). Their experimental set-up is shown in Fig. 6. The sample is situated within a magnet, producing a d.c. magnetic field, and additional Helmholtz coils produce an a.c. component ($\omega \simeq 50$ Hz). The modulation is of the order of 500 G for $B_o = 20$ kG. Under constant voltage conditions, the pulsed current through the sample (100 ns duration, repetition rate $\simeq 500$ Hz) is measured as a voltage drop in a 50Ω resistor. This voltage is fed into a sampling oscilloscope. The output of the sampling oscilloscope carries an a.c. modulation of the low frequency modulation of B (ω). With a lock-in amplifier, this a.c. component can be detected. For a linear relationship between the sample-resistance and B, the lock-in detec-

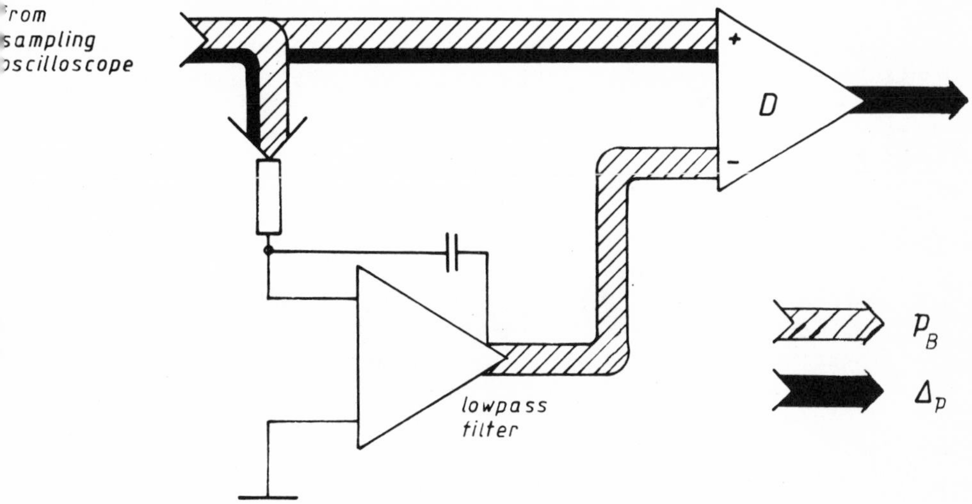

Fig. 7. Subtraction technique for a measurement of the oscillatory part $\Delta\rho$ of the magnetoresistance. (After Racek and Bauer, 1976).

tor signal will be constant, any nonlinearity changes the amplitude of the a.c. component of the sampling oscilloscope output and the first derivative dR/dB of the magnetoresistance is obtained. By setting the lock-in amplifier to the second harmonic frequency of the a.c. magnetic field modulation ($2\omega \simeq 100$ Hz), the second derivative is obtained. This method improves the signal/noise ratio obtainable with pulsed magnetophonon experiments. It is limited to not too high electric fields since the electric field pulse repetition frequency has to be of the order of 5 to 10 times the frequency of the a.c. magnetic field.

In the compensation technique, the oscillatory component $\Delta\rho$ is obtained by a subtraction of the non-oscillatory background (Stradling and Wood, 1968). The principle is shown in Fig. 7, due to Bauer (1974). The total magnetoresistance is monitored as a voltage drop along the sample. These voltage pulses are fed into a sampling oscilloscope and the output voltage is divided into two parts and fed equally into the non-inverting input of a differential amplifier and into the summing junction of an active low-pass filter. Here, it is cleared of the undesired oscillatory content and the non-oscillatory remainder is fed into the inverting input of the subtraction unit D, the output of which is the desired oscillatory signal $\Delta\rho$.

Hot Electron Effects

D.C. methods. To a larger extent, the precautions already mentioned in the preceding section on warm-carrier experiments have to be considered with higher electric fields. From the measured current density, the drift velocity $\vec{v}_d$ can only be determined if n remains constant, independent of $\vec{E}$. The carrier concentration can be affected by Joule heating, carrier injection from the contacts, or impact ionization.

Joule heating is minimized by applying short pulses (in the microsecond or nanosecond time duration range, depending on the total E and on σ). For an estimation of the heating effects, adiabatic heating of the samples can be assumed. Particularly at helium temperatures, the total energy input per pulse has to be restricted severely due to the relatively small specific heat of the semiconductors. In order to avoid injection effects, particular geometries have been used with large area contacts as shown e.g. by Ryder (1953) and Schmidt-Tiedemann (1962) to minimize any electric field gradients in the contact region, thus avoiding carrier injection.

The principal arrangement for pulsed d.c. measurements (Larrabee, 1959, Gunn, 1957b, Prior, 1959) is shown in Fig. 8. A d.c. power supply is connected to a pulse forming network, which consists of a coaxial cable, a stripline, or a L-C circuit. It is discharged with the aid of a switch which is either a Hg-wetted rela (for voltages up to 3000V) or a thyratron (for higher voltages). For measuring the voltage drop and the current through the sample, a voltage probe and a series resistance are used. Using strip line techniques, Gunn (1957b) performed experiments with n-Ge in fields up to 7×10^4 V/cm and pulse lengths of 2 ns. Especially for high conductivity samples, low impedance strip-line pulse generators ($Z_0 = 1.7\Omega$) with rise times below 1 ns have been developed by Bauer and Heinrich (1968) and by Heinrich and Jantsch (1969).

The most commonly used pulse generator with a rise time in the sub-nanosecond range now consists of a 50Ω coaxial cable, or a 50Ω air-line, which is connected with a carefully designed and mounted Hg-wetted relay (Elliott, 1968) and discharged into a 50Ω air-line which contains the sample and a current monitoring resistor in series. Overshoot and ringing on the top of the pulse can be diminished by carefully designed pulse-shaping filters (van Welzenis and Sens, 1971, van Welzenis, 1972) at the expense of rise-time degradation.

Jantsch and Heinrich (1970) and Fukui (1966) described a method which makes use of reflected pulses. According to the former authors, the sample is mounted at the end of a 50Ω transmission line

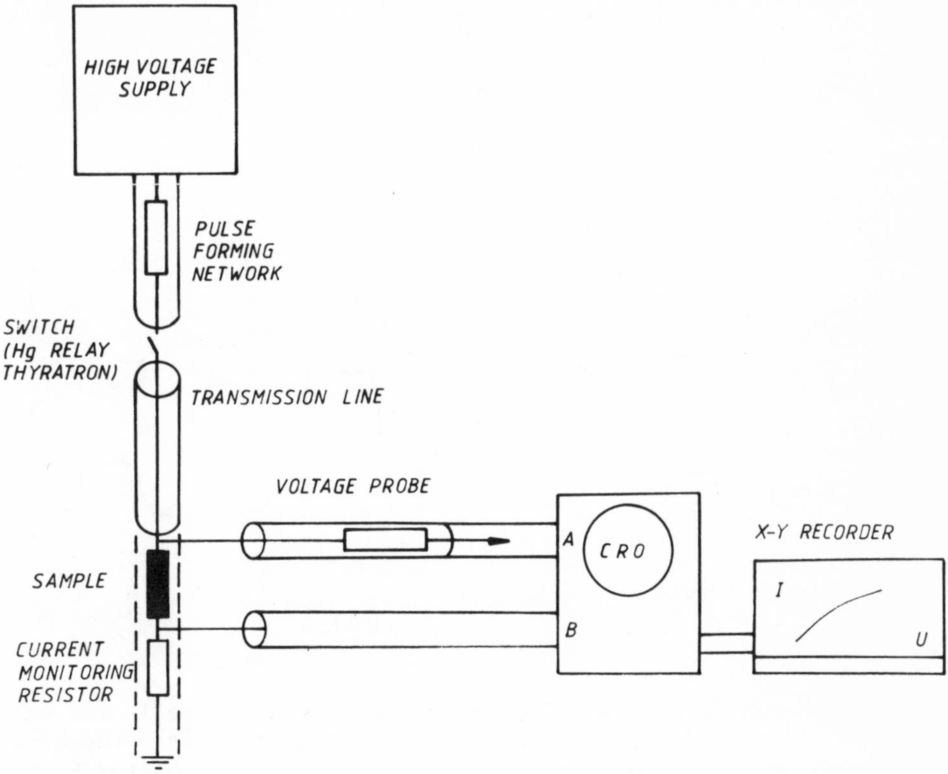

Fig. 8. Arrangement for pulsed d.c. measurements with nanosecond pulse risetime.

(see Fig. 9). The incident and the reflected pulses are observed with the aid of a sampling probe, which is further from the sample than the length of the charging line. This separates the incoming from the reflected pulse. The sample resistance is then determined by a comparison of the incoming voltage V_i and the reflected voltage, as V_r

$$R = Z\left(\frac{V_i + V_r}{V_i - V_r}\right),$$

where Z is the impedance of the transmission line. The sensitivity of this technique is better than about 6% when $0.1 < R/Z < 10$. Rise times of the order of 200 ps were achieved. Since the current monitoring resistor is abandoned, a possible temperature dependence and

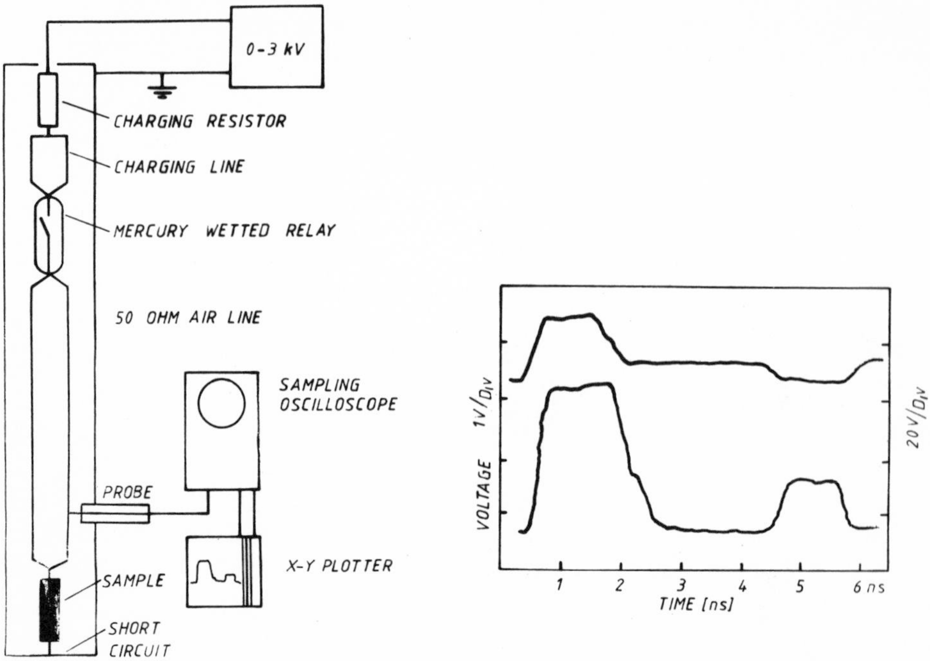

Fig. 9. Method for pulsed d.c. measurements using a reflected pulse technique. Left part: apparatus. Right part: incoming and reflected signal for R < Z (upper trace) and for R > Z (lower trace). (After Jantsch and Heinrich, 1970).

nonlinearity at high currents of this time, which occur with usual resistors, do not obscure the measurements.

Hot electron Hall effect, magnetoresistance, and the conductivity anisotropy in high fields. An experimental arrangement for pulsed Hall effect measurements is shown in Fig. 10. If a sampling oscilloscope with an input impedance of 50Ω is used, then series resistors in the kΩ range have to be used on the Hall probes (depending on sample resistance). This is not advantageous since the sensitivity is reduced. If the pulse length is longer than about 50 ns, commercially available differential amplifiers with input impedance of the order of 1MΩ can be used. A common-mode rejection of the order of $5 \times 10^{+4}$ is possible. Therefore, the difference of the voltage drops between the side arms can be measured directly.

Another possible experimental arrangement was used in measurements of the conductivity anisotropy in many-valley cubic

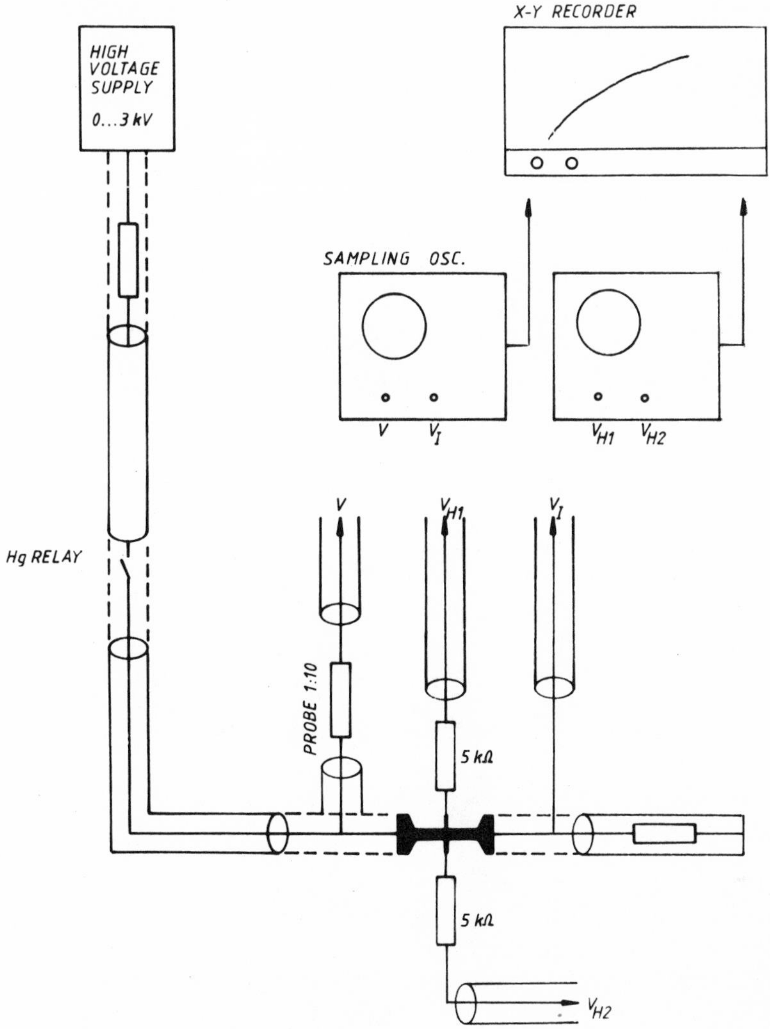

Fig. 10. Experimental arrangement for pulsed Hall effect measurements using series resistors with the Hall probes.

semiconductors. For arbitrary directions of the electric field with respect to the crystal axis, the current need not be parallel to the electric field. This leads to a build up of a transverse voltage, perpendicular to the direction of current flow (Sasaki and Shibuya, 1956, Sasaki et al., 1958, 1959). As with the Hall experiment, the electrodes are placed on opposite sides of the specimen. To eliminate the common voltage drop with respect to ground, the transverse

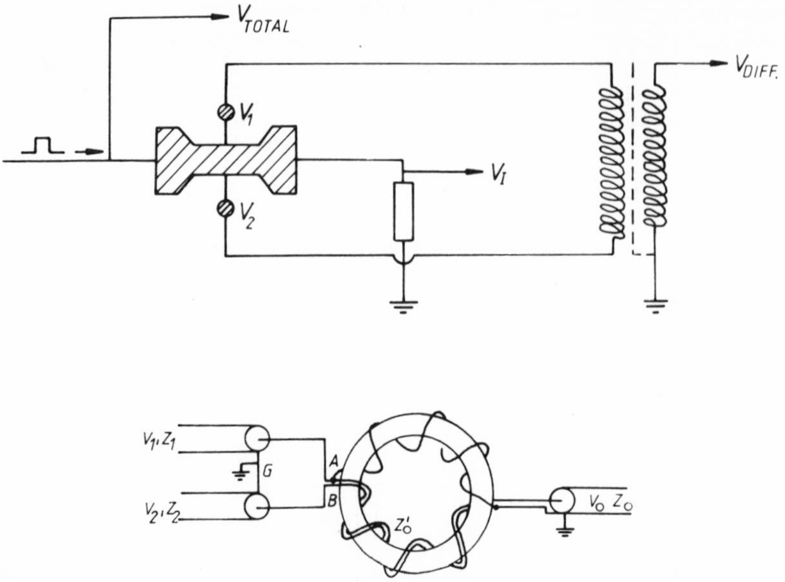

Fig. 11. Experimental arrangement for Hall effect or Sasaki effect studies using a pulse transformer. Lower part: pulse transformer for fast risetime measurements. (After Alberga, 1978).

voltage is fed into a transformer, which is connected to the oscilloscope (Fig. 11) (Sasaki and Shibuya, 1956). The geometry of the transverse electrodes has to be chosen carefully in order to reduce the distortion of the electric field lines in the neighborhood of these electrodes to a minimum.

Impurity and avalanche breakdown effects. Conductivity changes in high electric fields are caused not only by mobility variations. Especially at low temperatures, when there are neutral donors or acceptors, impact ionization due to the hot carriers may occur. These effects have been observed in a large number of semiconductors e.g. Ge, Si, GaAs, InP, InSb (Sclar and Burstein, 1957).

In small gap semiconductors like InSb, InAs, PbTe, the energy of the hot carriers may be sufficient to cause impact ionization of the electrons from the valance band. Thus, electron-hole pairs are created and an avalanche breakdown process, a dramatic increase of the current density, may occur beyond a certain breakdown field

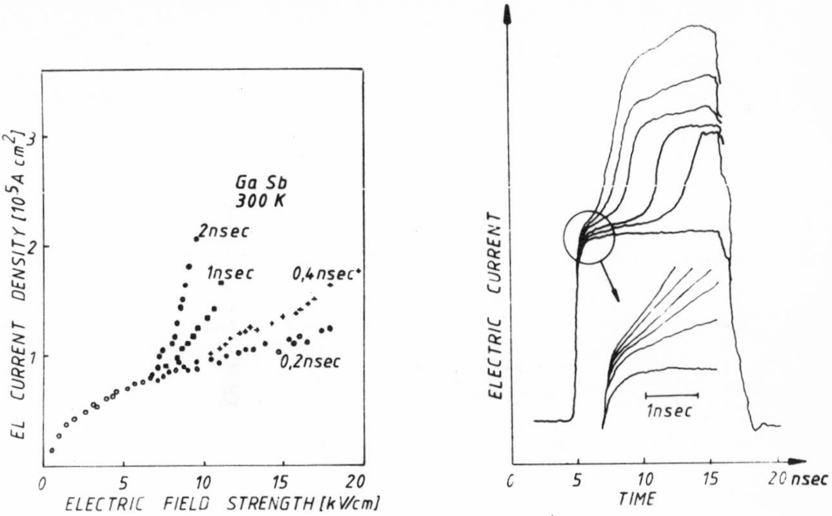

Fig. 12. Left part: time dependent current voltage characteristic of n-GaSb for various times after the rise of the voltage pulse. Right part: time dependent increase of the current under constant voltage conditions in the avalanche region. (After Jantsch and Heinrich, 1971).

(Glicksman and Steele, 1958, Glicksman, 1972, Ancker-Johnson, 1967). From the observation of the time-dependent increase of the current density under constant-voltage conditions, the carrier generation rate (in case of impurity breakdown) or the electron-hole pair generation rate, can be determined as shown in Fig. 12. The time-dependent j(E) characteristics of GaSb are shown, as well as the time-dependent increase of the electric current at various field strengths above the avalanche threshold electric field (Jantsch and Heinrich, 1971).

To get a full understanding on the impurity breakdown process and also of the avalanche breakdown across the fundamental gap, time-dependent Hall effect measurements are very helpful in addition to conductivity measurements. The most complete study has recently been made by Alberga (1978) on Hall effect measurements under avalanche conditions on n-type InSb at 77 K. He used a differential pulse transformer (see insert Fig. 11), the windings of which are made of bifilar wire (Ruthroff, 1959). The output of this transformer had a rise time in the subnanosecond range. This device is very useful in differentiating between mobility and carrier concentration effects under avalanche conditions.

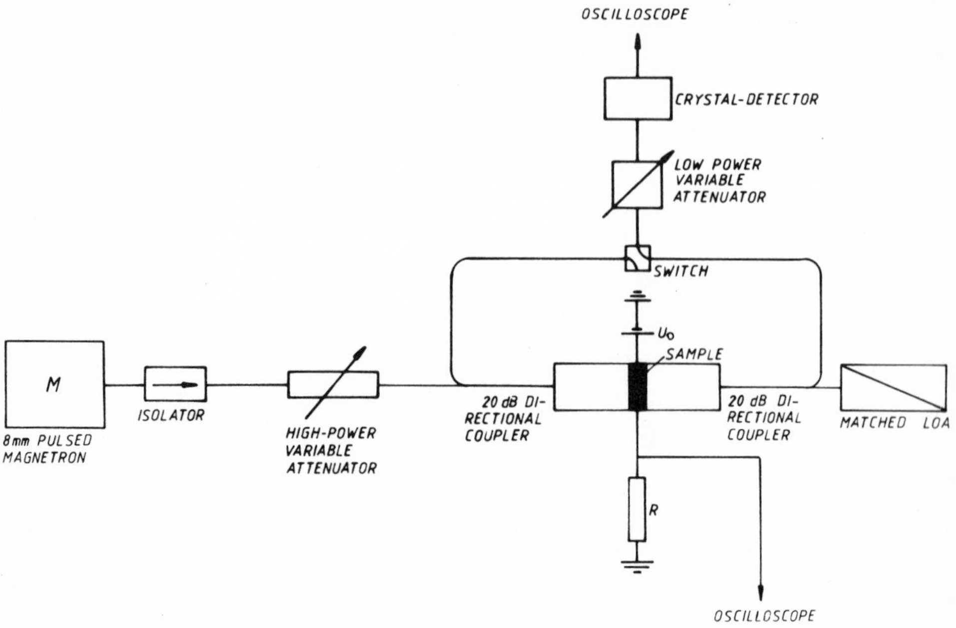

Fig. 13. Microwave circuit for measuring hot carrier conductivity by superimposing a small d.c. field on a high microwave heating field. Measurement of absorbed microwave power is also possible (transmitted and reflected power are measured). (After Acket and de Groot, 1967).

Microwave methods. In the investigation of hot electrons with microwave techniques, two methods are commonly used (similar to the warm electron case):

(i) measurement of the small signal d.c. conductivity as a function of the amplitude of the large microwave field.

(ii) the carriers are heated with a d.c. field and the small signal microwave conductivity is measured.

The first method was used by Morgan (1959), Seeger (1959), and Zucker et al. (1961) for an investigation of Ge and Si samples. Magnetrons were used as a source for large amplitude microwaves, usually pulsed with pulse lengths of the order of 200 ns and a repetition rate of 50 Hz. The experimental set-up is similar to the one described in the section on warm electrons. A typical set-up is shown in Fig. 13 (Acket and De Groot, 1967). If it is assumed that the current follows the field (which is true for not too high microwave frequencies), then the average current is give by

$$<j> = \frac{1}{2\pi} \int_{\omega t=o}^{2\pi} j(E_o + E_1 \cos\omega t) \, d\,(\omega t)$$

$$\simeq \frac{E_o}{2\pi} \int_o^{2\pi} \left(\frac{dj}{dE}\right)_{E_1 \cos\omega t} d(\omega t), \tag{21}$$

where j(E) describes the characteristic. The unknown function $(\frac{dj}{dE})$ is obtained from the experimental data $<j>/j_o$ ($j_o = \sigma_o E_o$) by solving the integral equation (21). Zucker et al. (1961) found that for Ge and Si, an approximate solution can be found if the experimental data can be fitted by the empirical relation

$$\frac{<j>}{j_o} = (1 - \frac{\sigma_\infty}{\sigma_o}) \frac{1}{\sqrt{1 + \frac{E_1{}^2}{E_o{}^2}}} \quad \frac{\sigma_\infty}{\sigma_o}, \tag{22}$$

where σ_∞ and E_c are chosen appropriately. Then it is found that

$$j(E) = (\sigma_o - \sigma_\infty) E_c \arctan (\frac{E}{E_c}) + \sigma_\infty \cdot E. \tag{23}$$

Acket and de Groot (1967) used a more general treatment, expanding j(E) in a polynomial series

$$j(E) = \sigma_o E (1 + \beta E^2 + \gamma E^4 + \delta E^6 + \ldots), \tag{24}$$

and

$$\frac{<j>}{j_o} = 1 + \frac{3}{2} \beta E_1^2 + \frac{15}{8} \gamma E_1^4 + \frac{35}{16} \delta E_1^6 + \ldots . \tag{25}$$

The coefficients β, γ, δ. . . are found by comparison from the experimental $<j>/j_o$ data as a function of E_1. In the experiments, however, one measures not E_1 but the absorbed microwave power of a sample with a resistance R_o:

$$P_{abs} = \frac{V}{2\pi} \int_o^{2\pi} j(E_1 \cos \omega t) \, E_1 \cos \omega t \, d\,(\omega t). \tag{26}$$

For j(E) given by

$$\frac{2\,R_o\,P_{abs}}{b^2} = E_1^2\,(1 + \frac{3}{4}\,\beta E_1^2 + \frac{5}{8}\,\gamma E_1^4 + \frac{35}{16}\,\delta E_1^6 + \ldots), \qquad (27)$$

where b is the wave-guide height and the coupled equations (27) and (21) have to be solved for j(E) and $E_1(P_{abs})$. Equation (27) shows that the measurement of P_{abs} as a function of E_1 is sufficient for determining j(E).

Information on the characteristic $j(E_1)$ can also be obtained from the variation of the microwave attentuation as a function of microwave amplitude (Acket and de Groot, 1967). If we define A as given by:

$$A = \frac{P_t}{P_o(1 - r_p)}\,, \qquad (28)$$

where P_t is the transmitted power, P_o is the incident power and r_p is the fraction of power reflected from the sample, then $\sigma(E_1)$ can be determined from A. This can be done by calibrating the relation between A and $\sigma(E_1)$ by measurements in the same geometry at low fields. This method is also applicable for samples with relatively high conductivity by making additional measurements of the absorbed microwave power $P_{abs} = \frac{1}{2}\sigma(E_1)\,E_1^2\cdot V$ (V being the volume of the semi-conductor sample).

At microwave power levels of several kW, microwave frequency multiplication occurs, as was first observed by Seeger (1963b). For a microwave field $E(t) = E_1 \cos\omega t$, j(t) is given by

$$j = j_1(E_1)\cos\omega t + j_3(E_1)\cos 3\omega t + j_5(E_1)\cos 5\omega t + \ldots \qquad (29)$$

Some information on the variation of j_3 or j_5 as a function of E_1 can be obtained from measuring the power radiated at the 3rd and 5th harmonic and j_3 and j_5 are proportional to $\sqrt{P_3}$ and $\sqrt{P_5}$, respectively.

In the method described in (ii), essentially a differential conductivity or mobility is determined (Gibson et al., 1961). There are different methods for this. Zucker et al. (1961) used a semi-conductor post in a RG-98/U waveguide. From a determination of the complex reflection coefficient, the complex high-frequency dielectric function is

$$\kappa(\omega) = \kappa_{lat.} + \frac{i\,\sigma\,(\omega)}{\kappa_o\,\omega}\,, \qquad (30)$$

where $\kappa_{lat.}$ is the lattice contribution to κ(without carriers) and $\sigma(\omega)$ is the complex conductivity, which is given for low frequencies, low fields, and simple bands by (Conwell, 1967)

$$\sigma(\omega) = \frac{ne^2}{m} \frac{\langle \varepsilon\tau_m/(1+\omega^2\tau_m)^2\rangle}{\langle\varepsilon\rangle} - i\omega \frac{\langle\varepsilon\tau_m/(1+\omega^2\tau_m)\rangle}{\langle\varepsilon\rangle} . \qquad (31)$$

(The brackets denote averages over the carrier distribution function). For a more realistic condition, the description of the complex $\sigma(\omega)$ is very complicated and details can be found in Nay (1972), Gibson et al. (1961), Hamaguchi and Inuishi (1966), Holm-Kennedy and Champlin (1972), Löschner (1972a), Bonek (1972), Bonek et al. (1970).

Seifert (1965) has suggested a method using a microwave cavity with the sample partially replacing the cavity walls, which has been used in the frequency range 10-85 GHz (Bonek, 1972) for studies on various III-V compounds (GaAs, GaSb, InAs, InSb). It uses a TE_{102} rectangular cavity and along a circular line of wall-current a hole is cut which is covered with the sample. Thus an arbitrary angle between the high d.c. field and the preferential direction of wall current can be chosen. The change of electron distribution due to the d.c. heating causes changes in the Q-factor and the resonant frequency of the cavity, and with perturbation theory the semiconductor impedance is obtained from the changes in the cavity parameters. By varying the measuring frequency, accurate information on the relaxation times τ_m and τ_ε and their dependence on the electric field can be obtained, in the case of GaSb, also for the intervalley relaxation $\Gamma \to L$. This method can also be adapted for measuring the warm-electron coefficient β as low as $10^{-8}cm^2/V^2$.

Energy relaxation times of GaAs and InP using a more conventional method of differential microwave conductivity measurements were determined by Glover (1973). The complex microwave conductivity was measured at 34.4 GHz as a function of d.c. fields up to 3kV/cm, using a hybrid-tee bridge with the sample mounted as an inductive post in the center of a tapered wave guide.

A new method for measuring electron drift velocities in the presence of impact ionization and even negative differential mobility using measurements of the average microwave conductivity has been suggested by Dargys et al. (1978). In addition, the non-equilibrium carrier concentration could be determined. The method is based on the detection of the change of the voltage drop across a sample with a sampling oscilloscope, caused by the application of a double microwave pulse, where the separation Δt of the two parts of the microwave double pulse can be changed. It is thereby possible to find out the lifetime of excess carriers and the time necessary for the non-equilibrium steady state concentration to be established.

Interesting applications of carrier heating effects are the use of nonlinear elements in heterodyne experiments at submillimeter frequencies as described by Whalen and Westgate (1970) for InSb mixers. These techniques have recently been applied to the detection of weak submillimeter radiation.

At high microwave frequencies, in the millimeter or even in the submillimeter wavelength range, the carrier distribution will no longer be able to follow the a.c. fields. From the relaxation phenomena which then occur, interesting information on the dominant scattering mechanisms of the carriers is obtained. The experimental techniques for $\nu > 100$ GHz are in principle similar to the ones used at lower frequencies. However, at very high ν, normal optical methods rather than wave-guides are used. Experiments at 136 GHz were described by Löschner (1972b) and at submillimeter waves by Löschner et al. (1972). These experiments allow a quantitative comparison with numerical solutions of the Boltzmann-transport equation. Anisotropic absorption of the electromagnetic radiation by hot carriers was well established by these authors and also was observed by Gershenzon (1974).

Negative differential mobility. The occurrence of negative differential mobility (ndm) due to transfer of hot carriers between non-equivalent valleys was first suggested by Ridley and Watkins (1961) and by Hilsum (1962). Gunn (1963, 1964) observed the generation of current oscillations in n-GaAs and n-InP for fields higher than a threshold field for ndm to occur.

In the region where $\mu' = dv/dE$ is negative, any infinitesimal perturbation of the electric field grows, and under certain conditions, a steadily travelling domain of high electric field will be formed. It originates near the cathode and travels towards the anode, where it disappears. The generation-repetition rate is given by v_d/L, where L is the sample length and v_d the drift velocity. For a given carrier concentration, there is a minimum sample length L_{crit} for domain formation, depending on μ', the drift velocity, and the dielectric constant. For n-GaAs this condition is

$$n_o \cdot L_{crit} = 3.10^{11} \text{ cm}^{-2}.$$

To probe the potential distribution along the sample, Gunn (1963, 1964) suggested the use of a capacitive probe (conductive probes are not possible with GaAs or InP since the contact resistance is too high). This is shown in Fig. 14, together with data on n-GaAs. The probe is moved a distance of approximately 5 μm (Ohtomo, 1968) parallel to the sample surface between cathode and anode. Its width is $x = 15$ μm, and it is about 300 μm long in the perpendicular direction. Depending on the mode of operation, it is

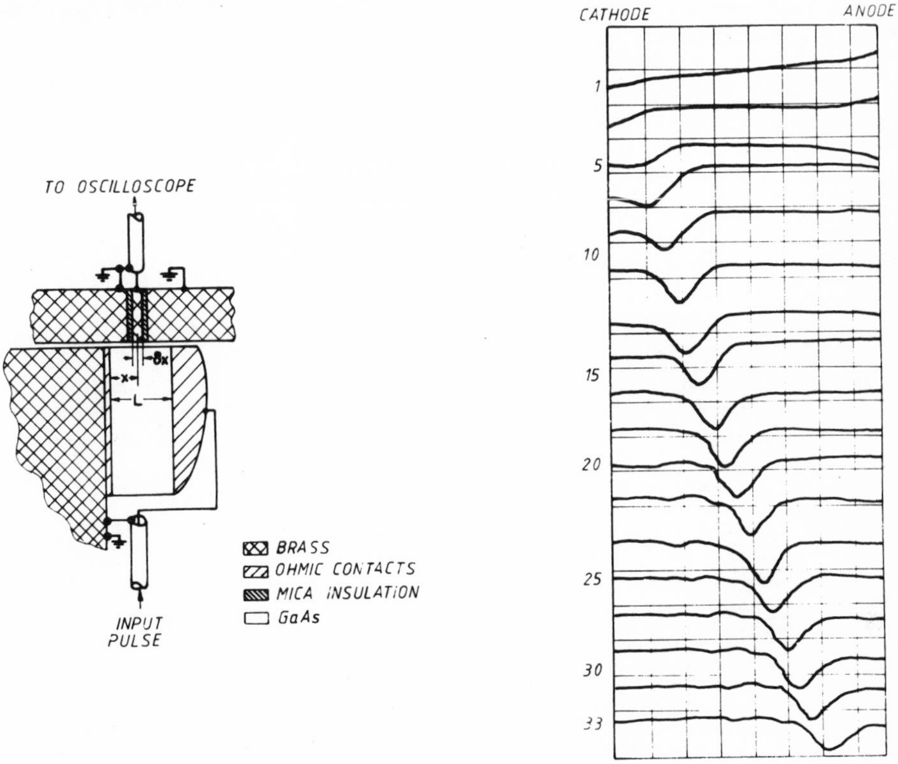

Fig. 14. Capacitive probe after Gunn to measure travelling voltage domains in ndm region of j(E) characteristics. Right part: experimental data on dV/dx in n-GaAs, 1 division corresponds to 26.5 μm. Top trace: measurement at a fixed time near the beginning of the applied pulse. Subsequent traces separated by intervals of 2 x 6.6 x 10^{-11}s. (After Gunn, 1964).

possible to display on the sampling oscilloscope $\frac{dV(x,y,t)}{dt}$ as a function of x. A resolution of the order of 20 μm is achieved. Similar techniques were employed to study the relatively weak ndm in n-Ge (McGroddy, 1970, McGroddy and Nathan, 1967) caused by the transfer of carriers from the L to the X valleys. Materials with a weak ndm may exhibit a region of saturated current with electric field, if the frequency of current oscillations is higher than the maximum frequency of the apparatus (McGroddy et al., 1969). It has been shown that for

$$n_o \cdot L > \kappa\kappa_o E_{thresh.}/e,$$

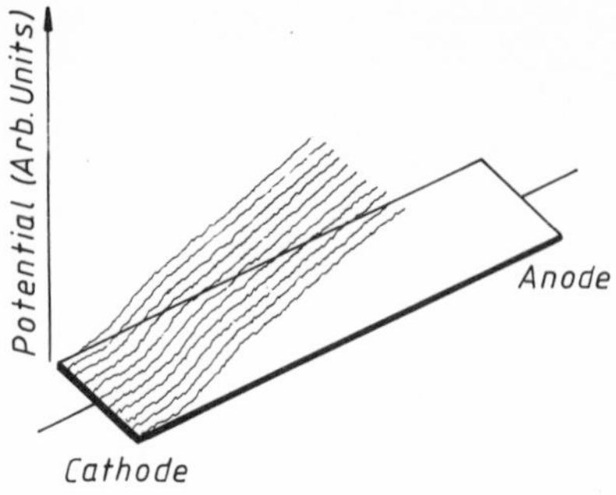

Fig. 15. Potential distribution of n-InSb between cathode and anode determined with a conductive probe. Time response of the potential probe: 1ns. (After Heinrich and Keeler, 1972).

the current density more or less saturates once $j > n_o \cdot e \cdot \max(v_d(E))$. This explains also the older experiments on n-Si for $\vec{E} || [100]$. In Si, there is a ndm due to repopulation effects between cold and hot <100> valleys for $\vec{E} || [100]$ at low lattice temperatures (Asche and Sarbej, 1971, Gram, 1972, Jorgensen et al., 1972, Jacoboni et al., 1977).

For materials like InSb, with a low contact resistance, it is possible to monitor the potential distribution along and across the sample with a conductive probe. Heinrich and Keeler (1972) developed this technique, using a 500Ω probe with a tip of 12 μm radius, connected to an x-y positioner. A resulting potential distribution exhibiting a domain is shown in Fig. 15.

With these conductivity probe techniques and the measurement of the amplitude and the frequency of current oscillations, it is possible to get directly the $v_d(E)$ curve, but the following information can be obtained (Foyt and McWorther, 1966, Butcher, 1967, Butcher et al., 1966): (i) the threshold field E_{thresh}, (ii) the electron drift velocity v_d for $E<E_{thresh}$, (iii) the domain velocity, (iv) the electric field inside and outside of the domain and (v) the peak-to-valley ratio (v_{max}/v_{min} in the $v_d(E)$ curve).

However, in order to obtain reliable $v_d(E)$ curves in ndm materials, the formation of high-field domains must be inhibited. This can be done if electric field pulses with a duration shorter than the negative dielectric relaxation time τ_d are used. This time determines the time interval which is necessary for the formation of field inhomogeneities if the applied field exceeds the threshold field. This time is given by (Kroemer, 1966)

$$|\tau_d^-| = \frac{\kappa_o \cdot \kappa}{e \cdot n_o |\mu_d^-|} \tag{32}$$

However, the time for which $E > E_{thresh}$ has to be large compared to τ_ε and to the intervalley relaxation τ_i. The first experiments with the pulse duration shorter than $|\tau_d^-|$ were performed by Gunn and Elliott (1966). The method was improved later by Elliott (1968). Since for n-GaAs, which was used, the dielectric relaxation time was of the order of 100 ps, the experimental set-up was rather complicated. The sample was mounted on a jig consisting of two crossed 50Ω micro-striplines separated by a common ground plane with a very low shunt capacitance (5.10^{-3}pF) and low lead inductance (≤ 0,1 nH). The pulse was generated by a charged 90Ω coaxial line through a mercury wetted relay. A superconducting delay line was used to pass the pulse to the sample on the jig. There is a problem with this method, which arises from the fact that the total current is the sum of

$$j = e\, nv(E) + \kappa\kappa_o \frac{dE}{dt},$$

and the displacement current vanishes only for $\frac{dE}{dt}$ (and thus $\frac{dV}{dt}$) = 0. This displacement current is a major source of experimental error and it turned out that the determined v(E) characteristics for n-GaAs exhibited too small a negative differential mobility.

Many determinations of the v(E) characteristic of samples with ndm were performed by using the microwave method of Acket and de Groot (1967). Also with this method, field inhomogeneities have to be avoided, which restrict the choice of the microwave frequency: it has to be high enough so that (Brauer et al., 1977)

$$\frac{1}{4 \cdot \nu} \lesssim |\tau_d^-|. \tag{33}$$

According to Glover (1971, 1972), this means that $\rho_o \cdot \nu \geq 20\Omega\text{cm}\cdot\text{GHz}$ (for n-GaAs), where ρ_o is the ohmic specific resistivity. An upper limit of the chosen microwave frequency is given by the energy relaxation time. The necessary corrections of the $v_d(E)$ characteristics due to τ_ε effects were discussed by Glover (1971, 1972). Using this microwave method, Inoue et al. (1973) have also determined velocity-field characteristics in the ternary compounds $GaAs_{1-x}\,Sb_x$ and $Ga_{1-x}\,In_x\,As$. Recently, the quaternary alloy $In_{1-x}\,Ga_x\,P_{1-y}\,As_y$ has also been the subject of extensive investigation in the search for high peak drift velocities (Hayes and Raymond, 1977, Houston et al., 1978).

A very reliable and often-used method in high field transport was proposed by Chang and Moll (1966) and by Ruch and Kino (1967,

1968). It has given very good results in the investigation of negative differential mobility. This is the time-of-flight technique, which consists of the measurement of the transit time t_r of carriers generated by irradiating a region near one of the electrodes, as they travel across a certain region under the influence of an applied field. This method has been described in detail and often used by the Modena group (Alberigi-Quaranta et al., 1971, Reggiani, 1979). The method has been modified for semiconductors with rather high conductivities by Neukermans and Kino (1973) and by Evans et al. (1972).

Fay and Kino (1969) proposed a method consisting of measuring the phase velocity of a growing space-charge wave, which can propagate in ndm material above the threshold electric field. The method was applied to high resistivity n-GaAs ($\simeq 1000\Omega$ cm) with $n_o \cdot L$ smaller than the critical product. For a sufficiently small carrier concentration, the dispersion relation of the space-charge wave leads to a velocity which is identical to the drift velocity of the electrons. The space-charge wave is excited in the cathode region and travels towards the anode. The space-charge wavelength is measured with a capacitive probe.

There are a large number of reviews on negative differential mobility in semiconductors and their applications. We refer the interested to those of Acket and Vlaardingerbrock (1969), Carroll (1970), Grubin et al. (1973), Copeland and Knight (1971), Hilsum (1972), Hobson (1973, 1974, Levinshtein and Shur (1975), Ridley (1977), Hartnagel (1968), Thim (1971), Thim and Pötzl (1976), Kino and Robson (1968), and Bosch and Engelmann (1975).

ELECTRON TEMPERATURE

The concept of the electron temperature has been widely used for the description of hot electron phenomena. In the early investigations, it was assumed that the carriers were distributed among the energy states according to a displaced maxwellian distribution function, centered at the drift momentum $\vec{k}_d = m\vec{v}_d/\hbar$ and with an electron temperature T_e instead of the lattice temperature T_L (Fröhlich and Paranjape, 1956, Stratton, 1957, 1958). Thus the distribution function $f(\vec{k})$ is given by

$$f(\vec{k}) \simeq \exp\,[-\,\hbar^2(\vec{k}-\vec{k}_d)^2/2mk_BT_e], \tag{34}$$

or for degenerate statistics,

$$f(\vec{k}) = \frac{1}{\exp\left[\frac{\hbar^2(\vec{k}-\vec{k}_d)^2-\varepsilon_F}{2mk_BT_e}\right]+1} . \tag{35}$$

If the drift velocity v_d is small compared to the average thermal velocity an expansion of (34) is made (diffusion approximation)

$$f(\vec{k}) \simeq \exp(-\varepsilon/k_BT_e)\ [1+\vec{k}\cdot\vec{E}(\hbar\mu/k_BT_e)] \tag{36}$$

where $\vec{v}_d = \mu\vec{E}$.

The mean electron energy is related to T_e, for MB statistics, as

$$\langle\varepsilon\rangle = \frac{\int \varepsilon\cdot g(\varepsilon)\ \exp(-\varepsilon/k_BT_e)\ d\varepsilon}{\int g(\varepsilon)\ \exp(-\varepsilon/k_BT_e)\ d\varepsilon} = \frac{3}{2}\cdot k_BT_e, \tag{37}$$

or for the displaced Fermi-Dirac distribution functions,

$$\langle\varepsilon\rangle = k_B\cdot T_e\ \frac{F_{3/2}\ (\eta)}{F_{1/2}\ (\eta)} , \tag{38}$$

where $\eta = \varepsilon_F/k_BT_e$ and $F_k\ (\eta)$ are Fermi integrals:

$$F_k\ (\eta) = \int_0^\infty x^k/[\exp(x-\eta)+1]\ dx. \tag{39}$$

The virtues of the drifted-Maxwellian or Fermian distribution function are the relatively simple approach and the possibility of solving the Boltzmann transport equation analytically for certain scattering mechanisms under additional simplifying assumptions. In addition, the knowledge of the mean carrier energy as a function of the applied field strength, together with the drift velocity, is useful. It is certainly of relevance when considering other non-linear effects such as impurity breakdown or avalanche multiplication. With the displaced MB or Fermi distribution functions, the Boltzmann equation is solved by the method of balance equations for momentum and energy:

$$\int \vec{k}\ \{(\frac{\partial f}{\partial t})_f + (\frac{\partial f}{\partial t})_c\}\ d^3k = 0,$$

$$\int \varepsilon\ \{(\frac{\partial f}{\partial t})_f + (\frac{\partial f}{\partial t})_c\}\ d^3k = 0, \tag{40}$$

where $(\frac{\partial f}{\partial t})_f$ and $(\frac{\partial f}{\partial t})_c$ are the field and collision terms of the Boltzmann equation. The coupled equations yield the dependence of $\vec{v}_d$ and of T_e on E.

For a justification of the above ansatz for the hot carrier distribution function, electron-electron scattering is taken into account. This scattering mechanism randomizes momentum and energy gained from the electric field. It is an inelastic scattering process, yet however, keeping the total momentum and energy of the carrier ensemble constant. Theoretical treatments of this process start from an appropriately screened Coulomb potential (Dykman and Tomchuk, 1961, 1962, Hasegawa and Yamashit , 1962, Hearn, 1965, Landsberg, 1965).

Although the electron temperature model is used less, and more refined numerical methods (Monte Carlo calculation of $\vec{v}_d(E)$, iterative solution of the Boltzmann equation) are used, this concept has given useful results. However, it can only be applied under conditions, where the inelasticity of the scattering mechanisms is not pronounced and the energy gained from the field is randomized. The calculations based on this model are relatively simple and physical insight on the dominant scattering mechanism is rapidly gained. In addition, this concept is still helpful under conditions where the electron-electron scattering is important or even dominant, since numerical solutions of the Boltzmann equation including this particular scattering mechanism are still tedious (Jacoboni, 1976). Therefore, there has been some interest in experimental methods which can be used for a direct determination of the mean electron energy $\langle\varepsilon\rangle$, or T_e, apart from the determination of the $\vec{v}_d(E)$ characteristic. In the following sections these methods will be discussed.

Mobility, Hall Effect and Magnetoresistance

A simple method for determining $T_e(E)$ is based on a comparison of the field dependence of the mobility $\mu(E)$ at a constant lattice temperature with the dependence of the Ohmic mobility $\mu_0(T_L)$ on the lattice temperature as shown in Fig. 16. For this comparison to be valid, it is necessary that ionized impurity scattering dominate the momentum relaxation of the carriers. This scattering mechanism has no explicit dependence on lattice temperature, but only depends on the carrier energy. The ionized impurity mobility depends on certain averages over the distribution function $[\mu_0(\eta) \sim F_2(\eta)/F_{\frac{1}{2}}(\eta)\cdot(k_BT_e)^{3/2}$ It is of course important for the applicability of this method, that the number of ionized impurities N_I remains constant in the range of electric fields applied. For $\eta \geq 10$ (moderately and highly degenerate materials), this method is not suitable at all, since then μ depends neither on T_L nor on E. In small magnetic fields, the comparison of electric field and lattice temperature dependences can be extende

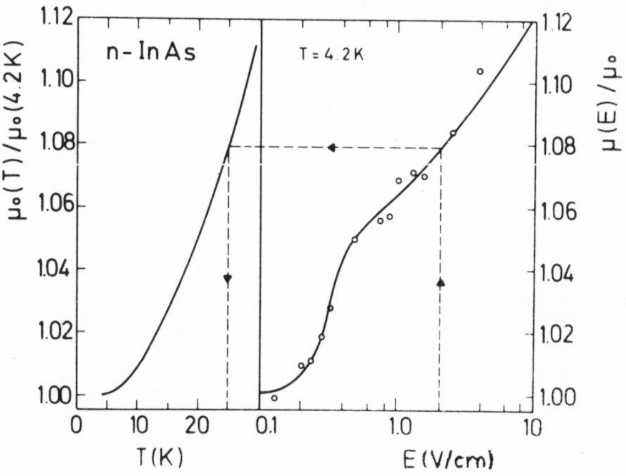

Fig. 16. Determination of electron temperature from a comparison of field dependent and lattice temperature dependent mobilities. (After Bauer, 1974).

to the Hall effect and to the magnetoresistance. The Hall coefficient factor r ($R_H = r/ne$) involves different averages over the carrier distribution function [$r \sim F_{7/2}(\eta)/F_{1/2}(\eta)/F_2^2(\eta)$] than μ_o. For degenerate material, r tends to 1 and the method is again not appropriate. However, extreme care has to be taken with the comparisons involving R_H and $\Delta\rho(B)/\rho_o$. At low lattice temperatuers (a necessary condition for dominant ionized impurity scattering), relative small magnetic fields may be sufficient to cause either impurity freeze-out effects (decrease of the free carrier concentration), or magnetic quantum effects might become important. This is of importance for small-gap materials with small effective masses.

The methods have been applied to III-V compounds, in particular to InSb, InAs and GaAs (Bauer, 1974, Miyazawa, 1969, Kinch, 1966, 1967, Crandall, 1970, Sandercock, 1970, Bauer and Kahlert, 1970). For dominant ionized impurity scattering, the time-dependent increase of the electron temperature, as a function of an applied step electric field, has been determined by several authors (Sandercock, 1965, Peskett and Rollin, 1963, Maneval et al., 1969, Yao et al., 1975). Maneval et al. (1969) used a double-step voltage pulse technique with the height of the second pulse $\Delta E \ll E$ and observed the current response. For small ΔE, the resulting energy variation $\Delta\varepsilon(t)$ is given (in the linear approximation) by

$$\Delta\varepsilon(t) = 2e\mu E\Delta E\tau_\varepsilon(1-\exp(-t/\tau_\varepsilon)).$$

For the time dependent increase of the current, the energy relaxatio time τ_ε can then be determined.

Hot Electron Faraday Effect

The Faraday effect is the rotation of the plane of polarization of linear-polarized light, for a propagation direction parallel to the applied magnetic field. The Faraday rotation δ is given by the different velocities for left-hand and right-hand circularly polarized light (Balkanski and Amzallag, 1968). In the hot electron case, only the free carrier contribution is of interest. Depending on the measuring frequency (in the infrared or microwave region), and on the applied magnetic field, two simple cases can be distinguished: (i) $\omega\tau >> 1$ and $\omega_c\tau << 1$: (infrared case), then the rotation is given by

$$\delta = \frac{2\pi e^3 nB.d'}{n_r c^2 m^2 \omega^2} , \tag{41}$$

with n_r the refractive index and d' the length of the sample. The rotation depends on the effective mass, but not on the relaxation time τ. (ii) $\omega\tau << 1$ and $\omega_c\tau << 1$ (microwave case):

$$\delta = \frac{2\pi e^2 nB\ d'\tau^2}{n_r\ c^2 m^2} . \tag{42}$$

The rotation here depends both on the effective mass and on the relaxation time.

For the determination of high electric field effects several authors (Wood, 1969, Almasov, 1973, Chattopadhyay and Nag, 1970, Asche, 1970, Ipatova et al., 1966) have considered the effects of repopulation (both equivalent and non-equivalent) and the effect of nonparabolicity of the bands. It has been shown by Vorobev et al. (1972a), that in the limit of (41), the Faraday rotation angle changes, if the carrier distribution is given by a displaced Maxwellian distribution function, as

$$\delta^* = \delta_o(1-10\ \frac{k_B T_e}{\varepsilon_g} - 4\ \frac{m\ v_d^2}{2\varepsilon_g}) . \tag{43}$$

Several experiments have been performed with n-InSb (Vorobev et al. 1972a, Heinrich, 1971) and the electron temperature has been determined for fields up to the avalanche breakdown field at 77 K. Sinc for E ≃ 200 V/cm, T_e was found to be about 200 K, these experiments show indirectly that the basic assumption of these experiments is not fulfilled very well. Electron-electron scattering has a strong influence on the average energy of the carriers (in samples with n ranging from 1.2 x 10^{15} to 4.9 x 10^{15} cm^{-3}), but the polar-optical scattering still introduces some streaming behaviour and is finally responsible for the avalanche breakdown at fields as low as 200 V/cm.

Investigation of the hot electron Faraday effect in semiconductors, where intervalley scattering is important were performed by Heinrich (1971) and Kriechbaum et al. (1972) in GaSb and n-Ge in the infrared region. In both materials, a decrease of the Faraday angle was found with increasing electric field due to the reduction of the number of carriers in the Γ (GaSb) or the L valleys (Ge) and the transfer of electrons to the higher L or X valleys, respectively. A rather moderate increase in T_e (300 to 330 K), at a field of more than 1 kV/cm in n-GaSb, causes a transfer of 10% of the electrons from the central Γ to the subsidiary L valleys. The change in Faraday rotation due to this transfer is about 0.15°.

Hot Electron Birefrigence

In strong electric fields, even in cubic semiconductors, the free carrier contribution to the dielectric function will be anisotropic and different for directions parallel and perpendicular to the applied electric field. This anisotropy arises from the anisotropy of the hot carrier distribution function, and will also depend on the non-parabolicity of the bands. It will give rise to a difference in the phase velocities (refractive indices), and the damping of light (causing elliptical polarization) which is polarized || or ⊥ to the applied d.c. electric field $\vec{E}_o$.

In many-valley semiconductors, this anisotropy in refractive index may also arise from differences in the populations of the equivalent valleys brought about by the applied field E. This effect was the original suggestion of Schmidt-Tiedemann (1961). In n-Ge, a difference in n_r occurs:

$$\Delta n_r = n_{r,\parallel} - n_{r,\perp} \text{ for } E \parallel \langle 111 \rangle$$

given by:

$$\Delta n_r = \frac{2\pi e^2 \, n_{\langle 111 \rangle}}{n_r \, m_l \, \omega^2} \; \frac{(n_c/n_h) - 1}{3 + \frac{n_c}{n_h}} \; (m_l/m_t - 1),$$

where $n_{<111>}$ is the total concentration, m_l and m_t are the longitudinal and transverse masses of carriers in the <111> valleys, n_c and n_h are the concentrations in the cold <111> and hot <111> valleys. The quantity which is actually measured is the phase shift δ of the two components polarized $||$ and $\perp$ to E:

$$\delta = \frac{2\pi d}{\lambda_o} \cdot \Delta n_r \ .$$

The experimental method is very sensitive, since changes Δn_r of the order of 1×10^{-5} can be detected. For n-Ge this corresponds to changes of the carrier concentrations in the hot and cold valleys of 1×10^{13} cm^{-3}. This method has also been applied to a study of n-Si at T = 80 K for E$||$[100](Vorobev et al., 1973). For the investigations of both n-Ge and n-Si, circularly polarized CO_2-laser radiation was used. With an analyzer inclined at 45° to the direction of E, the modulation of the circularly polarized light, due to the phaseshift δ, was determined. For electric field pulses of 200 ns, a fast photoconductive detector (Ge-Zn) had to be used.

The birefrigence due to hot carriers has been treated theoretically by Vasko (1973) and Brazis and Pozhela (1975). Hot electron birefrigence in single valley semiconductors like InSb and InAs, with appreciably non-parabolic conduction bands has been treated both theoretically (Almasov and Dykman, 1971, 1972, Almasov, 1972, 1973) and experimentally by Vorobev et a. (1971, 1972c). These latter authors have shown that even in degenerate InAs, this method yields information on T_e for fields up to 500 V/cm ($T_e \simeq 160$ K for $n = 1.5 \times 10^{16}$ cm^{-3}, $T_L = 80$ K). The experimental results were compared with calculations based on a displaced Fermi distribution function. Almasov (1972, 1973) has pointed out that, in hot electron Faraday experiments with non-parabolic semiconductors, not only the off-diagonal components of the dielectric tensor are changed, but also the diagonal ones, with the result that birefrigence occurs.

Burstein-Moss Shift

Shur (1969) has suggested an optical method for the determination of electron temperatures in degenerate semiconductors. It is based on the fundamental direct electronic absorption which is shifted towards higher energies than the gap energy ε_g, if the Fermi-level is within the conduction band (Burstein, 1954). At an energy of $4k_BT_e$ below the Fermi level, about 98% of the states are filled and therefore the absorption will only take place for energies higher than this threshold: the shift ε_B is given by

$$\varepsilon_B = (1 + m_c/m_v)(\varepsilon_F - 4k_BT_e). \qquad (44)$$

If the electrons gain energy from the field and T_e increases, the number of empty states below ε_F will increase due to the smoothing of the distribution function. As long as $\varepsilon_F - 4k_BT_e > 0$, this effect can be exploited and the change of the absorption spectrum can be used to determine T_e. This is done by a comparison with measurements of the Burstein-Moss shift as a function of lattice temperature. Another more direct influence of the electric field on the optical absorption edge, the Franz-Keldysh (1958) effect is small if the range of electric fields is restricted to some 100 V/cm. Heinrich and Jantsch (1971) have studied the effect of fields up to 150 V/cm on the Burstein-Moss shift in n-GaSb ($n=3.5\times10^{17}cm^{-3}$). An increase of T_e from 77 to 100 K was found. The direct comparison of the change of the optical transmission with lattice temperature and with electric field has to be corrected for the change of ε_g with T_L. Obiditsch and Kahlert (1976) used the same method to study n-GaSb at a lattice temperature of 6 K. In fields up to 200 V/cm, an electron temperature of 40 K was found. After an initial steep rise of T_e with E, above T_e = 15 K a kink is observed, which is most probably caused by a 2TA (simultaneous emission of 2 transverse acoustic phonons at the edge of the Brillouin Zone) process accompanied at higher electron temperatures by polar-optical phonon scattering. This method is of course only applicable for direct interband transitions when no phonons are involved in the absorption process of the light.

Hot Carrier Luminescence

The photoexcitation of electron-hole pairs by light with a photon energy $h\nu > \varepsilon_g$, the gap energy, has become a valuable tool for studying hot carriers. The photoexcited carriers interact with phonons and other carriers, lose their excess energy, and relax towards the band extrema. This leads to carrier heating and the generation of non-equilibrium phonons. The energy fed into the system is provided by the excitation instead of the electric field.

Whether a Maxwell-Boltzmann or Fermi distribution is established after excitation again depends on the self-interaction of the carriers, which has to be sufficiently strong. If electron-electron scattering dominates all other scattering mechanisms, then from simple energy balance considerations, a value of the electron temperature can be found. Under stationary conditions, the energy fed into the system is provided by the energy transferred from the photoexcited carriers to the carrier ensemble. This must be equal to the energy loss by phonon emission and photon emission. Information on the carrier temperature is obtained from the observation and analysis of luminescence due to electron-hole recombination, or electron-acceptor recombination.

Distinct differences occur with high and low excitation densities and reviews were given recently by Shah (1978) and Ulbrich (1978a). The variation of electron temperature of the carriers with the photon energy of the excitation source has been studied by Shah (1978) and by Goebel and Hildebrand (1979), Goebbel (1979), and Hildebrand et al. (1978). Most of this work on hot carrier luminescence has been done on GaAs and on the II-VI compounds CdS and CdSe.

Leheny and Shah et al. (1977, 1978) analysed not only the hot carrier luminescence but also the change of optical transmission above the band-gap, while the sample was excited simultaneously with a pump-beam generating electron hole pairs. These absorption experiments provide complementary information to the luminescence experiments and yield more than the electron temperature: the hot carrier distribution function. More details can be found in the above mentioned papers and in Ulbrich (1979). Recently, using techniques of picosecond spectroscopy, von der Linde and Lambrich (1979) have made a direct measurement of the time dependence of the optical absorption following electron-hole pair creation. The energy relaxation time in GaAs due to polar LO-phonon emission could be directly determined.

Hot Carrier Photoconductivity

Recently, Seiler et al. (1978) have suggested a new technique for the direct determination of electron temperatures, even in degenerate semiconductors. It is based on free carrier heating by optical absorption of infrared laser radiation. The carrier heating is observed as a change of the voltage drop across the sample, caused by the mobility changes. In Fig. 17, the experimental arrangement, and in Fig. 18, the results for n-InSb with $n = 1.4 \times 10^{15}$ cm^{-3}, are shown. The electron temperature for a given laser power and a certain laser wavelength is obtained from a comparison with the lattice temperature dependence of the Ohmic mobility and the non-Ohmic mobility due to the application of pulsed electric fields. The momentum relaxation is still determined by ionized-impurity scattering in this experiment and so this comparison is possible. The method also yields precise information on the free carrier absorption coefficient as a function of the CO_2 laser wavelength. It is possible to relate an effective absorption coefficient α_{Eff} to the incident laser power P_i per electron and the applied electrical power P_E per electron:

$$\alpha_{Eff}(\lambda, T_e) = \frac{P_E(T_e)}{P_i(T_e, \lambda)} \cdot \frac{1}{d}, \tag{45}$$

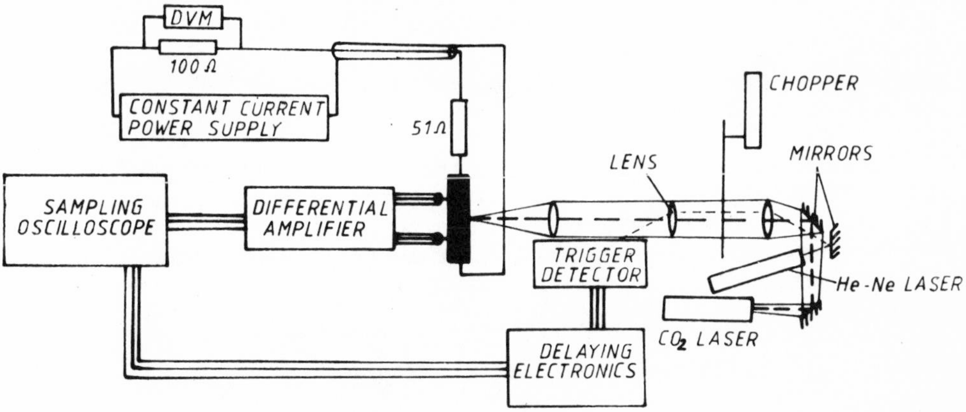

Fig. 17. Schematic diagram for measuring carrier heating induced by absorption of pulsed CO_2 laser radiation (pulse duration: 20μs, rep. rate: 1700 Hz) with a photoconductive technique. (After Seiler et al., 1978).

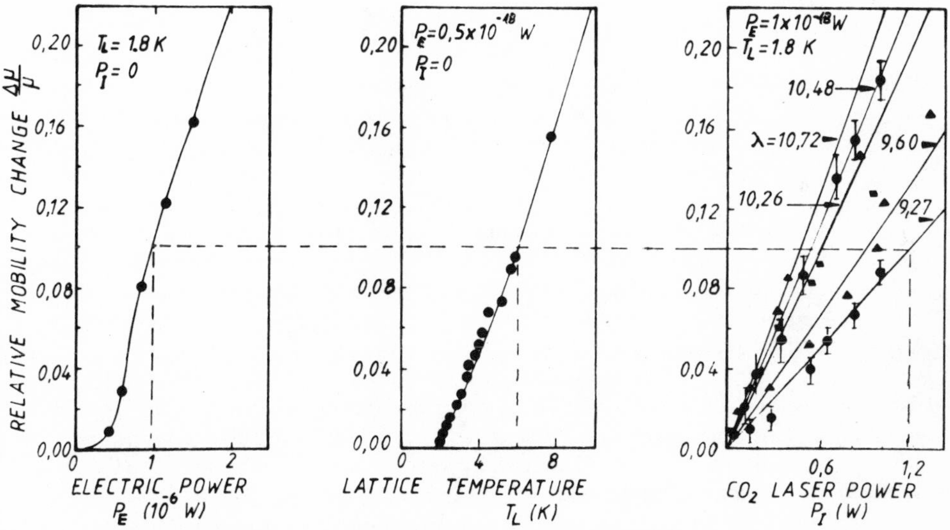

Fig. 18. Experimental results on carrier heating by optical absorption in InSb at T_L = 1.8 K with CO_2 laser radiation (λ = 10.72 . . . 9.27μm) comparison with d.c. heating and determination of electron temperatures are shown (after Seiler et al., 1978).

where $\alpha_{Eff} = \alpha(1-\beta)$. The loss factor β arises when a fraction of the original optical absorbed power per electron: $\alpha \cdot d \cdot P_i$ is transferred to the lattice by subsequent optical phonon emission before the threshold energy, at which the electron-electron scattering dominates the electron-phonon scattering, is reached. The experiments show that for CO_2 laser wavelengths with $P_I \simeq 1$ W at $T_L = 1.8$ K the electron temperatures rise up to 5 - 9 K (for $\lambda = 9.27 \ldots 10.72$ μm), with higher T_e for longer wavelengths. The loss factor β is not equal to zero. However, the finite width of the laser excitation pulse ($\Delta\varepsilon \simeq \varepsilon_F + 4k_BT_e$) made it impossible to observe any oscillatory behaviour related to the optical phonon energy $\hbar\omega_{LO}$.

Tunneling Experiments

Rowell and Tsui (1976) proposed a very elegant technique for the determination of electron temperatures in the range up to 20 K: on the semiconductor surface an oxide layer is deposited and onto this oxide layer a superconducting film is evaporated. Thus a superconductor-oxide-semiconductor sandwich, as shown in Fig. 19, is fabricated. The tunneling current through the junction now carries valuable information on the electrons in the semiconductor. The superconducting film is kept at very low temperature. The tunneling results from the electrons leaving the semiconductor for the empty states above the gap of the superconductor; preferentially into the high density-of-states region near the gap energy. By changing the diode voltage, the relative shift of the density-of-states curve of the semiconductor, with respect to the superconductor, can be altered. Thus the energy range of the hot electrons exposed to the empty states in the superconductor can be chosen, and any strong deviations of the hot carrier distribution function from a Fermi-function should be observable. In Fig. 19, the situation is shown for two different diode voltages at the same T_e.

The experiments were performed with n-InAs, a semiconductor which has an accumulation layer on its surface. For the superconductor, a $Pb_{0.8}$ - $Bi_{0.2}$ alloy was chosen (relatively high transition temperature). The sample configuration is shown in Fig. 19. One and 6 are the current contacts; 2, 5 and 7, 10 are contacts for the voltage drop, the Hall effect, and also a four-terminal connection to the tunnel junction. The other four contacts 3, 4 and 9, 8 are soldered directly on the $Pb_{0.8}$ $Bi_{0.2}$ stripes. The electron temperature in n-InAs, caused by the application of a field between the contacts 1 and 6 was determined by comparing the measured junction characteristics with calculated ones. For the calculations it was assumed that carriers in InAs are distributed according to a Fermi-distribution function and that the density-of-states of the Pb-Bi alloy is determined by the BCS density-of-states, with the

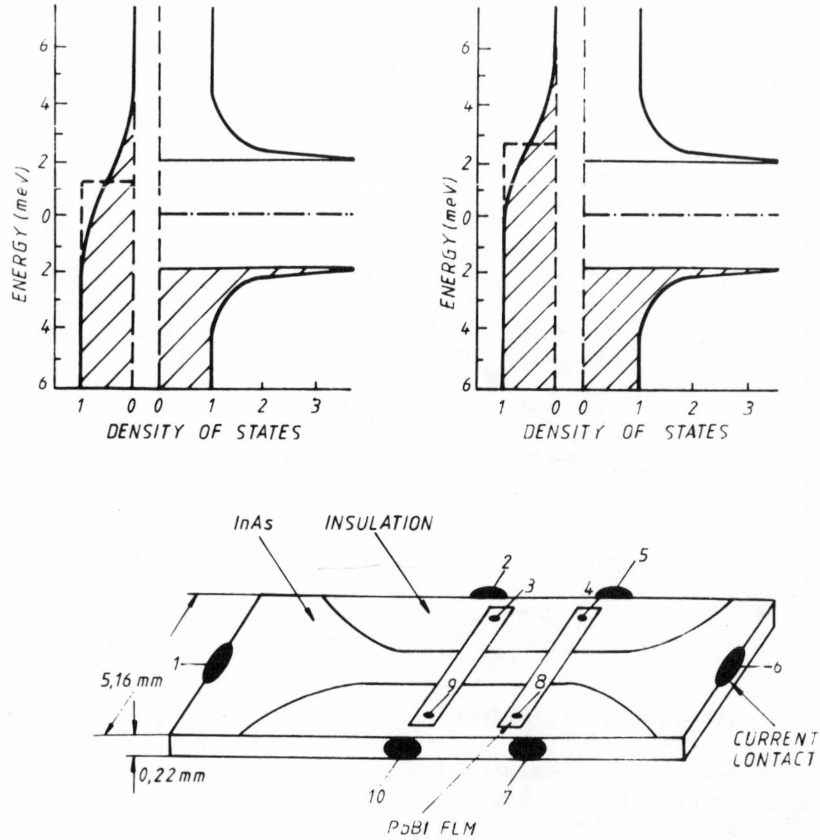

Fig. 19. Determination of electron temperatures with the aid of electron tunneling from a semiconductor through an oxide layer in a superconductor. Upper half: energy, density of states and occupancy of the possible states are shown. For two different bias fields at the semiconductor-oxide-superconductor structure, and the same carrier distribution in the semiconductor, the relative energy shifts are shown in the upper half; Lower half: typical sample geometry used for the tunneling experiments (after Rowell and Tsui, 1976).

superconducting gap fixed at its T = 0 value. These calculations were done for different bias voltages of the junction, thus providing the necessary test for the consistency of the technique (the same value of T_e was obtained for different bias fields).

Rowell and Tsui (1976) observed that the accumulation layer on n-InAs has some influence on the data. With the tunneling method,

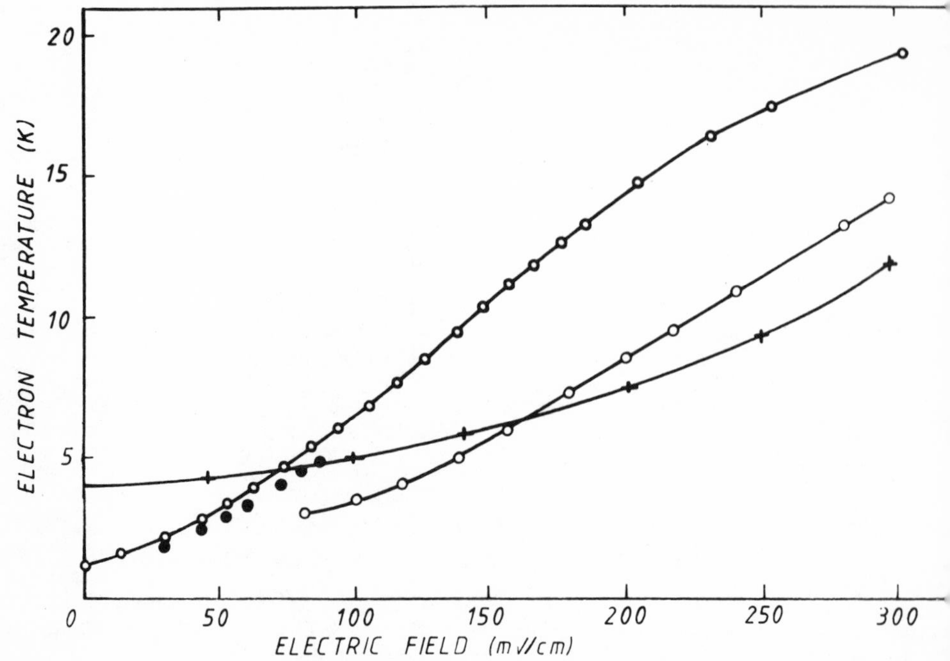

Fig. 20. Experimentally determined electron temperatures as a function of field for n-InAs. o,o: from tunneling experiments; o: from mobility measurements (yield smaller T_e-values) (after Rowell and Tsui, 1976). + from SdH experiments (after Bauer, 1974).

T_e is measured at the surface. Mobility methods measure T_e in the bulk of the sample. The measurements seem to indicate, that the surface and bulk electrons are not in thermal equilibrium with each other, but that there are 2 different electron temperatures, the electrons in the surface subbands being hotter (Fig. 20).

Hot-Electron Effects in Quantizing Magnetic Fields

In strong magnetic fields B, the electron motion in the plane perpendicular to B is quantized into Landau subbands. Each of these subbands is split into 2 states due to the two possible spin orientations of the carriers; as

$$\varepsilon = (N + \frac{1}{2})\ \hbar\omega_c + \frac{\hbar^2 k_z^2}{2m} \pm s.g^*.\mu_B.B, \tag{46}$$

where $\omega_c = eB/m$, N = 1, 2, 3 . . . , and μ_B denotes the Bohr magneton. For nonparabolic bands, a more general expression has to be used with $m = m(\varepsilon)$ and $g^* = g^*(\varepsilon)$. The density-of-states also changes dramatically; without damping it diverges as

$$g(\varepsilon) = \frac{eBV(2m)^{\frac{1}{2}}}{\hbar^2} \sum_{N,s} \{\varepsilon - (N+\tfrac{1}{2})\hbar\omega_c \pm s.g^*.\mu_B.B\}^{-\frac{1}{2}} \tag{47}$$

near the bottom of the Landau states ($k_z = 0$).

This significant change of the $\varepsilon(k)$ relationship, and of the associated density-of-states, considerably alters the electron-phonon and electron-electron relaxation rates (for a review, see e.g. Zlobin and Zaryanov, 1972). As far as the concept of the electron temperature is concerned, the important effect of the magnetic field is the reduction of the electron-electron scattering rate. This follows from the quasi-one-dimensional nature of the electron motion. The effect is particularly strong in the ultra-quantum limit, when just the lowest Landau level is occupied and no collisions between electrons belonging to different Landau levels occur. According to Zlobin and Zaryanov (1972), the ratio of the electron-electron to the electron-phonon collision rates in a quantizing magnetic field is approximately given by

$$\frac{\nu_{ee}}{\nu_{ep}} \sim \frac{(k_B T_e)^2}{(h\omega_c)^2} \exp\left(-\frac{\hbar\omega_c}{k_B T_e}\right). \tag{48}$$

[However, Kogan et al. (1976) have shown that in the ultra quantum limit, either 3-body collisions or 2-body collisions in the field of an impurity can occur. In addition, level broadening removes the sharp divergence of the density-of-states and restores some three-dimensionality to the problem.] The magnetic quantization can be used in various ways for an accurate determination of the electron temperatures of hot carriers, provided that $\omega_c\tau_m > 1$ and $\hbar\omega_c \gtrsim k_B T_e$. These restrictions impose an upper limit on the applied electric fields and on the range of lattice temperatures where experiments can be performed.

Most experiments have been performed with small gap semiconductors (InSb, InAs). For these materials, non-parabolicity has to be taken into account, and in a two-level scheme, instead of (46) the energy is given by (Zawadzki, 1973)

$$\varepsilon_{N,_z},^{\pm} = -\frac{\varepsilon_g}{2} + \{(\frac{\varepsilon_g}{2})^2 + \varepsilon_g \cdot \lambda_{N,k_z}\}^{\frac{1}{2}}, \tag{49}$$

where

$$\lambda_{N,k_z},\pm = \hbar\omega_c(N+\tfrac{1}{2}) + \frac{\hbar^2 k_z^2}{2m_o^*} \pm \frac{1}{2} g_o \mu_B \cdot B,$$

$$\omega_c = \frac{eB}{m_o^*}, \quad g_o^* = 2 + (1 - \frac{m_o}{m_o^*}) \frac{2\cdot\Delta}{3\varepsilon_g + 2\Delta},$$

and m_o^* and g_o^* are the effective band-edge mass and g-factor, respectively, m_o is the free-electron mass and Δ is the spin-orbit parameter (energy split).

There are striking differences for carrier heating in the transverse ($\vec{E}\perp\vec{B}$) and in the parallel ($\vec{E}||\vec{B}$) configuration. For $\vec{E}||\vec{B}$, the electron can move in the direction of $\vec{E}$ and between two subsequent collisions the work done by the field is $eE\ell$, ℓ being the mean free path. In the transverse configuration, the electric field determines the velocity in the direction perpendicular to both $\vec{E}$ and $\vec{B}$, and transport in the direction of $\vec{E}$ is due just to scattering. The work done by the electric field $\vec{E}$ is then $eE\lambda_B$, λ_B being the magnetic length ($\lambda_B^2 = \hbar/eB$).

It has been shown (Barker, 1972, Barker and Magnusson, 1974) that for small electric fields ($v_d \to 0$), the following expressions can be derived within a generalized electron temperature model (quantum transport approach):

$$T_e = T_L \; [1 + \frac{ne^2E^2}{mC_e k_B T_L} \cdot \tau_\varepsilon \; (\frac{\omega_\perp}{\omega_c^2 + \omega_\perp^2})] \text{ for } \vec{E}\perp\vec{B}, \tag{50}$$

$$T_e = T_L \; [1 + \frac{ne^2E^2}{mC_e k_B T_L} \cdot \tau_\varepsilon \; (\frac{1}{\omega_{||}})] \text{ for } \vec{E}||\vec{B}, \tag{51}$$

where C_e is the electronic specific heat, and $\omega_\perp$, $\omega_{||}$ are the transverse and longitudinal momentum relaxation rates, evaluated over the equilibrium density matrix. In the transverse geometry $\vec{E}\perp\vec{B}$, it is important to consider the connection between the applied electric field E_{ap} and the total field E. The Hall field E_H is only zero for a Corbino geometry ($E_{ap} = E$). In all other cases the total field will be given by

$$E = \sqrt{E_{ap}^2 + E_H^2} = E_{ap} \sqrt{1 + \tan^2\theta}$$

with

$$\tan\theta = \sigma_{xy}/\sigma_{xx}.$$

For high magnetic fields, $|\sigma_{xy}| >> \sigma_{xx}$ and $E \simeq E_H$. Thus, the total field can be much larger than the applied field, and E instead of E_{ap} has to be used in the energy balance equation.

Rohlfing and Prohofsky (1975) used a displaced Maxwellian with a different drift velocity v_N and electron temperature T_N assigned to each Landau level N. The momentum and energy balance equations are then given by:

$$eEn_N = \sum_{k_z} \hbar k_z \frac{\partial f_N(k_z)}{\partial t} = 0 , \qquad (52)$$

and

$$eE\, n_N\, v_N = \sum_{k_z} \left((N+\tfrac{1}{2})\hbar\omega_c + \frac{\hbar^2 k_z^2}{2m}\right) \frac{\partial f_N(k_z)}{\partial t} , \qquad (53)$$

where

$$n_N = n \exp\,[-(N+\tfrac{1}{2})\hbar\omega_c/k_B T_N]\ \mathrm{x}$$

$$\mathrm{x} \exp\left(\sum_{l=o}^{M} [\exp\,(-\,(M+\tfrac{1}{2})\hbar\omega_c/k_B T_M)\,]\right)^{-1},$$

(M + 1 is the total number of Landau levels which is considered);

$$f_N(k_z) = n_N\left(\frac{2\pi}{L}\right)\left(\frac{\hbar^2}{8\pi m k_B T_N}\right) \mathrm{x} \exp\left(\frac{-(\hbar k_z - m v_N)^2}{2k_B T_N}\right);$$

and $\frac{f_N(k_z)}{\partial t}$ is the rate-of-change of the distribution function due to scattering in the Nth and out of the Nth level. This generalized model is very suitable for the $\vec{E}||\vec{B}$ configuration and allowance is made for a distortion of the optic phonon population by coupling to the phonon Boltzmann equation. A comparison of this model to the more rigorous solution of the quantum kinetic equation is given in Calecki (1979).

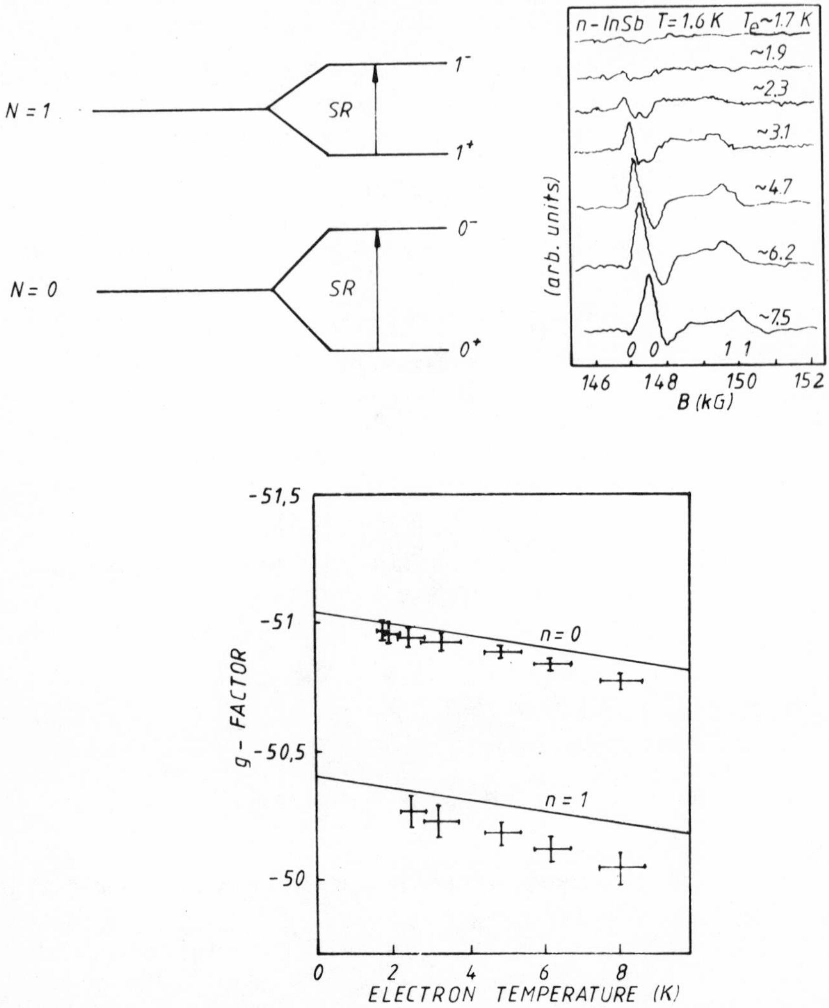

Fig. 21. Electron spin resonance of hot carriers. Upper right: level scheme in a nonparabolic semiconductor; upper left: experimental data on the derivative of the electron spin resonance of 100 GHz in InSb at various power levels (after Konopka, 1970); lower half: g-factor as a function of electron temperature in n-InSb as determined from ESR experiments. Upper curve: for $o^+ \to o^-$ transitions, lower curve for $1^+ \to 1^-$ transitions. (After Zawadzki, 1973).

Electron spin resonance of hot electrons. In a non-parabolic semiconductor, the electron g-factor will depend on energy, decreasing with increasing energy. Thus the energy difference between spin-split levels depends on k_z and on N. In an electron spin-resonance experiment, a g-factor will be obtained which is an average over the distribution function, and thus depends on T_e. Experiments with n-InSb with $n \simeq 1 \times 10^{14} cm^{-3}$ were performed by Gueron (1964), Kaplan and Konopka (1968) and Konopka (1970).

The heating of the carriers was achieved with microwaves ($\nu \simeq 100$ GHz) which were also used for the ESR measurement, keeping the sample at $T_L = 1.6$ K. At high microwave power levels, the N = 1 Landau level is also populated, as shown in Fig. 21. The data are analyzed by using, for the energy dependence of the effective mass and g-factor (Zawadzki, 1973):

$$m^*(k_z) = m_o^* \left(1 + \frac{2\varepsilon'(k_z)}{\varepsilon_g}\right) \tag{54}$$

and

$$g^*(k_z) = 2\left\{1 - \left(\frac{m_o}{m^*(k_z)} - 1\right) \frac{\Delta}{3\varepsilon'(k_z) + 3\varepsilon_g + 2\Delta}\right\}, \tag{55}$$

For a non-degenerate electron gas, the average energy in the z-direction is given by $\frac{1}{2}k_B T_e$. Using (54) and (55), and averaging over k_z, the full lines for the dependence of g on T_e were obtained (see Fig. 21).

Hot electron Shubnikov-de Haas effect. The Shubnikov-de Haas effect is an oscillatory magneto-resistance which is found in materials with the Fermi level within either the valence or the conduction band. Oscillations, which are periodic in 1/B, occur due to repopulation and scattering effects associated with the passage of Landau levels through the Fermi level. This effect can be observed under the following conditions:

$$\omega_c \tau \gg 1; \ \hbar\omega_c > k_B T_e; \ \varepsilon_F > \hbar\omega_c.$$

The magneto-resistance oscillations are damped with increasing temperature according to

$$\frac{\Delta\rho}{\rho_o} = a \sum_{s=1}^{\infty} \left[b_s \cos(2\pi s\varepsilon_F/\hbar\omega_c - \tfrac{1}{4}\pi) + R\right], \tag{56}$$

with

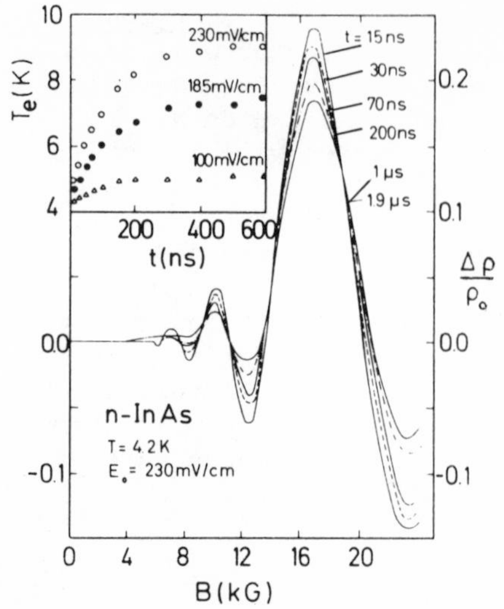

Fig. 22. Time dependence of SdH oscillations after voltage pulse E = 230 mV/cm in n-InAs. Insert: increase of T_e with time for three electric fields (after Bauer, 1974).

$$b_s \sim (\chi/\sinh \chi) \exp (- 2\pi k_B T_D^2/\hbar\omega_c) \tag{57}$$

and

$$\chi = 2\pi^2 k_B T_e/\hbar\omega_c . \tag{58}$$

(For details see Bauer, 1974).

At constant, low lattice temperatures, the electric field cause a reduction of the amplitudes of the Shubnikov-de Haas (SdH) oscilla tions due to the increase of T_e. For dominant ionized impurity scattering for the momentum relaxation, this $T_e(E)$ can be determined from a comparison with the damping of the oscillation amplitudes wit increasing lattice temperature under Ohmic conditions. It is preferable to perform the experiments in the $\vec{B}||\vec{j}$ configuration. In addition, the Dingle temperature T_D (57) must not change under the applied field E. The method is described in detail in Bauer (1974) and Bauer and Kahlert (1972) and has been used to determine electron temperatures in n-InSb, n-InAs, n-GaSb and p-Te (Kahlert, 1975). Using time-resolved measurements of the damping of the SdH amplitudes, after the application of a step voltage pulse, it was possibl to deduce the time-dependent increase of the electron temperature on a nanosecond time scale (Fig. 22) (Bauer and Kahlert, 1972). These measurements provided direct information on the relevant energy relaxation time for degenerate statistics. Moore et al.

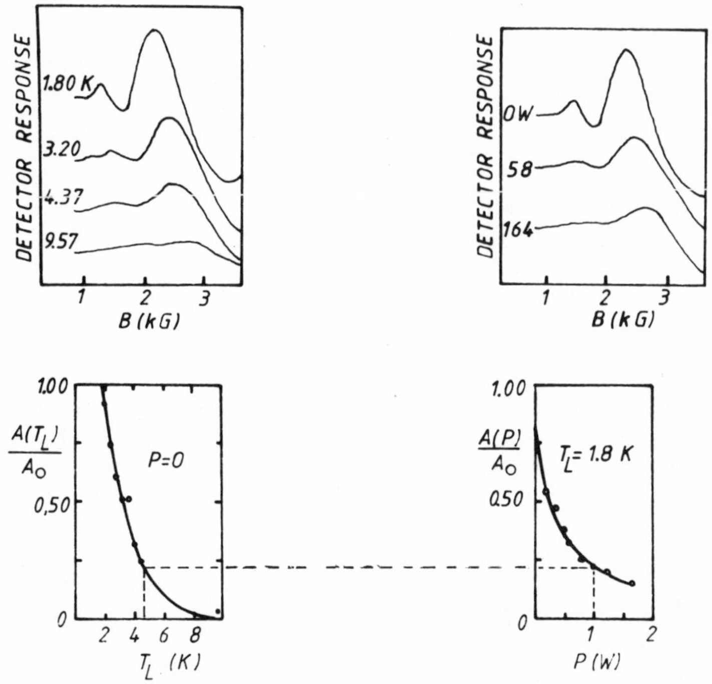

Fig. 23. Determination of electron temperatures in n-InSb with CO_2 laser heating using the damping of SdH oscillations. upper left: Ohmic SdH curves for different lattice temperatures. upper right: "hot" SdH curves at $T_L = 1.8$ K and laser powers 0 . . . 1.64 W, lower left: damping of SdH amplitudes as a function of T_L, lower right: damping of SdH amplitudes as a function of laser power (after Moore et al., 1978).

(1978) have shown that the effect of laser heating of the electrons can be monitored by making Shubnikov-de Haas effect measurements. Irradiating n-InSb with a pulsed CO_2 laser (at 10.2 μm, peak-power ≈ 2 watts) produced a damping of the Shubnikov-de Haas oscillation amplitudes which was observed simultaneously by a normal d.c. conductivity technique. In Fig. 23, the analysis of the data is shown, performed by a comparison with the damping of $\Delta\rho/\rho(B)$ with increasing T_L. A laser power of 2 watts increases T_e from 1.8 K to 6 K. The CO_2 laser pulse duration was about 3 μs. It could be shown that no significant lattice heating occurs. The SdH amplitude damping with these laser experiments is very similar to the damping produced by high d.c. fields. From time resolved measurements, τ_ε can also be obtained with this laser heating technique in degenerate InSb samples ($n = 1.4 \times 10^{15} cm^{-3}$).

It should be noted, that Lifschitz et al. (1972) and Romanovtsev and Shulman (1974) have discussed the influence of the disturbance of the phonon distribution function from the Bose-Einstein distribution, caused by the emission of phonons from the hot electrons at low lattice temperatures in InSb. These authors have also argued that the disturbance of the phonon distribution is enhanced in a quantizing magnetic field. The situation is however, still not clear since the experiments of Lifschitz et al. (1972) were carried out under conditions where magnetic freeze-out effects cannot be excluded. These effects may interfere and may make an unambiguous interpretation of the τ_ε-experiments in quantizing fields very difficult in InSb samples with $n \simeq 10^{14}\text{cm}^{-3}$.

Cyclotron resonance of hot electrons. As with spin resonance, in non-parabolic semiconductors cyclotron resonance can be exploited for a determination of electron temperatures (Kobayashi and Otsuka, 1972, 1974). With application of electric fields, the intensity of the $0^+ \rightarrow 1^+$ cyclotron resonance absorption changes due to population of higher Landau levels, and a second line $0^- \rightarrow 1^+$ appears. This can be observed with far-infrared cyclotron resonance either in absorption or in photoconductivity experiments (Bluyssen et al., 1978). Most experiments have been carried out so far with n-InSb. This material shows, in addition to the free carrier resonance, a resonance between bound states (000 → 110 in the notation as used in Johnson and Dickey, 1970). With increasing electric field, carriers from the impurity ground state are transferred by impact ionization into the 0^+ and 0^- Landau levels. Thus, the (000) → (110) line decreases with E. In the experiments of Kobayashi and Otsuka (1972, 1974), a chopped H_2O laser (30 Hz) has been used at $\lambda = 119$ μm. Electric field pulses ($\vec{E} \perp \vec{B} \| \vec{q}$) of 400 μs duration were applied at a frequency half of the chopping frequency. With a two channel boxcar integrator system connected to the Ge:Ga detector, the change of the absorption coefficient of the sample due to the applied d.c. field, was measured directly. An example is shown in Fig. 24. The electron temperature for a given electric field is determined from the change of the absorption strength under the assumption that the Landau levels 0^+ and 0^- are populated according to a Maxwell-Boltzmann distribution function:

$$n_{N,s}(E)/n \sim \exp\left\{-\frac{(N+\frac{1}{2})\hbar\omega_c + sg\mu_B . B}{k_B T_e}\right\}. \tag{59}$$

From this, the dependence $T_e(E)$ is determined. Additional measurements of $\mu(E)$ and $\mu_o(T)$ according to the method of Miyazawa (1969) yielded satisfactory agreement with $T_e(E)$ from the cyclotron resonance measurements.

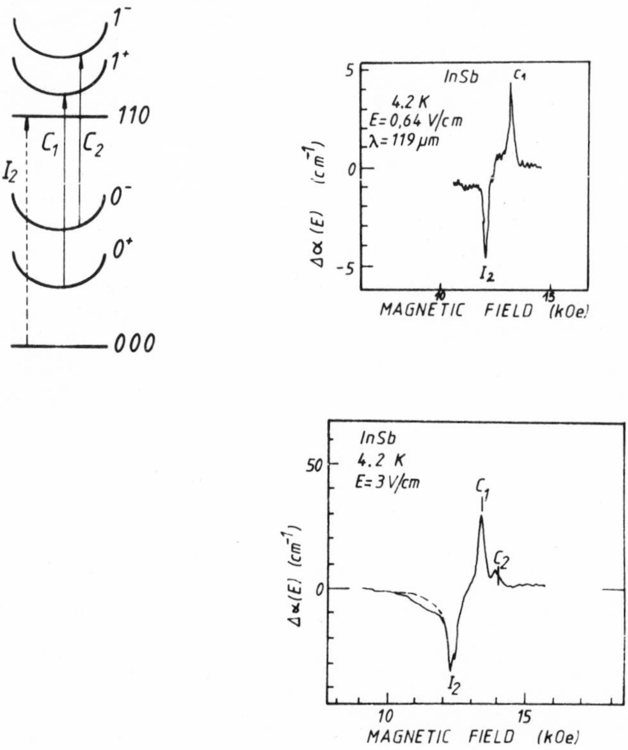

Fig. 24. Hot electron cyclotron resonance in n-InSb; upper left: level scheme with magnetic field split impurity and Landau states; upper right and lower right: change of absorption coefficient $\Delta\alpha(E)$ due to application of electric fields of 0.64 and 3 V/cm. At higher fields the $0^- \rightarrow 1^-$ transition appears. The change of absorption is directly measured as a difference between field-on and field-off detector response. (After Kobayashi and Otsuka, 1974).

HOT ELECTRON DISTRIBUTION FUNCTIONS

In contrast to the methods discussed in chapter II, which yield $\vec{v}_d(\vec{E})$, an integral average over the high field distribution function, or the methods of chapter III, which gave information on another average, the mean energy, in this chapter we discuss methods which yield details of the hot electron distribution function itself. Information on the occupancy of the possible states in energy or momentum space are obtained with the aid of optical methods: absorption, emission, and inelastic light-scattering. The band structure of the material has to be known exactly since a given photon energy

$\hbar\omega$ connects initial and final electron states $\varepsilon_i(\vec{k})$. For quantitative comparisons with calculated $f(\vec{k})$, the optical dipole-transition matrix-elements also have to be known. The great progress, which has been made in optical studies of hot electron distribution functions, has been reviewed recently by Bauer (1978) and Ulbrich (1978).

Optical Absorption

For the determination of the carrier distribution function, only direct optical transitions can be used. These are transitions between bands or transitions from localized electron states into bands. The frequency dependence of the absorption constant for direct band-band transitions is given by (Houghton and Smith, 1966)

$$\alpha(h\omega) = \frac{\pi c\mu_o e^2\hbar}{n_r\hbar\omega m_o^{\,2}} \sum_k |\vec{a}.\vec{p}|^2[f(\varepsilon_i(\vec{k})) - f(\varepsilon_f(\vec{k}))]\delta(\varepsilon_f(\vec{k}) - \varepsilon_i(\vec{k}) - \hbar\omega), \tag{60}$$

where $\vec{a}$ is the polarization vector of the radiation and $|\vec{a}.\vec{p}|^2$ is the transition matrix element squared. $f(\varepsilon_i(\vec{k}_i))$ is the probability for the occupation of the initial state $|i\rangle$. The difference in the distribution functions in (60) results from the fact that, apart from induced absorption, induced emission and the occupancy of the final states is also considered. Usually it is assumed that the applied electric field E does not affect the transition matrix element. An experimental situation is chosen where either the initial states are all filled ($f(\varepsilon_i(\vec{k})) = 1$) or the final states are all empty. The first case occurs in interband transitions from a filled valence band to the conduction band in an n-type semiconductor.

As (60) shows, a unique relationship between the photon energy and the energy of the initial and the final state $\varepsilon_i(\vec{k})$, $\varepsilon_f(\vec{k})$ is only obtained for parabolic and isotropic bands. Since the valence bands of the elemental group IV semiconductors and the III-V compounds are anisotropic and warped, for a given $\hbar\omega$ there will then be several, different $\vec{k}$ for which the δ function in (60) is fulfilled. Therefore, no direct determination of the distribution function is possible in this case. One has to assume, rather a distribution function, calculate the dependence $\alpha(\hbar\omega)$ for this function and a best fit to the experimental absorption data is sought by some numerical fitting procedures. For this purpose it is convenient to expand the carrier distribution function in a series of Legendre polynomials:

$$f(\vec{k}) = \sum_{l=o}^{\infty} f_l(\varepsilon)\, P_l(\cos\theta), \tag{61}$$

where θ is the angle between $\vec{k}$ and $\vec{E}$. Information on the anisotropy of the distribution function is obtained from experiments with light polarized ($||$) or perpendicular ($\perp$) to the applied field E. The absorption at a certain photon energy $\hbar\omega$ depends on the angular distribution of the carriers on a constant energy surface and on the angular form of the transition probability. For this purpose it is necessary to know, within the framework of a $\vec{k}.\vec{p}$ theory, the electron or hole wave functions. Calculations of these effects within a Kane model for InSb have been performed by Baumgardner and Woodruff (1968) and for intervalence band transitions in p-Ge by Christensen (1971, 1973).

Experimentally, the difference in absorption coefficient for light polarized $||$ and $\perp$ to E, $\alpha_{||}$ and $\alpha_{\perp}$ was first observed in p-Ge by Bray and Pinson (1963), Pinson and Bray (1964), Baynham and Paige (1964), and Baynham (1965) with intervalence band transitions. If the carrier distribution functions are expressed by (61), then the difference $\alpha_{||} - \alpha_{\perp}$ gives information on the even parts in the expansion (f_2, . . .), as

$$\frac{\alpha(\vec{E}_o, \vec{a})}{p} = A(\lambda) f_o(\varepsilon) \; \{1 + 0.1 \frac{f_2(\varepsilon)}{f_o(\varepsilon)} \cdot [1 - 3\cos^2(\vec{E}, \vec{a})]\}$$

and

$$\frac{f_2}{f_o} = 10 \; (1 - \alpha_{||}/\alpha_{\perp})/2 + \alpha_{||}/\alpha_{\perp}).$$

Baynham (1965) has also demonstrated, that from $f_2(\varepsilon)$ and $f_o(\varepsilon)$, $f_1(\varepsilon)$ can be derived if j(E) is known. If f_3 is negligible, $f_1(\varepsilon)$ determines the electric current. Baumgardner and Woodruff (1968) gave explicit expressions for $\alpha_{||}$ and $\alpha_{\perp}$ for valence-conduction band transitions in InSb-like semiconductors, based on the wave functions of Kane.

A review of this work up to 1977 was given in Bauer (1978). The main results are the following: due to the effectiveness of hole-optical phonon interactions in p-Ge, the distribution function shows a distinct kink near the optical phonon energy. If a two-temperature model for the distribution function is used (with T_1 for $\varepsilon_1(\vec{k}) < \hbar\omega_{LO}$ and T_2 with $\varepsilon_2(\vec{k}) > \hbar\omega_{LO}$), then T_2 does not differ much from the lattice temperature even in high electric fields. Experiments with p-Ge under uniaxial pressure were performed by Christensen

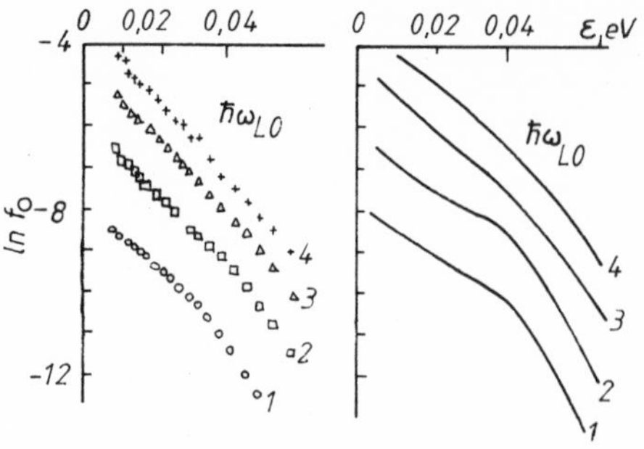

Fig. 25. Determination of hot heavy hole distribution function from intervalence band transitions in Ge. left part: experimental results on f_o for several carrier concentrations (1:p = 6.8 x 10^{14}; 2: 3.7 x 10^{15}; 3: 1.6 x 10^{16}; 4: 4.4 x $10^{16}cm^{-3}$ for μE^2 = const., ranging from 480 . . . 730 V/cm) ($\hbar\omega_{LO}$ = 0.037eV). right part: calculated distribution functions, taking h-h scattering into account. (After Vorobev et al., 1978).

(1971, 1973). Recently Vorobev et al. (1978) have performed experiments on p-Ge in fields up to 6 kV/cm. The kink near $\hbar\omega_{LO}$ in f_o decreases with increasing field and also $\alpha_{||} - \alpha_{\perp}$ decreases in fields in excess of 3 kV/cm. The experiments were compared with Monte Carlo calculations, allowing for optical, acoustical, ionized impurity scattering. For each of these scattering mechanisms, intra- and inter-band scattering was considered. Also, the influence of hole-hole scattering and of impurity concentrations of about 5 x $10^{16}cm^{-3}$ were considered on the distribution functions. As expected, hole-hole scattering also causes the kink in the distribution function f_o in the vicinity of $\hbar\omega_{LO}$ to disappear. At hole densities above $10^{16}cm^{-3}$, and fields below 1 kV/cm, the influence of h-h scattering and of ionized impurity scattering is very strong and f_o is nearly maxwellian. The calculations were performed (including hole-hole collisions) with the method of Matulenis et al. (1975) by simulating two equivalent random trajectories in the momentum space. Results are shown in Fig. 25.

Field induced changes in the fundamental electronic absorption have been used for a determination of the distribution function only in degenerate semiconductors. This method is based on the influence of the carrier heating on the Burstein-Moss shift and is explained in Fig. 26 (Jantsch and Brücker, 1977). In the conduction band of e.g. GaAs, a photon with energy $\hbar\omega$ creates electrons at two different energies ε_l and ε_h since the initial states are either in the light or heavy hole valence band. Due to the anisotropy and warping of the

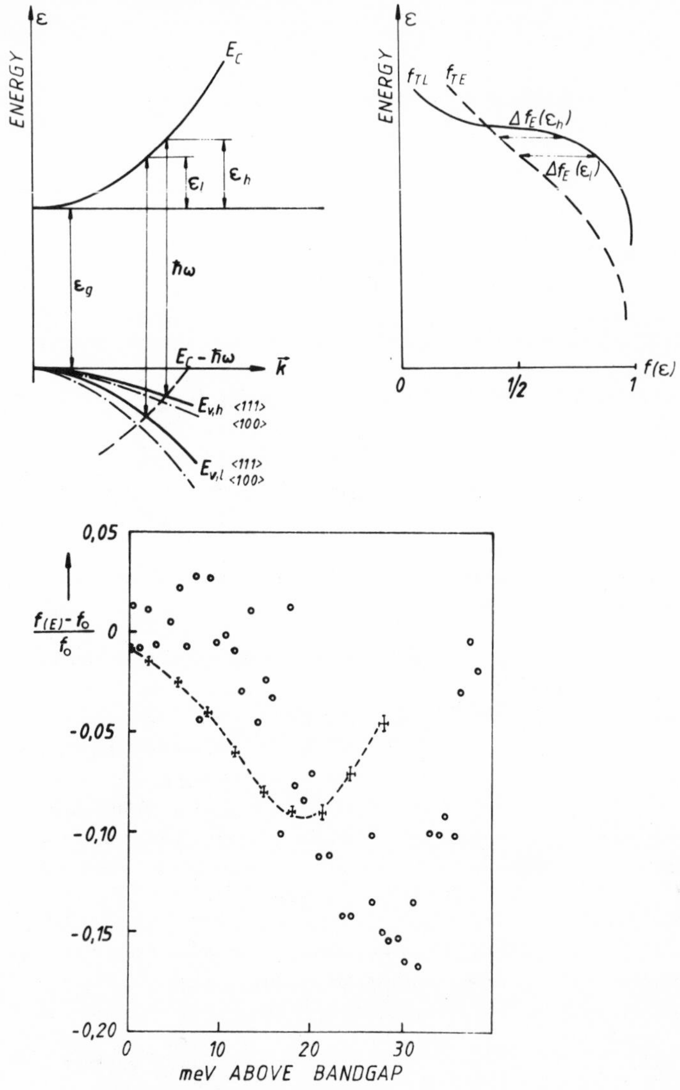

Fig. 26. Interband absorption in an electric field. Upper half: for constant $\hbar\omega$ at two energies ε_1, ε_h carriers in the conduction band are created. Upper right: change of distribution function from the thermal f_{TL} to the hot electron function f_{TE}. The change in distribution function $\Delta f_E(\varepsilon_h)$ and $\Delta f_E(\varepsilon_1)$ is indicated. lower part: relative change of the electron distribution function at 77 K in n-GaAs in an applied field of 900 V/cm (exp: + and dashed curve); circles: results of a Monte Carlo calculation by Bosi and Jacoboni (1976) (for $n = 5 \times 10^{17} cm^{-3}$) (After Jantsch and Brücker, 1977).

valence band, the energies ε_1 and ε_h are somewhat smeared out. For the parabolic approximation of the bands involved, the absorption is given by

$$\alpha(\hbar\omega) = \{\frac{e^2(2m_{r,h})^{\frac{3}{2}}\Phi_h}{4\pi n_r m_o \hbar^2 \varepsilon_o c}[1 - f(\varepsilon_h)] +$$

$$+ \frac{e^2(2m_{r,1})^{\frac{3}{2}}\Phi_1}{4\pi n_r m_o \hbar^2 \varepsilon_o c}[1 - f(\varepsilon_1)]\} \times (\hbar\omega - \varepsilon_g)^{\frac{1}{2}}, \qquad 62)$$

where $(m_{r,h,1})^{-1} = m_c^{-1} + (m_v(h,1))^{-1}$ is the reduced mass. From the field induced change $\Delta\alpha_E(\hbar\omega) = \alpha_E(\hbar\omega) - \alpha_o(\hbar\omega)$, the carrier distribution f_E in the electric field can then be obtained by an iterative procedure with the aid of (62) (Jantsch and Brücker, 1977). In n-GaAs samples with $n = 0.5\text{-}1 \times 10^{18}\text{cm}^{-3}$, it was found that the distribution function deviates considerably from a drifted heated Fermi distribution. Within experimental error, no difference between $\alpha_{||}$ and $\alpha_{\perp}$ was found.

Emission Experiments

The recombination radiation of hot electrons with donors, acceptors and with holes yields also information on the distribution function. Early experiments were performed by Shah and Leite (1964) and by Southgate et al. (1971). Since, for electron-hole recombination studies, the hole distribution function has to be known, the electron-impurity recombination is superior, since it involves only one distribution function. In addition, it is less affected by reabsorption effects. Nevertheless emission due to conduction-valence band recombination has been studied quite often, in addition to conduction band-acceptor or donor transitions, or in p-type materials hole-donor transitions. Reviews of these works can be found in Seiler (1978).

Recently, hot carrier recombination in n-GaAs (carriers excited with a He-Cd laser at 4416Å and subjected to electric fields up to 2 kV/cm) has been analyzed by Inoue et al. (1978) and Takenaka et al. (1979). The analysis was based on the assumption of a Maxwellian distribution for the heavy and light holes. The electron distribution function, derived from the emission spectra, shows distinctly non-Maxwellian behaviour and a kink near the $\hbar\omega_{LO}$. It can be approximated by a two-electron temperature Maxwellian distribution. The experiments were compared with Monte Carlo calcuations (see Fig. 27), but without inclusion of electron-electron scattering.

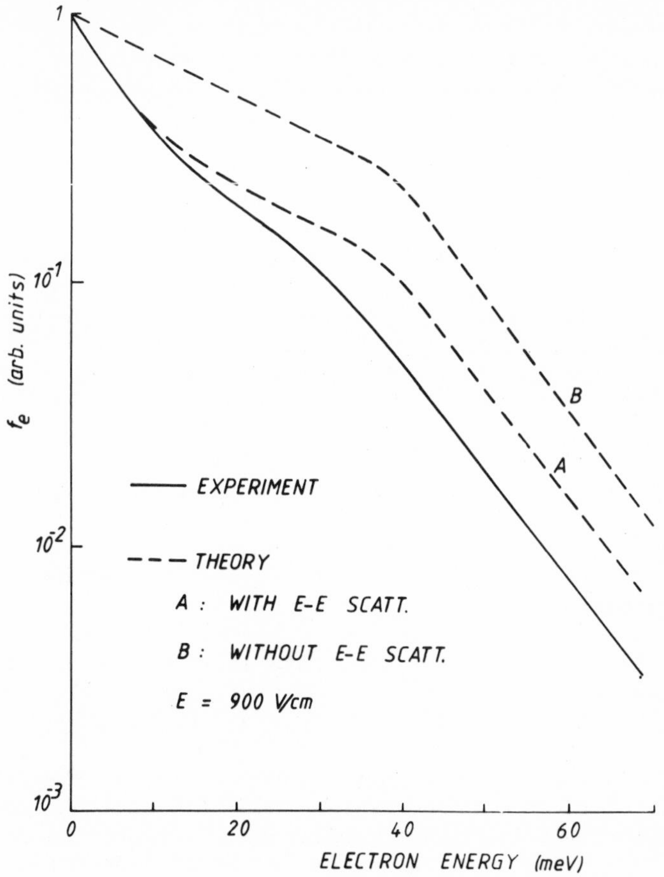

Fig. 27. Electron distribution function in n-GaAs (E = 900 V/cm, n = 5 x $10^{14}cm^{-3}$) derived from luminescence experiments and compared with Monte Carlo calculations. (After Inoue et al., 1978, and Takenaka et al., 1979).

The carrier concentrations of the samples investigated were of the order 0.5 - 5 x $10^{14}cm^{-3}$.

Inelastic Light Scattering

Free carriers make a contribution to the optical polarizability. A change of their properties due to an applied field results in changes of the light scattering. Scattering is caused both from collective excitations (plasmons, coupled phonon-plasmon excitations), but also from single particle scattering. It originates either from charge-density fluctuations, spin-density fluctuations, or, in non-parabolic semiconductors, from energy-density fluctuations (for a review, see Klein, 1969). The method has been used by Mooradian (1975) and Hamilton and McWorther (1969) to study the distribution function of electrons in n-GaAs in fields up to 1400 V/cm. In principle, it should be possible with this technique to study the asymmetry of a non-equilibrium carrier distribution function by varying the scattering geometry and taking advantage of the various possibilities offered by the direction of momentum transfer with respect to the applied field E. The experiments reported so far use just the geometry $\vec{q}_i || \vec{E}$ (where $\vec{q}_i$ denotes the photon wave vector of the incident light). Further experiments on single-particle scattering from hot photoexcited carriers and hot phonons were carried out by the Campinas group and we refer to the review by Leite (1978) and Levinson (1974).

Intraband Absorption and Emission in a Quantizing Magnetic Field

In the chapter on electron temperatures, the effect of a magnetic field on the orbital motion and on the carrier heating has already been discussed. The cyclotron resonance absorption is shown in Fig. 24 for a non-parabolic semiconductor like InSb. The presence of hydrogen-like donor states manifests itself in an additional transition between an impurity ground state (000) and an excited impurity state (110) $\equiv$ (N,M,λ) (Matsuda and Otsuka, 1978, Gornik, 1972, Gornik et al., 1976). An electric field causes a change of the carrier distribution and thus a repopulation of the impurity- and Landau states. The impurity line 000 $\rightarrow$ 110 will decrease in strength while the $0^+ \rightarrow 1^+$ and finally the $0^- \rightarrow 1^-$ transitions will increase. The carriers in the upper levels will lose their energy via non-radiative processes (phonon emission) and partly via a radiative recombination, the so called Landau-emission (Gornik, 1972), the inverse process to cyclotron-resonance absorption. Since the number of carriers in the upper levels (110), 1^+, 1^- will be small under usual experimental conditions ($T_L \simeq 4K$, $E = 0$ up to several V/cm), in the cyclotron-resonance absorption experiment, the occupancy of the (000), 0^+ and 0^- levels will determine the absolute value of the absorption constant, and its frequency dependence for a given B.

In the cyclotron-resonance emission experiment, the occupancy of the (110), 1^+ and 1^- levels determines the absolute emission intensities, since again the occupancy of the lower states is negligible for small carrier concentrations, ($n \approx 10^{14} cm^{-3}$) if the high density-of-states in the magnetic field near $k_z = 0$ is taken into account (Gornik et al., 1976). In a realistic model, the non-parabolicity of the conduction band has to be considered: it is not only responsible for the splitting of the $0^+ \rightarrow 1^+$ and $0^- \rightarrow 1^-$ lines but also causes asymmetries in the absorption or emission line shape: for different k_z-values, the energetic difference between two Landau levels is not the same. Cyclotron-resonance absorption and emission experiments performed together allow a check on the applicability of the electron temperature model (Partl et al., 1978, Muller et al., 1978a, Matsuda and Otsuka, 1979): absorption experiments probe the number of carriers in the 0^+ and 0^- levels, emission the number of carriers in the 1^+ and 1^- levels. If an electron temperature exists, then the number of carriers in the various levels should be determined by T_e and the appropriate energy differences. Additional mobility measurements, yield also an electron temperature which is mainly influenced by the carriers in the lowest level: 0^+ (Partl et al., 1978). However, the experiments usually yield different electron temperatures. There are now several definitions from the above mentioned authors:

(i) the average temperature T_{av} is the equivalent temperature, assuming Maxwellian statistics which corresponds to the average energy

$$\varepsilon_{av} = \int \varepsilon f(\varepsilon) g(\varepsilon) d\varepsilon / \int f(\varepsilon) g(\varepsilon) d\varepsilon,$$

(ii) the relative temperature or intersubband temperature which corresponds to the differences in population of the Landau levels

$$T_{e,1\to o} = \left(\frac{\hbar\omega_c}{k_B}\right) \ln[f(o)/f(\hbar\omega_c)]. \tag{63}$$

This is equivalent to the definition of a temperature T_e^{inter} (Matsuda and Otsuka, 1979) if $n(N^{\pm})$ is the population of the $N^{\pm}$ subband, and the electron distribution function in this subband is $f_{N^{\pm}}(k_z)$. Then

$$n(N^{\pm}) = A_1 \exp\,[-\varepsilon(N^{\pm}, k_z = o)/k_B T_e^{inter}]. \tag{64}$$

Finally, (iii):

$$f_{N^{\pm}}(k_z) = A_2 \exp[-\{\varepsilon(N^{\pm},k_z) - \varepsilon(N^{\pm},k_z = o)\}/k_B T_e^{intra}(N^{\pm})] \tag{65}$$

defines the temperature T_e^{intra}. The constants A_1 and A_2 are determined by the total electron concentration and the electron concentration within a Landau level $n(N^{\pm})$:

$$n = \sum_{N^{\pm}} n(N^{\pm}),$$

$$n(N^{\pm}) = \sum_{k_z} f_{N^{\pm}}(k_z).$$

Whether T_{av}, T_e^{inter}, T_e^{intra} are identical or different from each other, thus indicating that the distribution function is non-Maxwellian, depends not only on the carrier concentration but also on the geometry. For $\vec{E} \perp \vec{B}$, $T_{av} > T_e^{inter}$, $T_e^{intra} > T_e^{inter}$; for $\vec{E} || \vec{B}$: $T_e^{intra} < T_e^{inter}$. This can be deduced from the lineshape of the cyclotron resonances in the two geometries, which is related to T_e^{intra} where as the relative intensity yields T_e^{inter}. The temperatures are obtained by comparison (Matsuda and Otsuka, 1978) or from the total emitted power (Partl et al., 1978, Müller et al, 1978b). If no temperature model is used, then information on the number of carriers within a certain level is obtained from the integrated absorption α or emission P (Gornik et al., 1976, Müller, 1976):

$$\alpha_{i\to f} = \int \alpha(\nu)\,d\nu = \frac{\mu_o e^2 \hbar}{n_r \lambda_B^2 m_o^{*2}} \cdot n\,(N^{\pm})\ x$$

$$\int dk_z |M(k_z)|^2_{i\to f} \frac{f(\varepsilon_i(k_z))}{\omega_{i\to f}(k_z)} \Big/ \int dk_z f(\varepsilon_i(k_z)), \tag{66}$$

$$P_{i\to f} = \int P_{i\to f}(\nu)\,d\nu = \frac{\mu_o e^2 n_r}{4\pi c}(1-\cos\delta_{tot}) \cdot \frac{\hbar^2}{\lambda_B^2 m_o^{*2}}$$

$$x\ n(N^{\pm})\ \frac{V\cdot\Delta\Omega}{2\pi} \int dk_z |M(k_z)|^2_{i\to f}$$

$$x\ \omega^2_{i\to f}(k_z)\ f(\varepsilon_i(k_z))\,| \Big/ \int dk_z f(\varepsilon_i(k_z)), \tag{67}$$

where $|M(k_z)|^2$ denotes the non-parabolic correction to the transiti matrix element, δ_{tot} is the angle of total reflection, V the sample volume, ΔΩ the solid angle. For a comparison with theory, Matsuda and Otsuka (1978) used the model by Yamada and Kurosawa (1973), whi consists of a diffusion equation in energy space. The effect of frequent elastic scattering processes with small energy changes is regarded as a kind of Brownian motion in energy space. The effect of polar scattering is included by a boundary condition: each electron reaching $\hbar\omega_{LO}$ is immediately scattered back to $\varepsilon = o$. The electron heating predicted by this theory is however too strong. Also the deviations are stronger at high magnetic fields. Partl et al. (1978) and Müller et al. (1978b) improved this theory by taking Landau level broadening and electron-electron scattering int account. This scattering mechanism limits the heating effect at higher electric fields quite effectively. In the energy diffusion model of Yamada and Kurosawa (1973), electron-electron scattering i included using an iterative method, using the transition probabilities derived by Zlobin and Zyraynov (1970). It could be shown that electron-electron scattering is also the decisive mechanism which determines the lifetime of the carriers in the $N = 1^{\pm}$ Landau levels. This process leads to lifetimes as short as 10^{-10}s for $n = 10^{13}cm^{-3}$ in the first Landau level (Müller et al., 1978a). Indications for the effectiveness of electron-electron scattering in Auger-like processes are also given in Dornhaus et al. (1976) and were studied theoretically by Lewiner and Calecki (1978).

CONCLUSIONS

The study of hot carriers in semiconductors is by now a mature field. A basic understanding of the physics of nonlinear transport in bulk materials has been achieved. The new and exciting fields are certainly those associated with reduced dimensionality, picosecond phenomena, and optically-excited hot carriers. Studies of hot electrons in artificial structures (superlattices) are also very promising. There has been, for a long time, a field where experimental progress has been rather slow: the disturbance of the phonon distribution which is associated with the non-equilibrium carrier distribution. Recently, this subject has also been tackled successfully (Narayanamurti et al., 1978, Reupert et al., 1976) and a more complete understanding of non-equilibrium phenomena especially at low lattice temperature might emerge from these studies (theoretically, this problem has been treated by several authors. See e.g. Ferry, 1973, and references therein).

BIBLIOGRAPHY

Acket, G. A., and de Groot, J., 1967, IEEE Trans. Electron Dev. ED-14:505.
Acket, G. A., and Vlaardingerbrock, M. T., 1969, in "Festkorperprobleme IX," Ed. by O. Madelung, Vieweg, p. 280.
Alberga, G. E., 1978, Ph.D. Thesis, Tech. Univ. Eindhoven, unpublished.
Alberigi-Quaranta, A., Jacoboni, C., and Ottaviani, G., 1971, Rev. Nuovo Cim. 1:445.
Almasov, L. A., 1972, Phys. Stat. Sol.(b) 54:87.
Almasov, L. A., 1973, Phys. Stat. Sol.(b) 58:821.
Almasov, L. A., and Dykman, M. I., 1971, Phys. Stat. Sol.(b) 48:563.
Almasov, L. A., and Dykman, M. I., 1972, Phys. Stat. Sol.(b) 51:751.
Ancker-Johnson, B., 1967, J. Phys. Soc. Jpn. Suppl., 22:1156.
Asche, M., 1970, Phys. Stat. Sol. 41:67.
Asche, M., and Sarbej, O. G., 1969, Phys. Stat. Sol. 33:9.
Asche, M., and Sarbej, O. G., 1971, Phys. Stat. Sol.(a) 8:K61.
Balkanski, M., and Amzallag, E., 1968, Phys. Stat. Sol. 30:407.

Barker, J. R., 1972, J. Phys. C 5:1167.
Barker, J. R., 1978, Sol.-State Electr. 21:197.
Barker, J. R., and Magnusson, B., 1974, in "Proc. Intern. Conf. Phys Semicond., Stuttgart," Ed. by M. H. Pilkuhn, Teubner, Stuttgart p. 811.
Bauer, G., 1974, in "Springer Tracts in Modern Phys.," Ed. by G. Höhler, Springer-Verlag, Berlin, 74:1.
Bauer, G., 1978, Sol.-State Electr. 21:17.
Bauer, G., and Heinrich, H., 1968, J. Sci. Instr. 1:688.
Bauer, G., and Kahlert, H., 1972, Phys. Rev. B 5:566.
Bauer, G., et al., 1973, J. Phys. E 6:186.
Baumgardner, C. A., and Woodruff, T. O., 1968, Phys. Rev. 173:746.
Baynham, A. C., 1965, Sol. State Commun. 3:253.
Baynham, A. C., and Paige, E. G. S., 1963, Phys. Lett. 6:7.
Baynham, A. C., and Paige, E. G. S., 1964, in "Proc. Intern. Conf. Phys. Semicond., Paris," Ed. by M. Hulin, Dunod, Paris, p. 149
Bludau, W., and Wagner, E., 1976, Appl. Phys. Lett. 29:204.
Bludau, W., Wagner, E. and Quiesser, H. J., 1976, Sol. State Commun 18:861.
Bluyssen, H. J. A., Maan, J. C., van Ruyven, L. J., Williams, F., and Wyder, P., 1978, Sol. State Commun. 25:895.
Bonek, E., 1972, J. Appl. Phys. 43:5101.
Bonek, E., Pötzl, H. W., and Richter, K., 1970, J. Phys. Chem. Sol. 31:1151.
Bosch, B. G., and Englemann, R. W., 1975, "Gunn Effect Electronics," Pitman, London.
Bosi, S., and Jacoboni, C., 1976, J. Phys. C 9:315.
Braslau, N., and Hauge, P. S., 1970, IEEE Trans. Electr. Dev. ED-17:616.
Brauer, M., Kantelberg, G., Kliefath, K., and Petzel, B., 1977, in "Probleme der Festkörperelektronik," 9:166.
Bray, R., and Pinson, W. E., 1963, Phys. Rev. Lett. 11:502.
Brazis, R. S., and Pozhela, Yu. K., 1975, Sov. Phys.-Semicond. 9:11
Burstein, E., 1954, Phys. Rev. 93:632.
Butcher, P. N., 1967, Rept. Prog. Phys. 30:97.
Butcher, P. N., Fawcett, W., and Hilsum, C., 1966, Brit. J. Appl. Phys. 17:841.
Calecki, D., 1979, lecture in this volume.
Carroll, J. E., 1970, "Hot Electron Microwave Generators," Arnold, London.
Champlin, K. S., Armstrong, D. B., and Gunderson, P. D., 1964, Proc IEEE 52:644.
Chang, D. M., and Moll, J. L., 1966, Appl. Phys. Lett. 9:283.
Chattopadhyay, D., and Nag, B. R., 1970, Phys. Stat. Sol. 40:701.
Christensen, O., 1971, Ph.D. thesis, Tech. Univ. Denmark, unpublish
Christensen, O., 1973, Phys. Rev. B 7:763.
Conwell, E. M., 1967, "High Field Transport in Semiconductors," Academic, New York.

Copeland, J. A., and Knight, S., in "Semiconductors and Semimetals," Ed. by R. K. Willardson and A. C. Beer, Academic, New York, 7:3.
Crandall, R. S., 1970, Phys. Rev. B 1:730.
Dargys, A., Sedrakyan, R., and Pozhela, J., 1978, Phys. Stat. Sol.(a) 45:387.
Dienys, V., and Kancleris, Z., 1975, Phys. Stat. Sol.(b) 67:317.
Dienys, V., and Pozhela, Yu. K., in "Hot Electrons," Lith. Ac. Sci., Vilnius.
Dienys, V., Gintilas, Zh., and Martunus, Z., 1977, Sov. Phys. Semicond. 11:945.
Dornhaus, R., Müller, K. H., Nimtz, G., and Schifferdecker, M., Phys. Rev. Lett. 37:710.
Dykman, I. M., and Tomchuk, P. M., 1961, Sov. Phys.-Sol. State 2:1988.
Dykman, I. M., and Tomchuk, P. M., 1962, Sov. Phys.-Sol. State 3:1393.
Elliott, B. J., 1968, IEEE Trans. Instr. Meas. IM-17:330.
Evans, A. G. R., Robson, P. N., and Stubbs, M. G., 1972, Electr. Lett. 8:195.
Fawcett, W., 1975, in "Electronics in Crystalline Solids," Ed. by A. Salam, IAEA, Vienna, p. 531.
Fay, B., and Kino, G. S., 1969, Appl. Phys. Lett. 15:337.
Ferry, D. K., 1973, Phys. Rev. B 8:1544.
Ferry, D. K., Heinrich, H., Keeler, W., and Müller, E. A., 1973, Phys. Rev. B 8:1538.
Foyt, A. G., and McWorther, J., 1966, IEEE Trans. Electr. Dev. ED-13:79.
Franz, W., 1958, Z. Naturforsch. 13a:484.
Fröhlich, H., and Paranjap, B. V., 1956, Proc. Phys. Soc. (London) B69:21.
Fukui , H., 1966, Proc. IEEE 54:796.
Gershenzon, E. M., 1974, in "Proc. Intern. Conf. Phys. Semicond., Stuttgart," Ed. by M. H. Pilkuhn, Teubner, Stuttgart, p. 355.
Gibson, A. E., Granville, J. W., and Paige, E. G. S., 1961, J. Phys. Chem. Sol. 19:198.
Glicksman, M., 1972, in "Plasma Effects in Solids", Ed. by F. Seitz, D. Turnbull, and H. Ehrenreich, Academic, New York.
Glicksman, M., and Steele, M. C., 1958, Phys. Rev. 110:1204.
Glover, G. H., 1971, J. Appl. Phys. 42:4025; Appl. Phys. Lett. 18:290.
Glover, G. H., 1972, Appl. Phys. Lett. 20:224.
Glover, G. H., 1973, J. Appl. Phys. 44:1295.
Goebel, E. O., 1979, in "Festkörperprobleme," Ed. by J. Treusch, Vieweg, Braunschweig, 19:in press.
Goebel, E. O., and Hildebrand, O., 1979, Phys. Stat. Sol.(b) in press.
Gornik, E., 1972, Phys. Rev. Lett. 29:595.
Gornik, E., Bauer, G., and Müller, W., 1976, "Intern. Conf. Appl. High Magnetic Fields in Semiconductor Physics," Ed. by G. Landwehr, Wurzburg, p. 221.

Gram, N. O., 1972, Phys. Lett. A 38:235.
Grubin, H. L., Shaw, M. P., and Solomon, P. R., 1973, IEEE Trans. Electr. Dev. ED-20:63.
Gueron, M., 1964, in "Proc. Intern. Conf. Phys. Semicond., Paris," Ed. by M. Hulin, Dunod, Paris, p. 433.
Gulbrandsen, T., Meyer, N. I., and Schjaer-Jacobsen, 1965, Rev. Sci Instr. 36:743.
Gunn, J. B., 1957a, in "Progress in Semiconductors," Ed. by A. F. Gibson, Wiley, New York, 8:213.
Gunn, J. B., 1957b, J. Electr. 2:213.
Gunn, J. B., 1963, Sol. State Commun. 1:88.
Gunn, J. B., 1964, in "Plasma Effects in Solids," Ed. by J. Bok, Dunod, Paris, p. 199.
Gunn, J. B., and Elliott, B. J., 1966, Phys. Lett. 22:369.
Hamaguchi, C., and Inuishi, Y., 1966, J. Phys. Chem. Sol. 27:1511.
Hamilton, D. C., and McWorther, A. L., 1969, in "Light Scattering Spectra of Solids," Ed. by G. B. Wright, Springer-Verlag, Berlin, p. 309.
Harper, P. G., Hodby, J. W., and Stradling, R. A., 1973, Rept. Prog. Phys. 36:1.
Hartnagel, H., 1968, "Semiconductor Plasma Instabilities," Heineman London.
Hasegawa, A., and Yamashita, J., 1962, J. Phys. Soc. Jpn. 17:1751.
Hayes, R. E., and Raymond, R. M., 1977, Appl. Phys. Lett. 31:300.
Hearn, C. J., 1965, Proc. Phys. Soc. (London) 86:881; Phys. Lett. 20:113.
Heinrich, H., 1971, Phys. Rev. B 3:416.
Heinrich, H., and Jantsch, W., 1969, Sol. State Commun. 7:377.
Heinrich, H., and Jantsch, W., 1971, Phys. Rev. B 4:2404.
Heinrich, H., and Keeler, W., 1972, Appl. Phys. Lett. 21:171.
Hess, K., and Kahlert, H., 1970, J. Phys. Chem. Sol. 32:2262.
Hess, K., and Seeger, K., 1969, Z. Phys. 218:431.
Hess, K., and Seeger, K., 1970, Z. Phys. 237:252.
Hildebrand, O., Goebel, E. O., Romanek, K. M., Weber, H., and Mahler, G., 1978, Phys. Rev. B 17:4775.
Hilsum, C., 1962, Proc. IRE 50:185.
Hilsum, C., 1972, in "Proc. Intern. Conf. Phys. Semicond., Warsaw," Ed. by M. Miasek, Polish Sci. Publ., Warsaw, p. 585.
Hilsum, C., 1978, Sol.-State Electr. 21:5.
Hobson, G. S., 1973, J. Phys. E 6:229.
Hobson, G. S., 1974, "The Gunn Effect," Clarendon Press, Oxford.
Holm-Kennedy, J. W., and Champlin, K. S., 1972, J. Appl. Phys. 43:1878; 43:1890.
Houghton, J. T., and Smith, S. D., "Infrared Physics," Clarendon Press, Oxford, p. 131.
Hoult, R. A., 1975, Ph.D. Thesis, Oxford, unpublished.
Houston, B., Restorff, J. B., Allgaier, R. S., Burke, J. R., Ferry, D. K., and Antypas, G. A., 1978, Sol.-State Electr. 21:91.

Inoue, M., Ashida, K., Sugino, T., Shirafuji, Y., and Inuishi, Y., 1973, Jpn. J. Appl. Phys. 12:932.
Inoue, M., Takenaka, N., Shirafuji, Y., and Inuishi, Y., 1978, Sol.-State Electr. 21:1527.
Ipatova, I. P., Kazarinov, P. E., and Shubashiev, A. V., 1966, Sov. Phys.-Semicond. 7:1714.
Jacoboni, C., 1976, in "Proc. Intern. Conf. Phys. Semiconductors, Rome," Ed. by F. G. Fumi, Tipografia Marves, Rome, p. 1195.
Jacoboni, ., Canali, C., Ottaviani, G., and Alberigi-Quaranta, A., 1977, Sol.-State Electr. 20:77.
Jantsch, W., and Brücker, H., 1977, Phys. Rev. B 15:4014.
Jantsch, W., and Heinrich, H., 1970, Rev. Sci. Instr. 41:228.
Jantsch, W., and Heinrich, H., 1971, Phys. Rev. B 3:420.
Johnson, E. J., and Dickey, D. H., 1970, Phys. Rev. B 1:2676.
Jorgensen, M. H., Gram, N. O., and Meyer, N. I., 1972, Sol. State Commun. 10:337.
Kahlert, H., 1975, Phys. Stat. Sol.(b) 71:151.
Kahlert, H., and Seiler, D. G., 1977, Rev. Sci. Instr. 48:1017.
Kaplan, D., and Konopka, J., 1968, in "Proc. Intern. Conf. Phys. Semicond., Moscow," Ed. by S. Ryvkin, Nauka, Leningrad, p. 1146.
Kasai, K., Shirakawa, T., and Hamaguchi, C., 1978, J. Phys. Soc. Jpn. 44:216.
Keldysh, L. V., 1958, Sov. Phys.-JETP 7:788.
Kinch, M. A., 1966, Brit. J. Appl. Phys. 17:1257.
Kinch, M. A., 1967, Proc. Phys. Soc. (London) 90:819.
Kino, G. S., and Robson, P. N., 1968, Proc. IEEE 56:2057.
Klein, M. V., 1975, in "Light Scattering in Solids," Ed. by M. Cardona, Springer-Verlag, Berlin, p. 148.
Kobayashi, K. L. I., and Otsuka, E., 1972, Sol. State Commun. 11:815; also in "Proc. Intern. Conf. Phys. Semicond., Warsaw," Ed. by M. Miasek, Polish Sci. Publ., Warsaw, p. 903.
Kogan, Sh. M., Shadrin, V. D., and Shulman, Y. Ya., 1976, Sov. Phys.-JETP 41:686.
Konopka, J., 1970, Phys. Rev. Lett. 24:666.
Kriechbaum, M., Lischka, K., Kuchar, F., and Heinrich. H., 1972, in "Proc. Intern. Conf. Phys. Semicond., Warsaw," Ed. by M. Miasek, Polish Sci. Publ., Warsaw, p. 615.
Kroemer, H., 1966, IEEE Trans. Electr. Dev. ED-13:27.
Kubo, R., in "100 Years of the Boltzmann Equation," Ed. by M. H. Cohen and W. Thirring, Springer-Verlag, Vienna, p. 301.
Kuchar, F., Philipp, A., and Seeger, K., 1972, Sol. State Commun. 11:965.
Lagois, J., Wagner, E., Bludau, W., and Lösch, K., 1978, Phys. Rev. B 18:4325.
Landsberg, P. T., 1965, in "Lectures in Theoretical Physics," Boulder, Colo., 8A:313.
Larrabee, R. D., 1959, J. Appl. Phys. 30:857.
Leheny, R. F., and Shah, J., 1977, Phys. Rev. B 16:1577.
Leheny, R. F., and Shah, J., 1978, Sol.-State Electr. 21:167.

Leite, R. C. C., 1978, Sol.-State Electr. 21:177.
Levinshtein, M. E., and Shur, M. S., 1975, Sov. Phys.-Semicond. 9:411.
Levinson, I. B., 1974, Sov. Phys.-Semicond. 7:1121.
Lewiner, C., and Calecki, D., 1978, Sol.-State Electr. 21:185.
Lifschitz, T. M., Oleinikov, A. Ya., Romanovtsev, V. V., and Shulman, A. Ya., 1972, in "Proc. Intern. Conf. Phys. Semicond., Warsaw, "Ed. by M. Miasek, Polish Sci. Publ., Warsaw, p. 608.
Löschner, H., 1972a, J. Appl. Phys. 43:3585.
Löschner, H., 1972b, Ph.D. Thesis, Univ. of Vienna, unpublished.
Löschner, H., Zimmerl, O., Frank, K., Hillbrand, H., König, W., and Pötzl, H., 1972, in "Proc. Intern. Conf. Phys. Semicond., Warsaw," Ed. by M. Miasek, Polish Sci. Publ., Warsaw, p. 630.
Maneval, J. P., Zylberstejn, A., and Budd, H. F., 1969, Phys. Rev. Lett. 23:848.
Matsuda, O., and Otsuka, E., 1978, Sol. State Commun. 26:925.
Matsuda, O., and Otsuka, E., 1979, J. Phys. Chem. Sol. in press.
Matrilenis, A., Pozhela, Yu. K., and Reklaitis, A. S., 1975, Sol. State Commun. 16:1133.
McGroddy, J. C., 1970, IEEE Tran. Electr. Dev. ED-17:207.
McGroddy, J. C., and Nathan, M. I., 1967, IBM J. Res. Develop. 11:337.
McGroddy, J. C., Nathan, M. I., and Smith, J. E., 1969, IBM J. Res. Develop. 13:543.
Miyazawa, H., 1969, J. Phys. Soc. Jpn. 26:700.
Miyazawa, H., and Ikoma, H., 1967, J. Phys. Soc. Jpn. 23:290.
Mooradian, A., in "Laser Handbook," Ed. by F. T. Arecchi and O. E. Schulz-Dubois, North-Holland, Amsterdam, p. 1409.
Moore, B. T., Seiler, D. G., and Kahlert, H., 1978, Sol.-State Electr. 21:247.
Morgan, T. N., 1959, J. Phys. Chem. Sol. 8:245.
Morgan, T. N., and Kelly, C. E., 1965, Phys. Rev. 137:A1573.
Müller, W., 1976, Ph.D. Thesis, Techn. Univ. of Vienna, unpublished.
Müller, W., Gornik, E., Bridges, T. J., and Chang, T. Y., 1978a, Sol.-State Electr. 21:1455.
Müller, W., Kohl, F., Partl, H., and Gornik, E., 1978b, Sol.-State Electr. 21:235.
Nag, B. R., 1972, "Theory of Electrical Transport in Semiconductors," Pergamon, Oxford.
Narayanamurti, V., Logan, R. A., Chin, M. A., and Lax, M., 1978, Sol.-State Electr. 21:1295; Phys. Rev. Lett. 40:63.
Neukermans, A., and Kino, G. S., 1973, Phys. Rev. B 7:2693,2703.
Obiditsch, M., and Kahlert, H., 1976, Phys. Stat. Sol.(b) 77:677.
Ohtomo, M., 1968, Jpn. J. Appl. Phys. 7:1368.
Paige, E. G. S., 1964, in "Progress in Semiconductors," Ed. by A. F. Gibson, Wiley, New York, 2:213.
Parfenev, R. V., Kharus, G. I., Tsidilkovskii, I. M., and Shalyt, S. S., 1974, Sov. Phys.-Uspekhi 17:1.
Partl, H., Müller, W., Kohl, F., and Gornik, E., 1978, J. Phys. C 11:1091.

Peskett, G. D., and Roland, V. B., 1963, Proc. Phys. Soc. (London) 82:467.
Peterson, R. L., 1975, in "Semiconductors and Semimetals," Ed. by R. K. Willardson and A. C. Beer, Academic, New York, 10:221.
Philipp, A., Kuchar, F., and Seeger, K., 1977, Phys. Stat. Sol.(b) 79:115.
Pinson, W. E., and Bray, R., 1964, Phys. Rev. 136:A1449.
Price, P., 1978, Sol.-State Electr. 21:9.
Prior, A. C., 1959, J. Phys. Chem. Sol. 12:175.
Racek, W., and Bauer, G., 1976, Phys. Rev. Lett. 37:1032.
Reggiani, L., 1979, lecture in this volume.
Reupert, W., Lassmann, K., and de Groot, P., 1976, in "Phonon Scattering in Solids," Ed. by L. J. Challis, Plenum, New York, p. 315.
Ridley, B. K., 1977, J. Appl. Phys. 48:754.
Ridley, B. K., and Watkins, T. B., 1961, Proc. Phys. Soc. (London) 78:293.
Rohlfing, D. C., and Prohofsky, E. W., 1975, Phys. Rev. B 12:3242.
Romanovtsev, V. V., and Shulman, A. Ya., 1974, Sov. Phys.-Semicond. 8:364.
Rowell, J. M., and Tsui, D. C., 1976, Phys. Rev. B 14:2456.
Ruch, J. G., and Kino, G. S., 1967, Appl. Phys. Lett. 10:40.
Ruch, J. G., and Kino, G. S., 1968, Phys. Rev. 174:921.
Ruthroff, C. L., 1959, Proc. IRE 47:1337.
Ryder, E. J., 1953, Phys. Rev. 90:766.
Ryder, E. J., and Shockley, W., 1951, Phys. Rev. 81:139.
Sandercock, J. R., 1965, Proc. Phys. Soc. (London) 86:1221.
Sandercock, J. R., 1969, Sol. State Commun. 7:721.
Sasaki, W., and Shibuya, M., 1956, J. Phys. Soc. Jpn. 11:1202.
Sasaki, W., Shibuya, M., and Mizuguchi, K., 1958, J. Phys. Soc. Jpn. 13:456.
Sasaki, W., Shibuya, M., Mizuguchi, K., and Hatoyama, G. M., 1959, J. Phys. Chem. Sol. 8:250.
Schmidt-Tiedemann, K. J., 1961, Phys. Rev. Lett. 7:732.
Schmidt-Tiedemann, K. J., 1962, in "Festkörperprobleme," Ed. by F. Sauter, Vieweg, Braunschweig, 1:122.
Schmidt-Tiedemann, K. J., 1963, Philips Res. Repts. 18:338.
Schneider, W., and Seeger, K., 1966, Appl. Phys. Lett. 8:133.
Sclar, N., and Burstein, E., 1957, J. Phys. Chem. Sol. 2:1.
Seeger, K., 1959, Phys. Rev. 114:476.
Seeger, K., 1963a, Z. Phys. 172:68.
Seeger, K., 1963b, J. Appl. Phys. 34:1608.
Seeger, K., 1973, "Semiconductor Physics," Springer-Verlag, Vienna.
Seifert, F., 1965, Proc. IEEE 53:752.
Seiler, D. G., Editor, 1978, "Proc. 1977 Hot Electron Conf., Denton, Texas, in Sol.-State Electr. 21:1-323.
Seiler, D. G., Barker, J. R., and Moore, B. T., 1978, Phys. Rev. Lett. 41:319.
Shah, J., 1978, Sol.-State Electr. 21:43.

Shah, J., and Leite, R. C. C., 1964, Phys. Rev. Lett. 22:1304.
Shirakawa, T., Hamaguchi, C., and Nakai, J., 1973, J. Phys. Soc. Jpn. 35:1098.
Shockley, W., 1951, Bell Sys. Tech. J. 30:990.
Shur, M. S., 1969, Phys. Lett. 29A:490.
Southgate, P. D., Hall, D. S., and Dreeben, A. B., 1971, J. Appl. Phys. 42:2868.
Stradling, R. A., and Wood, R. A., 1968, J. Phys. C 1:1711.
Stratton, R., 1957, Proc. Roy. Soc. (London) A242:355.
Stratton, R., 1958, Proc. Roy. Soc. (London) A246:306
Takenaka, N., Inoue, M., Shirafuji, J., and Inuishi, Y., 1979, J. Phys. Soc. Jpn. in press.
Thim, H., 1971, in "Solid State Devices," Inst. of Physics, London, p. 87.
Thim, H., and Pötzl, H. W., 1976, in "Solid State Devices," Inst. of Phys., London, 32:73.
Ulbrich, R. G., 1978a, Sol.-State Electr. 21:51.
Ulbrich, R. G., 1978b, in "Proc. Intern. Conf. Phys. Semicond., Edinburgh," Ed. by B. L. H. Wilson, Inst. of Phys., London, p. 11.
Ulbrich, R. G., 1979, lecture in this volume.
van Welzenis, R. G., 1972, Ph.D. Thesis, Techn. Univ. Eindhoven, unpublished.
van Welzenis, R. G., and Sens, A. F. C., 1971, Rev. Sci. Instr. 42:722.
Vasko, F. T., 1973, Sov. Phys.-Solid State 15:1136.
von der Linde, D., and Lambrich, R. G., 1979, Phys. Rev. Lett. 42:1090.
Vorobev, L. E., Komissarov, V. S., and Stafeev, V. I., 1971, JETP Lett. 13:98.
Vorobev, L. E., Komissarov, V. S., and Stafeev, V. I., 1972a, Phys. Stat. Sol.(b) 52:25.
Vorobev, L. E., Stafeev, V. I., and Ushakov, A. Yu., 1972b, Phys. Stat. Sol.(b) 53:431.
Vorobev, L. E., Komissarov, V. S., and Stafeev, V. I., 1972c, Sov. Phys.-Semicond. 6:1009.
Vorobev, L. E., Stafeev, V. I., and Ushakov, A. Yu., 1973, Sov. Phys.-Semicond. 7:624.
Vorobev, L. E., Pozhela, Yu. K., Reklaitis, A. S., Smirnitskaya, E. S., Stafeev, V. I., and Fedortsov, A. B., 1978, Sov. Phys.-Semicond. 12:433, 440.
Whalen, J. J., and Westgate, C. R., 1970, IEEE Trans. Electr. Dev. ED-17:30.
Wood, V. E., 1969, J. Appl. Phys. 40:3740.
Yamada, E., and Kurosawa, T., 1973, J. Phys. Soc. Jpn. 34:603.
Yao, T., Inagaki, K., and Maekawa, S., 1975, J. Phys. Soc. Jpn. 38:1394.
Zawadski, W., 1973, in "New Developments in Semiconductors," Ed. by P. R. Wallace, Nordhoff, Leyden, p. 441.

Zawadzki, W., Bauer, G., Racek, W., and Kahlert, H., 1976, Phys. Rev. Lett. 37:1032.
Zlobin, A. M., and Zaryanov, P. S., 1970, Sov. Phys.-JETP 31:513.
Zlobin, A. M., and Zaryanov, P. S., 1972, Sov. Phys.-Uspekhi 14:379.
Zucker, J., Fowler, V. J., and Conwell, E. M., 1961, J. Appl. Phys. 32:2606.

TIME-OF-FLIGHT TECHNIQUES

Lino Reggiani

Gruppo Nazionale di Struttura della Materia
Istituto di Fisica, Università di Modena
Modena, Italy

INTRODUCTION

The basic ideas concerning time-of-flight techniques (ToF) as applied to the study of transport phenomena in semiconductors were first presented by Haynes and Shockley (1951) in their pioneer experiments. The most important quantity which results as output of the measurement is the value of the transit time T_R, that is the time taken by the charge carriers to travel across a given region of the sample under the influence of a known electric field. Thus, the ToF technique is usually synonymous with transit-time measurements.

After the Haynes and Shockley experiment, which was carried out at Bell Laboratories, other groups adopted this technique, and several refinements and improvements were introduced to avoid the limitations of the original set-up. Among these we cite, in chronological order, Spear (1957), Ruch and Kino (1968) and Sigmon and Gibbons (1969), and Alberigi-Quaranta, et al. (1970a).

At present, most of the current activity in this field comes from the Modena group, thus in the following, we will concentrate on the apparatus used at Modena University.

DESCRIPTION OF ToF TECHNIQUES

The schematic principle of ToF techniques is illustrated in Fig. 1. An ionizing radiation (of sufficient energy to generate electron-hole pairs) hits one side of a two terminal semiconductor device of length W with non-injecting contacts at its opposite sides (see Fig. 1a). The presence of a uniform electric field (see Fig. 1b)

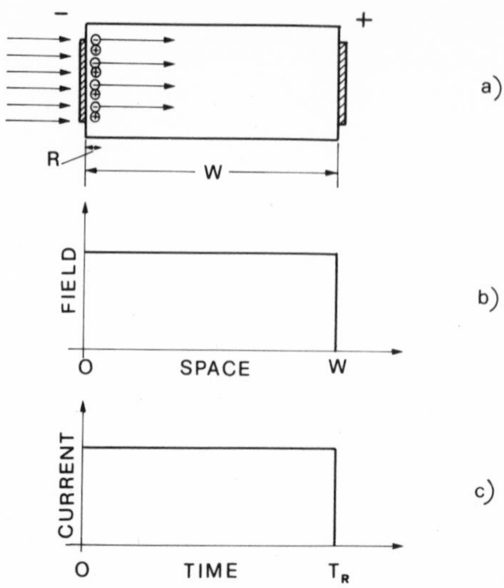

Fig. 1. Time-of-flight technique. a) An ionizing radiation creates charge pairs in a narrow region R of the sample close to one contact. b) Electric field inside the sample under ideal conditions. c) Current signal induced at the contacts by the carriers during their travel across the sample.

enables one type of carrier to travel across the region W of the sample while the other type is swept toward the opposite contact and gives a negligible contribution to the current. During the carrier travel, an induced current signal I(t) (current waveform) at the terminals of the sample is measured as a function of time (see Fig. 1c). The analysis of the current waveform enables one to achieve various kinds of information (as will be reported in next section).

The different types of radiation which have been reported in the literature and their main characteristics are summarized in Table 1. Table 2 summarizes the definitions of the different times of utility in analyzing the current signal.

Besides the requirement of a fast electronic apparatus characterized by a circuit rise time τ_c, the most severe operating conditions for ToF techniques concern: i) the characteristics of the ionizing radiation, and ii) the properties of the material under examination:

i) The ionizing radiation must be such as to create the charge pairs in a time much shorter than the transit time T_R, and in a range R much smaller than the sample length W.

Table 1. Ionized Radiations Used in ToF Techniques

Type	Duration (nsec)	Energy (keV)	Range (μm)
Low-energy heavy charged particles[1]	0.01 - 2	<5.5 10^3	30
-rays[2]		60	
Short light flashes[3]	2 10^3	10	0.1 - 1
Electron bursts[4]	0.07 - 1	7 - 50	3 - 10

1. Alberigi-Quaranta, et al. (1965)
2. Martini and McMath (1969)
3. Kepler (1960)
4. Alberigi-Quaranta, et al. (1970a)

Table 2. Definition of the times of interest in ToF. The order of magnitude is reported in parenthesis.

Electronic	τ_c = circuitry raise-time[1]	(~6 10^{-11} sec)
Ionizing radiation	τ_s = stopping time[2] τ_p = plasma time[1,3]	(10^{-11} - 10^{-12} sec) (10^{-11} - 10^{-7} sec)
Material	τ_ℓ = life-time[4] τ_ε = differential dielectric relaxation time[5]	($\geq T_R$) ($\gg T_R$)
Measured	T_R = transit time[6] τ_f = fall time[7] τ_r = rise time[7]	(5 10^{-9} - 10^{-7} sec) ($\geq$ 10^{-10} sec) ($\geq$ 10^{-10} sec)
Microscopic	τ^+ = mean drift time[6] τ_d = detrapping time[6] τ_e = energy relaxation time[8] τ_m = momentum relaxation time[8] τ_i, i = (ac., op.,...) = scattering time[8] τ_{th} = thermalization time[7]	(> 5 10^{-9} sec) (> 10^{-9} sec) (10^{-11} sec) (10^{-12} sec) (10^{-12} sec) (10^{-10} sec)

1. Taroni and Zanarini (1969a)
2. Alberigi-Quaranta, et al. (1969)
3. Taroni and Zanarini (1969b)
4. Smith (1959)
5. Gunn and Elliott (1966)
6. Alberigi-Quaranta, et al. (1971)
7. Reggiani, et al. (1978)
8. Conwell (1967)

ii) The lifetime τ_ℓ of the mobile carriers and the differential relaxation time τ_ε of the sample must be longer than the transit time T_R.

Provided conditions i) and ii) are satisfied, the current signa I(t) induced by the carrier motion is given by (see Fig. 1c):

$$I(t) = \begin{cases} Nqv_d/w, & \text{for } t < T_R. \\ 0, & \text{for } t > T_R. \end{cases} \qquad (1)$$

where N is the number of charge carriers of a given sign created by the radiation, q is the unit charge, and v_d is the drift velocity.

The duration in time of the signal is given by the transit time of the carriers across the sample. The current waveforms observed experimentally may be different from that predicted by Eq. (1), and this may be due to:

i) Non-uniform electric field inside the sample;

ii) Space charge limited currents present when the density of the carriers created is so high as to perturb the electric field inside the sample (plasma effects);

iii) Phenomena due to trapping and detrapping;

iv) Phenomena due to thermal diffusion.

Figure 2 shows the two situations when trapping, but not detrapping, occurs (see Fig. 2b), and when trapping and detrapping are active simultaneously with characteristic times shorter than the transit time. Figure 3 shows the effect of thermal diffusion on transit-time measurements. The spatial spreading of the initial she of carriers at increasing time (see Fig. 3a) leads to a fall time τ which differs from the rise time τ_r.

TYPES OF INFORMATION ATTAINABLE FROM ToF TECHNIQUES

The analysis of the current waveform enables several pieces of information to be obtained. This chapter will focus on the two mos important quantities which can be deduced, namely: the drift veloc ty v_d and the longitudinal diffusion coefficient D_ℓ. Other applica tions of ToF technique will be briefly surveyed in Sec. 5.

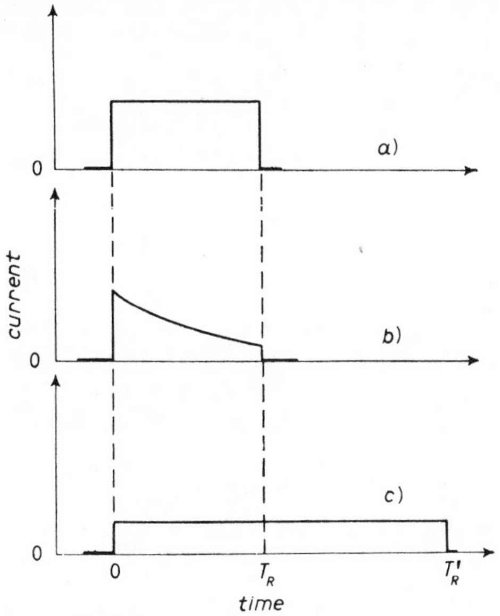

Fig. 2. Possible current waveforms obtained with ToF measurements. a) No trapping is present. b) Trapping but no detrapping occurs. c) Trapping and detrapping are present with characteristic times much shorter than the transit time.

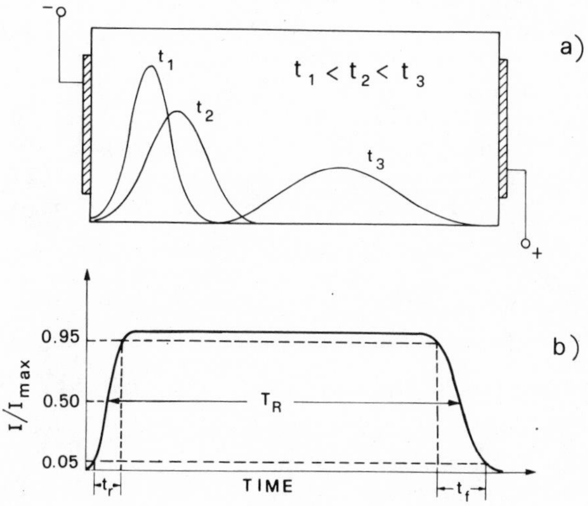

Fig. 3. ToF technique used for the determination of the longitudinal diffusion coefficient. a) Space distribution of carriers at three successive times as they travel across the sample. b) Current pulse induced by the carriers.

Drift Velocity

In this case the shape of the current waveform induced at the ends of the sample is determined only by the motion of the carriers which drift across W. Thus, by a measurement of T_R (for averaging purposes measured at half of the pulse height) and W, v_d for one kin of carriers is directly obtained as

$$v_d = W/T_R, \tag{2}$$

and since the electric field is given by $E = V_A/W$, V_A being the applied voltage, at low fields the Ohmic drift-mobility μ_d can be obtained as

$$\mu_d = v_d/E. \tag{3}$$

The total error in the drift velocity measurement can be estimated within ±5% at low fields and ±3% at high fields [Canali, et al. (1975)].

Diffusion Coefficient

The time and space evolution of the carrier-layer crossing the sample produces a fall time τ_f of the current pulse different from the rise time τ_r (for the purpose of normalization both are defined here as being between 0.05 and 0.95 of the pulse height). This difference originates from thermal diffusion spreading and space-charge repulsion effects. Apart from the contribution of the electronic rise time, which is common to both τ_r and τ_f, τ_f can be analyzed in terms of the times τ_D and τ_{sc}, which refer to diffusion and space-charge, respectively. Provided the following three conditions are satisfied, namely: (1) $X_o < X(T_R)$, where X_o is the initial and $X(T_R)$ the final thickness of the carrier layer; (2) $\tau_r < \tau_f$; and (3) the electric field is uniform throughout the sample, the diffusion time is

$$\tau_D = (21.6\ D_\ell\ W/v_d^3)^{\frac{1}{2}}. \tag{4}$$

When both space-charge and diffusion are present, the fall-time is

$$\tau_f = (\tau_{sc}^2 + \tau_D^2 + \tau_r^2)^{\frac{1}{2}} \tag{5}$$

with, in the dielectric relaxation-time approximation,

$$\tau_{sc} = C\ \exp(2\alpha I) \tag{6}$$

$$\alpha = (\mu' W\xi)/(\varepsilon\varepsilon_{\ell} v_d^2 S), \tag{7}$$

where C is a constant with the dimension of time, I is the current of the primary electrons, ξ is the primary electron energy, ε_{ℓ} is the energy necessary to create an electron-hole pair in the material, S is the useful area of the sample, ε is the dielectric constant and μ' is the differential mobility at the applied field.

For values of the current I sufficiently low, τ_f can be experimentally found to be approximately independent of I and (4) can be well approximated by

$$\tau_f = (\tau_D^2 + \tau_r^2)^{1/2} \tag{8}$$

Consequently, under the above conditions, from the measurement of τ_f and τ_r, τ_D is obtained from (8) and, in turn, D_{ℓ} from (4).

The total experimental error on D_{ℓ} (E), due to the experimental uncertainty regarding v_d, τ_f, τ_r and W, was estimated to be about 25% [Reggiani, et al., (1978)].

The choice of the crystallographic directions, of the value of the electric field strength and of the temperature, offers different possibilities of selecting the independent physical variable.

ALTERNATIVE TECHNIQUES

Tables 3 and 4 report a brief comparison of ToF with alternative techniques for the measurement of the drift velocity and of the diffusion coefficient, respectively.

Drift Velocity

Conductivity techniques are the most widely applicable for measuring hot-carrier currents. They do not yield directly the drift velocity so that an independent measurement of the carrier density is always necessary if v_d has to be determined. In order to avoid injection of the carriers from the contacts, conductivity techniques cannot be used when the resistivity is too high, in particular in very pure materials and at very low temperatures. It is also necessary to have some confidence that the carrier concentration remains constant during the experiment. Furthermore, only the drift velocity of the majority carriers can be measured.

Similar considerations hold for the microwave techniques with the additional drawback that the microscopic interpretation of the

Table 3. Comparison of ToF with alternative techniques for measurements of drift velocity.

	ToF	Conductivity	Microwave
Type of measurement	More direct measurement obtained by $v_d = W/T_R$ Electrons and holes	More indirect measurement obtained by $j = qnv_d$ $n = 1/q\ rR$ Majority carriers	More indirect measurements obtained by solution of integral equations. Majority carriers
Resistivity	High resistivity materials	Medium-low resistivity materials	Medium-low resistivity materials
Temperature	Limit at high temperatures	Limit at very low temperatures	Limit at very low temperatures
Contacts	Not-injecting	Not-injecting	Not-injecting

experimental results may be rather involved. In particular, the problem of avoiding distortion of the microwave field inside the sample, in order to obtain its amplitude, leads to supplementary difficulties. Microwave techniques, on the other hand, may yield information on microscopic times, when the frequency is appropriatel varied. Moreover, the problem of carrier injection at the contacts may be avoided when low d.c. fields and low currents are considered, since the carriers are heated by a large a.c. field.

ToF techniques have the advantage of giving more direct measurements of the drift velocity for both type of carriers (electrons and holes) within the same sample. They are not always applicable, however, since they require both the lifetime of the carriers and th dielectric relaxation time of the material must be longer than the transit time. Furthermore, with low resistivity materials the current signal due to the injected carriers may become too small when compared with the background current. ToF techniques also have severe limitations when low fields are applied. This occurs since the current signal is proportional to the drift velocity and, at low fields, the signal-to-noise ratio may become too small. It may be noted that conductivity techniques and ToF techniques are in a sense

Table 4. Comparison of ToF with alternative techniques for measurement of diffusion coefficient.

	ToF	White-noise	Beam-spreading
Type of measurement	Absolute determination of D_ℓ only. More direct method Electrons and holes	Relative determination of D_ℓ and D_t. More indirect method, it should be coupled with a conductivity measurement Majority carriers	Absolute determination of D_t only. More direct method Electrons and holes
Resistivity	High resistivity materials	Medium resistivity materials	High resistivity materials
Temperature	Limits at high temperature	Limits at low temperature	Limits at high temperatures
Field	Limits at very low fields	Limits at high fields	
Contacts	Not-injecting	Not injecting and free from 1/f and shot noise	Quite involved technology in making contacts

complementary, in fact they are applicable in low and high resistivity materials, respectively.

Diffusion Coefficient

Absolute longitudinal measurements of the diffusion coefficient can be obtained with ToF techniques. No "a priori" limits exist for the value of the field strength until breakdown of the sample or bulk. The accuracy of the measurement tends to improve with high fields owing to the higher signal-to-noise ratio.

A white noise technique [Nougier (1978)] enables a determination of both longitudinal and transverse diffusion coefficient to be carried out. This is a more indirect method since it requires coupling

Table 5. Different applications of ToF techniques

1 -	Measurement of ionization energy for electron-hole pairs
2 -	Study of carrier-carrier interaction
3 -	Study of plasma effects
4 -	Measurement of free drift time τ^+ and detrapping time τ_d
5 -	Study of negative dielectric relaxation phenomena

with a conductivity measurement; furthermore, only relative values of the diffusion coefficient can be obtained. The complementary of noise and ToF techniques has been demonstrated in a recent paper by the groups working at the Modena and Montpellier Universities [Nava, et al., (1979)].

Finally, absolute transverse diffusion coefficient can be obtained by the beam-spreading technique [Bartelink and Persky (1970), Persky and Bartelink (1971)]. The relevant complexity of the experimental set-up puts severe limits to this technique, in fact, to the author's knowledge, at present only the two mentioned experiment have been reported in the literature.

SUMMARY OF ToF MOST SIGNIFICANT APPLICATIONS

In concluding this lecture, I want to briefly summarize the most significant applications of ToF techniques which are besides the before mentioned cases of drift velocity and diffusion coefficient (see Table 5).

The creation energy of electron-hole pairs can be obtained by measuring the total collected charge, in absence of trapping phenomena, and normalizing its value to a known standard (for example Si) [Alberigi-Quaranta, et al., (1970b) and Canali, et al., (1972)].

By changing the energy and/or the intensity of the impinging radiation, the free carrier concentration can be varied, thus the effect of carrier-carrier interaction upon macroscopic quantities can be studied in detail [Jacoboni, et al., (1974)]. Furthermore, for the case of very high energetic radiation (for instance α-particles) plasma effects, related to the relaxation of the cloud of electron-hole pairs, can be evidenced [Taroni and Zanarini (1969a Canali, et al., (1971)].

When the material has a relevant number of trapping centers (for instance highly compensated materials), provided τ^+ and τ_d are shorter than T_R, τ^+ and τ_d can be obtained as a function of field strength by analyzing the decay in time of the current waveform (see Fig. 2b) [Ottaviani, et al., (1973)].

Finally, in materials exhibiting negative differential conductivity the strong space-charge effects which arise from the negative value of the differential dielectric relaxation time can be identified through a decreasing behavior of the fall time at increasing primary electron current [Nava, et al., (1978)].

BIBLIOGRAPHY

Alberigi-Quaranta, A., Cipolla, F., and Martini, M., 1965, Phys. Letters 17:102.
Alberigi-Quaranta, A., Martini, M., and Ottaviani, G., 1969, IEEE Trans. Nucl. Sci. NS-16:35.
Alberigi-Quaranta, A., Canali, C., and Ottaviani, G., 1970a, Rev. Sci. Instr. 41:1205.
Alberigi-Quaranta, A., Canali, C., Ottaviani, G., and Zauio, K. R., 1970b, Lett. Nuovo Cimento 4:908.
Alberigi-Quaranta, A., Jacoboni, C., and Ottaviani, G., 1971, Rivista del Nuovo Cimento 1:445.
Bartelink, D. J., and Persky, G., 1970, Appl. Phys. Lett. 16:191.
Canali, C., Martini, M., Ottaviani, G., Alberigi-Quaranta, A., and Zanio, K. R., 1971, Nucl. Instr. Meth. 96:561.
Canali, C., Martini, M., Ottaviani, G., and Alberigi-Quaranta, A., 1972, IEEE Trans. Nucl. Sci. NS-19:9.
Canali, C., Jacoboni, C., Nava, F., Ottaviani, G., and Alberigi-Quaranta, A., 1975, Phys. Rev. 12B:2265.
Conwell, E. M., 1967, "High Field Transport in Semiconductors," Academic Press, New York.
Gunn, J. B., and Elliott, B. J., 1966, Phys. Letters 22:369.
Haynes, J. R., and Shockley, W., 1951, Phys. Rev. 81:835.
Jacoboni, C., 1974, in "Proc. 13th Intern. Conf. Phys. Semicond., Rome," Ed. by F. Fumi, Marves, Rome, p. 1195.
Kepler, R. G., 1960, Phys. Rev. 119:1226.
Martini, M., and McMath, T. A., 1969, Appl. Phys. Letters 14:374.
Nava, F., Canali, C., and Jacoboni, C., 1978, Sol.-State Electr. 21:689.
Nava, F., Canali, C., Reggiani, L., Gasquet, D., Vaissiere, J. C., and Nougier, J. P., 1979, J. Appl. Phys., in press.
Nougier, J. P., 1978, in "Noise in Physical Systems," Ed. by D. Wolf, Springer-Verlag, Berlin.
Ottaviani, G., Canali, C., Jacoboni, C., and Alberigi-Quaranta, A., 1973, J. Appl. Phys. 44:360.
Persky, G., and Bartelink, D. J., 1971, J. Appl. Phys. 42:4414.

Reggiani, L., Canali, C., Nava, F., and Alberigi-Quaranta, A., 1978, J. Appl. Phys. 49:4446.
Ruch, J. G., and Kino, G. S., 1968, J. Appl. Phys. 174:921.
Sigmon, T. W., and Gibbons, J. F., 1969, Appl. Phys. Letters 10:320.
Smith, R. A., 1959, "Semiconductors", Cambridge Univ. Press, London.
Spear, W. E., 1957, Proc. Phys. Soc. 70:669.
Taroni, A., and Zanarini, 1969a, Nucl. Instr. Meth. 67:277.
Taroni, A., and Zanarini, 1969b, J. Phys. Chem. Solids 30:1861.

HOT-ELECTRON TRANSPORT IN QUANTIZING MAGNETIC FIELDS

R. J. Nicholas

Clarendon Laboratory
Oxford OX1 3PU, England

and

J. C. Portal

I.N.S.A., Department de Physique
31077 Toulouse Cedex, France

INTRODUCTION

The influence of a magnetic field upon the electrons and holes in metals and semiconductors is to quantize the energies of electrons into a set of Landau levels. This quantization can lead to several striking quantum effects in the electrical transport properties at high magnetic fields. When the carrier concentration is sufficiently high that a constant, well defined, Fermi energy exists within the conduction or valence band at low temperatures (e.g. in a heavily doped, degenerate semiconductor), then there is an oscillatory variation of the conductivity with magnetic field; the Shubnikov-de Haas effect. Oscillations occur as the Fermi level passes through the high density of states at the edge of each Landau level, while the amplitude of these oscillations is a strong function of temperature. Section 2 describes how this temperature dependence may then be used to investigate the temperature of the electron gas in cases where this is not the same as the lattice temperature. Provided that the electron gas may be described by a Maxwellian distribution function with a characteristic temperature T_e, then this method may be used to study the electron temperature as a function of the electric field, time, and also of the magnetic field.

For more lightly doped, non-degenerate semiconductors then it is not possible to observe Shubnikov-de Haas oscillations. However, when the energy relaxation mechanisms are monoenergetic, such as when L.O phonons are emitted, then a resonant cooling of the carriers can occ when the separation between one or more Landau levels is equal to th energy of the relaxation process. This resonant cooling may then be reflected in one of the transport coefficients. The positions of these resonances can thus be used to determine the energies of any strong energy relaxation process occurring. The most widely investi gated example of this is the 'hot-electron magnetophonon effect' whe oscillations are found to occur in the magneto-resistance of many different semiconductors due to the emission of both single, and pai of, phonons. Another example of such a resonant energy relaxation process is the 'magneto-impurity effect', where carriers are found t loose energy by exciting electrons or holes trapped on shallow impurity levels, into higher lying levels or into the conduction or valence band. Section 3 describes how the magnetophonon effect has been used to investigate the many energy relaxation processes which occur in semiconductors at low temperatures, both within and out of the conduction and valence bands. Section 4 will deal with the new and growing field of magneto-impurity effects.

THE SHUBNIKOV-de HAAS EFFECT

The application of a magnetic field to a semiconductor splits t continuous distribution of energy levels in the conduction and valen bands into sets of Landau sub-bands separated by $h\omega_c$. For degenerat semiconductors, this results in the observation of oscillations in t magnetic susceptibility, and in transport properties such as the magneto-resistance, which are periodic in the reciprocal of the magnetic field. The conditions necessary for the observation of these oscillations are the following:

$$\omega_c\tau >> 1, \quad \hbar\omega_c > kT, \quad \varepsilon_F > \hbar\omega_c$$

where τ is the relaxation or quantum-state lifetime, and ε_F is the Fermi energy in zero magnetic field.

The theoretical expressions for the amplitude of the longitudin and transverse oscillatory magnetoresistance have been given by Adam and Holstein (1959) and by Argyres (1959) and are

$$\frac{\Delta\rho}{\rho_0} = a \sum_{n=1}^{\infty} b_n \cos\left(\frac{2\pi n\varepsilon_F}{\hbar\omega_c} - \frac{\pi}{4}\right) + R \qquad (1)$$

with

$$b_n = \frac{(-1)^n}{\sqrt{1}} \left(\frac{\hbar\omega_c}{\varepsilon_F} \right)^{\frac{1}{2}} \frac{2\pi^2 nkT/\hbar\omega_c}{\sinh(2\pi^2 nkT/\hbar\omega_c)}$$

$$x \cos(\pi rn) \exp(-2\pi^2 kT_D/\hbar\omega_c), \qquad (2)$$

where ρ_0 is the zero-field resistivity, n is the harmonic, r is the ratio of the spin-splitting to the Landau-level spacing, and T_D is the Dingle temperature (Kubo, et al., 1965) which is related to the collision broadening relaxation time τ_c by $T_D = \hbar/\pi k\tau_c$. The term R in equation (1) represents an additional series of oscillatory terms in the transverse magnetoresistance which is generally much smaller than the term b_n and vanishes in the longitudinal case. The factor a is equal to 2.5 or 1 in the transverse and longitudinal magnetoresistance, respectively. Under the assumption of a constant Dingle temperature, the ratio of the amplitudes A of the oscillations at two temperatures is given by

$$\frac{A(T_1)}{A(T_2)} = \frac{\chi_1/\sinh \chi_1}{\chi_2/\sinh \chi_2}, \qquad (3)$$

where

$$\chi_i = 2\pi^2 kT_i/\hbar\omega_c,$$

which is clearly a strong function of temperature.

The influence of an electric field upon the carriers in a semiconductor is to heat the electrons out of thermal equilibrium with the lattice. One very simple and useful model used to describe the resultant change in the electron distribution function is to assume a Maxwell-Boltzmann or Fermi-Dirac distribution with the only adjustable parameter being the electron temperature T_e, which is higher than the lattice temperature T_L. This model is called the "electron temperature model" (Stratton, 1958). The use of this model is only justified where the carrier concentration is high enough that inter-electronic collisions are sufficiently frequent to establish the distribution function, and is thus most applicable for low electric fields. In order to determine the electron temperature in this model it is clear that (3) provides a direct method for the determination of T_e. This method was first used by Bauer and Kahlert (1970) to determine the electron temperature of degenerate n-InAs as a function of electric field. Figure 1 shows experimental recordings of the Shubnikov-de Haas oscillations as a function of both temperature and electric field. The electron temperature was deduced both from (3) and from a

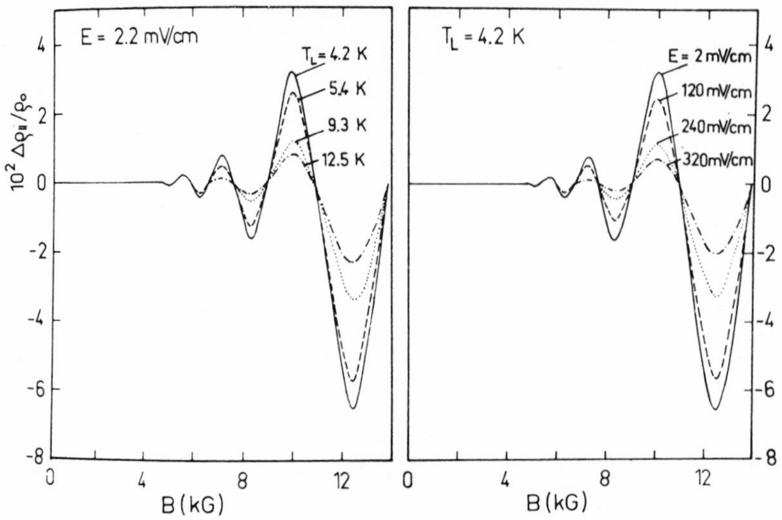

Fig. 1. Shubnikov-de Haas oscillations in degenerate n-InAs as a function of magnetic field, showing the influence of electric field and temperature (Bauer and Kahlert, 1972).

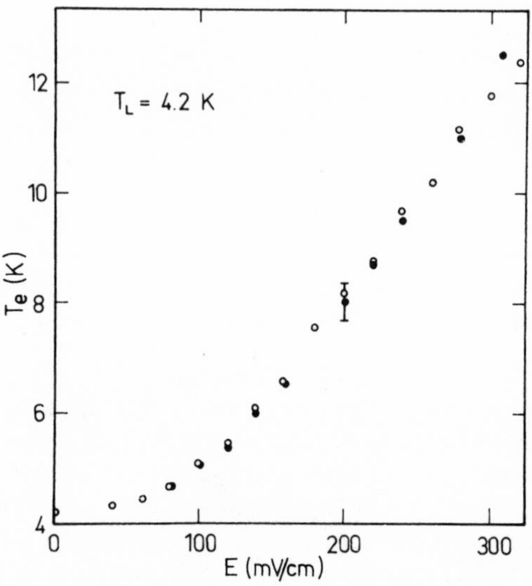

Fig. 2. The electron temperature of n-InAs as a function of electric field as deduced from equation (3) (Bauer and Kahlert, 1972).

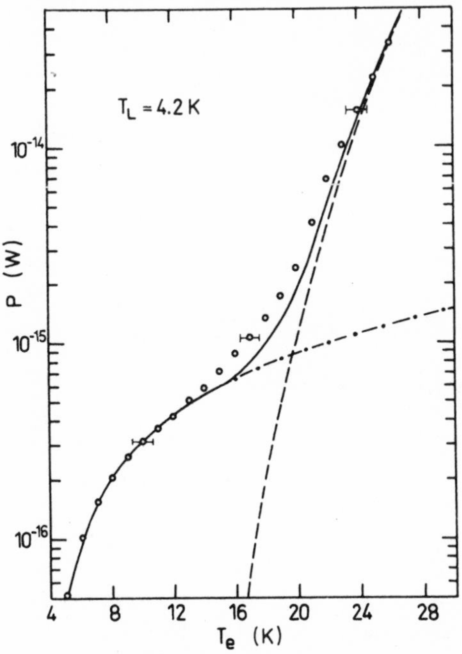

Fig. 3. The energy loss rate for electrons in InSb as a function of electron temperature. The solid lines are the calculated rate, taking account of acoustic phonon scattering and optic phonon emission. The experimental points were deduced from dependence of the electron temperature upon electric field (Kahlert and Bauer, 1973).

direct comparison with the experimentally determined temperature dependence and is illustrated in Figure 2. In addition, the use of rapid voltage pulses was used to measure the time dependence of the electron heating which was found to result from an energy relaxation time of the order of 10^{-7} sec.

Once the electron temperature is known it is possible to plot the energy loss rate, which is equal to the energy input = $e\mu E^2$ (where μ is the electron mobility), as a function of electron temperature. This is illustrated in Figure 3 for the case of InSb (Kahlert and Bauer, 1973) and is compared with the theoretical predictions for energy loss due to acoustic phonon scattering and optic phonon emission. As may be seen there is excellent agreement with the predictions except in the region of 14 K where a further energy relaxation process with a characteristic energy of the order of 100 K is required. Such a process has been found in magnetophonon experiments on InSb at 14 K (Stradling and Wood, 1970) where energy loss was found to result from the emission of pairs of band-edge T.A. phonons with a

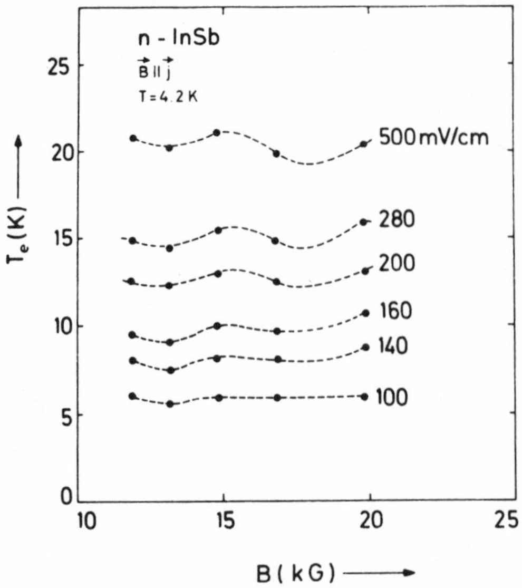

Fig. 4. The dependence of electron temperature upon magnetic field for n-InSb (Bauer et al., 1975).

characteristic energy $2\hbar\omega_{T.A.} = 110$ K. Further evidence of the importance of these two phonon processes will be given in Section 3.

It is also possible, by observing the temperature dependence of oscillations resulting from different harmonic numbers (n), to measu the electron temperature as a function of magnetic field. Results f InSb (Bauer, et al., 1975) are shown in Figure 4. It may be seen th there are pronounced dips in the electron temperature at magnetic fields of 13 kG and 18 kG. These dips result from a resonant coolin of the electron gas due to the emission of L.O. phonons. When the separation between one or more Landau levels is equal to the L.O. phonon energy then the emission rate for L.O. phonons is enhanced du to the large combined density of states for the initial and final states. Such a resonant cooling also gives rise to the magnetophono effect, as discussed in the next section.

The electric field dependence of the amplitude of Shubnikov-de Haas oscillations has also been used to investigate the energy loss rates (Hess, et al., 1977) in the two-dimensional electron gas forme in the inversion layer of MOS devices (Dorda, 1975). It was shown by the former, that the dominant energy loss mechanism is consistent with scattering by quantized surface waves or "surfons".

THE MAGNETOPHONON EFFECT

Higher Temperatures

For very pure samples of semiconducting materials, the carrier scattering time may be sufficiently long for well defined Landau levels to be formed even at high temperatures. If the carriers are heated out of equilibrium with the lattice by an electric field, then the rate of energy loss may be strongly affected by the quantization of the density of states. The electron temperature may thus be found to oscillate as a function of magnetic field, due to the presence of one or more strong monoenergetic energy relaxation processes. For pure materials where these effects are strongest, the donor or acceptor impurities are found to freeze-out at low temperatures and the material is non-degenerate.

The electron energies and densities of states are shown in Figure 5 for a fixed magnetic field. If, at this magnetic field, the relaxation energy is equal to the Landau level separation, or any integral multiple of it, then the energy relaxation process will both start and end at the high density-of-states regions at the bottom of each Landau level. The energy relaxation rate will thus be enhanced at magnetic fields given by the relation

$$E_{res} = N\hbar\omega_c = \frac{N\hbar eB_N}{m^*} \tag{4}$$

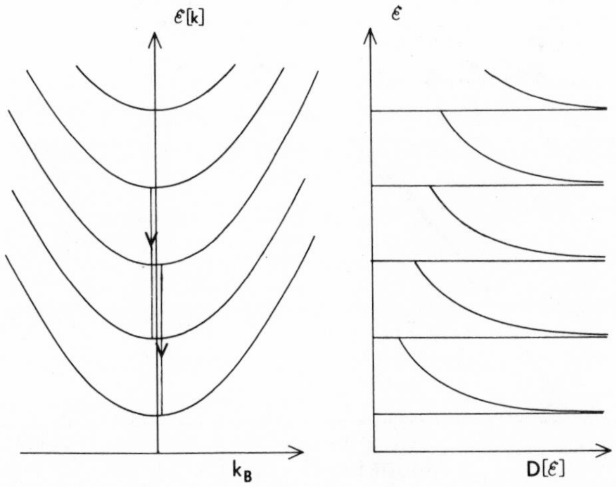

Fig. 5. A schematic view of the Landau sub-bands and density of states formed in a magnetic field. Magnetophonon resonance transitions are shown for the condition $2\hbar\omega_c = \hbar\omega_{L.O.}$.

where the product NB_N is known as the fundamental field, and there will be a resonant cooling of the electron gas resulting from transitions between Landau levels as shown in Figure 5. This resonant cooling of the electrons will then be reflected in any of the transport coefficients which are dependent upon electron temperature.

The most widely investigated example of such resonant energy relaxation occurs when L.O. phonons are emitted. Near the centre of the Brillouin zone the L.O. phonons show almost no energy dispersion with wavevector and can thus be regarded as monoenergetic. Resonance will thus occur whenever

$$\hbar\omega_{L.O.} = N\hbar\omega_c, \tag{5}$$

and are known as magnetophonon resonances. In order to detect this effect at higher lattice temperatures, care must be taken in order to distinguish between the emission and absorption of L.O. phonons, since the absorption of L.O. phonons can strongly relax the electron momentum and thus give rise to oscillations in the magnetoresistance (Harper, et al., 1973). The resonances may, however, be detected in the warm-electron coefficient β defined by the relation

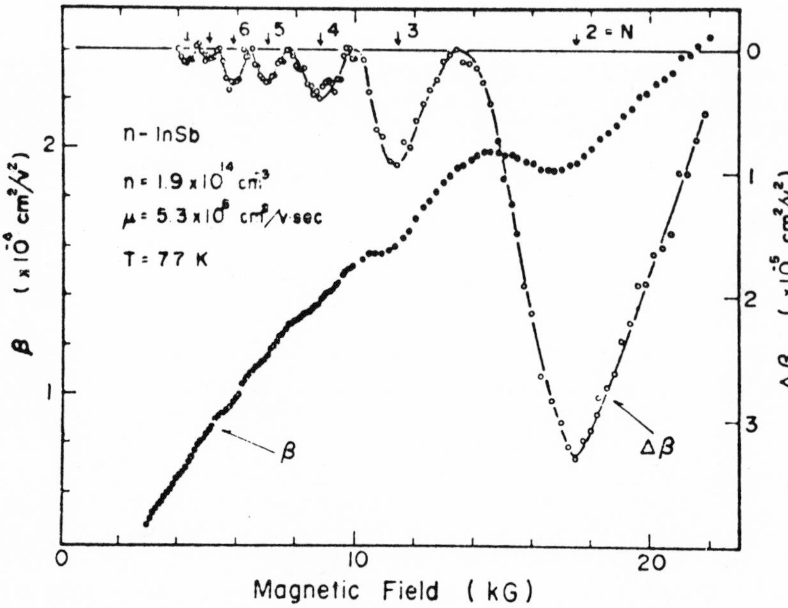

Fig. 6. The hot-electron coefficient β (given by equation (6)) for n-InSb as a function of magnetic field at 77 K (Hamaguchi et al., 1972).

$$\mu = \mu_o(1 - \beta E^2), \tag{6}$$

where μ is the carrier mobility, and β is proportional to the constant α which denotes the degree of electron heating in the formula

$$T_e - T_L = \alpha E^2, \tag{7}$$

where T_e and T_L are the electron and lattice temperatures respectively. Figure 6 shows plots of β as a function of magnetic field obtained by Hamaguchi, et al., (1972) for n-type InSb at 77 K. β is found to be an oscillatory function of magnetic field with minima occurring at magnetic fields determined by (5) above, showing that cooling of the electrons occurs at resonance. The amplitude of the dips in β increases as the harmonic number N falls, due to the higher magnetic field which produces better defined Landau levels.

Similar resonant cooling has been observed in n-type GaAs by Nicholas and Stradling (1976) and is shown in Figure 7. Since the magnitude of the oscillatory structure can be quite small, it was found to be necessary to use filters in order to detect the weak oscillations. The magnetic field is swept as a function of time, so that any oscillatory structure in the signal appears at a frequency which is relatively higher than that of the monotonic background onto which it is usually superimposed. If the signal is then fed into high-pass or band-pass filters, the monotonic background is strongly suppressed. The exact form of the resulting structure will depend upon the type of filters used and the manner in which the magnetic field is varied. Two different methods which have been commonly employed are:

(i) The magnetic field is swept linearly with time and the signal is fed into a filter with a frequency response $\alpha\ \omega^2$ (i.e. 12 dB/octave). In this case, the resulting output is proportional to the second derivative of the signal, and it is usual to plot $- d^2V/dB^2$ (where V is the signal voltage) so that maxima in the original signal also appear here as maxima. Due to the lower effective frequency of the oscillations at low harmonic numbers, this technique suppresses the amplitudes of the oscillations occurring at high magnetic fields. Several examples of recordings taken in this way are shown below.

(ii) The magnetic field is swept hyperbolically, so that equal intervals of 1/B appear at a constant rate. In this case any series of oscillations determined by a relation of the form of equation (4) will appear at a constant frequency and may be detected by feeding the signal into a "tuned" or band-pass filter. Such a method has the additional advantage that the oscillatory structure may be recorded digitally by sampling at equal time

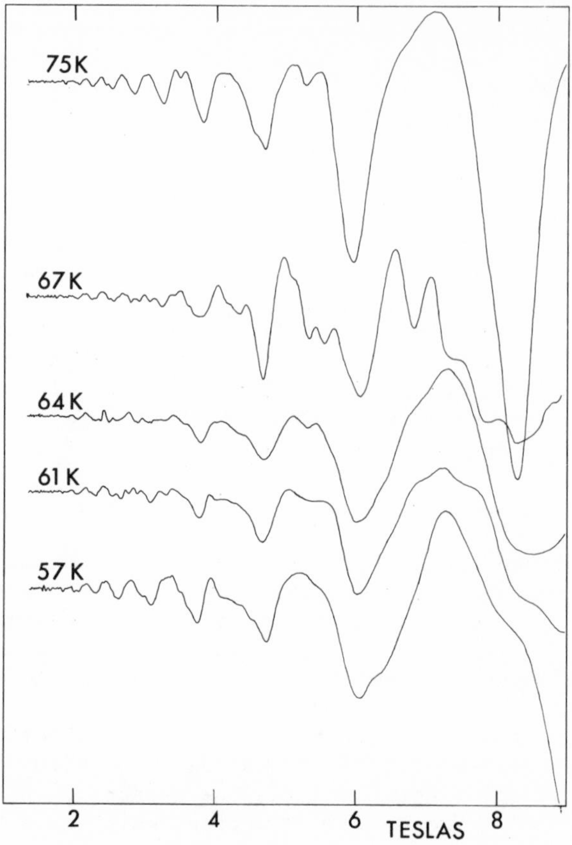

Fig. 7. Experimental recordings of the oscillatory structure in β for n-GaAs at several different lattice temperatures. The recordings are taken using a hyperbolic magnetic field swee as described in the text.

intervals, and may then be Fourier analysed (Eaves, et al., 19 by computer, in order to reveal the resonance energies (E_{res}) present. This method is particularly useful when more than one magnetophonon series is present and it is not clear from the original data exactly which series of resonances are occurring.

Many other techniques have also been used with success to detec magnetophonon oscillations, including compensation by a voltage proportional to the magnetic field, magnetic field modulation in order to obtain derivatives with respect to magnetic field, and direct accurate point-by-point measurements.

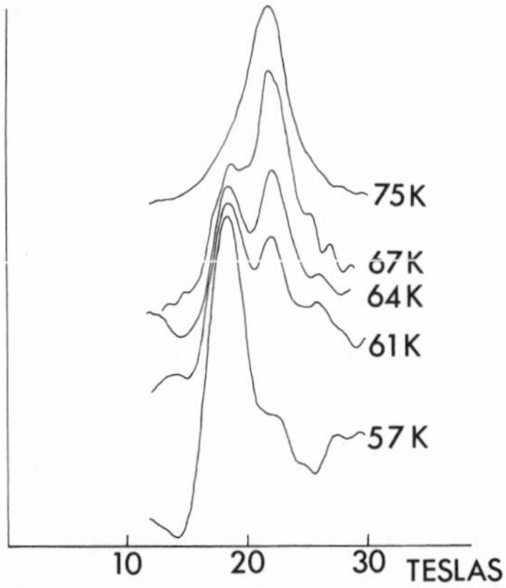

Fig. 8. Fourier transforms of the experimental recordings shown in Figure 7 (Nicholas and Stradling, 1976).

Typical examples of experimental recordings taken with a hyperbolic field sweep are shown in Figure 7 for magnetophonon resonances in β for GaAs at several temperatures. The resulting Fourier transforms are shown in Figure 8, and illustrate a very interesting phenomenon. At 77 K, one series of oscillations occurs with a fundamental field of 22.1 T which corresponds to the emission of L.O. phonons in a transition between Landau levels. As the temperature falls, the character of the oscillations changes and Fourier analysis reveals that the first series of oscillations is replaced by a second with a fundamental field of 18.3 T. The transition is extremely rapid and is complete by 55 K. The new series of oscillations was first observed by Stradling and Wood (1968) in n-GaAs at 20 K, and was interpreted as also resulting from electron energy relaxation by the emission of L.O. phonons but resulting from a transition from a Landau level to the ground state of a shallow hydrogenic donor impurity. Resonant cooling will thus occur whenever

$$N\hbar\omega_c = \hbar\omega_{L.O.} - E_I(B) \tag{8}$$

where $E_I(B)$ is the magnetic field dependent donor binding energy, which may be approximated by the relation

$$E_I(B) = E_I(0) + \tfrac{1}{2}\hbar\omega_c \tag{9}$$

which is valid at fields where $E_I(0) > \tfrac{1}{2}\hbar\omega_c$, so that

$$(N + \tfrac{1}{2})\hbar\omega_c = \hbar\omega_{L.O.} - E_I(0) \tag{10}$$

There are two striking features of this equation. Firstly, the resonance energy is decreased by $E_I(0)$ from the L.O. phonon energy, thus giving a predicted fall of 16% in fundamental field for GaAs in good agreement with the observed decrease of 17%. Secondly, resonance occurs for half integral values of $(N + \frac{1}{2})$ as opposed to the integral resonance condition (5). It is possible to verify this from the experimentally observed resonance positions, due to the non sinusoidal form of the resonances at high magnetic field which show distinct peak sharpening, and indicate that cooling of the electrons occurs at resonance.

A much more complex example of the hot-electron magnetophonon effect occurs for n-type silicon at 77 K. The situation is complicated by the equivalent many-valley nature of the conduction band, which means that scattering may occur within one valley (intra-valle or between different but equivalent valleys (inter-valley). Intervalley scattering occurs due to several different phonon modes which are of two types; g-phonons which scatter between ellipsoids whose principal axes are parallel, and f-phonons, which scatter between perpendicular valleys. Magnetophonon oscillations have been observe in the longitudinal magnetoresistance of n-Si under hot-electron conditions (Portal, et al., 1974; Eaves, et al., 1974, 1975) between 65 K and 77 K and were found to result from the emission of both g- and f-phonons. The resonance condition for the g-phonons is given b (5) with an appropriate phonon frequency ω_i; however, due to the ellipsoidal nature of the conduction band valleys up to three different cyclotron effective masses may be present, depending upon the orientation of the magnetic field relative to the crystal axes. The condition for the f-phonons is more complex: with B||(100) or B||(110) scattering between valleys with different cyclotron masses is determined by

$$(N + \tfrac{1}{2})\hbar\omega_{ci} - (M + \tfrac{1}{2})\hbar\omega_{cj} = \pm \hbar\omega_i \qquad (11)$$

where ω_{ci} and ω_{cj} are the cyclotron masses of the two valleys and thus two magnetophonon series result from each phonon, while scattering between valleys with the same effective mass is determined by (5). When B||(111) all the conduction band valleys are equivalent and the resonance condition for both f- and g-phonons is given by (5

Figure 9 shows experimental recordings of the oscillatory structure in the longitudinal magneto-conductivity of n-Si for B||(100), taken using the hyperbolic field sweep technique described above. The structure is extremely complex, so that Fourier analysis of the oscillations was used to analyze the recordings. The results are shown in Fig. 10 and 11 for B||(100) and B||(111). These show the presence of up to seven fundamental fields from 20.4 T to 216 T for B||(100). The simplest spectra occur for B||(111) where there is on one possible effective mass. There are five main peaks in the Fouri

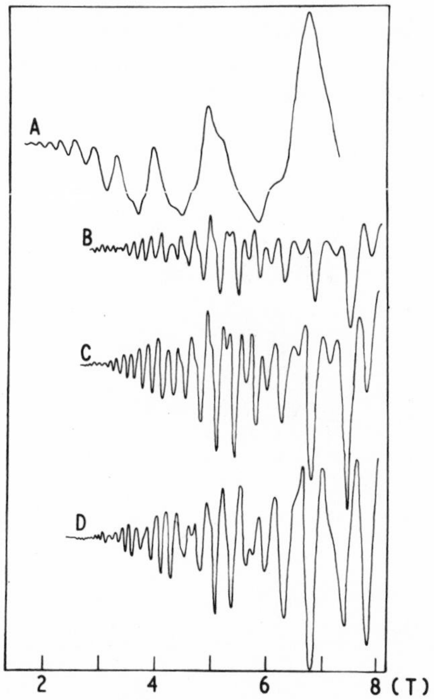

Fig. 9. The oscillatory magneto-conductivity of n-Si with j||B||(100) at a lattice temperature of 65 K, using a hyperbolic field sweep. Trace (a) is optimised for a fundamental field near to 100 T. The applied electric fields were (a) 11 V/cm; (b) 11 V/cm, (c) 17 V/cm; (d) 46 V/cm. (Eaves et al., 1975).

spectrum which result from the emission of phonons of energies of 130 K (g-type), 220 K (f-type), 535 K (f-type), 585 K (f-type) and 685 K (f-type), there is also weaker structure resulting from a 735 K (g-type) phonon. As the electric field is raised the series resulting from the emission of the higher energy phonons become much stronger due to the increased electron heating. When the magnetic field is parallel to (100), then two magnetophonon series are found to be observed from several of the different phonons, in agreement with (11).

The assignment of the observed phonon energies to f- or g-phonon scattering processes was made partly by use of the phonon dispersion curves measured by Dolling (1963), and partly from the interpretation of the stress dependence of the magnetophonon oscillations (Eaves, et al., 1974a). The application of a uniaxial stress in the (100) direction lifts the degeneracy of the conduction band valleys and

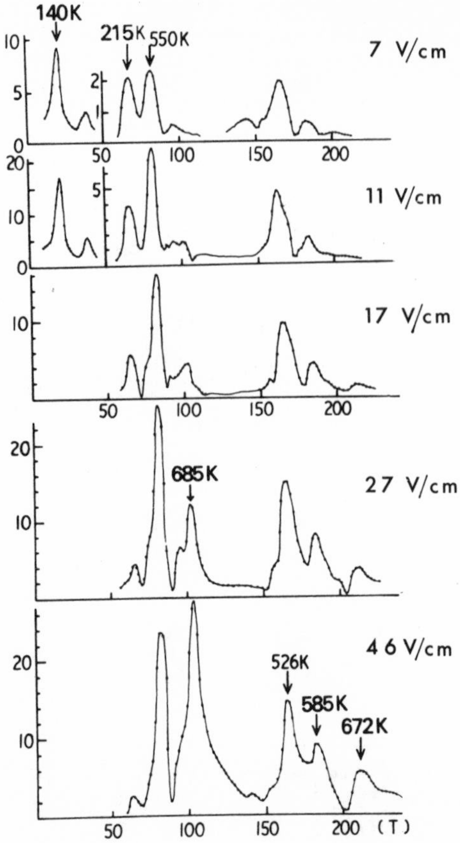

Fig. 10. Fourier transforms of the magnetophonon structure for n-S with j||B||(100) at a lattice temperature of 65 K. (Eave et al., 1975).

thus depopulates the four valleys perpendicular to the stress. In addition the change in relative position of the valleys alters the resonance condition (10) to

$$\pm \hbar\omega_i = (N + \tfrac{1}{2})\, \hbar\omega_{ci} - (M + \tfrac{1}{2})\hbar\omega_{cj} + \Delta E_s \quad (12)$$

where ΔE_s is the relative shift of the conduction band valleys. Th series resulting from the emission of f-phonons will be suppressed amplitude and shifted in fundamental field by the stress. The domi ance of inter-valley scattering in n-silicon arises due to the grou theoretical selection rules (Streitwolf, 1970) which forbid intra-valley scattering and also some inter-valley processes. Scattering by both the 140 K (g) and 220 K (f) phonons was predicted to be

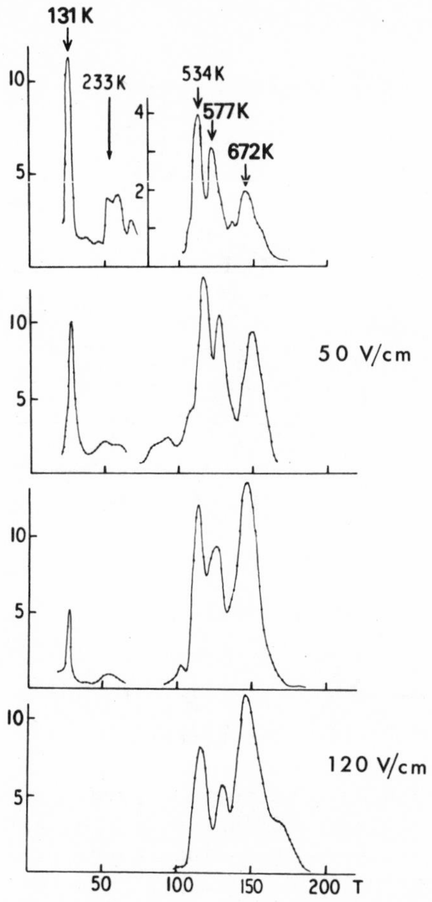

Fig. 11. Fourier transforms of the magnetophonon structure for n-Si with j||B||(111) at a lattice temperature of 65 K. (Eaves et al., 1975).

forbidden by Streitwolf (1970); however, it has recently been shown by Wallace and Joos (1978) that the presence of the magnetic field, impurities and other perturbations without the symmetry of the crystal, can make these processes quite strongly allowed.

Low Temperatures

When the lattice temperature is lowered to the region of 10-20 K it is found that there is a very considerable change in the magnetophonon structure observed, with many more features present. In

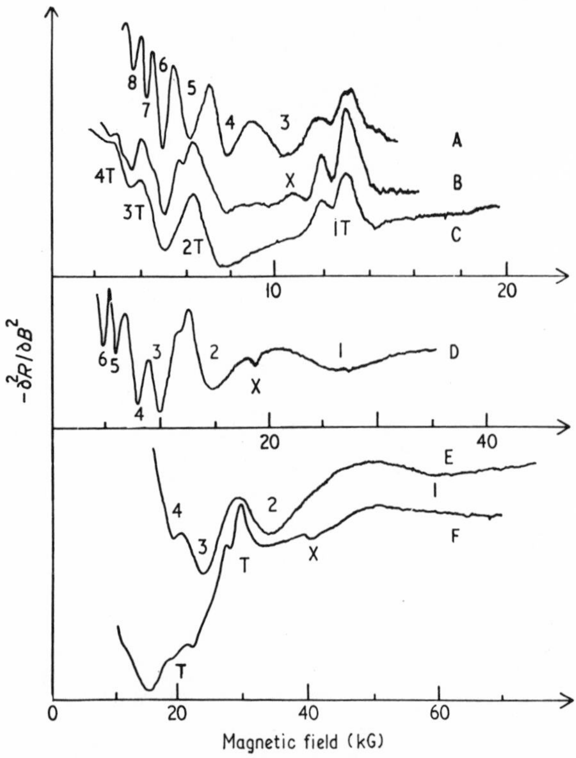

Fig. 12. Magnetophonon structure observed under hot electron conditions for InSb and InAs. Recordings show the second derivative of the longitudinal magnetoresistance as a function of magnetic field. The peaks marked T correspond to equation (12). Experimental conditions: A InSb, 11 K, 700 mV cm^{-1}; B InSb, 11 K, 170 mV cm^{-1}; C InSb, 11 K, 93 mV cm^{-1}; D InSb, 20 K, 650 mV cm^{-1}; E InAs, 11 K, 5 V cm^{-1}; F InAs, 11 K, 2 V cm^{-1}. (Harper et al., 1973).

this temperature range, the most commonly investigated transport coefficient has been the longitudinal magnetoresistance, since it is quite sensitive to electron temperature and does not show the strong monotonic magnetoresistance of the transverse case. Identical structure can, however, usually be observed in the transverse magnetoresistance providing confirmation that the resonances arise from energy relaxation. No trace of the "normal" magnetophonon effect due to relaxation of momentum by the absorption of phonons can be detected at these low lattice temperatures.

As the electric field applied to the sample is raised, the magnetophonon structure begins to appear. The first series of

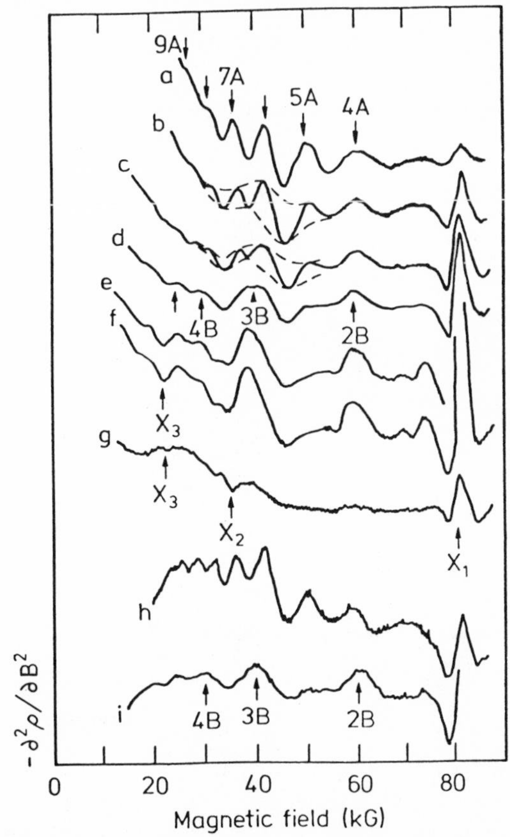

Fig. 13. Magnetophonon and magneto-impurity structure in the magnetoresistance (curves a-g $j \parallel B$, h and i $j \perp B$) of n-InP at 11 K. Peaks marked A are the impurity capture series (equation (10)) and peaks B are the 2TA(X) series (equation (12)). The peaks X_1 to X_3 result from magneto-impurity resonance (equation (14)). The electric fields applied are: a 12 V cm^{-1}; b 6.3 V cm^{-1}; c 4.5 V cm^{-1}; d 2.4 V cm^{-1}; e 1.4 V cm^{-1}; f 1.1 V cm^{-1}; g 0.2 V cm^{-1}; h 8.7 V cm^{-1}; i 2.4 V cm^{-1}. (Eaves et al., 1974b.)

resonances to appear is found to obey (4) with a resonance energy which is considerably below the L.O. phonon energy. The value of the resonance energy was found to correspond to a pair of band-edge T.A. phonons at the X-point, so that resonances occur when

$$2\hbar\omega_{T.A.} = N\hbar\omega_c. \tag{13}$$

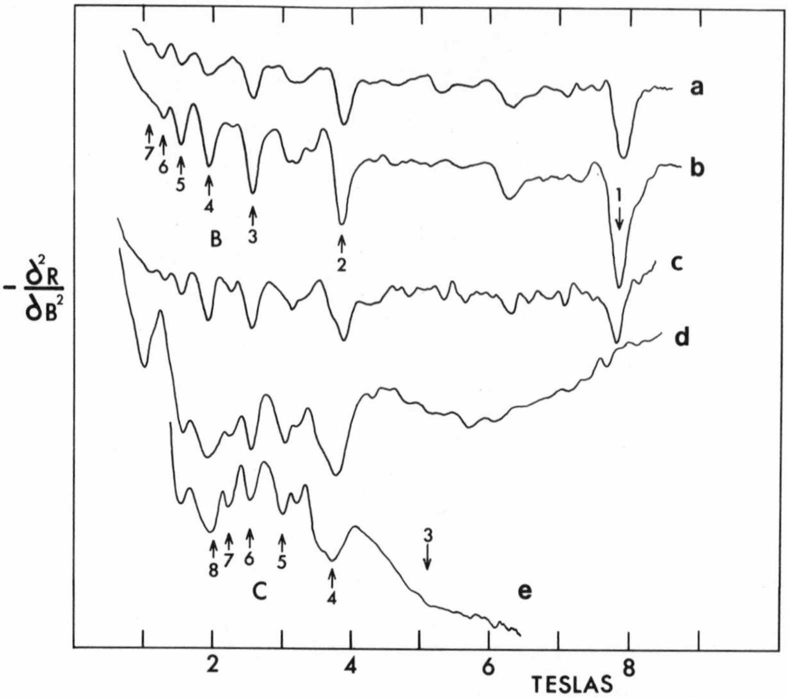

Fig. 14. Magnetophonon structure in n-CdTe. Series B corresponds t 2TA(X) emission (equation (13)) while series C results fr L.O. phonon emission accompanied by capture at 2s states (equation (14)). The lattice temperature were: a and b 20 K; c 17 K; d 14 K; and e 11 K.

This effect was first observed in InSb by Stradling and Wood (1970) and is illustrated in Figs. 12, 13, and 14, for InSb, InAs, InP and CdTe, and has also been observed in GaAs (Stradling, et al., 1970). The obvious importance of this energy relaxation mechanism is due partly to the very high density of phonon states near the edge of the Brillouin zone, where the T.A. phonon dispersion curves are almost flat. It is also partly due to the relatively higher electron population at the energies needed to emit T.A. phonon pairs since the emission of L.O. phonons requires electrons to be excited further into the band, with a consequently lower electron populatior

As the electric field is increased, and the carriers are heated still further out of equilibrium with the lattice, it is found that second series of oscillations occurs. This behaviour is illustrated in Figs. 12 and 13 for InSb and InP. The resonances for this series of oscillations are found to be well predicted by (10), which was

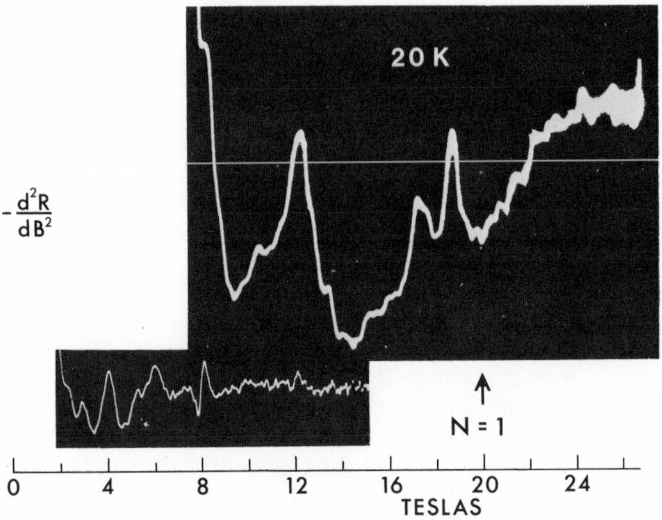

Fig. 15. Magnetophonon structure for InP at 20 K at high magnetic fields. The peak at 12 T is the fundamental (N = 1) of the 2TA(X) series, while the fundamental of the L.O. phonon-impurity capture series should occur at 20 T as shown.

shown above for GaAs to be the emission of an L.O. phonon accompanied by the capture of an electron from a Landau level into a shallow hydrogenic donor state.

Von Klitzing and Landwehr (1974) have shown that similar resonances occur for p-type tellurium due to the emission of an L.O. phonon accompanied by capture at a shallow acceptor level. Further complications arise in tellurium due to the "camels-back" nature of the valence band, which allows possible scattering at several different phonon energies, although the resonance positions seem to be well predicted by (10).

It is only at very high magnetic fields that (10) is found to break down. Even when the accurate magnetic field dependent donor binding energy $E_I(B)$ is used in (8), it is found that the observed structure is more complex than predicted and is shifted in magnetic-field position. This is illustrated in Fig. 15 for InP, where the N = 1 resonance is split into two and is shifted by $\sim$ 10% from the predicted position. Similar results have been observed by Hoult (1974) for GaAs.

The situation for n-CdTe is found to be more complex (present authors, to be published) with the observed resonances found to depend very strongly on both lattice temperature and impurity concentration. For more heavily doped samples, peaks were found to be described by (10), while for more lightly doped samples at temperatures below 14 K, L.O. phonon emission was found to be accompanied by capture at the 2s excited state of the donors. These resonances are shown in Fig. 14, and are determined by the relation

$$N\hbar\omega_c = \hbar\omega_{L.O.} - E_{2s}(B), \tag{14}$$

where $E_{2s}(B)$ is the magnetic-field-dependent binding energy for an electron in a 2s excited state. The origin of the dependence upon impurity concentration is thought to be due to the variation in strength of the phonon-emission rate to impurity states, which may greatly distort the electron distribution function for strong coupling. The low electron concentrations in the very pure samples used for magnetophonon studies are not sufficient for electron-electron collisions to maintain a Maxwell-Boltzmann-like distribution. It has been shown by Yamada (1973, 1974) and by Barker and Magnusson (1974) that this situation may be described by a process of diffusion through energy space until a very strong energy loss mechanism is reached, whereupon electrons rapidly lose energy and return to the bottom of the band. If the emission rate for L.O. phonons accompanied by electron capture to donors is sufficiently rapid, then electrons will be unable to reach an energy sufficiently high to emit L.O. phonons and remain in the band. Thus magnetophonon resonances given by (5) will not occur. In addition, the very strong distortion of the electron distribution function in the region of the Landau level makes it extremely difficult to predict whether magnetophonon resonances should give rise to maxima or minima in the resistivity. In practice, resonance positions are determined by the presence of peak sharpening.

As described above for higher temperatures, the hot-electron magnetophonon effect in n-silicon is considerably more complex than in materials with a single conduction band minimum located at the Γ-point. Recently impurity capture associated magnetophonon resonances have been observed in n-silicon (Portal, et al., 1979) at temperatures below 35 K, as shown in Fig. 16. Such effects are complicated still further, since the multi-valleyed conduction band results in a valley-orbit splitting of the 1s donor state into a triplet of 1s(A), 1s(E) and 1s(T) states. The possible resonance energies are shown in Table I, and as may be seen, the large donor binding energy results in a very large shift in possible resonance energy away from the original phonon energy. The appearance of the new resonances with falling temperature is shown in Fig. 16 for B||(100) and the spectra are shown in Fig. 17 as a function of electric field for B||(111). It was found to be possible to describe all of the observed magnetophonon series in terms of capture of

TABLE 1. The experimentally observed magnetophonon series fundamentals at zero stress.

Orientation	Series Fundamental field (Teslas)	m*	Resonance energy (K)	Calculated energy from equation (11) (K)	Capture process with phonon and final state
B||(100)	23.5	m_1	164	157	685(f) - 1s(A)
				142	535(f) - 1s(T)
				157	535(f) - 1s(E)
	31.8	m_1	220	207	735(g) - 1s(A)
				192	585(f) - 1s(T)
				207	585(f) - 1s(E)
	40	m_1	275	292	685(f) - 1s(T)
				307	685(f) - 1s(E)
	49	m_2	158	157	685(f) - 1s(A)
				142	535(f) - 1s(T)
				157	535(f) - 1s(E)
		m_1	333	342	735(g) - 1s(T)
	62	m_2	200	207	735(g) - 1s(A)
				192	585(f) - 1s(T)
				207	585(f) - 1s(E)
	93	m_2	299	292	685(f) - 1s(T)
				307	685(f) - 1s(E)
B||(110)	30.5	m_1	166	157	685(f) - 1s(A)
				142	535(f) - 1s(T)
				157	535(f) - 1s(E)
	64	m_1	206	207	735(g) - 1s(A)
				192	585(f) - 1s(T)
				207	585(f) - 1s(E)
B||(111)	34.5		153	157	685(f) - 1s(A)
				142	535(f) - 1s(T)
				157	535(f) - 1s(E)
	42		186	207	735(g) - 1s(A)
				192	585(f) - 1s(T)
				207	585(f) - 1s(E)
	61		265	292	685(f) - 1s(T)
	67		295	307	685(f) - 1s(E)

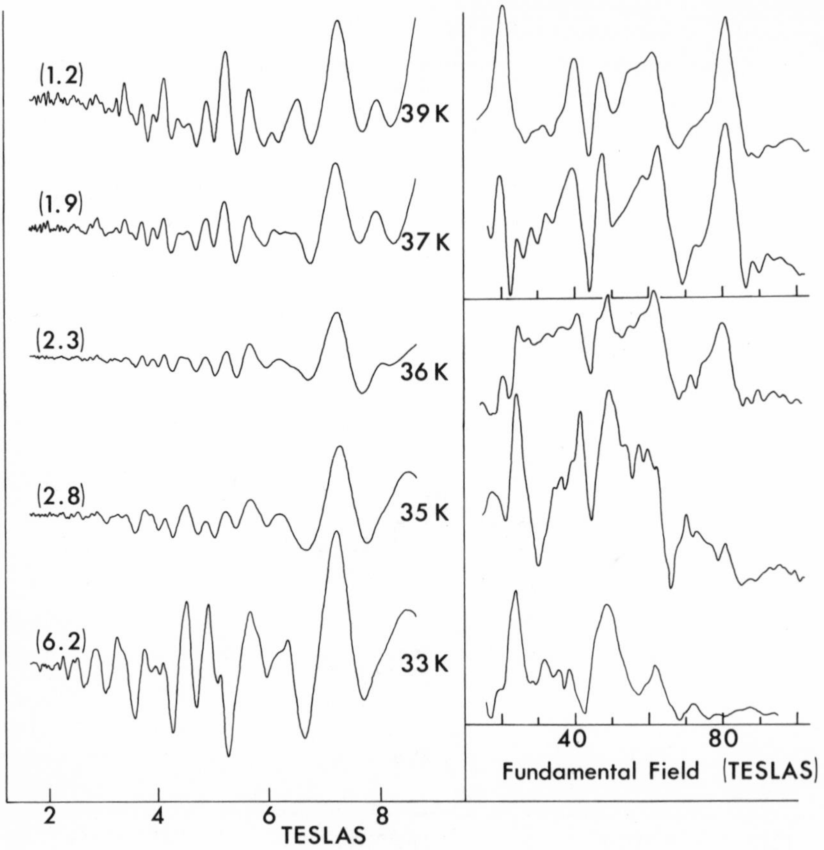

Fig. 16. The magnetophonon structure in n-Si as a function of temperature for j||B||(100). The recordings on the left show the original structure in the magnetoresistance while those on the right show the Fourier transforms of the originals. At 39 K the observed structure corresponds to the series observed at high temperatures.

either 735 K (g-type), 685 K (f-type) or 535 K (f-type) phonons at the three valley-orbit split 1s states of the shallow phosphorus donors which were known to be present in the samples investigated. The observed series of oscillations were well described by (10) with the appropriate cyclotron effective masses. The experimentally determined series fundamentals are also shown in Table I, together with the probable assignments of the processes occurring. In order to make accurate assignments of the processes, a study was made of the influence of uniaxial stress. The stress, in addition to

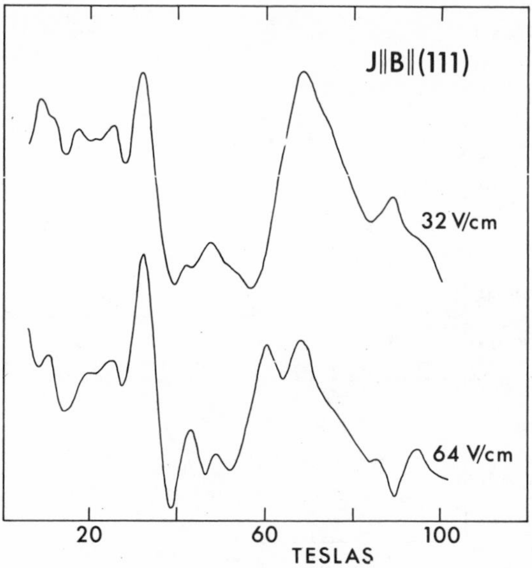

Fig. 17. Fourier spectra of the oscillatory magnetoresistance of n-Si at 30 K for j||B||(111).

splitting the conduction band valleys, shifts and splits the donor 1s states. The series fundamentals thus have a dependence upon stress (X) which is characteristic of the stress dependence of the donor binding energy, $E_I(0,X)$. Figure 18 compares the experimentally observed peaks in the Fourier spectrum with the calculated stress dependence of $E_I(0,X)$, giving excellent agreement and good confirmation of the assignments given in Table I. All f-phonon scattering is strongly suppressed by the stress, and the only features present at high stresses result from g-phonon scattering by 735 K phonons, previously undetected weak scattering by 195 K g-phonons, and some intra-valley scattering by 755 K optic phonons.

It has recently been found that one very sensitive technique for observing hot-electron energy loss mechanisms is to look at resonances in the magnetic field dependent photovoltage, which is produced by illumination with light of energy greater than the band gap. This technique was first employed by Parfen'ev, et al. (1968) to investigate n-InSb. Heating of electrons by sub-band gap radiation has also been employed by Morita, et al. (1973) who used a CO_2 laser to observe magnetophonon oscillations in n-InSb. A typical example of the use of above-gap radiation is shown in Figure 19 for a sample of semi-insulating GaAs at 2.2 K. The resonances observed correspond to the L.O. phonon-impurity capture process described above. The structure is independent of the excitation energy of the

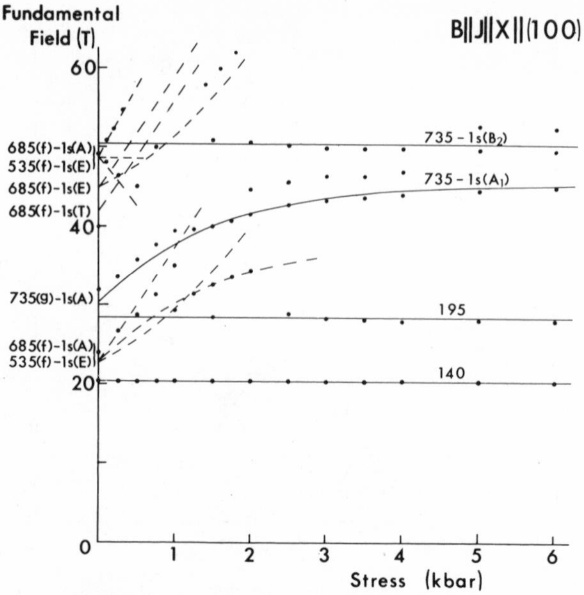

Fig. 18. The dependence of the peaks in the Fourier spectra of n-Si at 30 K upon the uniaxial stress applied to the sample. The solid lines represent the stress dependence of g-scattering processes calculated from equation (8) taking into account the stress dependence of $E_I(B)$. The dashed lines represent the calculated dependence for f-processes.

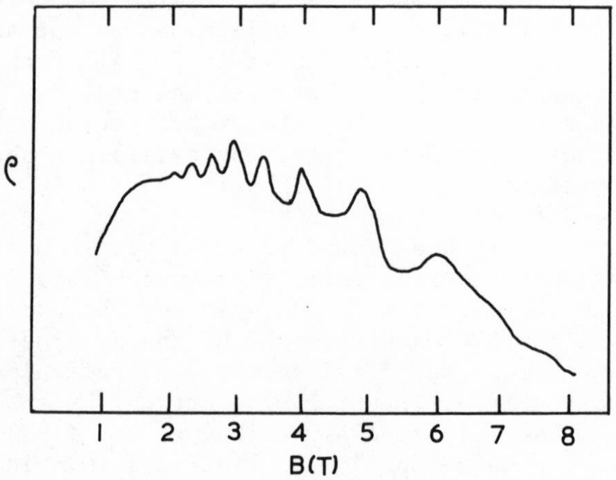

Fig. 19. Impurity capture associated magnetophonon resonances in the photo-magnetic E.M.F. of a sample of semi-insulating (Cr doped) GaAs at 2.2 K. (Eaves, Instone, Portal unpublished data.)

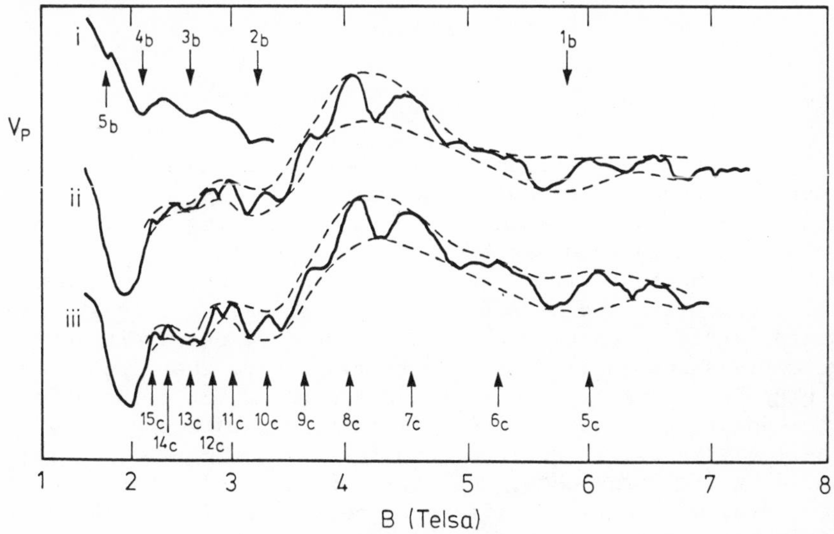

Fig. 20. The photomagnetic E.M.F. with the monotonic part subtracted for n-Ge doped with P and Sb. The magneto-impurity and magnetophonon resonances are shown by arrows b and c respectively (Instone et al., 1977).

carriers, identical results being found for excitation both by white and by monochromatic light of wavelength 633 nm and 514.5 nm. This technique was used by Instone, et al. (1977) to observe the first hot-electron magnetophonon resonance in n-type germanium. The magnetophonon structure observed in n-germanium is much simpler than for silicon since the group theoretical selection rules allow intra-valley scattering (Streitwolf, 1970), which is found to dominate the "normal" magnetophonon effect (Eaves, et al., 1970, 1975). Typical results observed for phosphorus doped germanium are shown in Figure 20. The oscillations result from an intra-valley process with the emission of L.O. phonons accompanied by electron capture at impurities. The oscillations are once again accurately described by (10), however a splitting of the peaks at high fields in both phosphorus (Fig. 20) and Antimony doped material is thought to result from the weak valley-orbit splitting of the donor 1s states which also occurs in n-germanium. It has so far proved impossible to detect these hot-electron magnetophonon resonances in the ordinary magnetoresistance of germanium.

Similar hot-electron magnetophonon resonances have also been observed in some materials in other transport coefficients. Dolat and Bray (1970) measured the acoustoelectric gain in InSb and found a strong resonant increase due to the emission of L.O. phonons.

Hot-electron magnetophonon resonances have also been observed in the Hall coefficient of InP (Eaves, et al., 1974b).

MAGNETO-IMPURITY RESONANCE (M.I.R.)

In the previous section it was shown that resonant cooling of hot carriers could occur due to the emission of phonons. In this section, a new type of resonant cooling or heating of electrons is described which is known as the magneto-impurity effect. Magneto-impurity resonance (M.I.R.) has been recently observed in n-GaAs, n-InP, p-tellurium and n- and p-type germanium, usually giving rise to sharp structure in the magnetoresistance or photovoltage. The effect arises due to inelastic scattering where there is an exchange of energy between a free carrier and another which is bound at a shallow donor or acceptor level. In a strong magnetic field the Landau levels are well defined, and the scattering will be strongly enhanced by the large densities of states at the bottom of each Landau level whenever the difference between Landau sub-bands is equal to the energy required for scattering. Thus the resonance condition is given by

$$N\hbar\omega_c = E_2(B) - E_1(B), \tag{15}$$

where $E_2(B)$ and $E_1(B)$ are the impurity energy levels, one of which is usually the ground state.

In order to observe M.I.R. it is necessary to be in the hot electron regime, which is realised either by heating the electrons with an electric field or by the photo-excitation of carriers across the band gap.

The magneto-impurity effect is thus usually observed at low temperatures, under conditions very similar to those where magnetophonon oscillations associated with electron capture are found to occur. This is illustrated in Fig. 13 for InP, where the structures marked X will now be attributed to M.I.R., and in Fig. 20 for n-Ge where the series b results from M.I.R.

The magneto-impurity effect has been recently reviewed in detail by Eaves and Portal (1979) and this section will give a brief summary of the results obtained in the four materials studied to date.

n-GaAs and n-InP

M.I.R. has been observed in high purity samples of n-GaAs and n-InP under conditions of weak heating by an electric field. Figure 21 shows a M.I.R. series (peaks y_1 to y_5) in the longitudinal

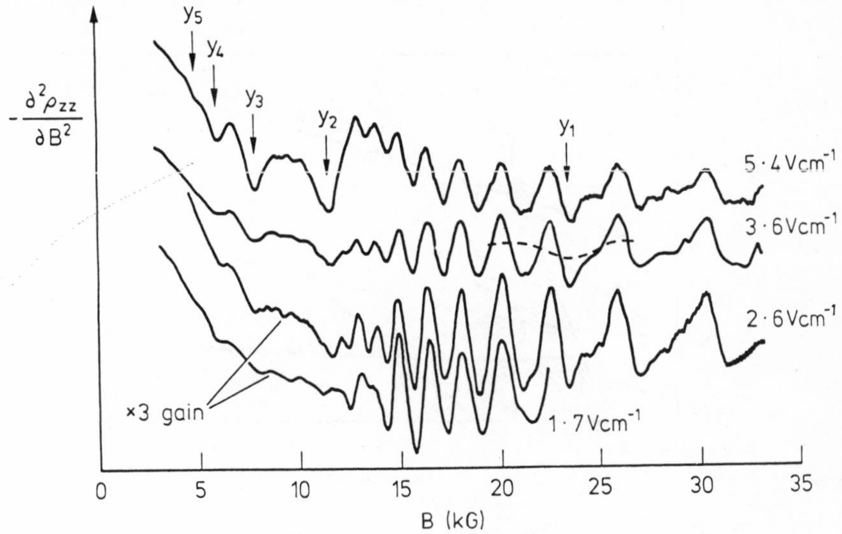

Fig. 21. The electric field dependence of the amplitude of the magneto-impurity series (peaks y_1 to y_5) in the longitudinal magneto-resistance of n-GaAs at 11 K. The peaks occurring at higher fields are due to magnetophonon resonance (Hoult, 1974).

magnetoresistance of n-GaAs at 11 K (Hoult, 1974). The oscillations are periodic in B^{-1} in agreement with (15) and give a fundamental field of 2.4 T. This series is due to resonant scattering of electrons between two Landau levels accompanied by a transition from the 1s to the 2p-level of a neutral donor. Figure 13 shows three sharp peaks x_1, x_2 and x_3 in the longitudinal magnetoresistance of InP at low applied electric fields, while at higher electric fields the structure becomes dominated by magnetophonon oscillations. The peaks correspond to M.I.R. with n = 1, 2 and 3. The process is different from that in n-GaAs, since for the impurity the transition goes from a 1s state to the bottom of the conduction band, i.e. to the N = 0 Landau level, and $E_2(B) = 0$ in (15). This series of oscillations has also been observed in the transverse magnetoresistance (ρ_{xx}), the Hall coefficient ρ_{xy} and the warm electron coefficient β, in which Nicholas and Stradling (1978) have reported resonances up to N = 6.

For InP, the fundamental (N = 1) M.I.R. peak occurs at 8.1 T and is extremely strong, being observable directly in the magnetoresistance. When studied carefully using a second derivative technique, this peak shows a splitting of 0.12 T due to a central-cell splitting of the two dominant shallow donors in InP (Nicholas and Stradling, 1978).

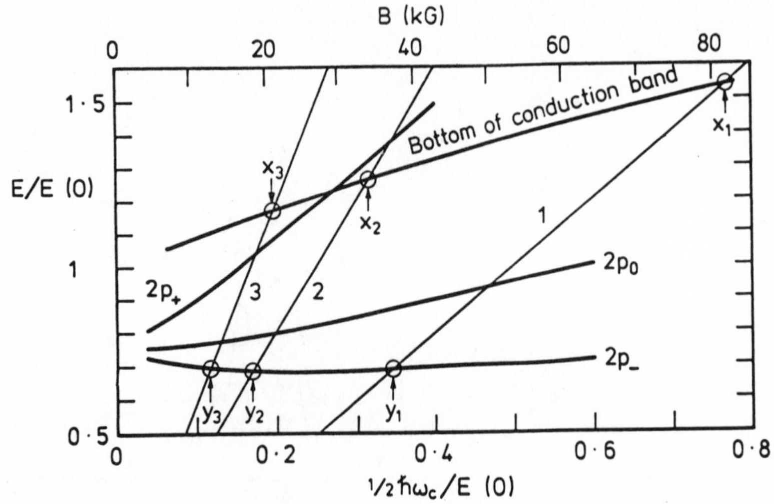

Fig. 22. Bold Lines: The energies of the $2p_-$, $2p_0$ and $2p_+$ bound states and the (N = 0, k_B = 0) lowest Landau state relative to the 1s state for shallow donors as a function of magnetic field. The vertical axis is in units of the zero field hydrogenic binding energy E(0) and the lower horizontal axis is in dimensionless units of $\frac{1}{2}$ $h\omega_c$/E(0). The upper horizontal axis (kG) corresponds to the case of InP only. Fine Lines: The energy intervals between Landau levels: Line 1 = $h\omega_c$; Line 2 = $2h\omega_c$; Line 3 = $3h\omega_c$. The intersections of the two sets of lines give the positions of the M.I.R. The x intersections are observed in InP whereas the y intersections are found in GaAs. (Eaves and Portal, 1979).

The different nature of the two M.I.R. series in n-GaAs and n-InP is shown in Fig. 22. It is difficult to tell whether the resonances correspond to cooling or heating (in an 'Auger-like' process) of the free carriers. This problem has been considered in detail by Nicholas and Stradling (1978) and by Eaves and Portal (197 who conclude that heating occurs in GaAs while both heating and cooling occur in InP depending upon the harmonic number N and the magnitude of the applied electric field. The effects may thus be considered to offer a good example of the complexity of hot-electron problems.

p-Tellurium

Von Klitzing (1978) has observed M.I.R. in the transverse magneto-resistance of p-tellurium at 2.2 K. The origin of the

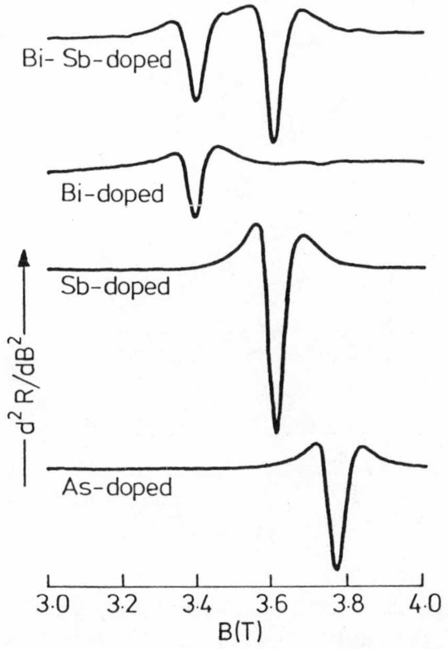

Fig. 23. Second derivative of the magnetoresistance for intentionally doped p-Te with B||i|c and T < 1.8 K (Von Klitzing, 1978).

resonances is similar to that in GaAs. However, in addition to this, there are several extremely sharp features which appear to be related to the chemical nature of the impurities. These are illustrated in Fig. 23 and are peculiar to tellurium, due to the "camel-back" form of the valence band. These peaks show a very pronounced splitting into four components, as the magnetic field is rotated away from B⊥c. It has been suggested by Eaves and Portal (1979) that these sharp structures arise from transitions between the shallow acceptor levels accompanied by the emission or absorption of an inter-valley acoustic phonon. Very recent measurements of the magnetic field dependence of the shallow impurity levels by von Klitzing (private communication), show clearly, however, that the true origin of these structures lies in resonant transitions only between impurity levels, with no phonon involved. The characteristic splitting into four components arises due to the g-factor splitting of the bonding and anti-bonding ground states, while the strong dependence upon the chemical nature of the impurity comes from the magnitude of the valley-orbit splitting.

n-type and p-type Germanium

The first observation of M.I.R. in p-germanium was made by Gantmakher and Zverev who have subsequently made a detailed study of

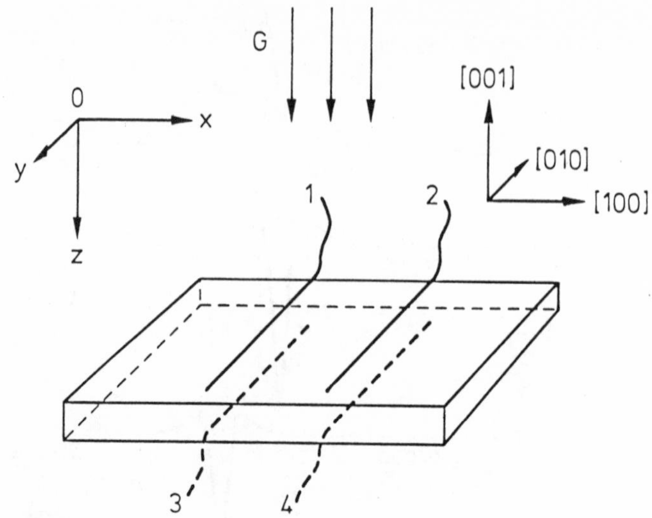

Fig. 24. Position of electric contacts 1, 2, 3, 4 and the direction G of the incident light in the experiments on p-Ge of Gantmakher and Zverev (1976a).

the effects of illumination, electric field, temperature, impurity concentration, and uniaxial stress on the resonant extrema (Gantmakher and Zverev, 1975-77; Zverev, 1977). The effect has also been observed recently in n-germanium by Instone, et al. (1977). In both n- and p-type material, the resonances arise from scattering of warm electrons which have been photo-excited across the band gap. Under these conditions both magnetophonon and magnetoimpurity structure are observed in n-germanium as shown in Fig. 20. In n-type samples scattering occurs from P and Sb donors, while for p-germanium interactions have been observed with B, Ga and In acceptors.

Figure 24 shows the experimental arrangement used by Gantmakher and Zverev, who illuminate one side of the sample, usually with a He-Ne laser. With the magnetic field $B||y$, the photomagnetic effect was observed by studying E_{GH}, the electric field proportional to the voltage between contacts 1 and 2. Measurements of the longitudinal magnetoresistance were made with $B||x$, and of the transverse magnetoconductivity with $B||z$. In all three configurations the magnetic field was usually parallel to a (100) axis.

Figure 25 shows magneto-impurity structure in E_{GH}, and illustrates that the structure disappears for impurity concentrations greater than 10^{15} cm^{-3}. The period of the oscillations is dependent upon the type of impurity due to their differing binding energies

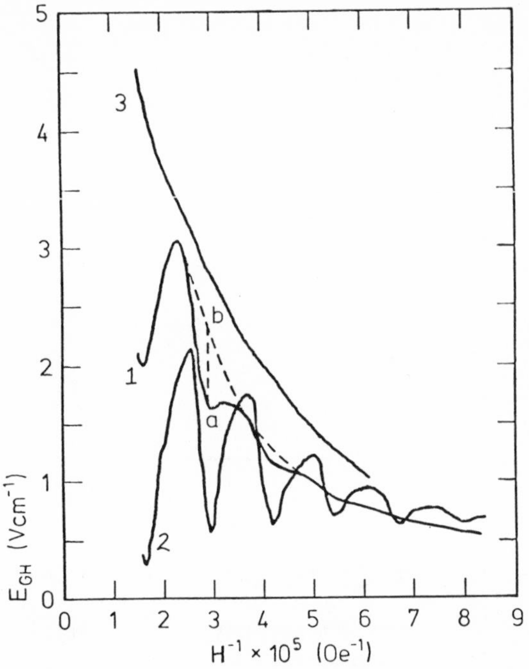

Fig. 25. Magneto-impurity structure in the electric field E_{GH} for Ga-doped Ge with $\underline{B}||(100)$. Curve 1: $\sim 3\ \Omega$ cm, T = 2.14 K; Curve 2: $\sim 40\ \Omega$ cm, T = 1.31 K; Curve 3: $\sim 1\ \Omega$ cm, T = 2.2 K. The resonances are not observed in Curve 3 for which the acceptor concentration is $3 \times 10^{15}\,cm^{-3}$. At these doping levels the excited states of the acceptors start to overlap. In each case the incident photon flux $\approx 10^{17}\,cm^{-2}s^{-1}$, $\lambda = 533$ nm. (Gantmakher and Zverev, 1976a).

(Figure 26). The resonance positions predicted by (14) correspond to the cyclotron energy of electrons in the lowest conduction band valley at the L-point of the Brillouin zone. The transition energy of the acceptors corresponds to the G transition (in the notation of Haller and Hansen, 1974; Skolnick, et al., 1974), which connects a $_8(2p_{3/2})$ excited state with the $_8(1s_{3/2})$ ground state. This transition energy is relatively insensitive to magnetic field.

Gantmakher and Zverev have shown that varying the intensity of illumination, the temperature, or the electric field may invert the extrema, and so turn maxima into minima at resonance. This extreme sensitivity of the extreme, together with the dependence upon uniaxial stress of the positions of the resonances (Zverev, 1977), has led Gantmakher and Zverev to conclude that the extrema arise from a

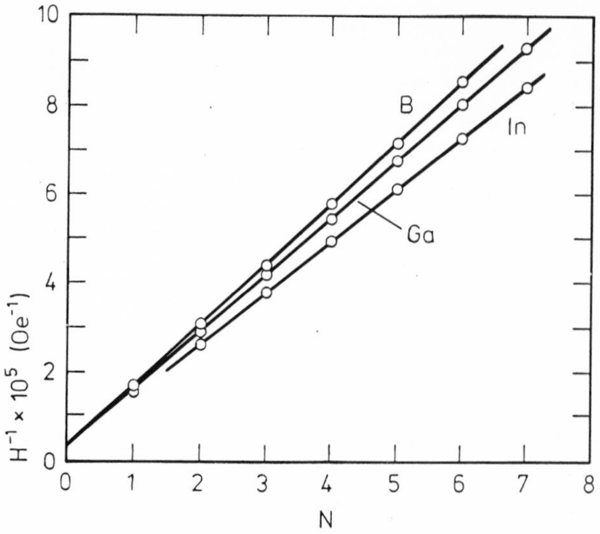

Fig. 26. Dependence of the position of the extrema on the integer N for different acceptors. The slope of the curves gives the acceptor transition energy using Equation (15) (Gantmakher and Zverev, 1976a).

resonant heating of cold electrons by de-excitation of acceptors (an Auger process) rather than from impact excitation of neutral acceptors.

Figure 20 shows magneto-impurity resonances observed in n-type germanium by Instone et al. (1977). The series of resonances 1b to 5b corresponds to M.I.R. given by equation (15) in which electrons in the L-valleys scatter inelastically off shallow donors causing transitions between the 1s and $2p_-$ states. It is not at present clear whether this process corresponds to a resonant heating or cooling of the conduction electrons.

ACKNOWLEDGEMENTS

The authors would like to thank S. Askenazy (Toulouse), L. Eaves (Nottingham), V. F. Gantmakher (Chernogolovk), K. von Klitzing (Würzburg) and R. A. Stradling (St. Andrews) for many useful discussions and for making available their experimental results prior to publication.

REFERENCES

Adams, E. N., and Holstein, T. D., (1959), J. Phys. Chem. Solids 10:254.

Argyres, P. N., (1959), J. Phys. Chem. Solids 4:19.

Barker, J. R., and Magnussen, B., (1974), "Proc. 12th Int. Conf. on Physics of Semiconductors," Stuttgart, p. 811.

Bauer, G., and Kahlert, H., (1970), "Proc. 10th Int. Conf. on Physics of Semiconductors", Boston, p. 65.

Bauer, G., and Kahlert, H., (1972), Phys. Rev. B 5:566.

Bauer, G., Kahlert, H., and Kocevar, P., (1975), Phys. Rev. B 11:968.

Bir, G. L., Krigel, V. G., Pikus, G. E., and Farbstein, I. I., (1974), Zh. Eksp. Teor. Fiz. Pis. Red. 19:48 (JETP Lett. 19-20: 29.

Couder, Y., (1969), Phys. Rev. Lett. 22:890.

Couder, Y., Hulin, M., and Thomé, H., (1973), Phys. Rev. B 7:4373.

Dolat, V., and Bray, R., (1970) Phys. Rev. Lett. 24:262.

Dolling, G., (1963), "Inelastic Scattering of Neutrons in Solids and Liquids", I.A.E.A., Vienna, p. 37 (Vol. 2).

Dorda, G., (1975), "Festkörperprobleme XIII" Springer-Verlag, Berlin, p. 215.

Eaves, L., Hoult, R. A., Stradling, R. A., Tidy, R. J., Portal, J. C., and Askenazy, S., (1975), J. Phys. C 8:1034.

Eaves, L., and Portal, J. C., (1979), J. Phys. C, to be published.

Eaves, L., Portal, J. C., Askenazy, S., Stradling, R. A., and Hansen, K., (1974a), Solid State Commun. 15: 1281.

Eaves, L., Stradling, R. A., Askenazy, S., Barbaste, R., Carrere, C., Leotin, J., Portal, J. C., and Ulmet, J. R., (1974b), J. Phys. C 7:1999.

Eaves, L., Stradling, R. A., and Wood, R. A., (1970), "Proc. 10th Int. Conf. on Physics of Semiconductors", Boston, p. 816.

Gantmakher, V. F., and Zverev, V. N., (1975), Zh. Eksp. Teor. Fiz. 69:695 (Sov. Phys. JETP 42:352).

Gantmakher, V. F., and Zverev, V. N., (1976a), Zh. Eksp. Teor. Fiz. 70:1891 (Sov. Phys. JETP 43:985).

Gantmakher, V. F., and Zverev, V. N., (1976b), "Proc. Int. Conf. on Physics of Semiconductors," Rome, pp. 1154-7.

Gantmakher, V. F., and Zverev, V. N., (1976c), Zh. Eksp. Teor. Fiz. 71:2314 (Sov. Phys. JETP 44:1220).

Gantmakher, V. F., and Zverev, V. N., (1977a), Pis'ma Zh. Eksp. Teor. Fiz. 25:44 (JETP Lett. 25:39).

Gantmakher, V. F., and Zverev, V. N., (1977b), Zh. Eksp. Teor. Fiz. 73:2337 (Sov. Phys. JETP 46:1223).

Haller, E. E., and Hansen, W. L., (1974), Solid State Comm. 15:687.

Hamaguchi, C., Shirakawa, T., Yamashita, T., and Nakai, J., (1972), Phys. Rev. Lett. 28:1129.

Harper, P. G., Hodby, J. W., and Stradling, R. A., (1973), Rep. Prog. Phys. 37:1.

Hess, K., Englert, T., Neugebauer, T., Landwehr, G., and Dorda, G., (1977), Phys. Rev. B 16:3652.
Hoult, R. A., (1974), Thesis, unpublished.
Instone, T., Eaves, L., Portal, J. C., Houlbert, C., Perrier, P., and Askenazy, S., (1977), J. Phys. C 10:1585.
Kahlert, H., and Bauer, G., (1973), Phys. Rev. B 7:2670.
von Klitzing, K., (1974), Solid State Comm. 15:1721.
von Klitzing, K., (1978), Solid State Electr. 21:223.
Kubo, R., Mikaye, S. J., and Hashitsume, N., (1965), "Solid State Physics", New York, Academic Press, 17:269.
Morita, S., Takano, S., and Kawamura, H., (1973), Solid State Comm. 12:175.
Nicholas, R. J., and Stradling, R. A., (1976), J. Phys. C 9:1253.
Nicholas, R. J., and Stradling, R. A., (1978), J. Phys. C 11:L783.
Parfenev, R., Farbshkin, I. I., and Shalyt, S., (1967), Sov. Phys. JETP 26:906.
Portal, J. C., Eaves, L., Askenazy, S., and Stradling, R. A., (1974), Solid State Comm. 14:1241.
Portal, J. C., Perrier, P., Houlbert, C., Askenazy, S., Nicholas, R. J., and Stradling, R. A., (1979), J. Phys. C, to be publishe
Skolnick, M. S., Eaves, L., Stradling, R. A., Portal, J. C., and Askenazy, S., (1974), Solid State Comm. 15:1403.
Stradling, R. A., Eaves, L., Hoult, R. A., Mears, A. L., and Wood, R. A., (1970), "Proc. 10th Int. Conf. on Physics of Semiconductors," Boston, p.369.
Stradling, R. A., and Wood, R. A., (1968), Solid State Comm. 6:701.
Stradling, R. A., and Wood, R. A., (1970), J. Phys. C 3:2425.
Stratton, R., (1958), Proc. Roy. Soc. (London) A246:406.
Streitwolf, N. H., (1970), Phys. Stat. Solidi 37:K47.
Wallace, P. R., and Joos, B., (1978), J. Phys. C 11:303.
Yamada, Y., (1974), Solid State Comm. 13:503.
Yamada, Y., and Kurosawa, T., (1973), J. Phys. Soc. Jpn. 34:603.
Zverev, V. N., (1977), Fiz. Tverd. Tela, 19:2015 (Sov. Phys. Solid State, 19:1179).

HOT ELECTRON DISTRIBUTION FUNCTION IN QUANTIZING MAGNETIC FIELDS

D. Calecki

Groupe de Physique des Solides
de l'Ecole Normale Supérieure
Université de Paris VII, France

INTRODUCTION

The successful generation of strong magnetic fields has made it possible to investigate experimentally hot electrons in semiconductors under conditions of quantization of their orbital motion. Such magnetic fields $\vec{B}$ are called "quantizing fields"; they alter significantly the field free electronic energy spectrum. The electron motion in the plane perpendicular to $\vec{B}$ is quantized in Landau states and the remaining degree of freedom of motion along $\vec{B}$ constitutes one-dimensional subbands, each of which is associated with one of the Landau states [Landau (1930)]. Moreover, the electronic density of states has an oscillatory structure with divergences characteristic of the one dimensional motion.

To observe quantizing effects, two conditions must be satisfied:

$$\hbar\omega_c >> kT_e$$

and

$$\omega_c\tau >> 1$$

where $\omega_c = eB/m$ is the cyclotron frequency, τ is the average lifetime in a Landau state, and T_e is the electron gas temperature.

When a high electric field $\vec{E}$ is applied, (in practice $\vec{E}$ is either parallel or perpendicular to $\vec{B}$) the energy acquired by electrons from the electric field can be larger than the energy given up to the lattice in the same time period. Thus electrons in a given subband can be heated to an energy sufficient to jump to higher subbands, and the electron distribution function (EDF) becomes very different from the thermal equilibrium situation.

The behaviour of hot electrons in quantizing magnetic fields is interesting for several reasons. The electron heating process is fa different from the B = 0 case because we are concerned with motion i a one-dimensional electronic band rather than the normal three-dimensional case. We expect that various non-linear I(V) characteristics with S and N shapes can occur, and these can be changed over wide range by the electric or magnetic fields and by the sample geometry (Hall geometry and Corbino disc). The control of the subband structure by the magnetic field leads to a resonance phenomena when it is coupled to some energy loss process with a definite threshold energy such as optical phonon emission. Such an effect, called the magnetophonon effect, is well known in the absence of electron heating, and has become a useful spectroscopic tool in the hot carrier regime. These questions are extensively discussed in tw excellent review papers by Zlobin and Zyrianov (1972) and by Barker (1978).

Despite extensive experimental and theoretical studies, hot electrons in quantizing magnetic fields are not quantitatively well understood. We cannot use the semi-classical Boltzmann equation, which is only valid in the range $\hbar\omega_c << kT_e$. We need a quantum transport theory based on the density operator ρ and a convenient master equation for its diagonal matrix elements. The influence of electron interactions with other electrons, acoustical and optical phonons, and impurities must be included in the expression for ρ. Its diagonal elements will give the EDF and the off-diagonal element will be used in the determination of the conductivity.

As in the absence of a magnetic field, the competition between carrier-carrier scattering and electron-phonon collisions is the central problem of hot electrons. It is evident that carrier-carrie scattering complicates matters if we want a rigorous treatment of th problem; but we also (over) simplify the hot electron problem if we merely assume that electron-electron collisions effectively redistribute energy among the electrons of the gas, and therefore the electrons are in a quasi-equilibrium state described by a Maxwell or a Fermi-Dirac function with an electric field dependent temperature T_e. In practice, the most important effect of a quantizing magneti field is to reduce strongly the ratio ν_{ee}/ν_{ep} (of the electron-electron collision frequency to the electron-phonon collision one); ν_{ee} is reduced by an enormous factor of order exp $(\hbar\omega_c/kT_e)$ whereas ν_{ep} is enhanced by a factor proportional to B^2[Zlobin and Zyrianov (1970)]. The origin of this exponential factor is qualitatively well understood; when $\hbar\omega_c >> kT_e$ in the non-degenerate electron gas, the majority of the carriers occupy the first Landau level (N = 0) and the number of electrons in the second level (N = 1) is smaller by the Boltzmann factor $\exp(\hbar\omega_c/kT_e)$. (In so doing, we assume the existence of an electron temperature T_e.) Furthermore, electron-electron collisions belonging to the same Landau level do not affect the distribution function because of the one-dimensional character of the electron motion parallel to the magnetic field. It therefore becomes necessary to consider collisions between electrons on different Landau levels such as N = 0 and N = 1. As ν_{ee} is proportional to the occupancy of the two levels, we are not surprised by the appearance of the exponential factor in its expression. In conclusion, the *a priori* use of an equilibrium distribution with an electron temperature is rather questionable, and this is well reflected in the literature on hot electrons in a quantizing magnetic field. We can classify the different papers on the subject in three main classes:

(1) The first set uses the simple concept of an electron temperature as a starting point [(Kogan (1963), Kalashnikov and Pomortsev (1964), Pomortsev and Kharus (1968), Pinchuk (1977), Licea (1971), Calecki (1971,1972), Cassiday and Spector (1974)]. The problem indeed is to determine the variations of T_e with the electric and magnetic fields, and also with the lattice temperature, taking into account the different scattering mechanisms of electrons, by phonons and by impurities. As in the B = 0 case, it is always the balance between the power received by electrons from the electric field and the power radiated by electrons to the phonons which give the expressions of T_e. The first power is expressed via the current density, whose expression must be correctly calculated. In what follows, we shall not present detailed calculations, but we shall establish qualitative results concerning T_e which will more readily enable us to discuss their interesting features.

(2) The second class of papers entirely neglects the electron-electron interactions--either the electron concentration is too small, or only the extreme quantum limit is treated. The first problem we are concerned with is to establish a master equation satisfied by the diagonal elements of the density operator and an expression for the electric current density. This has been conveniently done by Budd (1968) in the crossed-field configuration, and by Barker (1978) in the longitudinal one. Then, we have to search for solutions of the master equation to obtain the EDF and to determine for which conditions the concept of electron temperature is eventually meaningful. Kazarinov and Skobov (1962,1963) were the first to discuss the case of crossed fields. Their starting point is not entirely correct (in particular, their expression of the electrical current is questionable) and, later on, Yamashita and Inoue (1965,1969) considerably reduced the range of validity of their results. In an

entirely different spirit, Kurosawa and Yamada (1965,1973) have developed a phenomenological model describing the processes of energy exchange of an electron with the electric field and the phonons by a Brownian motion in energy space. Here again their starting point is not entirely rigorous, but the method is promising. It has recently been improved on by including both electron-electron scattering and collision broadening [Partl, et al., (1978)]. We shall present in Section 3 a method by Calecki (1969) that involves the Pauli master equation, derived by Budd (1968) and transform it into a Fokker-Planck equation. At the same time, we recover the arguments of the random walk approach of Kurosawa and Yamada (1973) on more rigorous grounds and finally we establish the restrictive conditions under which the concept of electron temperature can be used.

(3) The third set of papers is devoted to the effect of electron-electron collisions. Zlobin and Zyrianov (1970) were the first to calculate a critical density of electrons n_{cr}, such that for $n > n_{cr}$ one can use a thermal equilibrium EDF at temperature T_e However, their results seem applicable only in the case where no more than two Landau levels are occupied. In the extreme quantum limit, where only the first Landau level is populated, we have already said that binary electron-electron collisions do not contribute to the redistribution of energy between electrons. Nevertheless, a Maxwellian distribution has been found experimentally in InSb at T = 4°K, in a magnetic field of 10^4 Gauss and a concentration of 10^{14} cm^{-3} [Miyazawa (1969)]. This experimental fact led to the consideration in the extreme quantum limit of the possibility of scattering processes with three bodies taking part: electron-electron-charged impurity [Kogan, et al., (1976)] or electron-electron-phonon [Shadrin (1976)]. The interest of such studies lies in the fact that apparently the only real system for which three-body scattering processes determine gas kinetics is the electron gas in an ultra-quantizing magnetic field.

It would be necessary to complete this rapid review with many other interesting studies of quantizing magnetic fields: effect of collision broadening, influence of the electric fields in the Kane band structure, introduction of several valleys in a semiconductor, effect of heating of the phonon gas, impact of a high electric field on the screening in electron interactions . . . We have said no more on the distribution function of photoexcited electrons, which is also an exciting subject.

In what follows, we need to restrict ourselves to the main ideas and avoid a too lengthy technical development. After a qualitative introduction in the framework of the electron temperature concept, we shall discuss extensively the case of the EDF in the crossed-field configuration.

QUALITATIVE ESTIMATES OF ELECTRON TEMPERATURE

We shall begin with very classical results of electron motion in magnetic field and make use of the basic law of conservation of energy.

Crossed Electric and Magnetic Fields

When the high magnetic field condition: $\omega_c\tau_r \gg 1$ is fulfilled, where τ_r is the momentum relaxation time, the classical conductivity reduces to

$$\sigma_{xx} = \frac{n_e e^2 \tau_r}{m(\omega_c\,\tau_r)^2},$$

where n_e is the electron density. Assuming an infinite sample or a Corbino disc, the Hall field is zero and the energy balance equation takes the form

$$J_x E_x = \sigma_{xx} E^2 = \frac{n_e e^2 \tau_r}{m(\omega_c\,\tau_r)^2} E^2 = n_e \frac{\overline{\varepsilon}(E) - \overline{\varepsilon}_o}{\tau_\varepsilon} \tag{1}$$

where $\overline{\varepsilon}_o$ and $\overline{\varepsilon}(E)$ are equilibrium and non-equilibrium values of the electron mean energy and τ_ε is the energy relaxation time. All this is very approximate, as we have retained the linear expression for σ_{xx} and assumed the existence of a relaxation time for the energy; but let us go on.

Labelling R_L the Larmor radius of the cyclotron orbit, we have simply $m(R_L\omega_c)^2/2 = \overline{\varepsilon}_o$, and it follows from (1) that

$$\overline{\varepsilon}(E) = \overline{\varepsilon}_o[1 + \tfrac{1}{2}(\frac{eER_L}{\overline{\varepsilon}_o})^2 \frac{\tau_\varepsilon}{\tau_r}] \tag{2}$$

In the extreme quantum limit, where $\overline{\varepsilon}_o \ll \hbar\omega_c$, we can substitute for R_L its limiting value $\delta = (\hbar/m\omega_c)^{\frac{1}{2}}$.

Now let us estimate the ratio τ_r/τ_ε assuming that the electron energy and momentum are dissipated by acoustic phonons. The emission probablity per unit time of a phonon of momentum $\hbar q$ and energy $\hbar\omega_q = \hbar sq$ (s: sound velocity) is proportional to $(1 + N_q)$ and the same probability of absorption is proportional to N_q (the phonons distribution function at temperature T). The mean energy transferred to the lattice in one collision is

$$\Delta\varepsilon = \frac{(1 + N_q) - N_q}{1 + 2N_q} h\omega_q \tag{3}$$

But electrons interact preferentially with phonons whose wave vector $q \sim \delta^{-1}$; this results from the electron-phonon coupling and is a central point in the whole discussion about phonons in a quantizing magnetic field. In general, the phonon temperature T is such that $kT \gg \hbar s\delta^{-1}$ and N_q reduces to $kT/\hbar s\delta^{-1}$; thus

$$\Delta\varepsilon \simeq (\hbar s\delta^{-1})^2/kT$$

Now, remember that by definition, τ_ε is

$$\frac{d\bar{\varepsilon}}{dt} = -\frac{\bar{\varepsilon}}{\tau_\varepsilon} \Rightarrow \frac{\Delta\varepsilon}{\tau_r} \simeq \frac{\bar{\varepsilon}_o}{\tau_\varepsilon},$$

therefore

$$\frac{\tau_r}{\tau_\varepsilon} \simeq \frac{\Delta\varepsilon}{\bar{\varepsilon}_o} \simeq \left(\frac{\hbar s\delta^{-1}}{kT}\right)^2 \tag{4}$$

To assume a Maxwellian distribution with an electric field dependent electron temperature T_e implies that we replace $\bar{\varepsilon}(E)$ by kT_e; we finally get:

$$T_e \simeq T[1 + \tfrac{1}{2}(\frac{E}{sB})^2] \tag{5}$$

Indeed the factor ½ may be entirely spurious, but the dependence on E and B should be correct. In fact, such an expression for T_e has been obtained for the first time by Kazarinov and Skobov (1962,1963)

It is easy to take account of electron-impurity scattering. Impurities cannot exchange energy with electrons; so they do not change the value of τ_ε. But electron momentum is relaxed by impurities in a characteristic time τ_{ei}. The total momentum relaxation time is now τ_r such that:

$$\tau_r^{-1} = \tau_{ei}^{-1} + \tau_{ep}^{-1}$$

where τ_{ep} is due only to phonons. It follows that:

$$\frac{\tau_r}{\tau_\varepsilon} = \frac{\tau_{ep}}{\tau_\varepsilon}\frac{\tau_{ei}}{\tau_{ei}+\tau_{ep}} \simeq \left(\frac{\hbar s\delta^{-1}}{kT}\right)^2 \frac{\tau_{ei}}{\tau_{ei}+\tau_{ep}}$$

and finally:

$$T_e \simeq T[1 + \tfrac{1}{2}(\frac{E}{sB})^2(1 + \frac{\tau_{ep}}{\tau_{ei}})] \tag{6}$$

Optical phonons can be treated in the same way. Let $\hbar\omega_o$ be their energy, here taken as constant. We can distinguish the two cases of low and high phonon temperature. If $\hbar\omega_o >> kT$,

$$N_q \simeq \exp(-\frac{\hbar\omega_o}{kT}) << 1$$

and

$$\frac{\Delta\varepsilon}{\bar{\varepsilon}_o} \simeq \frac{\hbar\omega_o}{kT} .$$

We then obtain

$$T_e \simeq T[1 + \tfrac{1}{2}\frac{(eE\delta)^2}{kT\,\hbar\omega_o}], \tag{7}$$

and in the limit:

$$eE\delta >> \hbar\omega_o,$$

we find

$$kT_e \simeq \frac{(eE\delta)^2}{2\,\hbar\omega_o}$$

If $\hbar\omega_o << kT$,

$$N_q \simeq \frac{kT}{\hbar\omega_o} ,$$

and

$$\frac{\Delta\varepsilon}{\bar{\varepsilon}_o} \simeq (\frac{\hbar\omega_o}{kT})^2 ,$$

we then get

$$T_e \simeq T[1 + \tfrac{1}{2}(\frac{eE\delta}{\hbar\omega_o})^2]. \tag{8}$$

These last results will be obtained by very much more sophisticated methods in the next section. Moreover, we shall be able to determin their range of validity.

The Hall Field Problem

In experiments one usually employs conditions under which there is no current in the Y direction perpendicular to $\vec{E}$ and $\vec{B}$. In this case, the E_y Hall field is not zero. The preceding discussion can easily be extended. Let us write:

$$J_y = \sigma_{yx} E_x + \sigma_{yy} E_y$$

with $\sigma_{yy} = \sigma_{xx}$, $\sigma_{yx} = -\sigma_{xy}$ for homogeneous medium. The Joule power dissipated by hot electrons has the form:

$$J_x E_x = \sigma_{xx}[1 + (\frac{\sigma_{xy}}{\sigma_{xx}})^2]E_x^2$$

which is the same as in (1) if we replace σ_{xx} by $\sigma_{xx}[1 + (\frac{\sigma_{xy}}{\sigma_{xx}})^2]$. Finally we obtain for the general case of acoustic phonons and impurities:

$$T_e \simeq T[1 + \tfrac{1}{2}(\frac{E_x}{sB})^2(1 + \frac{\tau_{ep}}{\tau_{ei}})(1 + (\frac{\sigma_{xy}}{\sigma_{xx}})^2)] \qquad (9)$$

In (9) we must pay attention to the fact that T_e is also hidden in the expression on the right. Therefore T_e is given by a transcenden tal equation which does not always have a positive solution for all the values of the electric field. This point is the origin of certain instabilities that give rise to the S shape current-voltage characteristics.

Parallel Electric and Magnetic Fields

In this case the classical conductivity reduces to

$$\sigma_{zz} = n_e \frac{e^2 \tau_r}{m} ,$$

and the energy balance relation (1) is replaced by:

$$\frac{e^2 E^2 \tau_r}{m} \simeq \frac{\overline{\varepsilon}(E) - \overline{\varepsilon}_o}{\tau_\varepsilon} .$$

Expressed in terms of the mean free paths λ, $\overline{\varepsilon}_0 \simeq m(\lambda/\tau_r)^2/2$. Thus (2) is transformed into:

$$\overline{\varepsilon}(E) = \overline{\varepsilon}_0[1 + \tfrac{1}{2}(\frac{eE\lambda}{\overline{\varepsilon}_0})^2 \frac{\tau_\varepsilon}{\tau_r}]. \tag{10}$$

Comparison between (10) and (2) shows that the criterion for weak electric field is $eE\lambda << \overline{\varepsilon}_0$ for $\vec{E} \parallel \vec{B}$ and $eER_L = eE\delta << \overline{\varepsilon}_0$ for $\vec{E} \perp \vec{B}$. The inequalities differ by the factor $\lambda/\delta \simeq \omega_c\tau_r\ (kT/\hbar\omega_c)^{\frac{1}{2}}$ which should be greater than unity even in the extreme quantum limit. Thus we expect an easier heating of electrons in the longitudinal configuration.

STUDY OF THE ELECTRON DISTRIBUTION FUNCTION IN CROSSED FIELDS

In this section, we illustrate the kind of difficulties encountered in the determination of the EDF by treating the case of crossed fields, when carrier-carrier scattering is neglected.

Quantum Background

Let us consider an electron gas colliding only with phonons, and moving in the uniform, static fields $\vec{E} = (E,o,o)$ and $\vec{B} = (o,o,B)$. When interactions with phonons are neglected, the trajectories of electrons in a plane perpendicular to $\vec{B}$ are quantified cycloids, and a Hall current along the Y-axis perpendicular to $\vec{E}$ and $\vec{B}$ takes place due to the uniform drift velocity, $V_d = E/B$, of the carriers. Collisions with phonons produce an electric current along $\vec{E}$ allowing the electrons to receive a certain amount of energy from the electric field which is given to the phonons during collisions. We assume that by interacting with an outside bath, the phonons will always remain in thermal equilibrium at temperature T.

The advantage of the crossed-field configuration is that it enables us to diagonalize the electron Hamiltonian, including both the effects of $\vec{E}$ and $\vec{B}$. This remark is well known, since the pioneering work of Titeica (1935) on linear conduction in high magnetic field. We lose this opportunity when $\vec{E}$ and $\vec{B}$ are parallel.

The total Hamiltonian of the electron phonon system can be written:

$$H = H_e + H_p + H_{ep}, \tag{11}$$

where

$$H_e = \frac{(\vec{p} + e\vec{A})^2}{2m} + eEx. \tag{12}$$

Here we treat e > o and take the vector potential as A = (o,Bx,o) (it is easy to deal with the electron spin $\vec{S}$ by adding a term $g\mu_B S_z B$, but we shall disregard spin effects for simplicity). Also,

$$H_p = \sum_{\vec{q}} \hbar\omega_q (b_q^+ b_q + \tfrac{1}{2}), \tag{13}$$

and

$$H_{ep} = i \sum_{\vec{q}} (C(\vec{q}) b_{\vec{q}} e^{i\vec{q}\vec{r}} - c^*(\vec{q}) b_{\vec{q}}^+ e^{-i\vec{q}\vec{r}}) \tag{14}$$

where $C(\vec{q})$ depends on the nature of the electron-phonon coupling we consider.

Eigenvalues and eigenstates of H_e are specified by the three quantum numbers: n, which take any positive integral value, and k_y and k_z which are multiples of $2\pi/L$, where L represents a characteristic dimension of the electron enclosing sample. We summarize these three indices by ν and we can write [Budd (1968), Yamashita and Inoue (1965,1969)]

$$H_e|\nu\rangle = \varepsilon_\nu|\nu\rangle; \ \psi_\nu(\vec{r}) = \langle\vec{r}|\nu\rangle = \frac{1}{L} e^{ik_y y} e^{ik_z z} \phi_n(x - X_\nu), \tag{15}$$

and

$$\varepsilon_\nu = (n + \tfrac{1}{2})\hbar\omega_c + \frac{\hbar^2 k_z}{2m} + eEX_\nu + \tfrac{1}{2}mV_d^2, \tag{16}$$

where

$$X_\nu = -\left(\frac{\hbar k_y}{m\omega_c} + \frac{eE}{m\omega_c^2}\right) \tag{17}$$

is the mean abscissa of the electron in state ν. Therefore eEX_ν in (16) is the mean electrostatic energy of the electron in state ν, while $\tfrac{1}{2}mV_d^2$ is the kinetic energy part due to the electric field. Thus an electron that passes from state ν to state ν' during a collision with a phonon of wave vector q sees its mean abscissa change by the amount:

$$X_{\nu'} - X_\nu = \frac{\hbar(k_y - k_y')}{m\omega_c} = \frac{\hbar q_y}{m\omega_c}$$

since conservation of momentum implies $k_y - k_y' = q_y$.

We turn now to the EDF which is the diagonal matrix element of the density operator and satisfies the equation of motion

$$i\hbar \frac{\partial\rho}{\partial t} = [H,\rho], \tag{18}$$

while the EDF is defined by

$$f_\nu = \mathrm{tr} \langle\nu|\rho|\nu\rangle \tag{19}$$

and the trace is calculated over the phonon states.

The irreversible equation obeyed by f_ν was established in great detail a long time ago. The original treatment due to Van Hove (1955,1957) has been followed by many others [Kohn and Luttinger (1957), Zwanzig, et al., (1961)]. Neglecting the Pauli exclusion principle, they obtain the master equation

$$\frac{\partial f\nu}{\partial t} = \sum_{\nu'} [f_{\nu'}(t)W_{\nu'\nu} - f_\nu(t)W_{\nu\nu'}]. \tag{20}$$

Similarly, the current density component parallel to $\vec{E}$ calculated within the same approximations as in (20) is given by the famous Titeica expression

$$J_{||} = (-e) \sum_{\nu,\nu'} \frac{X_\nu - X_{\nu'}}{2} [f_{\nu'}W_{\nu'\nu} - f_\nu W_{\nu\nu'}]. \tag{21}$$

In (20) and (21), $W_{\nu\nu'}$ is the probability per unit time for an electron to be scattered from state ν to state ν' and is given by the Fermi golden rule:

$$W_{\nu\nu'} = \frac{2\pi}{\hbar} \sum_{\vec{q}} |\langle\nu|\ C(\vec{q})e^{i\vec{q}\vec{r}}|\nu'\rangle|^2 [N_{\vec{q}}\delta(\varepsilon_\nu - \varepsilon_{\nu'} + \hbar\omega_{\vec{q}})$$

$$+ (1 + N_{\vec{q}})\delta(\varepsilon_\nu - \varepsilon_{\nu'} - \hbar\omega_{\vec{q}})], \tag{22}$$

and

$$N_{\vec{q}} = [\exp(\frac{\hbar\omega_{\vec{q}}}{kT}) - 1]^{-1}$$

is the distribution function of the phonons assumed in thermal equilibrium at temperature T.

A very quick demonstration of (20) in the steady state and of (21) is given in the Appendix. It is remarkable that there is no driving term in the first member of (20); the electric field appears only in $W_{\nu\nu'}$. This result corresponds simply to the possibility of a steady state in the absence of collisions, because there is a magnetic field perpendicular to $\vec{E}$. In fact, all of the non-trivial physical information is contained in $W_{\nu\nu'}$ which has the following properties. We can easily verify that

$$|\langle\nu|C(\vec{q})e^{i\vec{q}\vec{r}}|\nu'\rangle|^2$$

depends only on the difference $X_{\nu'} - X_\nu$, is E-independent, and is different from zero only if $k_y - k'_y = q_y$ and $k_z - k'_z = q_z$. It follows that $W_{\nu\nu'}$ depends only on the difference $X_{\nu'} - X_\nu$; this corresponds to the homogeneity of the medium. Moreover, the electric field occurs only in the two δ-functions and only through the difference of the mean electrostatic potential energy: $eE(X_{\nu'} - X_\nu)$. Finally there is conservation of the y and z components of the momentum of the electron phonon system during the collision. Moreover the matrix element:

$$|\langle\nu|C(\vec{q})e^{i\vec{q}\vec{r}}|\nu'\rangle|^2$$

decreases quite rapidly as soon as q differs from δ^{-1}; this is the reason why we often say that electrons interact preferentially with phonons whose wave vector is of the order δ^{-1}. Finally, $W_{\nu\nu'}$ satisfies the principle of detailed balance

$$\frac{W_{\nu\nu'}}{W_{\nu'\nu}} = \exp\left(\frac{\varepsilon_\nu - \varepsilon_{\nu'}}{kT}\right) \tag{23}$$

which relies on our hypothesis concerning the thermal equilibrium state of phonons at temperature T.

Steady-State Solution of the Master Equation

We seek the solution of the apparently simple equation (20) in which $\frac{\partial f_\nu}{\partial t} = 0$:

$$\sum_{\nu'} [f_{\nu'} W_{\nu'\nu} - f_\nu W_{\nu\nu'}] = 0. \tag{24}$$

At first we see that any solution of (24) is independent of the strength of the coupling between electrons and phonons. Thus if the concept of electron temperature is meaningful, T_e only depends on the nature of the coupling with the phonons and not on the absolute coupling constant which is hidden in $C(\vec{q})$. Nevertheless when we assume that electrons interact with two types of scatterers, a solution of (24) must depend on the ratio of the strengths of the two scattering processes. This was apparent in the qualitative arguments of the preceding section, when we considered phonons and impurities; the expression of T_e included the ratio τ_{ei}/τ_{ep} which contains the ratio of the two scattering strengths. With the help of (23), it is evident that:

$$f_\nu = f^0_\nu \sim \exp(-\frac{\varepsilon_\nu}{kT}) \tag{25}$$

is a solution of (24). At the same time, this solution gives $J_{||} = 0$. Remarking that f^0_ν depends on X_ν (the mean abscissa of the electron in state ν) we are sure that f^0_ν describes an inhomogeneous distribution of electrons inside the sample. Without current in the x direction of the electric field, electrons accumulate in high electrostatic potential regions. Thus we are in a situation where electrons are not injected at one side of the sample and collected at the opposite side to produce a current. Such a steady state EDF is of no interest in our study. The desired solution of (24) must describe a situation where the electron density is homogeneous and therefore we must eliminate an X_ν or k_y dependence of f_ν. Furthermore f_ν must be an even function of k_z. To be convinced of that point, let us remember that the current density along $\vec{B}$ must be zero. But

$$J_z = \mathrm{Tr}\rho v_z = \sum_\nu f_\nu \frac{\hbar k_z}{m}$$

and we have $J_z = 0$ when f_ν is k_z^2 dependent.

In conclusion, labelling as simple ε that part of ε_ν that represents the total kinetic energy, index n remaining for the different Landau levels, the distribution function f_ν can be written as $f_n(\varepsilon)$.

Now we introduce a new transition probability per unit time $P_{nn'}(\varepsilon,\varepsilon')$ defined from $W_{\nu\nu'}$ by a k_y integration. More precisely,

$$P_{nn'}(\varepsilon,\varepsilon') = \int g_n(\varepsilon) g_{n'}(\varepsilon') W_{\nu\nu'} dk'_y, \tag{26}$$

where $g_n(\varepsilon)$ is the density of states associated with the nth Landau level. The master equation (24) takes the exact form

$$\sum_{n'} \int d\varepsilon' [\underline{P}_{n'n}(\varepsilon',\varepsilon) f_{n'}(\varepsilon') - \underline{P}_{nn'}(\varepsilon,\varepsilon') f_n(\varepsilon)] = 0. \tag{27}$$

In general a direct solution of (27) is impossible. But we can transform (27) into a set of differential equations of the Fokker-Plank type when $f_n(\varepsilon)$ varies slowly on the scale $\varepsilon' - \varepsilon$ of energy transfers. Thereby we hope to gain more insight. To see that, let us first assume that only the lowest Landau level n = 0 is occupied we shall say what happens when many levels are populated later.

Extreme Quantum-Limit Case. We can drop the n index in (27) which reduces to

$$\int d\varepsilon' [\underline{P}(\varepsilon',\varepsilon)f(\varepsilon') - \underline{P}(\varepsilon,\varepsilon')f(\varepsilon)] = 0 \tag{28}$$

It is remarkable that (28) is equivalent to setting

$$\frac{dJ(\varepsilon)}{d\varepsilon} = 0$$

where

$$J(\varepsilon) = \int_{-\infty}^{+\infty} d\varepsilon' \int_{-\infty}^{+\infty} d\varepsilon'' P(\varepsilon',\varepsilon'') f(\varepsilon') [\theta(\varepsilon - \varepsilon')\theta(\varepsilon'' - \varepsilon) - \theta(\varepsilon' - \varepsilon)\theta(\varepsilon - \varepsilon'')] \tag{29}$$

is exactly the current of electrons that cross the shell at energy ε per unit time. In (29) we expand $\rho(\varepsilon')$ in a Taylor series and we remember that ε' and ε'' are both very close to ε. Furthermore, we may assume that $P(\varepsilon',\varepsilon'')$, which is strictly a function of ε' and $\xi = \varepsilon' - \varepsilon''$, is considered as a function of ε and ξ in that range. Then (29) takes the simple form

$$J(\varepsilon) = -a(\varepsilon)f(\varepsilon) - b(\varepsilon)\frac{df(\varepsilon)}{d\varepsilon}, \tag{30}$$

with

$$a(\varepsilon) = \int_{-\infty}^{+\infty} \underline{P}(\xi,\varepsilon)\xi d\xi$$

$$b(\varepsilon) = \int_{-\infty}^{+\infty} \underline{P}(\xi,\varepsilon)\frac{\xi^2}{2} d\xi \tag{31}$$

The physical meaning of (30) is obvious; the current of electrons in energy space is due to:

(a) a drift term $-a(\varepsilon)f(\varepsilon)$ due to a systematic energy loss at each collision,

(b) a diffusion term $-b(\varepsilon)f'(\varepsilon)$ due to energy fluctuations, corresponding to a random walk in energy space.

This result can be considered as a justification of the phenomenological theory of Kurosawa and Yamada (1973). If we now introduce the net transition probability

$$\Gamma(\varepsilon) = \int d\varepsilon' \underline{P}(\varepsilon,\varepsilon') = \int d\xi \underline{P}(\xi,\varepsilon),$$

the coefficients a and b are related to the main energy loss per collision $\overline{\xi}$ and to the corresponding fluctuations $\overline{\xi^2}$ by

$$a(\varepsilon) = \Gamma(\varepsilon)\overline{\xi}\ , \qquad b(\varepsilon) = \tfrac{1}{2}\Gamma(\varepsilon)\overline{\xi^2} \tag{31}$$

The Taylor expansion used to go from (29) to (30) will be valid if we have

$$\overline{\xi^2} >> (\overline{\xi})^2$$

In steady state the current $\overline{J}(\varepsilon)$ satisfies the condition

$$\frac{dJ(\varepsilon)}{d\varepsilon} = 0$$

or $J(\varepsilon)$ = cst. As the flow is bounded at the lower value of ε, $J(\varepsilon)$ is necessarily zero and (30) has a trivial solution

$$f(\varepsilon) \sim \exp[-\int_0^{\varepsilon} d\varepsilon' \frac{a(\varepsilon')}{b(\varepsilon')}]. \tag{32}$$

The EDF in extreme quantum limit will be Maxwellian only if the ratio $a(\varepsilon)/b(\varepsilon)$ is ε-independent. If this is not so, there is no possibility of finding an electron temperature and $f(\varepsilon)$ is model dependent.

Several Landau levels are occupied. The rather simple result (32) breaks down as soon as we include several Landau levels in the picture. Let $J_n(\varepsilon)$ be the current in energy space for a given value of n. Again we can prove, in the same way as in the previous section that

$$J_n(\varepsilon) = - a_n(\varepsilon)f_n(\varepsilon) - b_n(\varepsilon)f_n'(\varepsilon), \tag{33}$$

where a_n and b_n are calculated with P_{nn} and depend on n as well as on ε. But now we must take account of the cross currents due to collisions that change $n \rightleftarrows n'$. For slowly varying $f_n(\varepsilon)$, we can perform the ε'-integration in (27). The condition for a steady solution becomes

$$\frac{dJ_n}{d\varepsilon} + \sum_{n' \neq n} [Q^{(1)}_{n'n} f_{n'} - Q^{(2)}_{nn'} f_n] = 0, \tag{34}$$

where

$$Q^{(1)}_{n'n} = \int d\varepsilon' \, \underline{P}_{n'n}(\varepsilon',\varepsilon)$$

$$Q^{(2)}_{nn'} = \int d\varepsilon' \, \underline{P}_{nn'}(\varepsilon,\varepsilon'), \tag{35}$$

The difficulty is now apparent: even if a_n/b_n is ε-independent, it will usually depend on n. The different Landau levels would thus be heated at different temperatures if $Q^{(1)}_{n'n}$ and $Q^{(2)}_{nn'}$ were zero. In fact, cross-transitions tend to equilibrate the various Landau leve[ls] but in so doing they spoil the Maxwellian nature of the distributio[n]. The steady solution is now model dependent.

In conclusion, we may expect a Maxwellian distribution with an electron temperature only in the extreme quantum limit. And even i[n] this simple case, the Maxwellian will not hold at the very bottom o[f] the band ($\varepsilon \lesssim \hbar\omega_q$, $eE\delta$) where our assumption of slow variation of $P(\varepsilon,\varepsilon')$ is not valid due to the divergent character of the density of states.

Application: study of the extreme quantum limit in the case where electrons interact with non-polar optical phonons. Calculations of $a(\varepsilon)$ and $b(\varepsilon)$ defined by (31) have been performed in the framework of an Einstein model for the phonons and when the couplin[g] constant $C(\vec{q})$ is q-independent. The ratio $a(\varepsilon)/b(\varepsilon)$ is found ε-independent so that $f(\varepsilon)$ is a Maxwellian with an electron temperatu[re]

$$T_e = T[1 + (\frac{eE\delta}{\hbar\omega_o})^2]$$

when $\hbar\omega_o << kT$ and

$$T_e \simeq \frac{(eE\delta)^2}{2k\hbar\omega_o}$$

when $\hbar\omega_o >> kT$ and $eE\delta >> \hbar\omega_o$.

These results are those obtained qualitatively in Section 2.1. Furthermore, we can recover part of them by solving a well-defined walk problem in energy space. All the details can be found in Calecki (1977).

Non-Heating Non-Linear Effect

The last effect we would like to point out is a non-linearity that appears in the dissipative current at low electric field, before heating of electrons can occur.

Titeica's expression (21) conveniently modified to take account of electron scattering on impurities has been extensively applied in the weak linear electric field region. It is well known that in the quantum limit the ohmic conductivity σ_{xx} diverges when collisions of electrons are assumed elastic, as it does for scattering on impurities. The divergence that arises from electrons of low energy can be removed by more careful calculations that consider non-Born scattering or collision broadening, Kubo (1965). However, it has been shown by Magarill and Savvinykh (1970), Elesin (1969), and Agaeva, et al., (1976) that the divergence is also removed in the case of elastic collisions and in the Born approximation, provided the theory is not entirely linearized with respect to the electric field. More precisely, let us write the expression for the dissipative current in the case of elastic impurity scattering as

$$J_{11} = (-e) \sum_{\nu,\nu',\vec{q}} \frac{X_\nu - X_{\nu'}}{2} |C(\vec{q})|^2 |\langle\nu|e^{i\vec{q}\vec{r}}|\nu'\rangle|^2 (f_{\nu'}-f_\nu)\delta(\varepsilon_\nu-\varepsilon_{\nu'}). \tag{36}$$

Here $C(\vec{q})$ is the Fourier component of the electron interaction energy with an impurity. We consider the extreme quantum limit and electric fields such that $eE\delta \ll \hbar s\delta^{-1}$; thus we do not expect any heating of electrons and in (36) f_ν can be replaced by its equilibrium value in absence of electric field. For a non-degenerate electron gas, we have:

$$f_\nu \sim \exp(-\frac{\hbar^2 k_z^2}{2mkT}).$$

We make no further assumption and we do not touch the δ-function that contains the electric field. After some algebra, we find a current $J_{||} \sim E\log E^{-1}$ in the limit $eE\delta \ll kT$. Thus there is no divergence except at the point $E = 0$ where we are faced with a non-analytic behavior of $J_{||}$. However, this singularity can be removed again if we forego the Born approximation or introduce collision broadening. Similar developments can be found for a degenerate electron gas [Elesin (1969)] and for semiconductors with a Kane band structure [Agaeva, et al., (1976)].

CONCLUSIONS

The example that we have just treated of hot electrons in crossed fields demonstrates that the behaviour of the EDF is rather difficult to be obtained analytically. Moreover the dissipative current depends on the knowledge of the EDF. If we assume that carrier-carrier scattering is not sufficiently effective to ascertain that an electron temperature can be introduced, we are face with the unavoidable task of solving a master equation. We have se that only in the quantum limit is there a possibility of encounteri a Maxwellian EDF. The simple solution with as many electron temper tures as Landau levels occupied cannot be justified. Thus we are rather pessimistic about the possibility of performing rigorous analytical calculations. We think that the field is open either to very crude approximations with qualitative deductions, or to numerical investigations based on a Monte Carlo method for example. This last direction now has an increasing importance in the interpretati of experimental data and particularly the magnetophonon effect.

APPENDIX. SHORT DERIVATION OF THE STEADY STATE MASTER EQUATION AND TITEICA'S EXPRESSION OF THE CURRENT

In steady state the density operator obeys the equation

$$[H_e + H_p + H_{ep}, \rho] = 0 \tag{A1}$$

We label $\lambda(\nu)$ and $\Omega_\lambda(\varepsilon_\nu)$ the eigenstates and eigenvalues of $H_p(H_e)$ and we calculate the diagonal and off-diagonal matrix elements of (A1). We get

$$\sum_{\nu'\lambda'} [(H_{ep})^{\lambda\lambda'}_{\nu\nu'} \rho^{\lambda'\lambda}_{\nu'\nu} - (H_{ep})^{\lambda'\lambda}_{\nu'\nu} \rho^{\lambda\lambda'}_{\nu\nu'}] = 0 \tag{A2}$$

and

$$(\varepsilon_\nu + \Omega_\lambda - \varepsilon_{\nu'} - \Omega_{\lambda'})\rho^{\lambda\lambda'}_{\nu\nu'} + (\rho^{\lambda'}_{\nu'} - \rho^{\lambda}_{\nu})(H_{ep})^{\lambda\lambda'}_{\nu\nu'}$$

$$+ \sum_{\substack{\nu''\lambda'' \\ \neq \nu'\lambda'}} (H_{ep})^{\lambda\lambda''}_{\nu\nu''} \rho^{\lambda''\lambda'}_{\nu''\nu'} - \sum_{\substack{\nu''\lambda'' \\ \neq \nu\lambda}} (H_{ep})^{\lambda''\lambda'}_{\nu''\nu'} \rho^{\lambda\lambda''}_{\nu\nu''} = 0 \tag{A3}$$

Following the ideas of van Hove (1955,1957) we know that off-diagon elements of ρ are of a higher order in the perturbation H_{ep} than th diagonal elements. Thus we retain in (A3) only the first two terms as the beginning of an iteration procedure. Dividing by $(\varepsilon_\nu + \Omega_\lambda - \varepsilon_{\nu'} - \Omega_{\lambda'})$ as in the distribution theory, we obtain the lowest order of $\rho_{\nu\nu'}^{\lambda\lambda'}$ in terms of the diagonal elements

$$\rho_{\nu\nu'}^{\lambda\lambda'} = (\rho_{\nu}^{\lambda} - \rho_{\nu'}^{\lambda'})(H_{ep})_{\nu\nu'}^{\lambda\lambda'}[i\pi\delta(\varepsilon_{\nu} + \Omega_{\lambda} - \varepsilon_{\nu'} - \Omega_{\lambda'}) + P(\frac{1}{\varepsilon_{\nu} + \Omega_{\lambda} - \varepsilon_{\nu'} - \Omega_{\lambda'}})] \tag{A4}$$

Inserting (A4) in (A2), the principal parts cancel each other and we have

$$\sum_{\lambda'\nu'} (\rho_{\nu'}^{\lambda'} - \rho_{\nu}^{\lambda})|(H_{ep})_{\nu\nu'}^{\lambda\lambda'}|^2\delta(\varepsilon_{\nu'} + \Omega_{\lambda'} - \varepsilon_{\nu} - \Omega_{\lambda}) = 0. \tag{A5}$$

We now assume that we can neglect correlation between electrons and phonons in the diagonal elements of ρ_{ν}^{λ}. It turns out that $\rho_{\nu}^{\lambda} = f_{\nu}P_{\lambda}$ where P_{λ} is the thermal equilibrium distribution of the phonon gas. Inserting this in (A5) and taking the trace over the phonon states, one obtains

$$\sum_{\nu'} [f_{\nu'} W_{\nu'\nu} - f_{\nu} W_{\nu\nu'}] = 0 \tag{A6}$$

with

$$W_{\nu\nu'} = \frac{2\pi}{\hbar} \sum_{\lambda\lambda'} |(H_{ep})_{\nu\nu'}^{\lambda\lambda'}|^2 \underline{P}_{\lambda}\delta(\varepsilon_{\lambda'} + \Omega_{\lambda'} - \varepsilon_{\nu} - \Omega_{\lambda}). \tag{A7}$$

Finally, when we introduce expression (11) in (A7), we obtain the explicit form (19) in the text.

In order to calculate the current, we need the particular off-diagonal element $\rho_{\nu\nu'}^{\lambda\lambda}$. This is obtained by the second step of the iteration procedure. We introduce the off-diagonal expression of ρ given by (A4) in (A3)

$$(\varepsilon_{\nu} - \varepsilon_{\nu'})\rho_{\nu\nu'}^{\lambda\lambda'} = \sum_{\nu''\lambda''} (H_{ep})_{\nu\nu''}^{\lambda\lambda''}(H_{ep})_{\nu''\nu'}^{\lambda''\lambda}\{(\rho_{\nu''}^{\lambda''} - \rho_{\nu'}^{\lambda})[i\pi\delta(\varepsilon_{\nu''} + \Omega_{\lambda''} - \varepsilon_{\nu'} - \Omega_{\lambda}) + \underline{P}\frac{1}{(\varepsilon_{\nu''} + \Omega_{\lambda''} - \varepsilon_{\nu'} - \Omega_{\lambda}}] - (\rho_{\nu}^{\lambda} - \rho_{\nu''}^{\lambda''})[i\pi\delta(\varepsilon_{\nu} + \Omega_{\lambda} - \varepsilon_{\nu''} - \Omega_{\lambda''}) + \underline{P}\frac{1}{(\varepsilon_{\nu} + \Omega_{\lambda} - \varepsilon_{\nu''} - \Omega_{\lambda''}}]\} \tag{A8}$$

Now let us look at the current parallel to the electric field

$$J_{||} = (-e)\ \mathrm{Tr}\ \rho v_x. \tag{A9}$$

It is easy to verify the useful relation

$$V_x = \frac{i}{\hbar\omega_c}[H_e, v_y].$$

Thus we get

$$J_{||} = (-e)\frac{i}{\hbar\omega_c}\sum_{\lambda\nu\nu'}(\varepsilon_{\nu'} - \varepsilon_\nu)\rho^{\lambda\lambda}_{\nu\nu'}(v_y)_{\nu'\nu} \tag{A10}$$

Inserting (A8) in (A10) and relabelling conveniently summation indices, we obtain

$$J_{||} = (-e)\frac{i}{\hbar\omega_c}\sum_{\nu\nu'}\sum_{\lambda\lambda'}(H_{ep})^{\lambda'\lambda}_{\nu'\nu}(\rho^{\lambda'}_{\nu'} - \rho^{\lambda}_{\nu})[i\pi\delta(\varepsilon_{\nu'} + \Omega_{\lambda'} - \varepsilon_\nu - \Omega_\lambda)$$

$$+ P(\frac{1}{\varepsilon_{\nu'} + \Omega_{\lambda'} - \varepsilon_\nu - \Omega_\lambda})]\{\sum_{\nu''}[(H_{ep})^{\lambda\lambda'}_{\nu''\nu'}(v_y)_{\nu\nu''} - (H_{ep})^{\lambda\lambda'}_{\nu\nu''}(v_y)_{\nu''\nu'}]$$

(A11)

The last factor is simply the matrix element of the commutator

$$<\nu\lambda|[v_y, H_{ep}]|\nu'\lambda'>,$$

but

$$[v_y, H_{ep}] = [\frac{p_y}{m} + \omega_c x, H_{ep}] = [\frac{p_y}{m}, H_{ep}].$$

As p_y is diagonal in the ν representation, with matrix elements

$$<\nu|p_y|\nu> = \hbar k_y.$$

we immediately get

$$<\nu\lambda|[v_y, H_{ep}]|\nu'\lambda'> = \frac{\hbar}{m}(k_y - k'_y)(H_{ep})^{\lambda\lambda'}_{\nu\nu'}$$

and

$$J_{||} = (-e)\frac{i}{\hbar}\sum_{\nu\nu'}\sum_{\lambda\lambda'}(H_{ep})^{\lambda\lambda'}_{\nu\nu'}|^2(\rho^{\lambda'}_{\nu'} - \rho^{\lambda}_{\nu})$$

$$(X_{\nu'} - X_\nu)[i\pi\delta(\varepsilon_{\nu'} + \Omega_{\lambda'} - \varepsilon_\nu - \Omega_\lambda) + P(\frac{1}{\varepsilon_{\nu'} + \Omega_{\lambda'} - \varepsilon_\nu - \Omega_\lambda})]$$

Again the principal part disappears for reasons of symmetry. The factorization of $\rho_\nu^\lambda = f_\nu P_\lambda$ and expression (A7) give the Titeica formula

$$J_{||} = (-e) \sum_{\nu\nu'} \frac{X_\nu - X_{\nu'}}{2} (f_{\nu'} W_{\nu'\nu} - f_\nu W_{\nu\nu'}) \quad \text{(A12)}$$

or equivalently

$$J_{||} = (-e) \sum_{\nu\nu'} (X_{\nu'} - X_\nu) W_{\nu\nu'} f_\nu . \quad \text{(A13)}$$

It is clear in (A13) that conduction along E is due to hopping of the mean electron abscissa at each phonon collision. The net current is the hopping probability $f_\nu W_{\nu\nu'}$ times the displacement $(X_{\nu'} - X_\nu)$ at each hop.

BIBLIOGRAPHY

Agaeva, R. G., Askerov, B. M., and Gashimzade, F. M., 1976, Sov.-Phys. Semicond. 10:1268.
Barker, J. R., 1978, Sol.-State Electr. 21:197.
Budd, H. F., 1968, Phys. Rev. 175:241.
Calecki, D., 1971, J. Phys. Chem. Solids 32:1835.
Calecki, D., 1972, J. Phys. Chem. Solids 33:379.
Calecki, D., Lewiner, C., and Nozieres, P., 1977, J. de Phys. 38:169.
Cassiday, D. R., and Spector, H. N., 1974, J. Phys. Chem. Sol. 35:957.
Elesin, V. F., 1969, Sov. Phys. JETP 28:410.
Inoue, S., and Yamashita, J., 1969, Prog. Theor. Phys. 42:158.
Kalashnikov, V. P., and Pomortsev, R. V., 1964, Fiz. Metal. Metallov. 17:343.
Kazarinov, R. F., and Skobov, V. G., 1962, Sov. Phys. JETP 15:726.
Kazarinov, R. F., and Skobov, V. G., 1963, Sov. Phys. JETP 17:921.
Kogan, Sh. M., 1963, Sov. Phys. Sol.-State 4:1813.
Kogan, Sh. M., Shadrin, V. D., and Shul'man, A. Ya., 1976, Sov. Phys. Sol.-State 41:686.
Kohn, W., and Luttinger, J. M., 1957, Phys. Rev. 108:590.
Kubo, R., Miyake, S., and Hashitsume, N., 1965, "Solid-State Physics", Academic Press, New York.
Kurosawa, T., 1965, J. Phys. Soc. Jpn., 20:937.
Landau, L. D., 1930, Z. Physik 64:629.
Licea, I., 1971, Z. Physik 249:115
Magarill, L. I., and Savvinykh, S. K., 1970, Sov. Phys. JETP 30:1128.
Miyazawa , H., 1969, J. Phys. Soc. Jpn. 26:700.
Partl, H., Muller, W., Kohl, F., and Gornik, E., 1978, J. Phys. C 11:1091.
Pinchuk, I. I., 1977, Sov. Phys. Sol.-State 19:383.

Pomortsev, R. V., and Kharus, G. I., 1967, Sov. Phys. Sol.-State 9:1150.
Pomortsev, R. V., and Kharus, G. I., 1968, Sov. Phys. Sol.-State 9:2256.
Shadrin, V. D., 1976, Sov. Phys. Semicond. 10:500.
Titeica, S., 1935, Ann. der Physik 22:129.
Van Hove, L., 1955, Physica 21:517.
Van Hove, L., 1957, Physica 23:441.
Yamada, E., and Kurosawa, T., 1973, J. Phys. Soc. Jpn. 34:603.
Yamashita, J., 1965, Prog. Theor. Phys. 33:343.
Zlobin, A. M., and Zyrianov, P. S., 1970, Sov. Phys. JETP 31:513.
Zlobin, A. M., and Zyrianov, P. S., 1972, Sov. Phys. Uspekhi 14:379.
Zwanzig, R., Montroll, E., and Argyres, P., 1961, in "Lectures in Theor. Phys." 3:106; 8a:183, Intern. Publishers.

HOT ELECTRON EFFECTS IN SEMICONDUCTOR DEVICES

H. L. Grubin

United Technologies Research Center
East Hartford, Connecticut 06108

INTRODUCTION

The original impetus for much of the hot electron work that has taken place in the last decade was the need to understand the contributions of hot electrons to semiconductor devices. The turning point in this area of study occurred when Gunn (1964a) published his results showing the presence of spontaneous and coherent oscillations in gallium arsenide and indium phosphide two terminal devices. Today, of course, hot electron phenomena are being studied for their own interest, and with ends that are ostensibly independent of device considerations. But, in fact, the two areas cannot be separated. Rather the emphasis is different, with hot-electron-device studies concentrating on situations where the electric field sustains large and fast transients and where the field is highly nonuniform.

Typically, when it is of interest to find an experimental situation that best illustrates hot electron device effects, the results of Gunn's early studies are shown. In Fig. 1, we display results (Gunn, 1964a) of a current instability in gallium arsenide. Here we show sampled current - voltage characteristics of a device thirty-six microns long. Mathematically, these results are plots of current as a function of voltage with time held constant. At low voltages the I-V relationship is nearly linear. At higher voltages the dc conductance first decreases slightly; then at some value of voltage there occur wide differences in current between successive samples. This appears as a large scatter of dots representing individual samples. In a subsequent series of experiments Gunn (1964b) established that these fluctuations in current were consequences of coherent oscillations associated with a moving distribution of potential within the semiconductor.

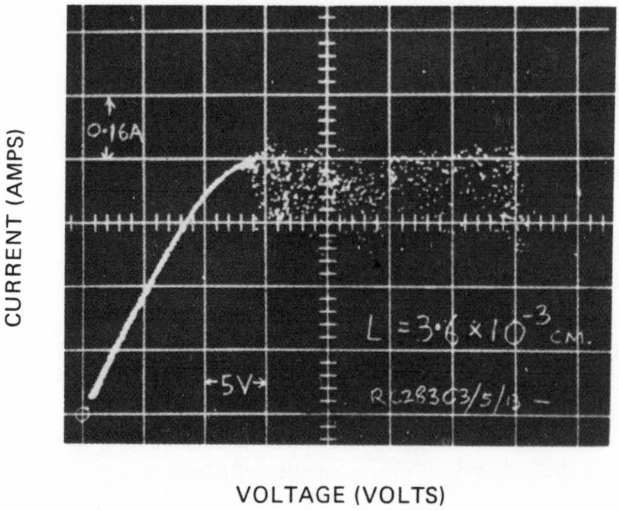

Figure 1. Sampled current-voltage characteristics of an n-type GaAs specimen. The active region lengths and scales are shown in the figure. From Gunn (1964a) with permission.

Gunn's experiments demonstrated that the semiconductor gallium arsenide was capable of converting dc to ac power. Our present explanation (Kroemer, 1964) for this conversion is that, through the mechanism of electron transfer (Ridley and Watkins, 1961) from the central to satellite valleys of the conduction band, gallium arsenide possesses a region of negative differential mobility (Hilsum, 1962) (see Fig. 2). As discussed by Ridley (1963), the presence of negative differential mobility can lead to unstable localized space charge regions. The observations of Gunn are then explained as arising from the periodic nucleation and disappearance of traveling space charge domains (Kroemer, 1964).

Figure 2, which is the drift velocity versus electric field curve for gallium arsenide (Butcher, 1967), represents what most device physicists regard as the basis of hot-electron transport effects in devices. Indeed, very sophisticated numerical programs have been designed around this curve and used to model the spatial and time-dependent dynamics of electrons in devices. Remarkable agreement with experiment has been obtained that has also included explanations of fine details of observations. I will illustrate some of these in a following chapter. Perhaps the most noteworthy point is that the velocity field curve, as represented by Fig. 2, derived

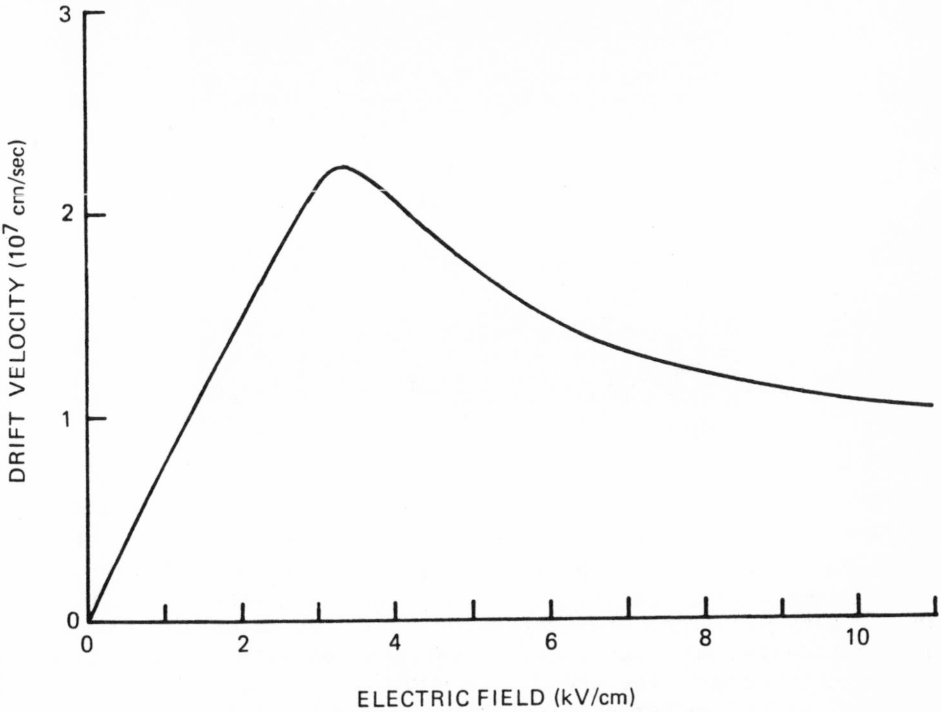

Figure 2. Calculation of the steady state drift velocity versus electric field curve for gallium arsenide. From Butcher (1967) Figure 3, with permisssion.

under steady-state conditions and spatially-uniform fields, is often used to explain transient phenomena and highly nonuniform situations. It is this latter use that is currently the subject of increased scrutiny and will occupy the rest of the current discussion. These topics properly come under the heading of transient hot-electron effects in semiconductor devices.

Before beginning the discussion we must lay a few ground rules. We will provide a microscopic description of carriers in a semiconductor device in terms of a displaced Maxwelliam

$$f(\vec{p}_i, T_i) = A(n_i/T_i^{3/2})\exp[-\hbar^2(\vec{k}-\vec{p}_i)^2/2m_i k_o T_i] \quad . \qquad (1)$$

The dynamics of the particles are then obtained from moments of the Boltzmann transport equation in the form (Bosch and Thim, 1974)

$$a_i \frac{\partial n_i}{\partial t} = - \frac{n_i}{\tau_{n_i}} + \frac{n_j}{\tau_{n_j}} \quad , \tag{2}$$

$$\frac{\partial n_i \vec{p}_i}{\partial t} = - en_i \vec{F} - \frac{n_i \vec{p}_i}{\tau_{p_i}} \quad , \tag{3}$$

$$\frac{\partial}{\partial t} \left\{ \frac{n_i p_i^2}{2} + \frac{3n_i k_o T_i}{2} \right\} = \frac{-en_i \vec{F}\cdot\vec{p}_i}{m_i} - \frac{n_i k_o T_i}{\tau_{E_i}} + \frac{n_j k_o T_j}{\tau_{i_j}} \quad . \tag{4}$$

In the above, a_i represents the number of equivalent satellite valleys, n_i, p_i and T_i are respectively the number density, the momentum and temperature of the ith valley. Further, we have ignored the spatial contributions to the moment equations. The τ's in the above equation are scattering times for intervalley and intra valley transfer and significant effects arise because of important differences between the energy and momentum relaxation time. Fig. 3 shows a set of scattering times for gallium arsenide calculated in 1974 by Bosch and Thim (1974). These calculations are for the Γ-X ordering in gallium arsenide in which the number of equivalent satel lite valleys is equal to 3. Calculations for the Γ-L ordering with four equivalent satellite valleys have not yet been published, although several groups are pursuing the problem. We note that at room temperature, the momentum scattering time for the central valle is significantly smaller than the energy scattering time. The significance of this will emerge below in connection with overshoot velocity.

TRANSIENT HOT ELECTRON EFFECTS IN SEMICONDUCTOR DEVICES

Velocity Overshoot

If a system of electrons is subjected to the combined influence of an electric field and scattering centers, ignoring for the moment electron transfer as a scattering mechanism, then the drift velocity of the particles asymptotically approaches the steady state value

$$v = -e\langle\tau\rangle\ F/m^* . \tag{5}$$

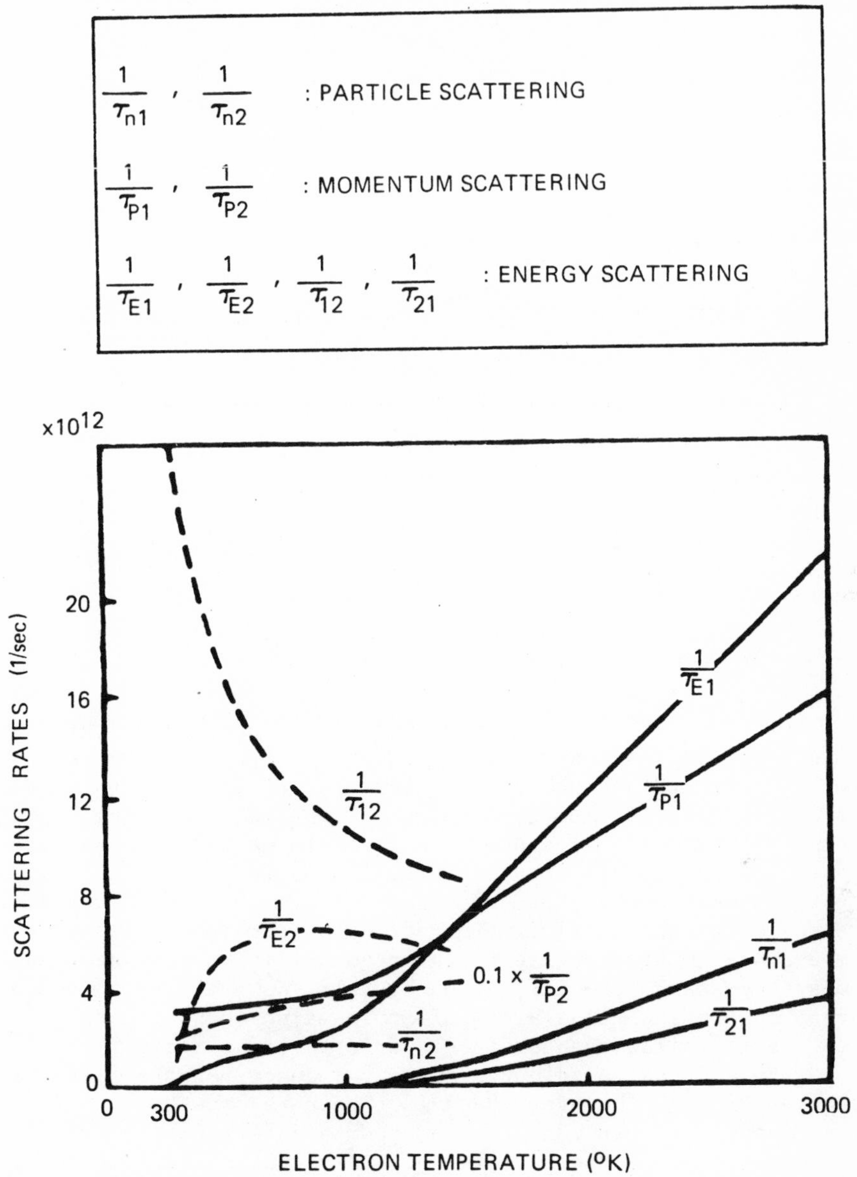

Figure 3. Calculated relaxation frequencies $(1/\tau)$ of GaAs versus the respective electron temperature. The symbols are listed in the figure. From Bosch and Thim (1974), with permission.

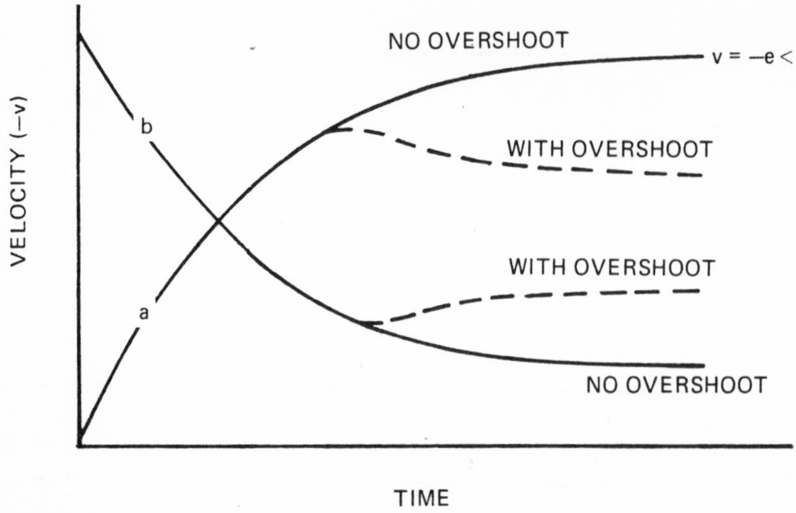

Figure 4. Approach to steady state, with and without transient relaxation effects. Part a denotes increasing velocity and part b denotes decreasing velocity values.

If we were just to consider the momentum balance equation, (2), we could be satisfied that the particles would approach this value in a time approximately equal to 2<τ> seconds, as illustrated in Fig. 4. But considering the momentum balance equation alone would give us an incomplete picture of events. Energy balance tells us that the electron temperature increases with increasing electric field and departs significantly from room temperature when the electric field exceeds a threshold value. The effect of the increasing electron temperature is to decrease the average collision time and to decrease the steady-state velocity, given by (5). Now, if the momentum and energy scattering times are similar in value then both momentum and energy will follow changes in electric field at approximately the same rate and the solid curve of Fig. 4 describes the approach to steady-state. On the other hand if the energy scattering time is significantly longer than the momentum scattering time, the average velocity of the carriers will find its value continually 'corrected' until steady state in the energy distribution is reached. The velocity will then relax in the manner shown by the dashed curve. A similar situation may be expected when the electric field is decreased, for here it also takes a finite time for the electric field to decrease and for the electron temperature to decrease.

We may now ask: where do we expect the overshoot transient effects to become significant? On the basis of the above discussion, substantial overshoot effects occur when the momentum scattering time is much shorter than the energy scattering time. Thus we would expect to detect these effects when the time rate of change of electric field falls somewhere within the limits

$$\frac{F_{th}}{\tau_p} \gtrsim \frac{dF}{dt} >> \frac{F_{th}}{\tau_E} \tag{6}$$

We are interested in the upper bound of the inequality and for this case we take τ_p to be represented by the momentum scattering time for LO intravalley phonons. Intervalley scattering is important as the electron temperature is elevated, and will reduce the momentum scattering time. For LO phonon intravalley scattering (Conwell, 1967),

$$\frac{1}{\tau_p} \approx \frac{F_0 e}{(2m^* k_o T_p)^{1/2}} \tag{7}$$

For the central valley of GaAs, with $F_0 = 5.6 \times 10^3$ V/cm (Butcher, 1967),

$$\frac{1}{\tau_p} \approx 4 \times 10^{12}/\text{sec} \tag{8}$$

which is consistent with the more complete calculations of Fig. 3 ($\frac{1}{\tau_{p_i}} \sim 3.5 \times 10^{12}$/sec for temperatures below 1000 K). Transient overshoot effects are then expected to be significant when the carriers experience changes in the electric field of the order of F_{th} (3kV at the doping levels of 10^{15} / cm^3) occurring during a time interval of the order of; or somewhat larger than, τ_p.

In Fig. 5 we show (Rees, 1969) the response of the central valley velocity, the satellite valley velocity, the satellite valley population and the drift velocity, when the electric field is stepped down from 6kV/cm to 5kV/cm. The drift velocity is computed from the equation

$$v = \frac{n_c v_c + a_s n_s v_s}{n_c + n_s} \tag{9}$$

For gallium arsenide, the satellite valley scattering times are significantly shorter than the central valley scattering times and so the satellite velocity readjusts itself to the change in electric

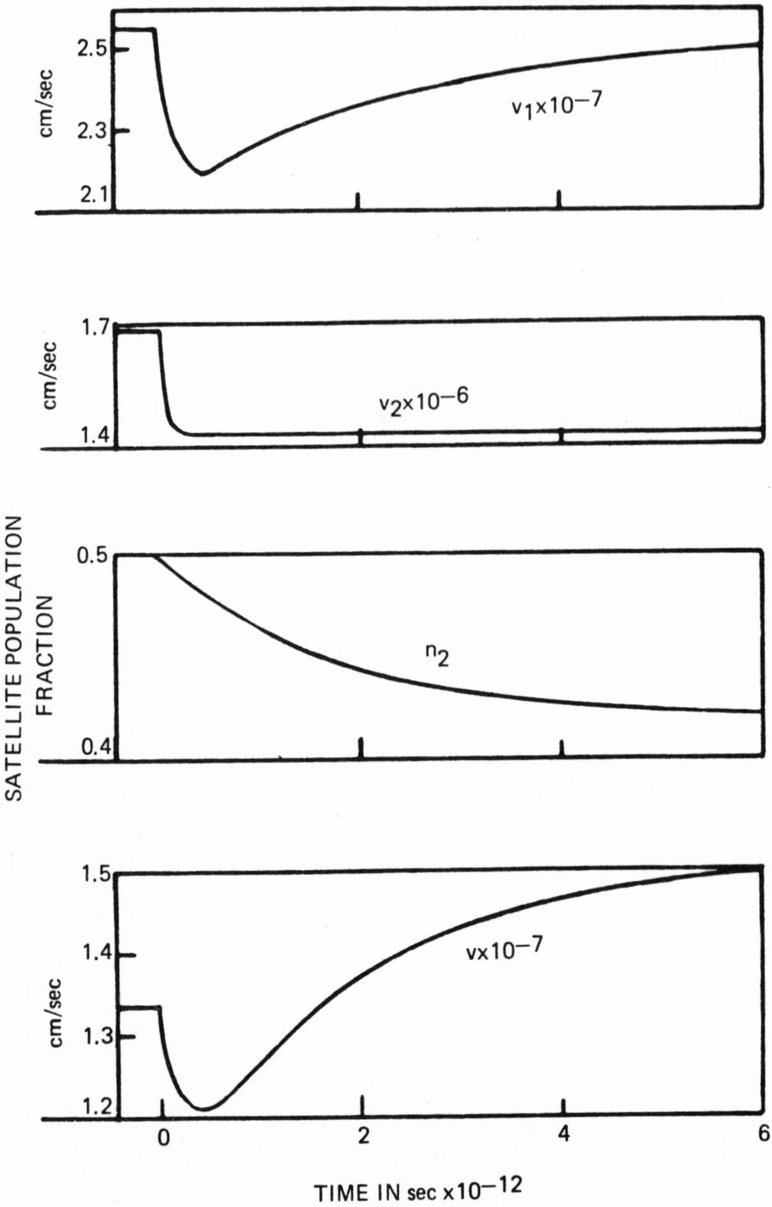

Figure 5. Response time of electrons to a field stepped down from 6 to 5 kV/cm at time t = 0. (a) and (b) are respectively the central and satellite valley velocities. (c) is the satellite population fraction. (d) is the drift velocity of the carriers. From Rees (1969) with permission.

field in less than 10^{-13} sec. The central valley velocity takes somewhat longer to reach its minimum value and to then recover. Here v falls for about 4×10^{-13} sec and then begins to recover over a period of approximately 5 picoseconds. The transfer of electrons from the central valley to the satellite valley is also slow. The net response of the drift velocity shows an initial drop due to the central and satellite valley velocities followed by a recovery due partly to the central valley velocity and partly due to differential electron transfer. The final velocity exceeds the initial velocity and illustrates the presence of negative differential mobility.

Transient calculations of the above type are important for determining the conditions under which a material like gallium arsenide will remain an active device. Rees (1969), using the results of Fig. 5, determined the frequency dependence of the differential mobility. These results are shown in Fig. 6. For this calculation, the Fourier transform of the drift velocity was taken, yielding v (ω). The mobility was defined as the ratio of $v(\omega)/5.5$ kV/cm (for a change from 6kV/cm to 5kV/cm); 5.5 kV/cm is the average of the initial and final values of the electric field. What we notice here, is that the real part of the mobility is *negative* at frequencies below 80 GHz. It is positive above 80 GHz, reflecting the short time response of the central valley electrons.

Spatial Dependence

The above arguments are for uniform fields. We can use this information to estimate the extent to which spatial changes in the electric field will affect the transient response of the carriers. To carry out this argument, we must first calculate the mean free path of electrons injected, with zero initial drift velocity, into a region of high electric field. The mean free path is calcualted prior to thermalization assuming a scattering time given by (7). From the momentum balance equation, we have

$$v = p/m^* = -(e\tau_p/m^*)F\left[1 - \exp\left(-t/\tau_p\right)\right] \quad , \tag{10}$$

and in a time $t = 2\tau_p$, v will have reached approximately 87% of its peak value. During this time the drifting electrons will have travelled a distance

$$\ell_o \cong (e\tau_p^2/m^*)\, F\, (1 - e^{-2}) \quad , \tag{11}$$

before thermalization occurs. We estimate that if the spatial variation in the electric field is such that large changes in the electric

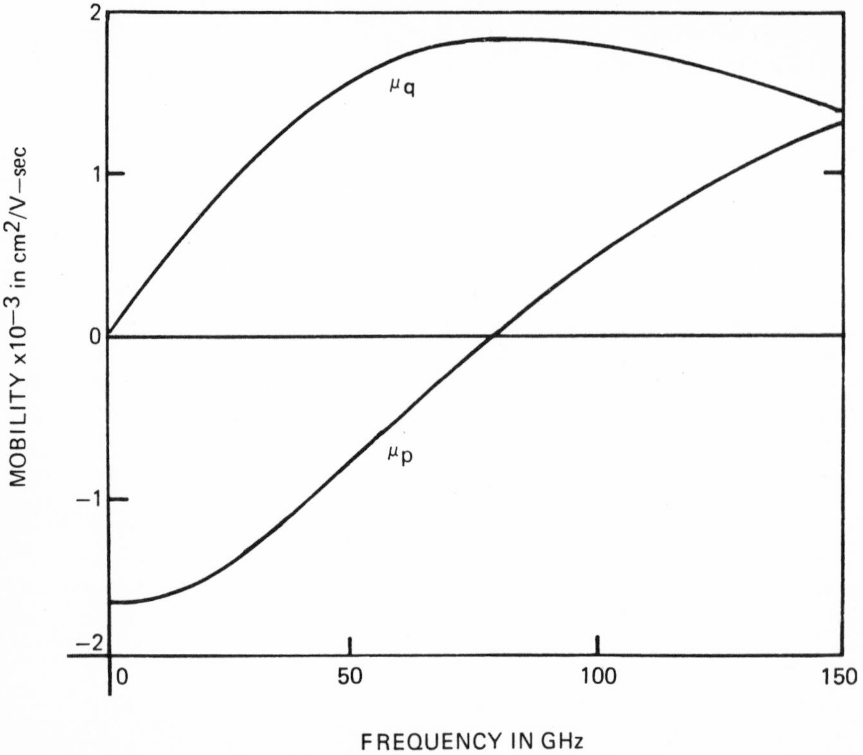

Figure 6. Frequency dependence of the differential mobility for a bias field of 5.5 kV/cm. μ_p and μ_p denotes respectively the components in phase and quadrature with the ac field. From Rees (1969) with permission.

field occur in distances less than that given by (11), then significant relaxation effects may be expected to take place. As an example, for an electric field of 5kV/cm and a cold mobility of $6500cm^2/V$-sec, electrons in the central valley will have attained a velocity of $2.8x10^7$cm/sec in a time $2\tau_p$, and will have travelled a distance of approximately 0.1 microns before the steady state value is reduced by increases in the electron temperature. Figure 7 illustrates this situation (Ferry and Barker, 1979). Here the velocity response of electrons in GaAs is shown for a field of 5kV/cm. These calculations include intervalley transfer and contributions from the satellite valley, for a Γ-L ordering.

The above results suggest that transient overshoot effects will be important when the electric field undergoes a change of the order

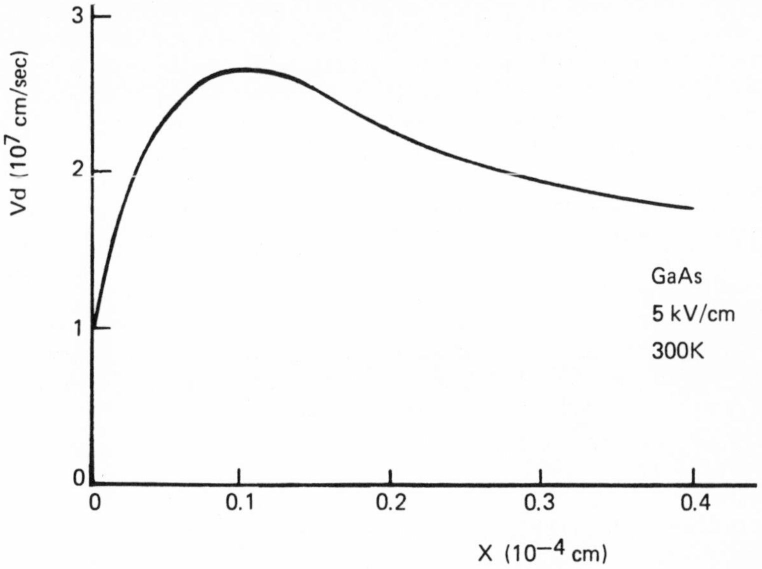

Figure 7. The velocity response of electrons in GaAs for an electric field of 5 kV/cm. From Ferry and Barker (1979), with permission.

of F_{th} (or greater) within a distance of ℓ_0. In other words overshoot effects will be pronounced when

$$\frac{dF}{dx} \gtrsim \frac{F_{th}}{\ell_o} \quad . \tag{12}$$

Using Poisson's equation this can be translated into the condition

$$\frac{N-N_0}{N_0} \gtrsim \frac{\tau}{\tau_p} \tag{13}$$

where τ is the dielectric relaxation time. At $N_0 = 10^{15}/cm^3$, the right-hand-side of the above equation is equal approximately to 4. Thus, when the ratio of N/N_0 exceeds 5 we may expect relaxation effects to enter in an important way. In an early discussion, Butcher, et al., (1967) estimated the maximum value of N/N_0 for the accumulation portion of a stable travelling domain. As is well known from domain theory, as the outside field begins to decrease the domain increases in size and the accumulation portion of the domain increases dramatically. Butcher, et al., (1967) found that when the outside field is below 2kV/cm, the accumulation ratio exceeds 5, and that for outside fields approaching the sustaining field the ratio is substantially higher. The implication is that a sizable portion of the accumulation layer is not accurately modeled by steady-state velocity-electric-field relationships. Rather, transient effects must be included for a complete picture. The degree to which this particular transient contribution will affect the output at the terminals of device has not yet been determined.

As we go to higher carrier concentrations, τ decreases and the criteria expressed by (13) becomes more restrictive. For example at 10^{17} cm^{-3}, accumulation layers in excess of 4% of N_0 will introduce significant relaxation effects.

Relaxation Effects Within a Travelling Accumulation Layer: The Jones-Rees Effect

The discussion above for nonuniform fields was based on concepts derived from a uniform field analysis. While this approach is useful for estimating some of the relaxation contributions, it is not wholly satisfying. Jones and Rees (1973) have examined some of these limitations. For this discussion, we refer to the accumulation layer profile sketched in Fig. 8.

The accumulation layer profile consists of two regions, a high-field and a low-field region separated by a region of substantial charge accumulation. The downstream portion of the accumulation layer consists mainly of satellite electrons. In analyzing the propagation of the accumulation layer, Jones and Rees (1973) observed a phenomena that appeared in all of their simulations. The fall of the satellite electron density from the high-field region to the low-field region was quite sharp. This behavior was not expected. In the spatially uniform analysis in which the field decreased rapidly to a value below threshold, a large fraction of the satellite electrons was initially retained. About 5-10 psec was required for their thermalization (see Fig. 5). Since the layer velocity in their computer simulations was above the satellite electron velocity, a long trailing edge of satellite electrons was expected. But in fact this did not occur. Rather a time constant of the order of 1 psec represented the rate of transfer. The mechanism proposed by Jones and Rees (1973) to explain the difference between the relaxation time for uniform fields and nonuniform fields is described below.

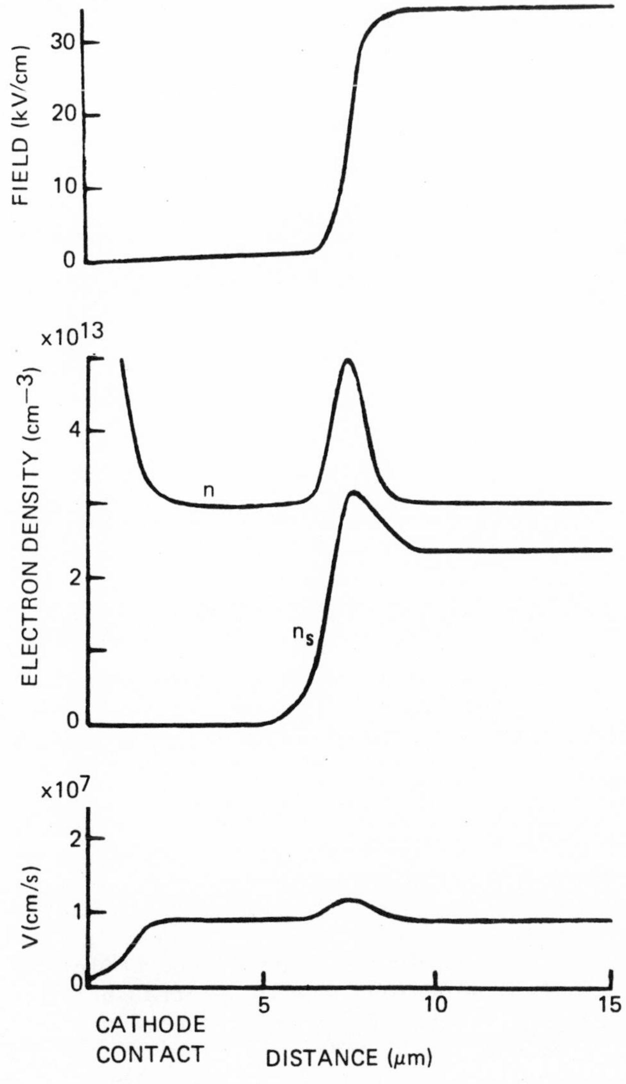

Figure 8. Profiles of electric field, total electron density and satellite valley density, and drift velocity for a mature propagating domain. From Jones and Rees (1973) with permission.

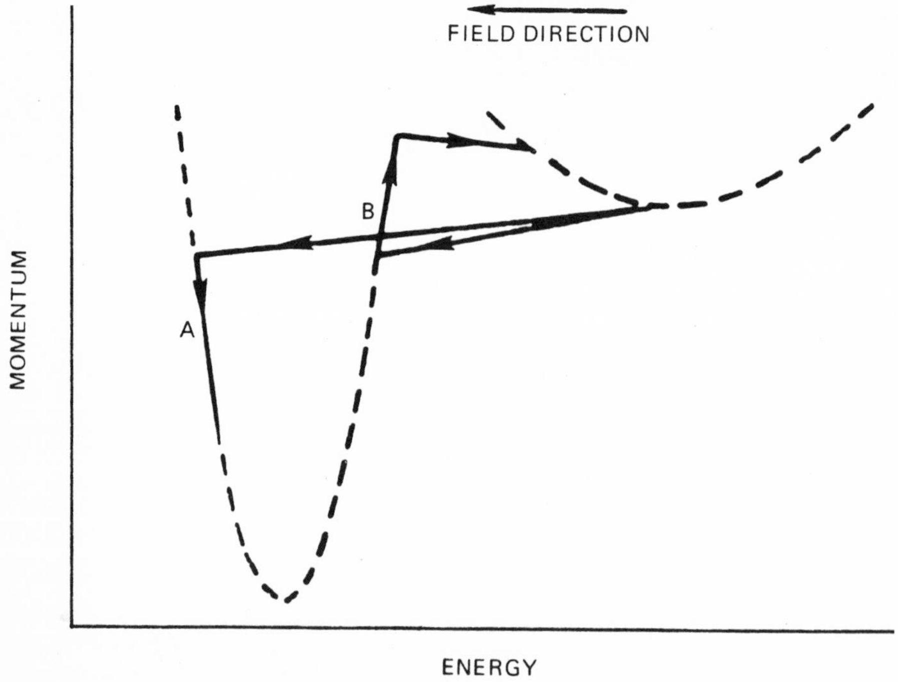

Figure 9. k-space representation of carrier cooling behind an accumulation layer. 'A' electrons scattered to Γ valley-backward momenta states lose energy and fall behind the layer. 'B' electrons scattered to Γ valley-forward momenta states gain energy and remain within the layer.

Figure 9 illustrates possible k-space trajectories for electrons initially within the layer. We will assume, for illustrative purposes, that the electrons are initially occupying the satellite valley states. These electrons rapidly scatter between the satellite and central valley states with high energy and momenta. Some of these electrons will scatter into k-states whose direction is opposite to that of the moving distribution. Those scattered electrons will lose energy and will not scatter back into the satellite valley. They are then concentrated behind the accumulation layer and account for the lower satellite valley population. In contrast, electrons scattered to high forward momenta states gain energy from the field and are concentrated in the front of the accumulation layer. At least part of the propagating accumulation layer is due to this scattering mechanism. Thus the traveling accumulation layer segregates the electrons into two classes and, in leaving a cold distribution behind the layer, it overrides the tendency to leave electrons heated, which characterizes a rapid fall of field in a spatially uniform distribution.

CONCLUSIONS

In the above, I have outlined several hot electron transient effects in transferred electron device. As of this writing, it is not at all clear what role these specific effects will have on the operation and output of small electron devices. What is clear is that it is incorrect to try to discuss the behavior of very-small, highly-doped devices without examining transient relaxation phenomena.

BIBLIOGRAPHY

Bosch, R., and Thim, H.W., 1974, IEEE Trans. Electron Dev. ED-21:16.
Butcher, P., 1967, Repts. Progr. Phys. 30:97.
Butcher, P.N., Fawcett, W., and Ogg, N.R., 1967, Br. J. Appl. Phys. 18:755.
Conwell, E.M., 1967, "High Field Transport in Semiconductors," Academic Press, New York. We use here (3.623) with the assumptions $\hbar\omega \approx k_oT_p$ and $\varepsilon = m^*V^2/2 < k_oT_p$. Note the comment following (3.623) that the scattering time does not represent a relaxation time. In our discussion, we nevertheless use τ_p for intravalley optical scattering as though it were a relaxation time.
Ferry, D.K., and Barker, J.R., 1979, Sol.-State Electr., in press.
Gunn, J.B., 1964a, IBM J. Res. Develop. 8:141.
Gunn, J.B., 1964b, in "Proc. 7th Intern. Conf. Phys. Semiconductors," Dunod Cie., Paris, p. 199.
Hilsum, C., 1962, Proc. IRE 50:185.
Jones. D., and Rees, H.D., 1973, J. Phys. C. 6:1781.
Kroemer, H., 1964, Proc. IEEE 52:1736.
Rees, H.D., 1969, IBM J. Res. Develop. 13:537.
Ridley, B.K., 1963, Proc. Phys. Soc. (London) 82:954.
Ridley, B.K. and Watkins, T.B., 1961, Proc. Phys. Soc. (London) 78:293.

OPTICAL EXCITATION OF HOT CARRIERS

R. G. Ulbrich

Institut für Physik
Universität Dortmund
46 Dortmund, Fed. Rep. Germany

INTRODUCTION

The study of charge transport in semiconductors and the domain of applied optical spectroscopy have promoted each other in various ways in the last two decades: efficient photo-cathodes and photo-conductors, extremely fast and sensitive detectors, and especially taylored light emitting diodes and injection lasers were all the result of applied research in semiconductors. All of these opto-electronic devices are based on the optical excitation of mobile carriers, or its time-reversed process, radiative recombination. They could not develop to full maturity before the basic mechanisms of optical absorption, carrier transport, and electron-hole-pair recombination in the relevant semiconductor materials were understood (for a good survey of the relevant phenomena see Pankove, 1971). On the other hand, the impact of laser spectroscopy, improved experimental techniques (among them optical probes with subpicosecond time resolution) and the availability of well-characterized group IV, III-V and II-VI semiconductor crystals have helped to keep the field of "hot", i.e. non-equilibrium, carrier phenomena so attractive. High density electron-hole plasma experiments, picosecond relaxation studies, and the direct access to carrier densities and energy distribution functions f(E) via light scattering and application of emission spectroscopy became possible in the last ten years. These were initially triggered by spectroscopic studies aimed at fundamental static properties of semiconductors: band structure, impurity levels, phonon frequencies. But soon the interest also centered on the dynamic aspects of interactions between the elementary excitations, like e.g. electron-phonon coupling, electron-electron collisions, capture of carriers by impurities, exciton interactions, etc. A great number of recent optical experiments has led to direct insight into microscopic scattering processes of mobil carriers, which were hitherto only theoretical postulates and had been used phenomenologically in formal transport theory (See, e.g., Seiler, 1978).

This article will outline the basic phenomena in the context of optical excitation of electron-hole-pairs in semiconductors. A short explanatory list of terms will be given in the second paragraph on optical excitation (one-electron band structure, description of excited states, correlated electron-hole-pairs and exciton picture, polaritons and optical absorption). We shall then discuss explicitly three important experiments involving optical excitation: (i) carrier heating by optical injection, (ii) band filling, and (iii) "oscillatory" photoconductivity and photoluminescence.

OPTICAL EXCITATION OF ELECTRON-HOLE-PAIRS

The ground state of a semiconductor at $T = 0$ can be described in the one-electron approximation in terms of an average, self-consistent crystal potential seen by the electrons. The eigenvalues $E(\vec{k})$ of the corresponding effective one-electron Hamilton operator represent the possible energies for a (fictional) extra electron added to the (existing) N-electron crystal (Friedel, 1972, Bassani and Pastori-Parravicini, 1975). The available eigenstates of the crystal are filled according to the Pauli-principle up to the Fermi-energy E_f, which in turn depends on the total number of electrons contained in the crystal in its neutral ground state.

Figure 1 shows as an example the theoretical one-electron band structure of GaAs (Zucca, et al., 1970), the energy zero is E_f. The lowest lying excited states of this specific crystal are separated from the ground state by the gap energy E_g of Fig. 1 minus a small correction of ~4 meV due to correlation and exchange between the excited crystal electron (of Γ_6 symmetry in the example of Fig. 1) and the remaining Γ_8 "hole" state, i.e. the short form for the N-1 valence electron system. The corresponding elementary excitation is the "exciton". In the effective-mass-approximation (EMA), the dynamic properties of the single constituent "electron" and "hole" of these low lying discrete exciton states are easily understood - they are described by the E(k) band structure mentioned above. The total energy of the e-h-pair, which includes the mutual Coulomb interaction, is given by

$$E(\vec{k}) = E_g - \frac{R^{*2}}{n^2} + \frac{\hbar^2 K^2}{2M^*_{Ex}}, \quad n = 1, 2, \ldots, \tag{1}$$

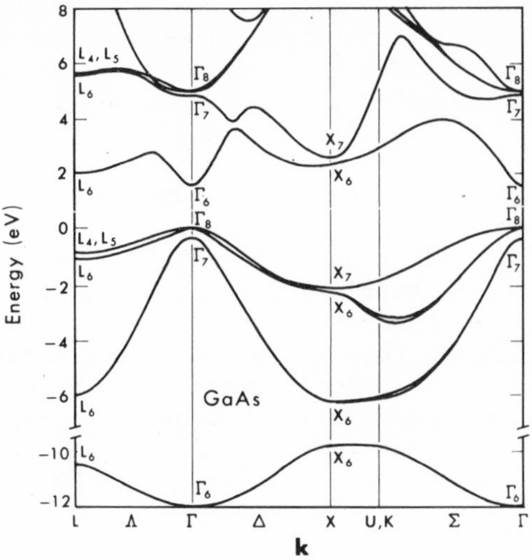

Fig. 1. One-electron band structure of GaAs, calculated by Zucca et al. (1970), including the spin-orbit interaction.

where $\vec{K} = \vec{k}_e - \vec{k}_h$ (k_e, k_h - electron and hole wavevector), $M^*_{Ex} = m^*_e + m^*_h$ is the sum of effective electron and hole mass, and n is the exciton orbital quantum number. In analogy with the hydrogen problem the exciton Rydberg is given by

$$R^* = \frac{\mu q_e^4}{2\varepsilon^2\hbar^2} = \frac{q_e^2}{2\varepsilon a_B} , \tag{2}$$

and is of the order of 2-20 meV for typical semiconductors. $\mu^{-1} = m_e^{*-1} + m_h^{*-1}$ denotes the reduced exciton mass resulting from the separation of translational and relative motion of both particles and a_B is the exciton Bohr radius (~20 Å to several hundred Å).

Above this discrete exciton series (i.e. for pair energies greater than E_g) one finds a continuum of correlated e-h-pair states represented by coulombic wave functions (Bassani and Pastori-Parravicini, 1975). If one neglected Coulomb correlation, these states would represent the "free electron", "free hole" states of the naive one-electron picture. However, due to the existence of e-h correlation, excitonic effects are of importance and are not at all negligible, at least for excitation energies close to the band edge.

The coupling of the crystal electrons with the electromagnetic field at optical frequencies can be treated on different levels of sophistication: using the uncorrelated states of the one-electron scheme, we obtain in first order perturbation theory - applying Fermi's Golden Rule - the dipole transition, if there is a non-vanishing dipole (or momentum) operator matrix element between the initial and final plane-wave electron states. This is the essence of most elementary treatments of optical properties of semiconductor (Pankove, 1971). The resulting spectra describe only the gross features of actually observed band-to-band optical transitions: Allowed direct ("vertical" transitions in Fig. 1) and forbidden indirect transitions (with participation of momentum-transfer by other excitations, like phonons, impurities or other mobile carriers both reflect the combined conduction and valence band density-of-states weighted with the $\vec{k}$-dependent optical dipole matrix element. The uncorrelated pair scheme fails to describe properly the shape of absorption and reflectance spectra close to the band edges of direct and indirect gap materials (Bassani and Pastori-Parravicini, 1975).

The extension towards a description in the exciton scheme is straightforward and is able to account for the main features of exciton (and continuum) absorption spectra close to E_g. However, the coupling of the exciton to the light field, which is generally strong in the case of direct allowed dipole transitions, requires a more rigorous treatment than a perturbation approach. Consequently, one has to introduce the polariton concept, i.e. the description in terms of mixed exciton-photon states within the crystal (Hopfield, 1969, Friedel, 1971). They represent the proper eigenstates of the (otherwise unperturbed) crystal in the close vicinity of discrete exciton resonances. The polariton approach is useful as long as the exciton-photon coupling is strong compared wit all other damping mechanisms of the electron (and hole) by phonon or impurity scattering. Exciton absorption, or polariton damping, and also the resonance behaviour of inelastic and elastic light scattering and related effects around the bandgap energy can be consistentl described in the polariton picture (Hopfield, 1969). If the inciden photon energy $\hbar\omega$ is sufficiently above E_g, excitonic effects are less pronounced, but the Coulomb enhancement in optical transitions is still appreciable even for $\hbar\omega > E_g + R$ (Bassani and Pastori-Parravicini, 1975).

The symmetries of the initial and final wave functions involved in the optical transition and the directional property of the $\vec{p}\cdot\vec{A}$ coupling term lead, in general, to anisotropic electron and hole distributions in $\vec{k}$-space. The specific case of direct-gap zinc-blende crystals (like GaAs, InP, CdTe etc.) excited with circularly or linearly polarized light of energy $h\omega$ such that $E_g < \hbar\omega < E_g + \Delta$ has recently been treated in detail (Dymnikov et al., 1976). The distributions reflect essentially the Γ_8 symmetry of the degenerate valence band orbitals and are of central importance for the interpretation of optical spin pumping experiments.

Apart from these spin-dependent effects in optical excitation of e-h-pairs, the related cubic warping of the heavy- and light-hole energy surfaces in $\vec{k}$-space causes a finite width of final electron and hole energies even under conditions of mono-energetic illumination. The valence band parameters in the common III-V and II-VI semiconductors discussed here are such that the warping effect leads to an energy broadening between 5% and 30% of the mean kinetic energy of the photoexcited electrons (and holes) and may thus be neglected in many cases of practical interest.

Almost all of the experimental work in optical excitation has been performed by using laser light sources. The recent progress in tunable cw dye lasers has extended the accessible spectral range from 400 nm to more than 1 μm wavelength (for an overview see e.g. Weisbuch, 1979). Further development of color center lasers (Welling, 1979) in the near infrared spectral range (1 μm to ~4 μm) looks promising and will considerably improve the experimental possibilities. Resonant excitation of narrow-band-gap semiconductors, and high resolution spectroscopy of deep impurity states will soon become feasible. Both laser variants, dye and color center crystal, can be operated in mode-locking conditions for picosecond light pulse generation and open fascinating possibilities for studying fast kinetics of photoexcited carriers (a survey of the present status of picosecond excitation spectroscopy is given by Shank et al., 1978).

The outstanding advantage of optical excitation methods in studying hot carrier phenomena are evidently the inherent selectivity in excitation energy (as compared with thermal or electric field excitation) <u>and</u> the enormous extension of time resolution into the 10^{-12} sec range! Even though the kinetic energies of optically generated e-h-pairs are not necessarily as monoenergetic as the incident laser light, there is no easier way to inject carriers of well defined density in semiconductor crystals. The following three sections will treat three specific experiments which represent the current status of optical studies of hot electrons.

CARRIER HEATING BY OPTICAL INJECTION

The principle of carrier heating by injection of additional hot electrons and holes is simple: due to Coulomb interaction among the carriers, the initial excitation of non-equilibrium states is balanced by relaxation towards an equilibrium among the carriers. If the carrier system is also coupled - via the e-phonon interaction- to an external thermal bath, the steady-state optical injection of e-h-pairs will cause, in general, a change in the mean energy of the carrier populations. Heating (or cooling) with respect to the bath is observed. The process represents a complex transport problem, sometimes far from equilibrium, and with highly non-uniform, i.e. spatially dependent variables.

In most cases the penetration depth of optical excitation is small, typically less than 1 μm for above-band-gap illumination in direct gap semiconductors, and several tens of μm in indirect materials, depending on $\hbar\omega - E_g$. For this reason, drift and diffusion of the carriers play a major role in determining the steady state density profile in the sample. Of crucial importance are als the radiative and nonradiative recombination channels, which are, in general, strongly dependent on the carrier energy. The quantitative description of the observed phenomena is obviously difficult Even the use of thin samples would not help because of the known effective surface recombination (Hwang, 1972). The problem has bee analyzed in the context of photoconductivity (de Vore, 1956). In spite of the problems of a proper description, carrier heating experiments have given a lot of direct information on hot carrier relaxation mechanisms (Shah, 1978).

The first experiment was performed by Shah and Leite (1969) on GaAs (E_g = 1.52 eV) with fixed pump photon energy 2.41 eV. The results of this cw heating experiment were in qualitative agreement with a description in terms of a simple power balance equation, taking into account the most prominent dissipative scattering mechanism (coupling to LO phonons) and assuming energy randomization among the electrons via interelectronic collisions. Their work was followed by a series of more detailed experiments in GaAs and other semiconductors (Shah, 1978). The regime of carrier energy relaxati by interaction with acoustic phonons at low temperatures and the ro of e-e-scattering at much lower densities ($n_e \approx 10^{11}cm^3$ to $10^{14}cm^{-3}$) was investigated in GaAs (Ulbrich, 1973, 1978). The electron energ distribution function f(E) was probed by emission spectroscopy of a well-defined conduction band → acceptor (BA or e,A°) transition. The principal scheme of this experiment is shown in Fig. 2. The emission spectrum gives directly the product of f(E) with the conduction band density-of-states, the energy dependence of the matrix element is negligible in the kinetic energy range discussed here. Figure 3 shows a comparison between an observed and theoretical lineshape under resonant excitation conditions ($\hbar\omega_{ex} = E_g + 3/2\ kT_L$) and medium electron density $n_e \approx 10^{14}cm^{-3}$. The good agreement over a wide dynamic range of more than two orders of magnitude in f(E) makes this spectroscopic probe an ideal tool for the investigation of electron distributions in direct gap materials.

Figure 4 shows the result of a transient experiment performed with 200 psec, 1.92 eV light pulses incident on GaAs at T_L = 4.2 K. Within the available time resolution of ~0.5 nsec of the (e,A°) spectrum probe, the injected carriers, with ~0.4 eV excess kinetic energy, relax towards mean energies around 3 meV. The subsequent relaxation close to thermal lattice (resp. bath) energies occurs on a 10^{-8} sec scale, in good agreement with theoretical estimates from a power balance equation with the inclusion of piezo-electric

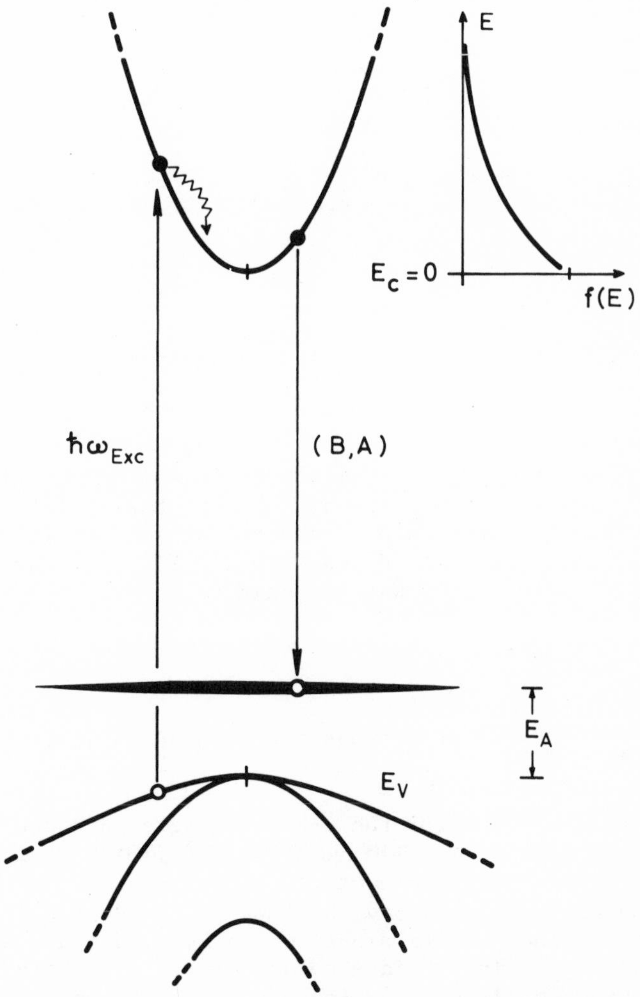

Fig. 2. Scheme of the optical excitation of non-equilibrium electron-hole-pairs by incident light of energy $\hbar\omega_{Ex}$ and simultaneous probing of electron energy distributions f(E) by radiative band-to-acceptor (BA) transitions. E_A is the acceptor hole binding energy.

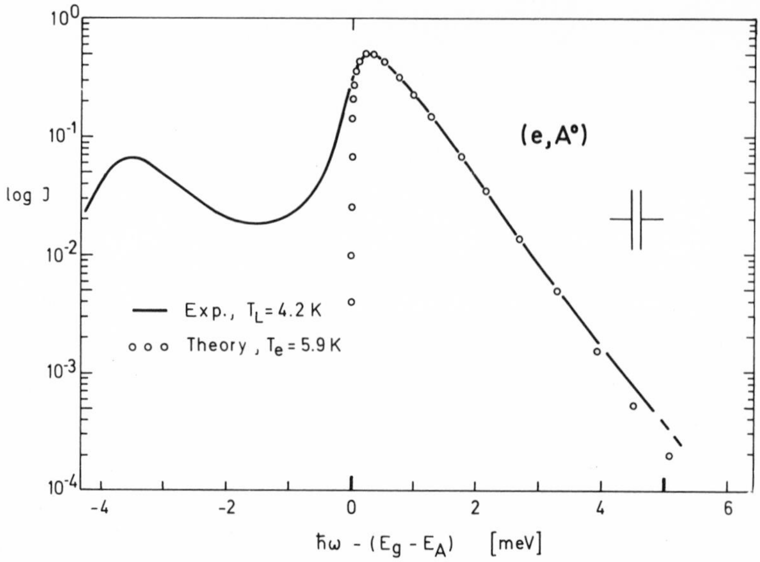

Fig. 3. Comparison between measured (e,A°) band-to-acceptor emission line shape and theory, taking into account the energy dependence of the transition matrix element. The sample was GaAs doped with 2 x $10^{14}cm^{-3}$ carbon acceptors, the excitation was resonant ($\hbar\omega_{ex} = E_g + 3/2\ kT_L$, carrier density $n_e \approx 10^{14}cm^{-3}$.

and deformation potential coupling with acoustical phonons (Ulbrich 1973).

In their search for the e-h-plasma ground state properties in GaAs, inP and GaSb, Hildebrand et al. (1978) have investigated, *inter alia*, the dependence of plasma temperature on the optical injection excess energy and the pump power under pulsed excitation conditions. The results are shown in Figs. 5 and 6 and can be summarized as follows: in accordance with cw excitation experiments (Weisbuch, 1978) they found a monotonic and linear increase of mean carrier energy with increasing injection energy. The increase is less pronounced than expected from a simple model and indicates energy relaxation rates above the known electron and hole one-phonon dissipation rates which have been calculated (Goebbel and Hildebrand, 1978) and are shown in Fig. 6. The reason for the discrepancy is possibly the assumed power balance equation, which considers only the injection excess energy and energy dissipation into the cold lattice. Further dissipation mechanisms and shortening of pair life-time by stimulated emission might also play a role.

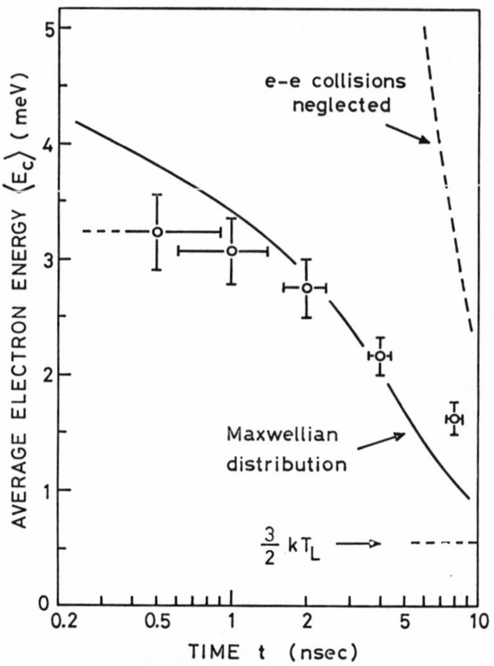

Fig. 4. Average electron energy as a function of time after pulsed photoexcitation at t = 0 with 200 psec light pulses, and $\hbar\omega_{ex}$ = 1.92 eV in GaAs (E_g = 1.52 ev, T_L = 4.2 K). The full line is the theoretical dependence, taking into account the relevant optical and acoustical interactions, and assuming the electron distribution to be Maxwellian. (After Ulbrich, 1973).

After picosecond light pulse excitation and probing techniques had been developed (Shapiro, 1977), a series of experiments was performed to study the time evolution on that scale of (i) carrier energy relaxation, (ii) plasma density and (iii) exciton screening after the pulsed injection of high density plasmas with relatively high mean energy (Shank et al., 1979, von der Linde and Lambrich, 1979). The quoted results were all in quite good agreement with theoretical descriptions known from the theory of electron transport in semiconductors.

The above cited carrier heating experiments were representative examples - much more work than reported here has been done (see e.g. Yassievich and Yaroshetskii, 1975, and references contained therein).

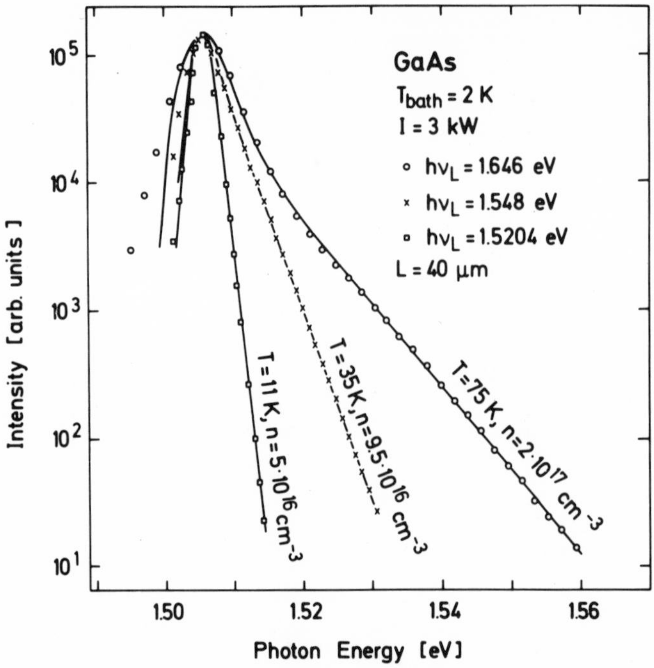

Fig. 5. Band-to-band emission spectra from a high density electron-hole-plasma in GaAs excited by 10 nsec laser pulses with the indicated energies $h\nu_L$. The plasma temperatures were obtained from a fit to a model including amplification by stimulated emission and reabsorption losses. (After Goebbe and Hildebrand, 1978).

BAND FILLING BY OPTICAL EXCITATION

The energy dependence of the electron-density-of-states close to a band edge is proportional to $m_e^{*3/2} \cdot E^{1/2}$ and can be quite small, especially in the case of small band gap semiconductors. For $m_e^* = 0.07\ m_o$, one has 2.5×10^{15} states per cm^{-3} in an energy interval of 1 meV above E_g. Typical lifetimes of optically created electron-hole-pairs at low carrier densities are controlled by non-radiative recombination and can range from fractions of 10^{-9} sec to 10^{-6} sec, depending on the number of recombination centers and impurities present in the crystal. Under steady-state conditions and at low temperatures one needs therefore generation rates between 10^{24} and 10^{21} cm^{-3} sec^{-1} to reach the limit of carrier degeneracy! This corresponds to light power densities of the order of 0.1 to 100 W/cm^2 (at $\hbar\omega$ = 1 eV), if the pair recombination time itself does not decrease with increasing pump power, e.g. through the onset of stimulated emission in the case of direct gap materials.

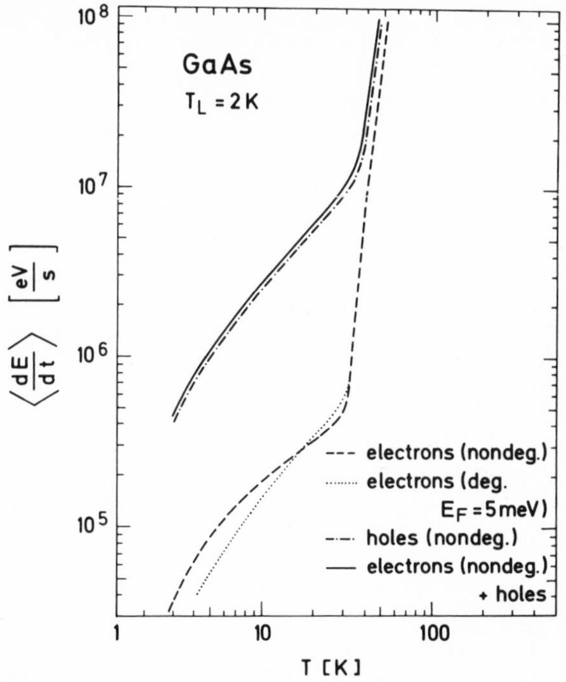

Fig. 6. Theoretical energy dissipation rates of degenerate and nondegenerate quasi-equilibrium distributions of electrons and holes of temperature T in a GaAs crystal of lattice temperature T_L = 2K. (After Goebbel and Hildebrand, 1978).

Recently Lavallard et al. (1977) and Bichard (1979) have undertaken a careful and exhaustive study of saturation phenomena of the optical absorption by an electron-hole-plasma in InSb under cw quasi-resonant excitation with a CO laser. From the excitation power dependent transmission data and analysis of the band-to-band emission spectra, they obtained the first consistent picture of saturation and bleaching in InSb under cw photoexcitation. Strong evidence for a photon-recycling effect (caused by reabsorption of the band-to-band recombination radiation and related to the exciton-polariton problem discussed above) was found and had to be taken into account in their analysis. The steady state density of carriers in their experiment was well above the theoretical plasma density corresponding to minimum total energy, so that the observed effects indicate the existence of highly non-equilibrium plasma phases under the given pumping conditions. Detailed information on the renormalization of the band gap due to exchange and correlation in the e-h-plasma was also obtained.

Observations of a dynamic Burstein shift of the absorption edg in GaAs, and saturation effects in excitonic absorption spectra of GaAs in the presence of intense optical excitation, were reported recently (Shah et al., 1976, Shah and Leheny, 1977, Gibbs et al., 1979). In Shah and Leheny (1979), the persistent effect of exciton contributions to the absorption coefficient above E_g even at high densities (where the discrete exciton peak below E_g disappears) has been clearly observed. The effects described by Gibbs et al. (1979 belong to non-linear phenomena in the context of light-matter inter- action: the basic mechanisms of electron-hole-correlation and carrier scattering in many cases generate non-linear sample trans- mission characteristics which can be employed for optical switching or bistable device applications.

The interest in band-filling effects originates from the need for solid-state saturable absorbers suited for applications in lase modulation, e.g. passive mode-locking or Q-switching in high power lasers.

OSCILLATORY PHOTOCONDUCTIVITY AND PHOTOLUMINESCENCE

After optical excitation with monochromatic light of energy $\hbar\omega > E_g$, the initial electron and hole distributions are more or le sharp in energy and anisotropic in k-space. Any subsequent scatter- ing process on either the other mobile carriers, or on impurities and phonons, will tend to randomize this initial "hot" non-equili- brium state towards equilibrium with the surrounding bath to which the sample is coupled. We shall discuss here only one example of such a counterplay between mono-E, directional-$\vec{k}$ distributions with relaxed quasi-equilibrium carriers under conditions of optical pumping: the so-called oscillatory photoconductivity and -lumines- cence effect, which has up to now been observed in almost all impor- tant group IV, III-V, and II-VI compounds.

The quoted periodicity in $\hbar\omega$, for which "oscillatory" stands, is in principle due to the strongest inelastic coupling mechanism with well-defined amount of energy transfer (which is typically the Fröhlich LO-phonon coupling in the reported photo-conductivity, photo-luminescence, and photo-excitation spectra (Mazurczyk and Fan, 1970, Shaw, 1971, Collins et al., 1969, Kovaleskaja et al., 1967, Legros and Marfaing, 1973, Stocker et al., 1966, Ulbrich, 1971, Hei and Hiesinger, 1974, Lüth, 1972, Ibach and Mönch, 1973) are simple consequences of spatial, energetic and wavevector deviations from homogeneity and quasi-equilibrium, which in turn depend critically on the injection energy of the electron-hole-pairs.

There has been controversy in the literature on the basic orig of these oscillatory effects - whether lifetime, mobility, or carri density are responsible for the structures seen in the spectra. It seems clear now that all three parameters play a role, but that one has to be cautious not to oversimplify the description. In addition to the three factors mentioned, we stress here the importance of spatial inhomogeneity. This fact has been almost completely neglected in all interpretations, though it is known quite well that non-radiative surface recombination (Hwang, 1972) together with density profile effects (de Vore, 1956) strongly influence photoconductivity and photoluminescence data. The balance between excitation energy dependent relaxation (or "hot carrier diffusion") into the bulk of the crystal, and the surface-related decay of e-h-pairs certainly plays a key role in a realistic description. Linked with this balance is the net lifetime of the injected distribution, which is mainly controlled by the rate at which carriers reach the surface. The energy dependent mobility also plays a role: its influence is proportional to the ratio of the energy relaxation time over the total carrier lifetime.

Folland (1970) has analyzed a theoretical model neglecting electron-electron scattering, and taking polar-optical and acoustical phonon scattering into account. He solved the transport equation for small applied electric fields and with the assumption of a constant carrier lifetime. The solutions obtained had all the qualitative features of actually observed oscillatory photo-conductivity spectra: asymmetric "sawtooth"-like dependence of the conductivity on $\hbar\omega$, and a decrease of the effective modulation depth with increasing temperature. Inclusion of interelectronic collisions would have smeared out the sharp structures, in accordance with experimental findings at carrier densities around $n_e = 10^{14}$ cm^{-3} (Weisbuch, 1978).

More detailed descriptions of the transport problem with strong LO-phonon coupling would start from a polaron picture - treating the coupling to phonons not as a perturbation, but rather incorporate it fully by means of proper diagonalization.

CONCLUSION

The few principles, experimental observations, and insights presented in these notes on the complex field of optical excitation of electron-hole-pairs may easily be generalized to cover other specific situations, e.g. indirect gap semiconductors, more complicated spatial structures (like superlattice crystals), or even actual semiconductor devices. Further development and consequent application of the techniques discussed here will contribute substantially to progress in the field of hot carrier phenomena in semiconductors.

REFERENCES

Bassani, F., and Pastori-Parravicini, G., 1975, in "Electronic State and Optical Transitions in Solids," Ed. by R. A. Ballinger, Pergamon Press, Oxford, Ch. 3.
Bichard, R., 1979, Ph.D. thesis, Université Paris VI, unpublished.
Collins, A. T., Lightowlers, E. C., and Dean, P. J., 1969, Phys. Rev. 183:725.
de Vore, H. B., 1956, Phys. Rev. 102:86.
Dymnikov, V. D., D'yakonov, M. I., and Perel, N. I., 1976, Zh. Eksp. Theor. Fiz. 71:2373 [transl. in Sov. Phys. JETP 44:1252].
Folland, N. O., 1970, Phys. Rev. B 2:418.
Friedel, J., 1972, in "Optical Properties of Solids," Ed. by F. Abeles, North Holland, Amsterdam.
Gibbs, H. M., Gossard, A. C., McCall, S. L., Passner, A., and Wiegmann, W., 1979, Solid State Commun. 30:271.
Goebel, E. O., and Hildebrand, O., 1978, Phys. Stat. Sol.(b) 88:645.
Heim, U., and Hiesinger, P., 1974, Phys. Stat Sol.(b) 66:461.
Hildebrand, O., Goebel, E. O., Romanek, K. M., Weber, H., and Mahler G., 1978, Phys. Rev. B 17:4775.
Hopfield, J. J., 1969, Phys. Rev. 182:945.
Hwang, C. J., 1972, Phys. Rev. B 6:1335.
Ibach, H., and Mönch, W., 1973, Phys. Stat. Sol.(b) 55:243.
Kovalevskaja, G. G., Nasledov, D. N., and Slobodchikov, S. V., 1967, Phys. Stat. Sol. 23:755.
Lavallard, P., Bichard, R., and Benoit á la Guillaume, C., 1977, Phys. Rev. B 16:2804.
Legros, R., and Marfaing, Y., 1973, Phys. Stat. Sol.(a) 19:635.
Lüth, H., 1972, Phys. Rev. Letters 29:1377.
Mazurczyk, V. J., and Fan, H. Y., 1970, Phys. Rev. B 1:4037.
Pankove, J. I., 1971, "Optical Processes in Semiconductors," Prentice Hall, Englewood Cliffs, N.J.
Seiler, D., Ed., 1978, "Proceedings of the International Conference on Hot Electrons in Semiconductors, Denton, 1977," Sol.-State Electronics 21:1-324 [This reference contains a collection of review articles and original papers which document the recent progress in the field of optical studies of hot electrons].
Shah, J., 1978, Sol.-State Electr. 21:43.
Shah, J., and Leheny, R. F., 1977, Phys. Rev. B 16:1577.
Shah, J., and Leite, R. C. C., 1969, Phys. Rev. Letters 22:1304.
Shah, J., Leheny, R. F., and Lin, C., 1976, Solid State Commun. 18:1035.
Shank, C. V., Ippen, E. P., and Shapiro, S. L., Eds., 1978, "Picosecond Phenomena," Springer-Verlag, Berlin.
Shank, C. V., Fork, R. L., Leheny, R. F., and Shah, J., 1979, Phys. Rev. Letters 42:112.
Shapiro, S. L., Ed., 1977, "Ultrashort Light Pulses: Picosecond Techniques and Applications," Springer-Verlag, Berlin.
Shaw, R. W., 1971, Phys. Rev. B 3:3283.

Stocker, H. J., Levinstein, H., Stannard, C. R., Jr., 1966, Phys. Rev. 150:613.
Ulbrich, R., 1971, Phys. Rev. Letters 27:1512.
Ulbrich, R., 1973, Phys. Rev. B 8:5719.
Ulbrich, R., 1978, Sol.-State Electr. 21:51.
von der Linde, D., and Lambrich, R., 1979, Phys. Rev. Letters 42:1090.
Weisbuch, C., 1978, Sol.-State Electr. 21:179.
Weisbuch, C., 1979, in "Proc. 14th Intern. Conf. Phys. Semiconductors," Inst. of Phys., London, p. 93.
Welling, H., 1979, in "Advances in Solid State Physics, XIX," Ed. by J. Treusch, Vieweg, Braunschweig, in press.
Yassievich, I. N., and Yaroshetskii, I. D., 1975, Fiz. Tekh. Poluprovodn. 9:857 [transl. in Sov. Phys. Semicond. 9:565].
Zucca, R. L., Walter, J. P., Shen, Y. R., and Cohen, M. L., 1970, Solid State Commun. 8:627.

THEORETICAL CONCEPTS OF PHOTOEXCITED HOT CARRIERS

C. J. Hearn

Department of Physical Oceanography
Marine Science Laboratories, U.C.N.W.
Menai Bridge, Gwynedd, U.K.

INTRODUCTION

Most of the extensive literature on hot carriers refers to carrier heating by an electric field. This lecture concerns a totally different mechanism for producing a non-equilibrium carrier distribution. Carriers are photoexcited into a band at a steady rate with a mean energy in excess of the thermal energy. If the recombination lifetime τ_r is sufficiently short in relation to the thermalization time τ_{th} the carriers spend insufficient time in the band to completely thermalize via the electron-phonon interaction prior to their recombination and a hot steady-state carrier distribution is obtained. The thermalization time is not a very well defined quantity and like the recombination time is dependent on carrier energy, but broadly speaking we can use these quantities to classify the types of heating that we can expect to find. If $\tau_r \ll \tau_{th}$ the distribution is wholly recombination-controlled with effectively no thermalization. If $\tau_r \gg \tau_{th}$ the majority of carriers are thermalized and the distribution function only differs from equilibrium by having a high-energy tail containing a small minority of unthermalized carriers. The tailed-thermal distribution approaches the equilibrium limit for large τ_r by the height of the tail decreasing to zero. The presence, and shape, of the tail can be inferred from the optical recombination spectrum. The intermediate case where $\tau_r \sim \tau_{th}$ shows distribution functions with a more substantial high-energy tail and the low-energy carriers are not thermalized and can be described as either warm or moderately heated.

The theoretical overlap between studies of optical and high-field heating of carriers lies in the dependence of the distribution functions on the electron-phonon interaction. In some materials this interaction is still only understood in a fairly rudimentary manner and in principle experimental hot carrier studies can help provide additional information on the parameters of the interaction. Optical heating experiments can doubtlessly be of great value in this respect, but any such potential cannot be realized without greater experimental exploration of the topic. Optical heating considered as such an investigative tool has the advantage that a wide variety of different distribution functions can be obtained by choosing different optical excitation spectra and varying the recombination lifetime by preparing samples with different doping concentrations. The ability of the electron-phonon interaction to move carriers between states of widely differing dynamic properties leads to important non-linear field effects for high-field heated systems such as the Gunn effect in multi-valley conduction bands. Similar non-linearities must occur when electric fields are imposed on optically heated carriers and indeed some non-linearities can be predicted which are unique to the distribution functions resulting from monoenergetic photoexcitation. The investigation of the macroscopic consequences of such non-linearities in photoexcited systems will be of great interest and may help extend our presently very limited knowledge of the optimum form of phenomenological equations to describe the dynamics of non-linear transport in semiconductors.

THE BARKER-HEARN MODEL

The model employs the steady-state Boltzmann equation modified to include recombination and generation terms. The linear electric field properties of the photoexcited carriers are obtained by finding the distribution function in zero electric-field and then using conventional low-field perturbation theory to deduce the transport parameters of this distribution. We shall outline an iterative method of finding the zero-field distribution in the next section. The non-linear field properties can be derived by solving the modified Boltzmann equation directly using, for example, a Monte Carlo method. The model assumes that the system is linear in the strength of photoexcitation which implies low photocarrier densities, so that such non-linear effects as statistical degeneracy, trap emptying and filling, inter-carrier interaction and non-equilibrium phonon populations are excluded. The model has only been used for a symetric band structure so that assuming isotropic excitation the distribution function f is a function of energy ε only. If $\underline{k}$ is the wave-vector of a state of energy ε then $f(\varepsilon)$ is the statistical carrier occupancy of state $\underline{k}$ and satisfies the steady-state rate equation

$$W\omega_e(\varepsilon) - f(\varepsilon)/\tau_r(\varepsilon) = f(\varepsilon)Z(\varepsilon) - \int Y(\varepsilon,\varepsilon')f(\varepsilon')d\varepsilon' \qquad (1)$$

which is a linear integral equation. $W\omega_e(\varepsilon)$ is the rate of photoexcitation into a state $\underline{k}$ and W is the total rate of excitation into the band which, in view of the linearity, can be assigned any arbitrary value. $\tau_r(\varepsilon)$ is the energy-dependent recombination time. $Z(\varepsilon)$ and $Y(\varepsilon,\varepsilon')$ are defined in terms of the inelastic scattering probabilities per unit time derived from the electron-phonon interaction

$$Z(\varepsilon) \equiv \sum_{\underline{k}'} P(\underline{k},\underline{k}') \tag{2}$$

$$\int Y(\varepsilon,\varepsilon')f(\varepsilon')d\varepsilon' \equiv \sum_{\underline{k}'} P(\underline{k},\underline{k}')f(\varepsilon') \tag{3}$$

The ideal choice of excitation spectrum is the monochromatic form

$$\omega_e(\varepsilon) = \delta(\varepsilon-\overline{e}) \tag{4}$$

and the minimum width of a spectral line which can be modelled by a δ-function will be mentioned later. Monoenergetic excitation has four advantages over broad spectrum excitation. First, it avoids any uncertainties as to the shape of the spectrum which are introduced by our lack of knowledge of the matrix elements and selection rules involved in the topical excitation process. Second, if $\overline{\varepsilon} \gg kT$, the distribution function for a given recombination lifetime is furthest removed from a thermal distribution because no low-energy carriers are excited into the band. Third, the distribution obtained from monoenergetic excitation exhibits features which are simple consequences of the physics of the electron-phonon interaction and although these may be obscured experimentally by various broadening effects, they are certainly of interest in theoretical understanding. Four, the monoenergetic distribution is the Green function of the scattering-recombination operator and if denoted by $f(\varepsilon,\overline{\varepsilon})$ the distribution function $f(\varepsilon)$ for any excitation spectrum $\omega_e(\varepsilon)$ can be synthesized via

$$f(\varepsilon) = \int_0^\infty f(\varepsilon,\overline{\varepsilon})\omega_e(\overline{\varepsilon})d\overline{\varepsilon} \tag{5}$$

Another spectrum which is used, for experimental convenience, derives from the exposure of a low temperature sample to room temperature (T_r) ambient radiation. If we assume that the radiation is black-body and excitation takes place from localized states which are in excess of kT_r below the band then

$$\omega_e(\varepsilon) \propto \varepsilon^{\ell} \exp(-\varepsilon/kT_r) \tag{6}$$

where ℓ is a (positive) shape parameter originating from the matrix elements of the transition process. Fortunately, the actual value assigned to ℓ does not seem to be of critical importance. The spectrum is preferably terminated at high energies to prevent impact ionization of localized states.

Of relevance to the form of ω_e is the optical phonon emission process. This process is strictly contained in the scattering terms on the right-hand side of the rate equation. However, the process is very much faster than any other occurring in our problem and it is computationally simpler to use the instantaneous emission approximation which means that at low temperatures a carrier photoexcited into a state in the range $n\hbar\omega < \varepsilon < (n+1)\hbar\omega$, where $n\hbar\omega$ is an integral multiple of the optical phonon energy, is instantaneously transferred to a state of energy $\varepsilon - n\hbar\omega$. Hence the effective excitation spectrum is

$$\omega_e(\varepsilon) \rightarrow \sum_n \tilde{\omega}_e(\varepsilon - n\hbar\omega) , \qquad \tilde{\omega}_e(\varepsilon') \equiv \omega_e(\varepsilon')\theta(\varepsilon')\theta(\hbar\omega - \varepsilon') , \tag{7}$$

θ being the unit step function. This is not a complete description of the instantaneous emission approximation and we continue the matter in a later section.

The energy dependence of the recombination lifetime is the least understood factor in the problem. Experimentally one can only measure the mean reciprocal lifetime $\langle 1/\tau_r \rangle$ which corresponds theoretically to

$$\langle 1/\tau_r \rangle = \int_0^\infty (1/\tau_r)\, f\, \varepsilon^{\frac{1}{2}}\, d\varepsilon \Big/ \int_0^\infty f\, \varepsilon^{\frac{1}{2}}\, d\varepsilon \, . \tag{8}$$

The reciprocal lifetime is assumed to be proportional to the density of capture centres N,

$$1/\tau_r = \sigma(\varepsilon)\, v(\varepsilon) N \; , \tag{9}$$

where $v(\varepsilon)$ is the velocity of a carrier of energy ε and $\sigma(\varepsilon)$ the capture cross-section. The mean $\langle\sigma\rangle$ is defined as $\langle \sigma v \rangle / \overline{v}$ where $\overline{v}$ is the mean thermal velocity. This is a conventional definition which is rather unsatisfactory for heated carriers. By using appropriate thermal distributions one can infer the dependence of τ_r upon upon ε in the lower portion of the band provided there is no explicit dependence on lattice temperature T. Another approach is to employ a long lifetime (low n) and to heat the carriers by a variable electric field and assume an effective carrier temperature, which can be deduced from mobility measurements.

From the comments made in the introduction we know that the form of the function $\tau_r(\varepsilon)$ is of prime importance in determining the shape of f in the recombination-controlled regime $\tau_r \ll \tau_{th}$ where $f \sim \tau_r{}^{\omega}e$. However, our main concern, with the lifetimes available to us, is with the warm or moderately heated regime $\tau_r \sim \tau_{th}$ and the tailed-thermal regime $\tau_r \gg \tau_{th}$. Most models of recombination processes suggest that τ_r is a fairly monotonic function of ε and often there is a pronounced tendency for τ_r to reduce as ε decreases. In this case the form of $\tau_r(\varepsilon)$ only significantly affects the heated and tailed distributions in a narrow region at the edge of the band. In this edge region the distribution is essentially recombination-controlled, because the acoustic phonon scattering becomes increasingly ineffective towards the band-edge, due to the constraints of energy and momentum conservation, and the recombination probably becomes more effective. The width of the edge region decreases as τ_r/τ_{th} increases and in the recombination-controlled distributions the edge region has effectively invaded the entire carrier system. If τ_r actually vanishes for $\varepsilon = 0$ then the same must be true of f and although this is strictly unphysical it is probably indicative of the type of behavior to be expected in the edge region.

The best understood recombination mechanism is cascade capture into shallow ionized centres via the acoustic phonon interaction. According to Lax (1960)

$$\sigma(\varepsilon) = \sigma_1(\varepsilon_s/kT)^4(kT/\varepsilon)\Phi(\varepsilon,T) , \tag{10}$$

where $\varepsilon_s = \frac{1}{2}m^*s^2$ and s is the velocity of sound, σ_1 is a constant and

$$\Phi(\varepsilon,T) \sim kT/\varepsilon, \ \varepsilon \geq \varepsilon_s \tag{11}$$

For $\varepsilon < \varepsilon_s$ Lax suggests that Φ smoothly tends towards a constant as ε decreases to zero. However, the energy dependence of σ only affects f in the edge region and since ε_s is equivalent to a temperature of less than 1°K this modification is not important. The cascade model was developed in order to explain the giant capture cross-sections observed for group V impurities in group IV semiconductors at low temperatures. For thermalized carriers, the Lax cross-section varies as T^{-4} and this is in some agreement with experiment at moderately low temperatures. However, below a few degress Kelvin the capture cross-section ceases to increase as rapidly as T^{-4} and there is an effective cut-off in the increase of cross-section. Improved calculations of the capture process have not produced any very clear explanations as to the cause of this cut-off. Following Mattis (1960) it seems evident that at least a part of this behavior must stem from a tendency of the carriers to become increasingly non-thermal as the magnitude of the cross-section increases. This will be true for any process of carrier injection into the band and of course an increase of mean carrier energy above 3/2kT will reduce $\langle\sigma\rangle$. Barker and Hearn (1968, 1973) made a calculation of $\langle\sigma\rangle$ for electrons photoexcited by room-temperature radiation into the conduction band of silicon below 50°K. Unfortunately, there seems to be some dispute over the interpretation of the results of this calculation arising from the use of a Hall number r appropriate to a low magnetic field so that the value of r is dependent on the shape of the distribution function.

Noguera and Hearn (1978) considering the capture of holes in germanium by copper impurities have employed $\tau_r \propto \varepsilon^{\frac{1}{2}}$ on the basis of somewhat scant measurements reported for the variation of carrier concentration with electric field. This much weaker energy-dependence of τ_r does mean that a Maxwelliam distribution with an effective temperature, in excess of the lattice temperature, can be employed as a simple artifice without the need for any contrived analytical procedures to avoid the edge region imparting a divergence on the total recombination rate. To conclude our comments on the energy dependence of τ_r it seems that unless we are interested in the recombination-controlled regime $\tau_r \ll \tau_{th}$ or in the edge-region the energy-dependence of τ_r is not of vital importance and indeed the use of a constant recombination lifetime is probably adequate. However, systems may exist in which the energy-dependence shows some dramatic change within the bulk of the band due to thresholds for different recombination processes and as such becomes or prime importance.

SOLUTION OF THE INTEGRAL EQUATION

For monochromatic excitation (4) can be substituted into (1) to give

$$f(\varepsilon) = A\delta(\varepsilon-\overline{\varepsilon}) + \overline{f}(\varepsilon), \quad A = W\tau_o$$
$$1/\tau_o \equiv Z(\overline{\varepsilon}) + 1/\tau_r(\overline{\varepsilon}) \quad , \tag{12}$$

where $\overline{f}$ contains a delta-function whose magnitude is controlled by a balance between the total rate of excitation W and the rate $A\tau_o$ at which carriers exit due to scattering out and recombination. The spectrum of this scattering rate from the delta-function is then the exciation rate for the remaining non-singular portion of the distribution. We shall henceforth assume that for monoenergetic excitatio this transformation to the non-singular form of (1) has been made.

The equality of the sum of f/τ_r taken over the band, which is the total recombination rate, to the total excitation rate is a consequence of the zero sum of the right-hand side of (1) due to particle conservation. This property can be used as an aid within an iterative solution of (1), and is referred to as normalization of the distribution function, or it can be used just as a check upon the validity of a final iterated solution. A better checking procedure is of course to make a comparison of the final values of the homogeneous terms with the excitation spectrum for all energies.

The integral equiation is solved by an iterative method based on a finite-interval mesh extending sufficiently beyond the tail of of the distribution. The mesh size is mainly dictated by considerations of the acoustic phonon scattering term. Let us denote by x the quantity $\varepsilon^{\frac{1}{2}}$. The maximum change in x for acoustic scattering of a carrier is

$$\Delta x = 2\varepsilon_s^{\frac{1}{2}} \quad \text{if } x > \varepsilon_s^{\frac{1}{2}} \quad .$$

For a typical band $\Delta x \sim 1$ $\text{deg}^{\frac{1}{2}}$K so that the scattering is nearly elastic for T ~ 300K but for low temperatures it is clearly not feasible to use any quasi-elastic approximation to treat the inelastic acoustic scattering. Now it is important that the integral over the small range Δx is accurately evaluated otherwise particle conservation is violated and cumulative errors occur as iteration proceeds. In order to achieve a sufficiently fine mesh in x at low ε it is found that a constant mesh size in ε would be too small for practical computation. This is overcome firstly by using a constant mesh size in x. Secondly, the mesh size is chosen as an exact sub-multiple of Δx and an unweighted sum is used to evaluate the integral. This means that particle conservation is automatically satisfied exactly and effectively the integral equation has been replaced by a (sparse) system of algebraic equations for a discrete set of states uniformly dispersed in k-space. The number of mesh points in Δx can now be made comparatively small (~6) without affecting the form of the distribution function.

The only successful iteration scheme which has been employed is

$$f_{n+1}(\varepsilon)[Z(\varepsilon) + 1/\tau_r(\varepsilon)] = W\omega_e(\varepsilon) + \int Y(\varepsilon,\varepsilon')f_n(\varepsilon')d\varepsilon' \qquad (14)$$

where f_n is the n-th approximation to the solution. It is fairly easy to see than f_n represents the distribution of carriers after n scattering events from the starting distribution f_0 and this can be formalized analytically. As such the iterative scheme fits into the framework of iterative methods developed for high-field carrier heating. Furthermore, it is clear that convergence increases as τ_r/τ_{th} decreases and fewer scattering events occur. The method was originally developed by Hearn (1966) with renormalization of the distribution after each iteration. In terms of the finite set of algebraic equations used to represent the integral equation the scheme (14) is a Jacobi iteration and details of its convergence criteria are given in standard texts together with procedures for accelerating the convergence. These procedures have not proved very reliable in the present case and because we are usually working with distributions which are well removed from the recombination-controlled regime the convergence is necessarily very slow requiring from fifty to in excess of several hundred iterations.

PHOTOEXCITED HOLES IN Cu-DOPED GERMANIUM

We shall discuss in this section the properties of carriers which are photoexcited by room temperature black-body radiation. The first study of this type was made by Barker and Hearn (1968, 1973) prior to any really reliable experimental work and was later developed by Noguera and Hearn (1978) to specifically analyze the detailed measurements of Norton and Levinstein (1973). These measurements concern photoexcited holes in Cu-doped Ge from 4 to 15°K. Five samples were very carefully characterized in terms of the densities of acceptors and compensating donors and these are detailed in the table. Very detailed studies of conductivity and Hall coefficient in the dark establish the parameters of the elastic scattering cross-sections of the impurities using an assumed deformation potential. The capture cross-section was measured directly for samples with low compensation density and we shall assume that lifetimes in all samples can be simply derived from this value. The photomobility is derived from photoconductivity and photohall measurements. The theoretical task which is posed by these measurements would be considerably simpler if they involved electrons rather than holes because of the greater complexity of the valence band structure. Firstly, the anisotropy of the valence bands is ignored by considering spherically averaged bands. Secondly, we have two valence bands and as such the rate equation (1) is replaced by a pair of integral equations which are coupled by the inter-band acoustic scattering

$$W\omega_e^i(\varepsilon) - f^i(\varepsilon)/\tau_r^i(\varepsilon) = f^i(\varepsilon)\, Z^i(\varepsilon) - \sum_j \int Y^{ij}(\varepsilon,\varepsilon')f^j(\varepsilon')d\varepsilon' \quad (15)$$

$$Z^i(\varepsilon) = \sum_j \sum_{\underline{k}'} P^{ij}(\underline{k},\underline{k}') \quad (16)$$

$$\int Y^{ij}(\varepsilon,\varepsilon')f^i(\varepsilon')d\varepsilon' = \sum P^{ji}(\underline{k}',\underline{k}) \quad , \quad (17)$$

Doping densities for Ge samples used by Norton and Levinstein and summary of one-band mobility calculations by Noguera and Hearn

Sample	Doping densities in cm^{-3}		Temperatures in °K		Mobilities in $10^4 cm^2V^{-1}s^{-1}$		
	Donor	Acceptor	T	T_e	μ_{exp}	μ_{cal}	$\mu(T_e)$
S_1	$1.9x10^{11}$	$2.7x10^{13}$	4 8 15	4.9 8.5 15.4	158 108 58.2	155 112 58.5	166 112 58.2
S_2	$2.5x10^{12}$	$5.9x10^{13}$	4 8 15	9 11 17	42.8 40.0 29.5	37.5 40.2 32.1	54.5 44.2 32.3
S_3	$1.7x10^{13}$	$2.2x10^{15}$	4 8 15	19 19 23	17.8 15.5 14.0	17.7 15.6 15.0	22.4 18.6 15.8
S_4	$1.1x10^{14}$	$2.5x10^{15}$	4 8 15	42 38 40	9.8 8.7 7.1	13.3 10.0 8.4	13.8 10.7 8.69
S_5	$2.2x10^{14}$	$3.0x10^{15}$	4 8 15	56 51 51	7.1 6.6 5.6	11.3 8.7 7.0	11.2 8.7 7.0

where i and j take values 1 and 2 corresponding to the two bands. The excitation rates are assumed to be given by (6). Instantaneous intra-band optical phonon emission is included via (7) but in the 'coupled' model of excitation inter-band optical phonon emission is also included.

$$\omega_e^i(\varepsilon) \rightarrow \sum_n \sum_j \alpha_{ijn} \, \tilde{\omega}_e^j(\varepsilon - n\hbar\omega) , \qquad (18)$$

where α_{ijn} is calculated from the relative probabilities $\beta(k,j)$ for intra and inter-band transitions and involves a sum over all possible n-phonon chains,

$$\alpha_{ijn} = \sum_{m_1} \sum_{m_2} \cdots\cdots \sum_{m_{n-1}} \beta(j,m_1)\beta(m_1,m_2) \cdots\cdots \beta(m_{n-1},i) . \qquad (19)$$

A further compliation arises from the greater complexity of the acoustic scattering theory which involves three deformation potentials. It is possible to simplify this theory somewhat to include only the dilatation potential a and the effective shear potential b. Whilst the value of b can be measured as -2.4eV the value of a is extremely uncertain in spite of various attempts to fix its value from normal low field transport measurements. Noguera and Hearn (1978) therefore hoped that the hot photocarrier measurements would provide a means of uniquely fixing the value of the dilatation potential and indeed their methodology does illustrate the use of such measurements as an investigative tool as discussed in the Introduction. Unfortunately, the results in this case seem rather inconclusive. It could be that the deformation interaction suffers unexpectedly heavy screening by electrons in both impurity and band states or that under the conditions of the experiment some of our model assumptions, particularly those concerning the light-hole band, had failed. It is probable, however, that the three-parameter deformation theory of Bir, Normantas and Pikus (1962) which was employed here is rather inadequate and attention should be directed towards improving this part of the analysis.

We shall therefore only consider Noguera and Hearn's approximate analysis based on a simple one-band model. This implies very strong inter-band coupling so that $f^{(1)}$ and $f^{(2)}$ are identical. In Fig. 1, we have shown the two distributions, as calculated from the two-band model, and in this case the density-of-states factor $m_i^{3/2}$ is included and the variable $K^2 \equiv \varepsilon/kT$ is used. The strong coupling approximation does appear to be equite good in agreement with some high-field studies (but see Bosi et al., 1979). The one-band model employed a simple deformation potential of 10eV and as a first approximation a Maxwellian distribution (truncated at $\varepsilon = \hbar\omega$) was

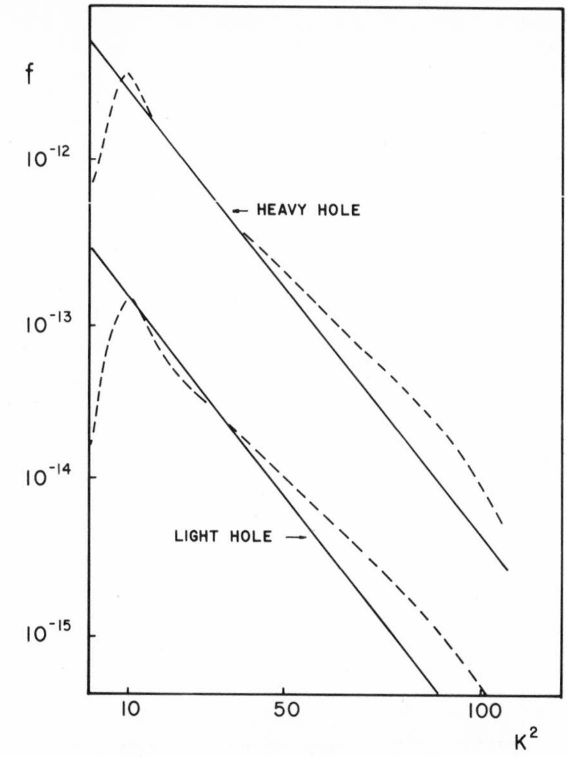

Fig. 1. Numerical (broken lines) and Maxwellian distributions for light and heavy holes in Ge sample S_5 at 4 K, with T_e=56°K.

used. The carrier temperature is readily calculated from the energy balance equation obtained by multiplying both sides of (1) by ε and summing over the band. Hence the Maxwellian mobility $\mu(T_e)$ is obtained from the effective temperature T_e of the carriers. The rate equation (1) was also solved exactly leading to the calculated rate equation mobility

$$\mu_{Cal} = -(e/3m_c)\int_o^\infty K^3(\partial f/\partial K)\tau_s(K)dK/\int_o^\infty K^2 f\ dK \ , \tag{20}$$

m_c being the conductivity mass and τ_s the harmonic sum of the various momentum relaxation times for impurities and the acoustic deformation scattering (which is open to minor objections at the lowest values of T_e); the integral in (20) is evaluated numerically from f after integration by parts.

In the Table, we show the details of the one band calculations, the experimental mobility being denoted by μ_{exp}. The mean recombination lifetime τ_r for samples S_1 to S_4 is of order 10^{-7}, 10^{-8}, 10^{-9}

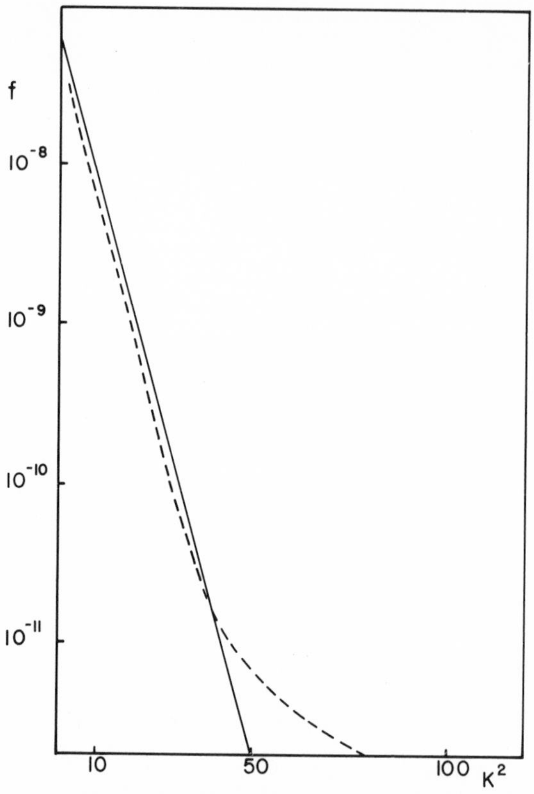

Fig. 2. Numerical (broken line) and Maxwelliam distributions for homes in Ge sample S_1 at 4°K.

and 10^{-10} sec respectively and τ_{th}, the thermalization time, can be taken as 10^{-9} sec. Thus moving vertically down the table we note that $\delta T \equiv T_e - T$ increases as τ_r/τ_{th} decreases. When T_e is close to T we note that δT increases as T decreases, because the relevant phonon transition rates (at kT) decrease with kT. However, when δT is large there is no such trend because T_e is largely determined by the spontaneous phonon emission rate and this is independent of T. In Figs. 2 and 3, we show calculated distribution functions together with the Maxwellian approximations. Figure 3 is an example of a tailed-thermal distribution for $\tau_r >> \tau_{th}$. For small values of K^2 there is an edge region (which is not shown) where f tends to be zero because of the energy-dependence of the lifetime $\tau_r \propto \varepsilon^{\frac{1}{2}}$. This is followed by a low energy region where the carriers are essentially thermalized, and this region contains the vast majority of carriers. At higher energies there is a tail which represents the only significant departure from equilibrium and this tail closely resembles a Maxwellian at the radiation temperature T_r. The Maxwellian at T_e is a first-moment straight-line fit to log f (note that the carrier densities differ because of the energy dependence of τ_r). Figure 3

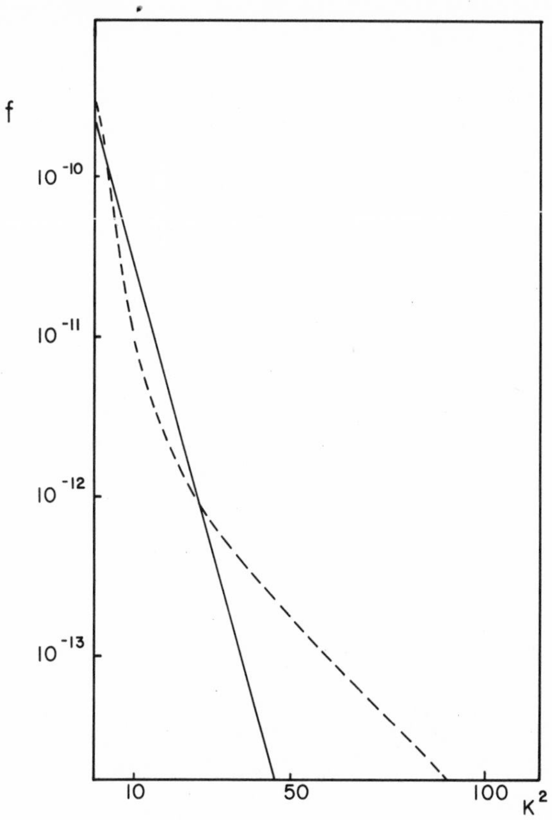

Fig. 3. Numerical (broken line) and Maxwelliam distributions for holes in Ge sample S_3 at 4°K.

is an intermediate type of distribution $\tau_r \sim \tau_{th}$ with a more substantial tail, corresponding to T_r, and a low-energy region which is markedly distorted from an equilibrium distribution. The values of μ_{cal} and $\mu(T_e)$ are in reasonable agreement except in the lower doped samples at very low T. The mobility calculations do of course involve only the bottom of the low-energy region and the form of the distribution at higher energies could only be confirmed by direct observation techniques such as radiative recombination spectroscopy. The agreement between the calculated and experimental mobilities is rather irregular but is fairly satisfactory for samples S_1 and S_3. The temperature dependence of the mean lifetime is strongest in S_1 and the calculated distributions lead to $T^{0.75}$ which is only slightly lower than the photohall measurement. In the heavily-doped samples the dependence is much weaker because the distribution function is weakly dependent on T and in S_4 the calculated and experimental dependence is $T^{0.2}$ to $T^{0.3}$.

MONOCHROMATIC EXCITATION OF CARRIERS

We have an excitation spectrum given by (4) and using the simultaneous optical phonon emission approximation $\bar{\varepsilon}$ is effectively less than $\hbar\omega$ so that we are concerned only with acoustic phonon scattering. Detailed numerical calculations by Noguera (1978), and Maksym (1979) show that the distribution consists of an edge-region, a low-energy region and a tail plus the image (12) of the excitation line. For a tailed-thermal distribution the tail itself has three distinct regions. Provided that $\bar{\varepsilon} >> kT$ there exists above the low-energy region a plateau which extends almost to $\bar{\varepsilon}$ and consists of carriers cascading down in energy from $\bar{\varepsilon}$ by acoustic phonon emission. Immediately below $\bar{\varepsilon}$ there is a structured region consisting of several peaks or shoulders in the distribution. In the 'anti-Stokes' region above $\bar{\varepsilon}$ the distribution drops very sharply to zero in a step which has the shape of a Maxwellian at the lattice temperature T and some structure may also be superimposed on the Maxwellian slope of this step. These features are shown in Figs. 4 to 8 calculated by Maksym for electrons in GaAs at 1.7 K with an energy-independent lifetime $\tau_r = 10^{-7}$ sec and since $\tau_{th} \sim 10^{-9}$ sec they represent tailed-thermal distributions.

The quantity α shown on the captions to these figures is the fraction of carriers remaining in the singular part of the distribution which for a constant lifetime is

$$\alpha = \tau_o/\tau_r \tag{21}$$

and since $\alpha << 1$, $1/\tau_o \sim Z(\bar{\varepsilon})$. Figure 9 is a calculation by Noguera for holes in sample S_3 of Ge discussed above, where $\tau_r \sim \tau_{th}$ so that the low-energy carriers are now sufficiently heated to almost obscure the plateau region.

The anti-Stokes region contains carriers which have absorbed phonons while in the region $\varepsilon \leq \bar{\varepsilon}$ and its shape directly reflects the Boltzmann factor in the Planck distribution describing the equilibrium phonon population at the lattice temperature. Optical phonon absorption can also contribute to any part of the anti-Stokes Maxwellian which extends beyond $\hbar\omega$. The structured region immediately below $\bar{\varepsilon}$ is a direct consequence of the finite energy range of acoustic phonon emission events. Thus a carrier initially at $\bar{\varepsilon}$ cannot descend beyond $(\bar{\varepsilon}^{\frac{1}{2}}-\Delta x)^2$ by a single emission or $(\bar{\varepsilon}^{\frac{1}{2}}-2\Delta x)^2$ by two emission events. Hence one can expect structure in the distribution which is evenly spaced in $\varepsilon^{\frac{1}{2}}$ with a period of Δx. The structure is not discernible after a few periods due to convolution of the emission paths and the amplitude and shape are dependent on the variation of the transition probability $P(\underline{k},k')$ in (3) with $(\underline{k}'-\underline{k})$. The figures show that deformation scattering, which favours

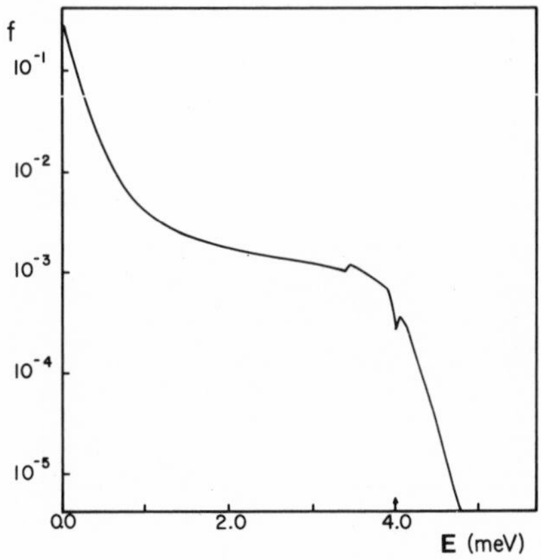

Fig. 4. Distribution in GaAs with only longitudinal piezoelectric scattering $\bar{\varepsilon}$ = 4meV (arrow) α = 2.47%.

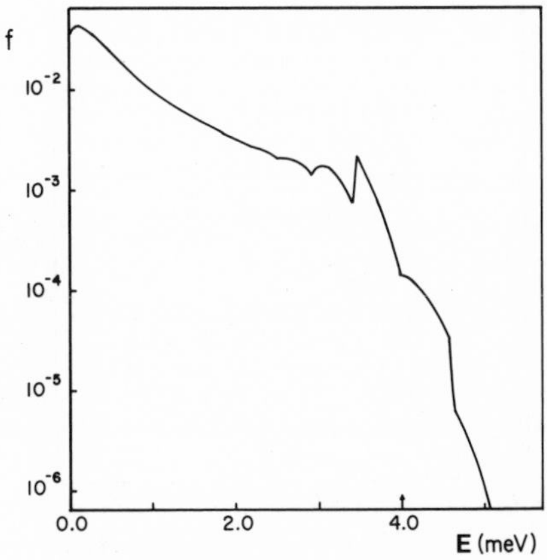

Fig. 5. Distribution in GaAs with only deformation potential scattering $\bar{\varepsilon}$ = 4meV (arrow) α = 0.41%

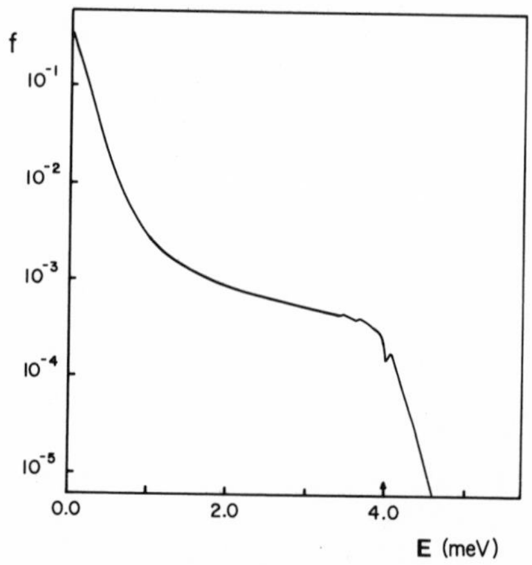

Fig. 6. Distribution in GaAs with longitudinal and transverse piezoelectric and deformation potential scattering $\bar{\varepsilon}$ = 4meV (arrow) α = 0.28%.

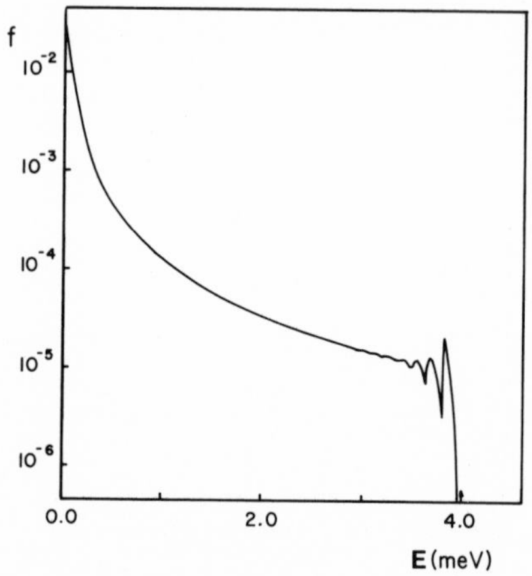

Fig. 7. Distribution in GaAs with only deformation potential scattering $\bar{\varepsilon}$ = 40meV (arrow) α = 0.28%.

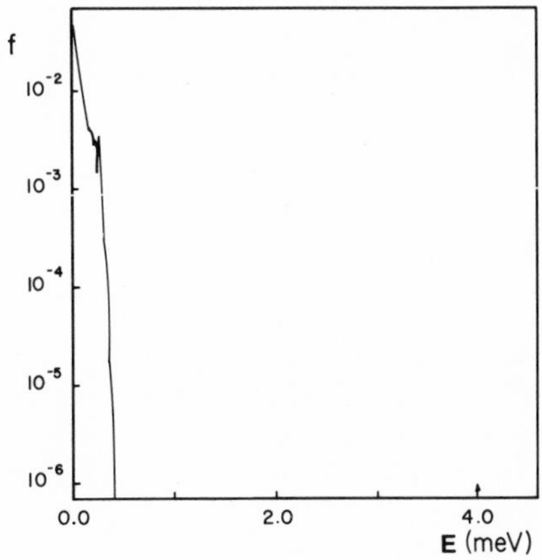

Fig. 8. Distribution in GaAs with deformation potential and longitudinal optical scattering $\bar{\varepsilon}$ = 40 meV (arrow).

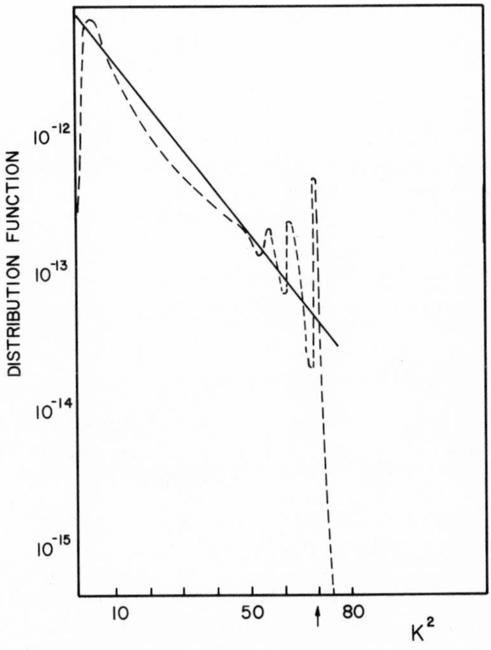

Fig. 9. Distribution of holes in Ge with deformation potential scattering in one band model.

large momentum change, gives more pronounced peaks than peizoelectr scattering which is also affected by screening as mentioned in Section 7. Similarly, the amount of structure occurring in the anti-Stokes region is dependent on the type of acoustic scattering includ Whether this structure is physically observable for a very narrow excitation line is dependent on the presence of broadening mechanis not included in our model. The presence of structure due to acoust phonons in unique to the monoenergetic photocarrier problem and doe not occur for high field heating.

To further illustrate our interpretation of the shape of the distribution we show in Fig. 10 the deformation potential distribution at various times ε after switching-on a square pulse excitation spectrum at $t = 0$. This time evolution is calculated by inclu ing a term $\partial f/\partial\varepsilon$ on the right-hand side of (1) and representing it numerically by finite differences. The time step was 0.1 nsec. Fo the first few frames taken at intervals of 0.5 nsec one observes an increasing number of peaks as more scattering events occur but as the number of scattering events increases so the magnitude of the structure diminishes. It is observed that after 50 nsec the platea and anti-Stokes Maxwellian are approaching completion. The low energy thermal region has not even started to develop but compariso of the 25 and 50 nsec frames shows that the cascading carriers are close to reaching the bottom of the band. This adds force to our comment in the Introduction that it is extremely difficult to chara terize the carrier scattering by a simple thermalization time τth.

If the delta-function input is broadened to a square pulse or narrow Guassian the only significant topological effect is that the acoustic phonon structure is destroyed if the width of the line $\Delta\varepsilon$ is of order $\Delta x\bar{\varepsilon}^{-\frac{1}{2}}$. This smoothing out of the structure can be investigated in detail by using (5). The importance of the ratio of $\Delta\varepsilon$ to $\Delta x\bar{\varepsilon}^{-\frac{1}{2}}$ in determining certain features of the distribution has bee emphasized by Ridley and Harris (1976).

EFFECTS OF OPTICAL PHONONS

The effect of optical phonon emission is very accurately described, for lifetimes of practical significance, by the instantaneous emission approximation. This approximation assumes that the chain of states $\varepsilon + nh\,\omega(n=0,1,2,\ldots)$ have a carrier distribution $f(\varepsilon)\exp(-nh\omega/kT)$ since they are infinitely tightly coupled together by the optical phonon interaction. The Boltzmann factor arises fro optical phonon absorption and stimulated emission and in the effective excitation method (7) of applying the approximation this facto is ignored because at low temperatures there is only a very small fraction of carriers above the optical phonon emission threshold.

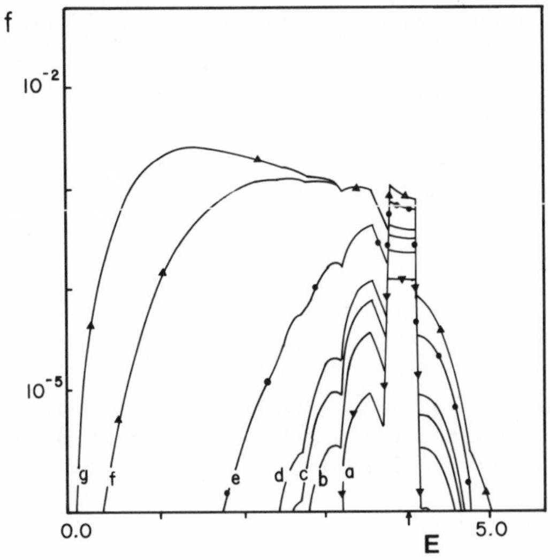

Fig. 10. Time evolution of the distribution with square-pulse excitation corresponding to Fig. 5. Each curve represents one frame with times in nsec as follows: t = 0.5(a), 1.0(b), 1.5(c), 2.0(d), 2.5(e), 25(f), 50(g).

Note that the top of the square pulse becomes higher as time progresses and curves appear superimposed along its sides whereas in reality the slopes of its sides do decrease very slowly with time.

One can avoid the approximation completely at the cost of extra computional effort by including the optical phonon processes in the scattering terms. A calculation of this type is shown in Fig. 8, with $\hbar\omega$ = 36me and a comparison with Fig. 5, which would correspond to the use of the effective excitation demonstrates the validity of the approximation; Fig. 7 represents the distribution which would occur with no optical phonon interaction. There are, however, two important points to make about the failure of the effective excitation method. The method gives a modified thermal Maxwellian decay to the tail at $\varepsilon = \hbar\omega$ if the tail of the distribution reaches the threshold. This is due to acoustic absorption by carriers below the threshold and although the shape is broadly correct the detailed form of the distribution will be entirely incorrect. In reality carriers crossing the threshold by acoustic absorption immediately emit optical phonons to enter states at the bottom of the band. Furthermore, even if only a small fraction of carriers cross the threshold in this way during their lifetime the neglect of the subsequent energy loss by optical phonon emission can lead to a significant under-estimate of the total energy loss to the lattice by the heated carrier system.

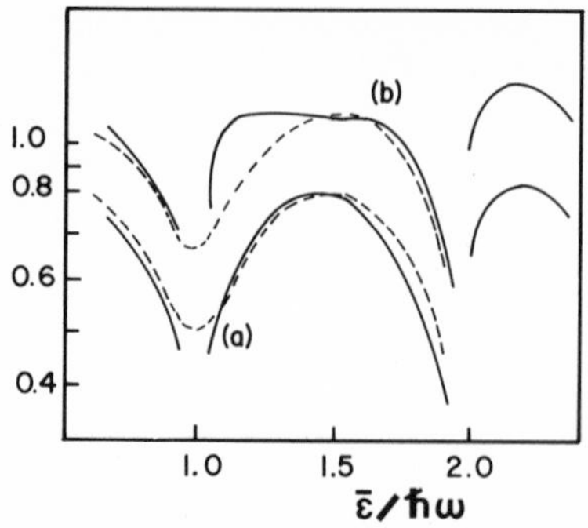

Fig. 11. Comparison of Monte Carlo (full lines) and measured oscillatory photoconductivity in InSb at 8°K. Curves (a) and (b) refer to an electric field of 1.2 and 2.5 Vcm^{-1} respectively.

A consequence of the instantaneous emission approximation is that the distribution for monoenergetic excitation has a periodic dependence on $\bar{\varepsilon}$ with period $\hbar\omega$ and even if the approximation fails there will be a residual periodic component. Hence, if the resultant distribution is heated the photoconductivity is periodic and this is the so-called oscillatory photoconductivity effect. Figure 11 shows the observed (Stocker et al. 1966) and calculated oscillation in InSb at 8°K for two different electric field strengths. The calculations were made by Barker and Hearn (1969a) to confirm a reported non-linear dependence of the photoconductivity on field strength. The iterative technique cannot be used in non-vanishing fields whereas the Monte Carlo method becomes quite successful when the field produces a substantial asymmetry in the distribution (Barker and Hearn, 1969b). The results shown were based on a Lax recombination model (10) with a mean lifetime of 10^{-10} sec so that considerable heating occurred. Good agreement is obtained with the experimental results except when $\bar{\varepsilon}$ is close to a multiple phonon threshold $n\hbar\omega$. A possible origin of this discrepancy is mentioned later.

In the previous section, we showed that the distribution for monoenergetic excitation has a thermal Maxwellian decay to zero at $\bar{\varepsilon}$. When an electric field of non-vanishing magnitude is applied this sharp step in the distribution is removed and the distribution is elongated above $\bar{\varepsilon}$ due to some carriers being accelerated by the field to energies greater than $\bar{\varepsilon}$. In $\underline{k}$-space this elongation of the distribution $f(\underline{k})$ lies in the region of particle current $e\vec{v}$ parallel to $\vec{F}$, and represents an increase in the positive current component given by this region. There is a corresponding reduction in the negative current component given by carriers in the region of $e\vec{v}$ anti-parallel to $\vec{F}$. If $\bar{\varepsilon}$ is fairly close to $\hbar\omega$ the elongated distribution may extend across the emission threshold and these accelerat-

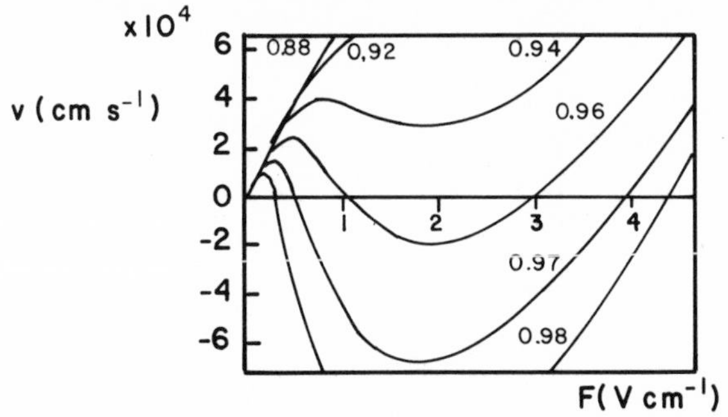

Fig. 12. Velocity-field characteristic for electrons in InSb obtained by the one-dimensional model. The numbers on the curves are $\bar{\varepsilon}/h\omega$.

carriers immediately lose all velocity by entering states at the bottom of the band. Hence, increasing the field reduces the current increase in the parallel region and the velocity-field characteristic of the entire carrier system shows a negative differential slope above some threshold field (see Fig. 13). If $\bar{\varepsilon}$ is sufficiently close to $\hbar\omega$ it is possible that at higher fields the parallel region may show a current decrease as a substantial fraction of the carriers in this region are transported to zero-velocity states. If this decrease exceeds the current increase in the antiparallel region a negative current is obtained. Thus in Fig. 13, the velocity-field characteristic becomes negative and then is restored at higher fields to positive values as the carriers in the zero-velocity states are re-accelerated by the field. These non-linear field effects can only occur if the carriers are at least moderately well heated optically because otherwise the number of carriers in the tail below $\bar{\varepsilon}$ which can enter the elongated region is too small to make the effects discernible. Furthermore, the effects disappear if the momentum relaxation time is significantly less than the lifetime because a carrier then reverses its velocity too frequently during its lifetime for any elongation to occur.

Barker and Hearn (1969c) showed that if inelastic scattering is totally excluded $\tau_{th} \to \infty$, so that the zero-field distribution (which lies in the recombination-controlled region) is simply a delta-function, then the velocity-field characteristic for a one-dimensional space model can be derived analytically. The results are shown in Fig. 12 for parameters appropriate to InSb at 10°K, and in Fig. 13 we show a characteristic derived by the Monte Carlo technique for the same parameters. Apart from the expected differences in the values of the various critical fields, the curves are broadly similar although there has clearly been some error in deriving the correct momentum relaxation time for the one-dimensional model and this results in the

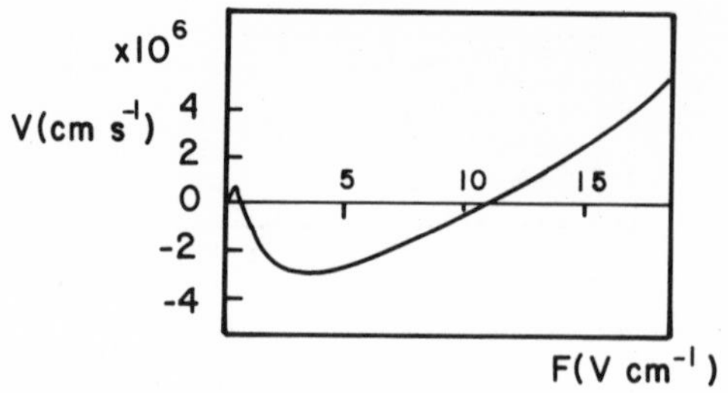

Fig. 13. Velocity-field characteristic for electrons in InSb at 10°K obtained by Monte Carlo for $\bar{\varepsilon}/h\omega = 0.978$.

drift-velocity being too low. One can conclude with some confidenc that the non-linearity and negative current predictions are real effects provided the carriers are sufficiently well heated opticall The negative current state involves a fraction of the optical energ being transferred to the electric field so that the sample is actin as a power supply to the external electrical circuit.

Normally the electric field in a semiconductor is uniform over the bulk of the sample so that the observed current can be deduced directly from the velocity-field characteristic. However, if the characteristic has a negative slope region and the fields in the sample lie within this region one finds that the uniform field situ tion does not occur because of instability against space-change for tion. It is possible in these circumstances to find either time-independent non-uniform field configurations or uniformly propagati field-domain solutions to the phenomenological equations governing the system. The domain solution is observed in the Gunn effect. Barker and Hearn (1969c) and later Maksym and Hearn (1979) have mad detailed studies of the possible solutions of those equations for the velocity-field characteristic discussed here. Their conclusion are that both domain solutions and negative current states are poss ble. This clearly represents a fruitful area for experimental inve tigation. The discrepancy mentioned earlier between theoretical an experimental values for photoconductivity near a multiple phonon threshold needs detailed investigation. A possible explanation of the discrepancy lies in some non-uniformity of the field, and a pro measurement of the field distribution would be of value.

FURTHER PHYSICAL EFFECTS

The Barker-Hearn model is limited to effects which are linear the rate of excitation. It therefore excludes carrier-carrier inte actions because these depend upon the magnitude of the photocarrier concentration. The model can, however, be readily adapted to inclu such interactions and this is discussed in a separate lecture. A subsidiary non-linear effect is the screening of the electron-phonon interaction. This is a poorly understood topic for non-equilibrium carriers but the polar interactions are considered to be screened by the inclusion of an inverse screening length λ such that $(\lambda^2 + q^2)$ replaces q^2 in the matrix element derived from the Coulomb interaction. Contributions to the screening come from both band electrons and electrons localized on impurities. Considering just the contribution from the hot carriers, it can be shown that the relation

$$\lambda^2 = (e^2 n/2\mu) \langle 1/\varepsilon \rangle$$

is probably valid where μ is the permittivity of free space multiplied by the low frequency dielectric constant for the piezoelectric interaction and the high frequency dielectric constant for the optical polar interaction.

For a thermal Maxwellian $\langle 1/\varepsilon \rangle = 2/kT$, and since λ is dominated by low-energy carriers this form can be adopted for tailed-thermal distributions, or T replaced by an estimated effective temperature T_e. Hence, the carrier distribution appropriate to a screened polar interaction can be derived for a particular value of η. This approach was used in the GaAs calculations above where η was taken as $10^{14}cm^{-3}$. The only discernible effect upon the shape of the distribution occurs in the structured region where the peaks due to piezoelectric scattering are greatly suppressed. The screening of the deformation potential interaction is less understood and is worthy of further study for the hot hole system in Ge discussed above. The screening of the ionized impurity interaction is relevant to the mobility calculations discussed in that section and involves the Brooks-Herring formula. However, λ in this case is dominated by the compensated acceptors and does not involve η.

Another non-linearity comes from possible departures from equilibrium of the phonon system since this is dependent on W. The Planck distribution assumed in our model is disturbed by the net phonon creation by the heated carriers. The existence of such effects is clearly established by laser excitation on semiconductors such as CdS where an excess population of optical phonons can be directly measured. Clearly a coupled model of the two non-equilibrium systems is required. Other non-linear effects arise from the disturbance of the filling factors for the generation and recombination states. This process is of considerable significance for non-linear high-field transport in the presence of generation-recombination (see Maksym and Hearn, 1978). For across-the-gap excitation one is led to a coupled electron-hole heated system which should be tractable.

A very important area for all hot carrier studies is that of transport in conditions of non-uniform field, carrier concentration etc. In the high-field case our understanding is limited to the us of rather crude phenomenological equations and this is also true of the studies of non-linear field effects for photoexcited carriers (see appendix to Maksym and Hearn, 1979). For photoexcited carrier there is an immediate problem of the influence of non-uniformity in the excitation rate, particularly with reference to a limited volum of the sample being illuminated. Although an internal electric field will oppose diffusion terms to produce zero net current, the distribution must be changed within the illuminated volume and a ne transport of carrier energy could occur. It should be possible to investigate the significance of any such processes by using a spatially dependent heated Maxwellian distribution in a similar mann to the high-field work of Cheung and Hearn (1972a,b).

BIBLIOGRAPHY

Barker, J.R., and Hearn, C.J., 1968, Phys. Letters, 26A: 148.
Barker, J.R., and Hearn, C.J., 1969a, Phys. Letters, 29A: 215.
Barker, J.R., and Hearn, C.J., 1969b, "Proc. Conf. on Computation Phys.," U.K.A.E.A., Culham, 34: 1.
Barker, J.R., and Hearn, C.J., 1969c, J. Phys. C., 2: 2128.
Barker, J.R., and Hearn, C.J., 1973, J. Phys. C., 6: 3097.
Bir, G.L., Normantas, E., and Pikus, G.E., 1962, Sov. Phys.-Solid St 4: 869.
Bosi, S., Jacoboni, C., and Reggiani, L. 1979, J. Phys. C., 12: 152
Cheung, P.S., and Hearn, C.J., 1972a, Electron. Lett., 8: 79.
Cheung, P.S., and Hearn, C.J., 1972b, J. Phys. C., 5: 1563.
Hearn, C.J., 1966, Proc. Phys. Soc., 88: 407.
Lax, M., 1960, Phys. Rev., 119: 1502.
Mattis, D.C., 1960, Phys. Rev., 200: 52.
Maksym, P.A., 1979, Ph.D. Thesis, Univ. of Warwick, U.K., unpublishe
Maksym, P.A., and Hearn, C.J., 1979, J. Phys. C., in the press.
Noguera, A., 1978, Ph.D. Thesis, Univ. of Warwick, U.K., unpublishe
Noguera, A., and Hearn, C.J., 1978, Solid-State Electronics, 21: 17
Norton, P., and Levinstein, M. 1973), Phys. Rev. B, 6: 478.
Stocker, H.J., Levinstein, M., adn Stannard, C.R., 1966, Phys. Rev. 150: 613.

THE PHYSICS OF NONLINEAR ABSORPTION AND ULTRAFAST CARRIER RELAXATION IN SEMICONDUCTORS

Arthur L. Smirl

North Texas State University
Denton, Texas 76203

INTRODUCTION

The absorption of light quanta of energy greater than the band gap in a semiconductor induces an electron to make a transition from the valence band to a state high in the conduction band, leaving behind a hole in the valence band (Fig. 1). After such an absorption process, the photoexcited electron is left with an excess energy ΔE_e that is given by

$$\Delta E_e = (h\nu - E_g)(1 + m_e/m_h)^{-1}, \tag{1}$$

where m_e is the electron effective mass and m_h is the hole effective mass. The excess energy of the photogenerated hole is

$$\Delta E_h = (h\nu - E_g) - \Delta E_e. \tag{2}$$

These energetic electrons (holes) will quickly relax through various collisional processes to the bottom (top) of the conduction (valence) band, where eventually they will recombine. It is well established that if the photoexcitation is sufficiently intense this relaxation process results in the generation of hot electron and phonon distributions.

The intraband relaxation kinetics of a photoexcited electron located energetically above the conduction band minimum are complex; moreover, as we shall see, the evolution of many of these processes occurs on a time scale too rapid for *direct* measurement by present

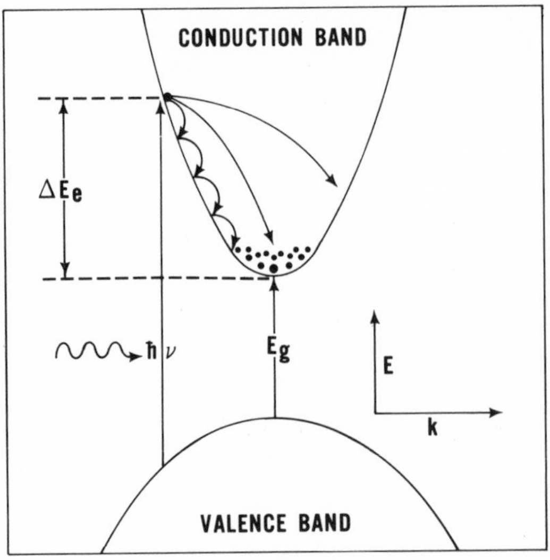

Fig. 1. Direct optical transitions in a semiconductor.

electronic detection systems. In some cases, characteristic relaxation times for these ultrafast processes have been determined *indirectly* from transport measurements or from steady-state photoluminescence measurements.

In general, the measurement of a transport property such as th mobility or drift velocity always represents an integrated or average effect over the distribution function. Consequently, it is usually necessary to assume the form of the distribution function. Once the distribution function has been assumed or determined, measurement of one of a number of transport properties such as the mobility, Hall effect, magnetoresistance, or the Shubnikov-de Haas effect will yield the carrier temperature. Details of these studie have been provided by Bauer in earlier chapters. For additional discussion of the methods for determining hot electron temperatures from transport measurements the reader may find the reviews by Conwell (1967) and Bauer (1974) of interest.

By employing steady state optical techniques such as photoluminescence or optical absorption, information concerning the *form* of the distribution function as well as the carrier temperature can be obtained. This additional information concerning the distributi function is obtained primarily because a direct optical transition (absorption or emission) occurs between a well defined initial energy state E_i and a well defined final state E_f, where $E_f - E_i = h\nu$. The transition rate between these two states depends on the occupancy of the initial state $f(E_i)$ and the availability of the final state $1 - f(E_f)$. As a result, we can measure the entire energy dependence of the distribution function by monitoring all possible optical transition energies. The methods for determining hot electron distribution functions and hot electron temperatures from radiative recombination, absorption, or inelastic light scattering have been reviewed by Hearn and by Ulbrich in earlier chapters. Additional details are provided by the reviews of Bauer (1976), Shah (1978), Ulbrich (1978), Leite (1978), and Weisbuch (1978).

Generally, these detailed studies of hot electron and phonon distributions, both those employing transport techniques and those employing photoluminescence techniques, have been conducted under steady state conditions. That is, the knowledge of the electron temperature and the form of the distribution function are determined by assuming that the rate at which the electron-hole gas receives energy from the electrical or optical field is equal to the rate at which the carriers lose that energy to the lattice or impurities. This detailed knowledge of the steady-state distribution function then provides information concerning electron-electron and electron-phonon interactions; from this information, ultrafast relaxation rates are *indirectly* assigned. Direct time resolved studies have been performed (Ulbrich, 1973), but they have, previously, been limited in resolution to nanosecond time scales. With the advent of mode-locked lasers and the picosecond optical pulses that they produce, *direct* measurement of many of these relaxation times is now possible. The remainder of this lecture and a seminar to follow will describe the procedures, results, and interpretation of experiments that directly measure the nonlinear, nonequilibrium optical properties of semiconductors on a picosecond time scale in an effort to obtain information concerning ultrafast carrier relaxation processes.

As a rule, these direct experimental studies of ultrafast relaxation processes have employed a variation of the "excite-and-probe technique". This method was first used on a picosecond time scale by Shelton and Armstrong (1967). In this technique, a single ultrashort (0.2 psec - 10 psec) optical pulse is switched by an electro-optic shutter from a train of such pulses produced by a mode-locked laser. This single pulse is split into two by a beam splitter, and a relative delay is introduced between the two pulses. The intensity of the second (probe) pulse is attenuated to a small fraction (e.g. 2%) of the corresponding intensity of the first (excite) pulse. Both pulses are focused onto the semiconductor sample as shown in Fig. 2. Intense ultrashort pulses (typically having peak powers on the order of 10^8 watts) whose photon energy

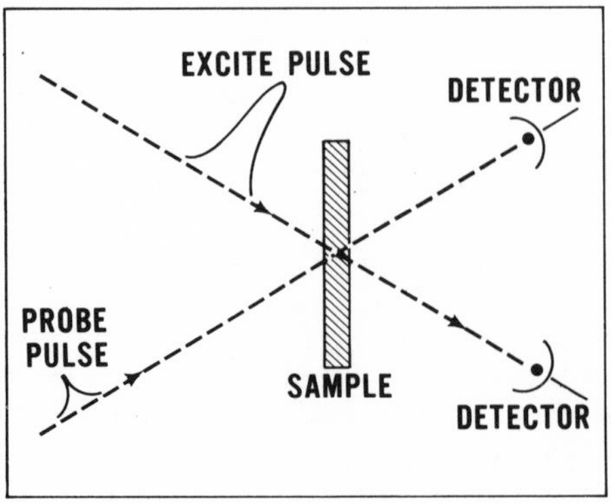

Fig. 2. Experimental technique for measuring ultrafast relaxation times in semiconductors.

$h\nu$ is greater than the band gap E_g of the semiconductor, when tightl focused, can produce an enormous number (10^{19} - 10^{20} cm^{-3}) of electron-hole pairs on a time scale that is short compared to many of the kinetic processes involved in the evolution of the carrier distribution. Thus, the absorption of the exciting pulse creates a large, rapidly evolving, nonequilibrium carrier distribution that changes the transmission properties of the sample. This initial pulse is then followed at various time delays by the weak probe pulse that monitors the evolution of the enhanced sample transmissic with time as it returns to its equilibrium condition. There are any number of variations of this technique. For example, the probe puls can be monitored in reflection as well as transmission, the probe wavelength can be purposefully chosen to be different from the excit wavelength, and the probe polarization can be varied with respect tc the polarization of the excite pulse. Another variation of the technique is to hold the time delay between excite and probe pulses constant while varying the energy of the excite pulse. Of course, i addition to varying the optical parameters, one can vary the environ ment of the optical sample as well (e.g. vary the sample temperature or subject the sample to hydrostatic pressure). We shall review the application of many of these techniques to the time-resolved measure ment of the optically induced changes in the transmission and

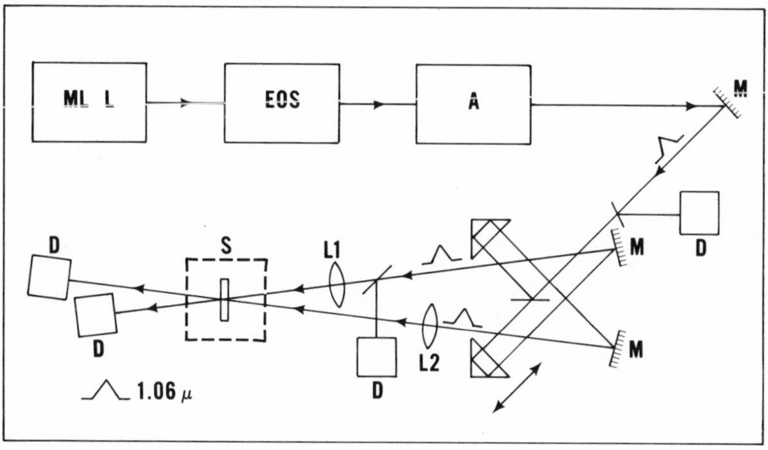

Fig. 3. Block diagram of the experimental configuration for excite and probe measurements at 1.06 μm, where MLL denotes the mode-locked laser, EOS the electro-optical switch, A the laser amplifier, M a mirror, D a detector, L1 and L2 lens, and S the germanium sample.

reflection spectrum of Ge and GaAs in a seminar to follow. In the remainder of this chapter, for pedagogical purposes, we shall restrict our discussion to a single application of the "excite-and-probe technique" to germanium.

Specifically, we intend to discuss an early excite and probe experiment performed in germanium using intense picosecond optical pulses with a wavelength of 1.06 μm as depicted in Fig. 3. In this particular application of the excite-and-probe technique, the exciting pulses were selected by a laser-triggered spark gap and a Pockel's cell from trains of pulses produced by a mode-locked Nd-glass laser. The pulses were approximately 10 psec in duration and had peak powers of approximately 10^8 watts at a wavelength of 1.06 μm. They produced a measured irradiance of approximately 10^{-2} J/cm^2 when focused on the crystal surface (to a spot size 2 mm in diameter). The electron-hole plasma produced by the absorption of the excitation pulse was probed using weak probe pulses at 1.06 μm.

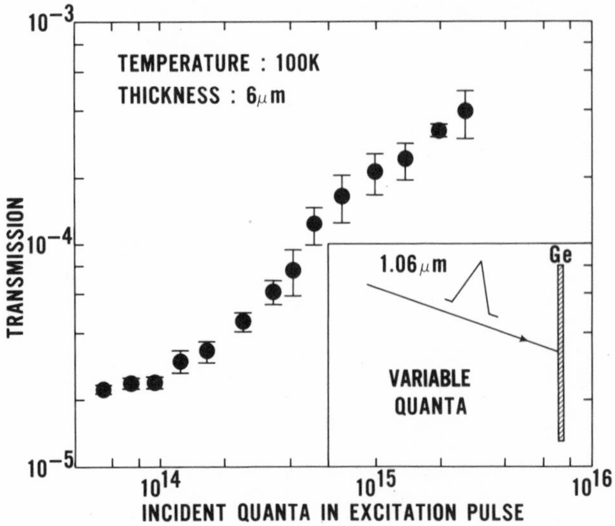

Fig. 4. Change in transmission of a 6 μm-thick germanium sample as a function of incident excite pulse energy at 1.06 μm. Not a pulse energy of 2 x 10^{15} quanta corresponds to a surface irradiance of approximately 10^{-2}J/cm^2.

The germanium sample was a high purity (ρ_{min} = 40 Ωcm) single crystal cut with the (111) plane as the face. The sample was polished and etched with Syton to a thickness of 6 μm as determined by interferometric techniques.

The results of a measurement of the change in transmission of the thin germanium crystal at 1.06 μm as a function of increasing carrier number (created by the direct absorption of the excitation pulse) are shown in Fig. 4. These data were obtained in the following manner (see inset of Fig. 4). The crystal was illuminated by variable energy excitation pulses with a wavelength of 1.06 μm, and the transmission of each pulse was measured. The crystal transmission at 1.17 eV (1.06 μm) is seen to increase by a factor of approximately 30 at high photoexcitation levels. These measurements were performed at 100 K. Thus, we see that our excitation pulse can be made energetic enough to alter the optical properties of the germanium!

The results of measurements of the temporal evolution of the enhanced transmission of this thin germanium sample at a wavelength

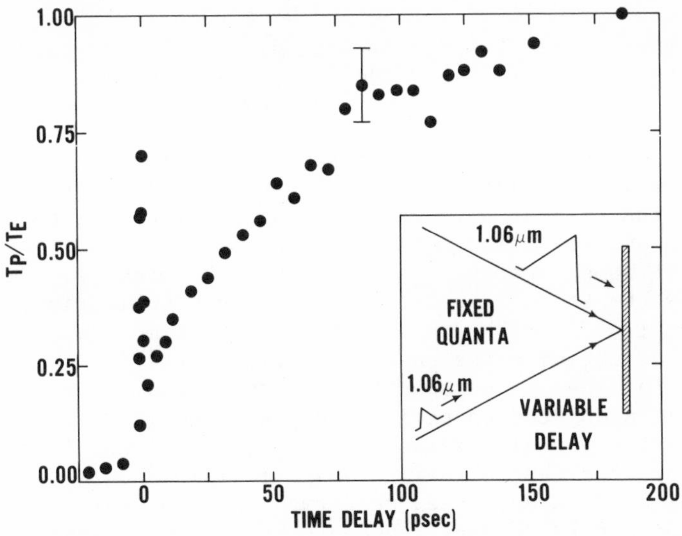

Fig. 5. Probe pulse transmission vs. delay between the excite pulse at 1.06 μm and the probe pulse at 1.06 μm for a sample temperature of 100 K. The data are plotted as the normalized ratio of probe pulse transmission to excite pulse transmission, T_P/T_E, in arbitrary units. The error bar represents twice the typical statistical standard deviation.

of 1.06 μm following the creation of a dense electron-hole plasma by the excitation pulse are shown in Fig. 5. The measurements were performed in the following manner (see inset Fig. 5). The sample was irradiated by 1.06 μm excite pulses containing approximately 2×10^{15} quanta, and the transmission of each excite pulse was measured. Each excite pulse was then followed at various delays by a weak probe pulse at 1.06 μm. These measurements were performed for a sample temperature of 100 K. The data are plotted as the ratio of the probe pulse transmission T_P to excite pulse transmission T_E, in arbitrary units, versus time-delay in picoseconds. The arbitrary units are chosen so that the peak of the probe transmission at 100 K is unity. The actual value of the ratio T_P/T_E was observed to be as large as six; however, this value strongly depends on the quality of the spatial overlap of the focused excite and probe pulses on the sample surface. Notice that the graph of the probe transmission versus time exhibits two distinct features. The first is a rapid rise and fall in the probe transmission. This narrow spike is approximately two picoseconds wide and is centered about zero delay. This spike is followed by a gradual rise and fall of the probe transmission lasting hundreds of picoseconds. These measurements are identical to those reported by Kennedy et al. (1974), Shank and Auston (1975), and Smirl et al. (1976). However, when comparing the different sets of data, one must realize that the sample thicknesses and focused optical spot sizes are not identical

We are now in a position, for the first time, to clearly define our goals for the remainder of this lecture. We wish to enumerate and discuss the various effects and physical processes that can occur during and following the absorption of an *intense, ultrashort* optical pulse whose energy is greater than the bandgap. Our intent is to follow the temporal evolution of the nonequilibrium, optically created electron distribution on a picosecond time scale in the hope of *directly* determining the characteristic scattering rates of these processes. In order to be specific, we shall use the excite and probe transmission studies at 1.06 μm presented in Fig. 4 and Fig. 5 as the basis for our discussion. We choose this experiment and this semiconductor for the following reasons: (1) To this point, more experimental studies have been performed on germanium than on any other semiconductor, primarily because it is a well characterized material with a bandgap energy that is comparable to, but less than, the photon energy of the 1.06 μm Nd-glass laser. (2) More effort has been expended theoretically attempting to model the picosecond optical response of germanium than any other semiconductor. In particular, much of this theoretical work has been an attempt to interpret the data of Fig. 4 and Fig. 5. (3) Finally, we choose to emphasize this data because to this point in time there is no single accepted explanation for the slow rise in probe transmission displayed in Fig. 5. As a result, these particular early experiments graphically illustrate problems that can be encountered in the interpretation of the data obtained with such intense ultrashort pulses--unless care is taken in designing the experiment. During early experiments, freedom in this regard was limited by the scarcity of picosecond sources at wavelengths other than 1.06 μm.

We emphasize, once again, that the picosecond optical studies described here, potentially, have all of the advantages of the studies described in earlier chapters by Hearn and by Ulbrich. They differ from these studies, however, in two ways. First, picosecond clocks, such as the one pictured in Fig. 2, provide picosecond time resolution. Secondly, absorption of such intense optical pulses in a semiconductor create carrier densities much larger than those encountered in steady-state optical experiments. This allows the study of processes that become significant only at high carrier densities. Traditionally, these phenomena have been studied in the presence of strong impurity effects caused by high concentrations of donors and acceptors in heavily-doped materials.

The remainder of this chapter is organized as follows. In the next section, we summarize the pertinent features of the germanium band structure. In the third section, we qualitatively discuss the physical processes that occur during and after the nonlinear absorption of the excite pulse by germanium. Finally, we discuss an initial attempt at modeling the picosecond optical response of germanium, as depicted in Fig. 4 and Fig. 5. In doing so, we shall have to invoke or consider rather a large number of physical processes. We shall review experimental attempts to separate the roles of these processes in determining the picosecond temporal evolution of the optically-created carrier distribution in germanium in a later seminar of this ASI.

Finally, the author notes that this is not a review article and no attempt has been made to provide a comprehensive survey of the area of picosecond spectroscopy or the application of picosecond techniques to semiconductors. Nor has any attempt been made to provide the reader with an exhaustive bibliography. For further information on the techniques and applications of picosecond optical pulses, the reader is referred to recent works edited by Shapiro (1977) and Shank et al. (1978). Our intent here is to provide the reader with an introduction to picosecond optical interactions in semiconductors. In this regard, we have relied heavily on our own work on germanium as well as work by others which is related to our own. Remarks that we make concerning areas of general agreement, controversy, and future study (the reader should be warned) reflect our own point of view and are by definition subjective. The reader should also be warned that our point of view has been known to change from time-to-time as these studies progress.

REVIEW OF THE GERMANIUM BAND STRUCTURE

The energy band structure of Ge is well known [Cardona (1963), Fawcett (1965), Cardona and Pollak (1966), van Borzeszkowski (1974)], and the relevant features are shown in Fig. 6. The significant features of the conduction band are the locations of the conduction-band valleys. The minimum which is located at Γ is separated from the top of the valence band by 0.805 eV at 300 K. This separation increases to 0.889 eV at 77 K. This central conduction-band valley is highly nonparabolic; however, an enormous simplification in calculations is obtained by replacing the actual central valley structure by a parabolic structure with effective mass $m_0 \simeq 0.04$ m, where m is the electron rest mass. The consequences of such a simplification are discussed by Latham et al. (1978). The indirect gap, from the valence band top to the minima at L, has a separation of 0.664 eV at 300 K and a separation of 0.734 at 77 K. There are four valleys like the one shown for the entire set of [1 1 1] directions. The minima for these valleys are located exactly at the Brillouin-zone

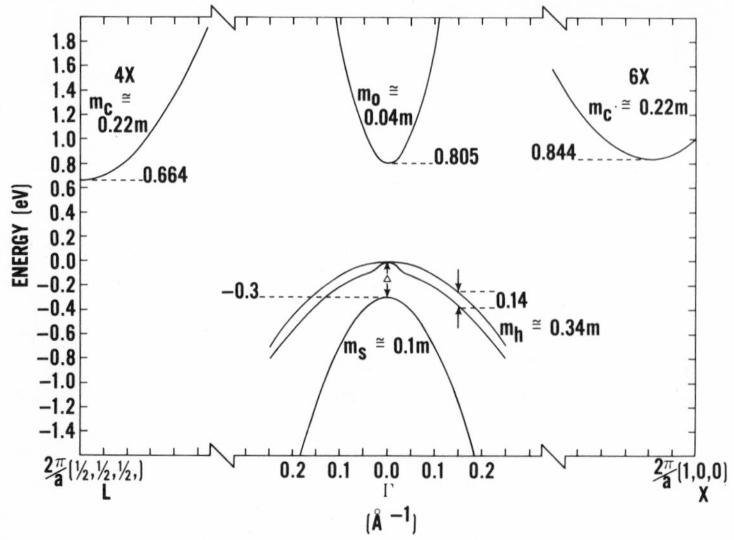

Fig. 6. Approximate germanium band structure at 300 K.

boundary. The band minimum at X is 0.18 eV higher than that at L. There are six equivalent valleys like the one in the [1 0 0] direction, located in from the zone boundary. The energy surfaces in the valleys at L and X (along [100]) are elongated ellipsoids; the density-of-states effective mass for the side valleys is taken to be $m_c \simeq 0.22$ m.

There are three valence bands that are split by the spin-orbit interaction. The valence-band maximum is at Γ. At the center of the Brillouin zone, the heavy-hole band and the light-hole band are degenerate, and the third band is separated from them by an energy $\Delta \simeq 0.3$ eV. Near the center of the zone, the light-hole band can be approximated by an effective mass of $m_l \simeq 0.04$ m. However, except for this small region near the center of the Brillouin zone, the structure introduced by the spin orbit coupling is minor, and we can treat the heavy hole and light hole bands as having the same curvature, separated in energy by $\Delta' \simeq 0.14$ eV. The effective mass of both heavy-hole and light-hole valence bands away from k = 0 is taken to be $m_h \simeq 0.34$ m.

PHYSICAL PROCESSES

The transient properties of a dense electron-hole plasma created in a semiconductor by the interband absorption of an intense ultra-short optical pulse are determined by the simultaneous interaction of a large number of electronic processes. Thus, if we are to follow the dynamics of the carrier distribution on a picosecond time scale, we require a detailed knowledge of which processes occur, their rates, and their effects on the evolving plasma. In this section, we list the fundamental processes believed to be important, and we estimate their rates and effects on the carrier distribution. We use the word "estimate" because some of these processes occur on a time scale too rapid for direct measurement even by picosecond techniques, and others have yet to be measured in a clear and concise manner at these large carrier densities. We list only those processes or effects that have been observed by experimentalists or invoked by theorists in their interpretation of picosecond studies in germanium. As a result, we recognize from the outset that our list contains only a few of the myriad of possible electronic interactions. We discuss these processes in the context of understanding and interpreting the data of Fig. 4 and Fig. 5.

When an intense excite pulse is incident on a thin germanium sample, a fraction of the pulse is reflected; the unreflected portion enters the bulk of the crystal where most of it is absorbed. The light entering the bulk of the crystal is absorbed primarily by direct optical transitions. In this process (see process "a" of Fig. 7), a quantum of light from the excitation pulse is absorbed inducing an electron to make a transition from near the top of the valence band to the conduction-band valley near Γ, leaving behind a hole in the valence band. As shown in Fig. 7, such a transition can occur between any of the three valence bands and the conduction band, since the energy of the light quanta $h\nu$ is substantially greater than the direct band gap energy E_0. However, we shall ignore the contribution of the split-off band to the direct absorption coefficient because of its small contribution to the hole density of states (in fact, at temperatures below room temperature, e.g. 77 K, we are not energetically coupled to the split-off band at all). The linear, direct absorption coefficient α_0 for germanium at 300 K and 1.06 μm is 1.4×10^4 cm^{-1}, yeilding an absorption length of approximately 1.0 μm. The direct absorption of an intense optical pulse generates carriers in a single, optically-coupled state in the conduction band at a rate that is given approximately by

$$G(t) \simeq \alpha_0 I(t)/[\eta(E)\Delta E h\nu], \tag{3}$$

where I is the incident optical intensity, $\eta(E)$ the density of states at the optically-coupled energy, and ΔE is the spread in optically-coupled energies caused by the finite bandwidth of the incident

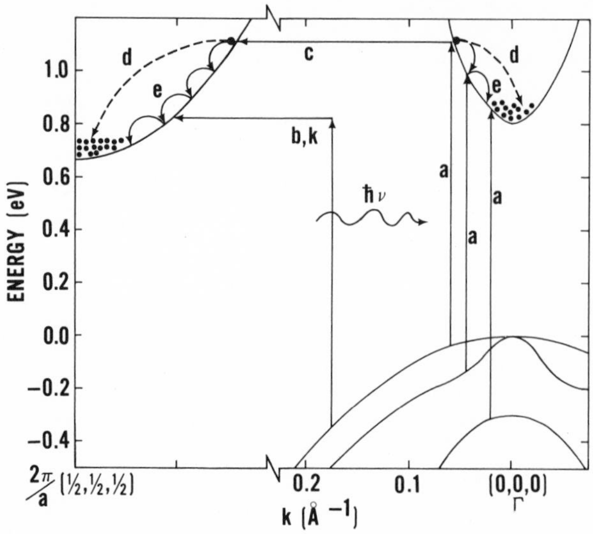

Fig. 7. Schematic representation of (a) direct interband absorptio (b) phonon-assisted indirect absorption, (c) long-wavevect phonon-assisted intervalley electron scattering, (d) electron-electron scattering, (e) phonon-assisted intravalley electron relaxation, and (k) Coulomb-assisted indirect absorption processes in germanium.

optical pulse. For typical excitation pulses of 10 psec width at 1.06 μm, with an energy density of 10^{-2}J/cm^2 and a bandwidth in the range of 10Å to 100Å, the generation rate into a single state i approximately 10^{13} to 10^{14} sec^{-1}. The absorption of such a pulse can create carrier densities between 10^{19} and 10^{21} cm^{-3} in a time perio of 10 psec. In addition, each electron is deposited into the conduction band with an excess energy $\Delta E_e \simeq 0.33$ eV, with respect to t bottom of the central conduction-band valley, and an excess energy $\Delta E_e \simeq 0.47$ eV, with respect to the valley at L. Consequently, the effect of direct absorption is the creation of a large number of carriers with excess energy ΔE_e.

Indirect phonon-assisted interband absorption processes, which involve the transition of an electron from the valence band near Γ to either the L or X conduction-band valleys by the simultaneous absorption of a photon and the absorption or emission of a phonon, are also allowed (process b, Fig. 7). These processes are not important in our problems. As we shall see, the probability that an electron will reach the L valley by means of a real optical transition to the Γ valley followed by a phonon-assisted scattering to one of these side valleys is much greater than the probability that the electron will reach the same valley by means of a second-order phonon-assisted optical transition. The indirect-absorption coefficient for germanium (Dash and Newman, 1955), at 1.06 μm and room temperature, is approximately $3.5 \times 10^2\ cm^{-1}$. For this reason, we ignore phonon-assisted indirect absorption effects in the remainder of our discussions.

As we have already stated, direct absorption of the excitation pulse deposits electrons high in the conduction band, leaving behind holes in the valence band. Since the excitation pulse is approximately monochromatic, a very narrow set of states in the valence band is optically coupled to a narrow set of states in the conduction band. The optically-excited electrons and holes are initially deposited in these states and initially occupy very localized energy ranges in the conduction and valence bands, respectively. Because of the small number of optically-coupled electron (hole) states available in the conduction (valence) band, one might be tempted to conclude that the direct transitions are saturated at very low pulse energies. In fact, this need not be the case. The number of carriers occupying the optically-coupled states at any given time is determined by the relative strengths of the direct optical generation rate into the states, given by (3), and the combined scattering rates out. If the generation rate exceeds the scattering-out rates, then the optically-coupled states are partially filled, and the transmission of the germanium will be enhanced. This process is called *state filling* and is distinct from *band filling* (to be discussed later). This state filling, if significant, results in a delta-function-like spike in the distribution function located at the excitation excess energy. We now consider the mechanisms by which the photoexcited electrons can be scattered from the optically-coupled states and loose their excess energy ΔE_e.

One of the primary processes that removes the nonequilibrium electrons from their localized initial states is long wavevector phonon-assisted intervalley electron scattering (process c, Fig. 7). Such transitions are energetically allowed since $h\nu > E_{L,X} + \hbar\omega_{\mu\vec{q}}$, where $E_{L,X}$ refers to the indirect gaps at L and X and $\omega_{\mu\vec{q}}$ to the phonon frequency of mode μ and momentum $\vec{q}$. Elci et al. (1977) have calculated the intrinsic state lifetime τ_0 of an electron initially in a state k in the central valley of the conduction band as it is scattered by long-wavevector optical and acoustic phonons to available states k' in all side valleys. That is, they calculate

$$1/\tau_o = \sum_{\vec{k}'} R(\vec{k},\vec{k}'), \qquad (4)$$

where $R(\vec{k},\vec{k}')$ is a scattering rate calculated from first-order perturbation theory using Fermi's "Golden Rule". They estimate the scattering rate $1/\tau_0$ to be greater than 10^{14} sec^{-1}. Consequently, electrons are emptied from the central to the side conduction-band valleys at a rate that is comparable to, or even larger than, the optical generation rate.

The electrons scattered to the side valleys by optical and acoustic phonons are deposited there with an excess energy $\Delta E_e \simeq 0.47$ eV. These nonequilibrium carriers might be thought to occupy very localized regions within the side valleys, and the holes expected to still occupy a localized set of states in the valence band. These energetic carriers can give up their excess energy to the distribution as a whole via carrier-carrier collisions (process d), or they can lose their excess energy by intravalley optical phonon emission (process e), as shown in Fig. 7. Carrier-carrier scattering events (including electron-electron, electron-hole, and hole-hole collisions) occur because each carrier must move in the screened Coulomb field of the other carriers. The rate for such collisions is very large. Elci et al. (1977) have estimated this rate to be greater than 10^{14} sec^{-1} at carrier densities of 10^{20} cm^{-3} These collisions ensure that the electron and hole distributions will be Fermi-like. They also ensure that the Fermi distribution for the holes and the Fermi distribution for the electrons will reach a common temperature, which is different from the lattice temperature. This initial temperature can be obtained by equating the total optical energy absorbed to the total energy of the electron-hole distribution,

$$\sum_c \sum_{\vec{k}} E_{ec}(\vec{k})\ f_{ec}(\vec{k}) + \sum_v \sum_{\vec{k}} E_{hv}(\vec{k})\ f_{hv}(\vec{k}) = nh\nu, \tag{5}$$

where $f_{ec}(\vec{k})$ represents the electron-Fermi-distribution function in the conduction band c, $f_{hv}(\vec{k})$ the hole-distribution function in the valence band v, $\sum_c$ denotes a summation over all conduction band valleys, and $\sum_v$ a summation over all valence bands. Finally, n represents the number of photogenerated electron-hole pairs as determined by integrating the total optical generation rate over the optical pulse width. The approximate result of such a calculation i

$$T \simeq \frac{1}{3k_B}(h\nu - E_L), \tag{6}$$

where k_B is the Boltzmann constant. Consequently, we obtain an initial carrier temperature of approximately 1800 K!

Electrons located high in the tail of this hot Fermi distribution can relax by intravalley optical phonon emission. Similar comments also apply to the holes in the valence band. The effect of this relaxation mechanism (process e, Fig. 7) is to reduce the carrier temperature and increase the lattice temperature. The rate at which the carrier distribution loses energy to the lattice is of fundamental importance in determining the temporal evolution of the germanium transmission. Unfortunately, the electron-optical phonon coupling constant D for germanium is uncertain by a factor of 3. Experimental measurements and theoretical estimates of this value range [Conwell (1967), Meyer (1958), deVeer and Meyer (1962), Reik and Risken (1962), Jorgensen et al. (1964), Ito et al. (1964), Jorgensen (1967), Fawcett and Paige (1971), Herbert et al. (1972), Seeger (1973), Costato et al. (1973)] from 6.4 x 10^{-4} erg/cm to 18.5 x 10^{-4} erg/cm. Since the carrier energy relaxation rate is proportional to D^2, the time required for the carrier distribution temperature to reach that of the lattice is uncertain by an order of magnitude. We shall return to this controversy later in this lecture and again in a later seminar. For the moment, it is sufficient to note that an individual nonequilibrium carrier in the side conduction band valley will emit optical phonons at a rate of approximately 10^{12} sec^{-1}. In other words, a carrier at 1800 K will *initially* loose its energy at a rate of roughly 0.03 eV/psec. We emphasize that these numbers represent only a rough order of magnitude, and they are uncertain by at least a factor of 10. In any event, it is clear that the phonon relaxation rate is much slower than the carrier-carrier thermalization rate discussed in the previous paragraph.

Also notice (Fig. 7) that a single electron will emit approximately 15 optical phonons while relaxing to the conduction band minimum. As 10^{20} carriers/cm^3 cool to lattice temperature, an enormous number of such short wavevector optical phonons are created. These optical phonons eventually decay into two long wavevector acoustic phonons. If the rate at which the optic phonons decay into acoustic phonons is smaller than the rate at which optic phonons are created by hot carrier relaxation, the result will be a nonequilibrium phonon distribution with a temperature T_p greater than the lattice temperature. According to Safran and Lax (1975), the optical phonons decay with a characteristic time of 10 psec at 77 K and 5 psec at 297 K.

Once there are electrons in the conduction band and holes in the valence band, as a result of direct absorption of the excitation pulse, then free-carrier absorption is possible. Free-carrier absorption (process f, Fig. 8) denotes a process where an electron in any one of the conduction band valleys is induced to make a transition to a state higher in that same valley by the simultaneous absorption of a photon and the absorption or emission of a phonon (optical or acoustic). An identical process occurs for holes in the valence band. The rate for direct absorption is usually larger than that for free-carrier absorption; however, the rate at which direct absorption events occur decreases as the number of occupied

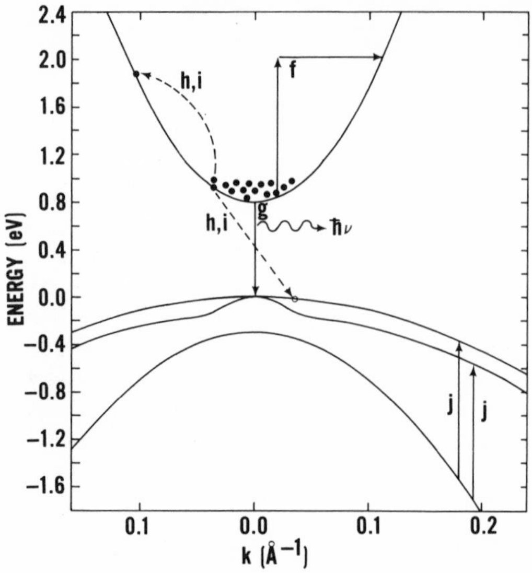

Fig. 8. Schematic representation of (f) free carrier absorption, (g) radiative recombination, (h) Auger recombination, (i) plasmon-assisted recombination, and (j) direct intervalence-band absorption processes in germanium.

states in the Γ valley increases. On the other hand, the rate for free-carrier absorption events increases as the number of electrons (holes) in the conduction (valence) band increases. The total number of free-carrier events occuring per unit time per unit volume in our experiments can be estimated from

$$R_{FCA}^{Total} = \alpha_{FCA} I/h\nu, \qquad (7)$$

where α_{FCA} is the free carrier absorption coefficient and I the optical intensity. The rate per carrier can then be obtained by dividing (7) by the number of carriers/volume present in the sample:

$$R_{FCA} = \alpha_{FCA} I/h\nu n. \qquad (8)$$

The free-carrier absorption coefficient α_{FCA} is directly proportional to carrier density and is estimated (Elci et al., 1977) to reach values between 3×10^2 and 3×10^3 cm^{-1} at the optically-created carrier densities encountered in the experiments of Fig. 4 and Fig. 5. Since α_{FCA} is directly proportional to n, the rate/carrier R_{FCA}

is independent of carrier concentration. For optical energy densities of 10^{-2}J/cm^2 and optical pulsewidths of 10 psec, the rate/carrier is between 10^{10} and 10^{11} sec^{-1}. As a result of a free-carrier absorption event, a carrier will gain an excess energy $h\nu$. This excess energy is quickly redistributed to the distribution as a whole through carrier-carrier collisions. As a result, free-carrier absorption serves to further elevate the carrier temperature.

It is important to notice that, of the processes discussed to this point, only direct and indirect absorption events will increase the carrier number. Free-carrier absorption and phonon-assisted relaxation serve only to elevate or reduce, respectively, the carrier temperature. Various recombination processes can reduce the carrier number, as discussed below.

The recombination processes can be divided into two general categories: radiative and nonradiative. Radiative recombination can be of two types: direct and indirect. The recombination of an electron in the Γ-valley of the conduction band with a hole in the valence band by means of emission of a photon is termed direct (process g, Fig. 8); the recombination of an electron in the L or X-valley with a hole in the valence band by means of the simultaneous emission of a photon and emission (or absorption) of a phonon is termed indirect (not shown). Direct gap recombination is the faster of the two processes. The transition probability is energy and distribution function dependent. At the high carrier densities under consideration here, the rate has been estimated (Elci et al. 1977; van Driel et al. 1976) to be approximately 10^9 sec^{-1}. Much shorter effective lifetimes have been predicted (Ferry, 1978); however, we presently have no direct experimental evidence to substantiate these claims. Consequently, we assume that these processes occur on nanosecond time scales and ignore them for the remainder of our discussion.

At very high carrier densities, such as those produced here, third-order, nonradiative Auger-recombination can become important. Auger recombination is a Coulombic three-body interaction conserving energy and momentum. In this process (h, Fig. 8), an electron recombines with a hole, and the excess energy is transferred to another electron (or hole) in the form of kinetic energy. The Auger equation for electron-hole recombination has the form

$$dn/dt = -\gamma_A n^3, \qquad (9)$$

where γ_A is defined as the Auger rate constant. This rate constant has been estimated (Conradt and Aengenheister, 1972) to be approximately 10^{-31} cm^6 sec^{-1}. We can use (9) to determine an estimate of the initial decay rate *immediately* following carrier creation by direct absorption of the excitation pulse:

$$\left.\frac{dn}{dt}\right|_{0^+} \simeq -(\gamma_A n_o^2)n, \tag{10}$$

where n_o is the initial photogenerated carrier density and $\gamma_A n_o^2$ the initial rate. For typical carrier densities generated here (10^{20} - 10^{21} cm^{-3}), this yields an initial recombination rate of 10^9 - 10^{11} sec^{-1}. Obviously, this rate is strongly dependent on the carrier density; an uncertainty of an order of magnitude in carrier density results in two orders of magnitude error in the initial carrier loss rate. We shall report on picosecond optical measurement of these Auger rates in a later seminar. Because of the small magnitude of γ_A, Auger events are only observed at very high carrier densities. They serve to reduce the carrier number and heat the carrier distribution.

In passing, we note that the inverse of the Auger process, the so-called avalanche process is also allowed. This process (not shown in Fig. 8) is also a three-body Coulombic interaction. Here, however, an electron located high in the conduction band makes a transition to an energy lower in the band, and the excess energy is used to *create* an electron-hole pair. For this process to be significant, a substantial number of carriers must be located sufficiently high in the conduction band to possess an excess kinetic energy larger than the direct band gap energy E_o. As a result, this process depends strongly on carrier temperature. The rate for this process has been estimated (Leung, 1978) to be less than 10^8 sec^{-1} for typical distribution temperatures encountered here.

As the carrier density builds up (primarily as a result of direct absorption of the excite pulse), the plasma frequency of the carriers increases. At sufficiently high plasma frequencies, an electron in the Γ valley can recombine with a hole near the top of the valence bands via emission of a plasmon. The plasma frequency is given by

$$\omega_p^2 = \frac{e^2 n}{\varepsilon}\left(\frac{1}{m_c} + \frac{1}{m_h}\right), \tag{11}$$

where e is the elementary electron charge, ε is the dielectric constant, n is the carrier density, and m_c and m_h are the electron and hole effective masses, respectively. Since most of the electrons are located in the side valleys, the side-valley electron effective mass occurs in (11). Normally, an electron near the conduction-band edge (at Γ) can recombine with a hole by emission of a plasmon *only* if the plasma frequency ω_p is larger than the direct gap frequency E_o/h. However, in our problem, according to Elci et al. (1977), the plasma resonance is substantially broadened during the period the excitation pulse is passing through the sample. This is because direct absorption populates only the Γ valley. As a result, the Fermi energy of the Γ-valley electrons is perturbed relative to the Fermi energy of the L,X valley electrons when the excitation pulse is present in the sample. This relative perturbation is rapidly damped as the two Fermi energies try to rapidly equalize by means of phonon-assisted intervalley scattering. This rapid damping causes the broadening of the plasmon resonance. As a result of the broadening, the plasmon lifetime is short compared to a picosecond. The energy lost in the decay of the plasma oscillations is rapidly transferred to single electron and hole states and ultimately increases the temperature of the carrier distribution. Consequently, the end result of plasmon assisted recombination is the same as that of Auger recombination; it reduces the carrier number and raises the distribution temperature. As a matter of interest, a carrier density of 10^{21} cm^{-3} would result in a plasma energy $\hbar\omega_p$ equal to the band gap energy E_0. The effect of plasmons in early theoretical models may have been overestimated because of errors in early estimates of the carrier density.

Once a large population of holes has been created in the valence band (again by the direct absorption of the excitation pulse), we must consider the importance of another process that occurs only at high carrier densities--direct intervalence-band absorption. These band-to-band transitions occur between either the light- or heavy-hole valence band and the split-off valence band. While quantum selection rules forbid direct transitions between valence subbands at $\vec{k} = 0$, they are allowed at $\vec{k} \neq 0$. The energetically allowed direct intervalence-band transitions are indicated by arrows (process j) in Fig. 8. These transitions occur relatively far from the center of the Brillouin zone. As a result, unless the hole concentration is large, the intervalence-band transition rate will be small compared to that for direct interband absorption, since both initial and final states will be occupied. The availability of the final state for absorption depends on both the hole concentration and temperature. The direct intervalence-band absorption coefficient in germanium for a lattice temperature of 300 K, a carrier temperature of approximately 1800 K, and a carrier number of 10^{20} cm^{-3} has been estimated (Leung, 1978) to be 10^3 cm^{-1}. Intervalence-band absorption does not change the carrier number but, like free-carrier absorption, serves only to elevate the distribution temperature.

At high carrier densities, Coulomb-assisted indirect transitions might enhance absorption as well. A Coulomb-assisted indirect transition is one in which an electron makes an indirect transition from the valence band near Γ to a conduction-band side valley by the absorption of a light quantum. In contrast to phonon-assisted indirect transitions, here the momentum required for the virtual scattering of the electrons from the central valley to a side valley is provided by electron-electron scattering. This process is shown as process k of Fig. 7. The expected importance of this process in our experiments is based on observations of enhanced indirect absorption in heavily doped n-type germanium by Haas (1962). By extrapolating his results on heavily-doped samples, we can obtain a rather crude estimate of the importance of this process at our photon energies and our carrier densities. His measurements suggest that at a wavelength of 1.06 μm and at carrier densities of 10^{20} cm^{-3}, the Coulomb-assisted indirect absorption coefficient might be in the range of 10^3-10^4cm^{-1} One of the problems associated with this extrapolation is the presence of the large hole densities in our experiments; these holes can partially fill the initial states required for the transition. The effect of this process is to increase the carrier number. Since the momentum required for the virtual transition is provided by carrier-carrier scattering, this process is sensitive to carrier density.

The diffusion of carriers from and within the interaction region (focused spot size times sample thickness) will reduce the carrier density seen by the probe pulse. Because of the size of the focused spot for the optical pulse (typically 1 - 2 mm in diameter), diffusion transverse to the direction of light propagation is entirely negligible on picosecond time scales. However, diffusion of carriers from the region near the surface of the sample into the crystal bulk in the direction of light propagation can be significant. We term the diffusion in the direction of light propagation longitudinal. Under the assumption that the photogenerated carriers (10^{20} cm^{-3}) that are created by the excitation pulse are initially deposited in an exponential absorption depth of approximately 1 μm, the time required for the carrier plasma layer to double its thickness has been estimated (Leung, 1978; Auston and Shank, 1974) to be 75-100 picoseconds. Consequently, longitudinal diffusion effects can be significant in our problems.

One final effect of the huge carrier densities deserves mention here: band-gap renormalization. At high carrier concentrations, exchange contributions and free-carrier induced shifts in phonon frequencies cause a narrowing of the energy gap. Ferry (1978) has estimated this narrowing to be approximately 32 meV when the carrier density in the central Γ-valley is 2 x 10^{19} cm^{-3}. Thus, we see that all of the processes and interactions discussed above are complicated by the presence of a dynamic energy-gap narrowing as the carrier densities evolve with time. Since this energy-gap narrowing is electronic in origin, we expect the gap to instantaneously (10^{-15} to 10^{-14} sec) reflect any change in carrier density.

This concludes our brief survey of the basic physical processes that could be important in describing the picosecond optical response of germanium. The key features of this review are summarized in Table I. This table illustrates, once again, the large number of processes that must be considered if we are to describe the evolution of a large photogenerated carrier distribution on a picosecond time scale. Although many of these processes are important in low intensity experiments as well, their rates are drastically altered at the high photoexcitation levels and large carrier densities present here. Others (e.g. Auger recombination, avalanche processes, plasmon recombination, direct intervalence band absorption, Coulomb-assisted indirect absorption, and band-gap narrowing) are only significant at high excitation levels and huge carrier densities. In the following section, we present an early model that attempts to account for the excitation pulse transmission and probe pulse transmission as displayed in Fig. 4 and Fig. 5 in terms of some of these processes.

INITIAL MODELS

In Fig. 5, we have presented a graph of the probe pulse transmission versus time delay between the excitation pulse at 1.06 μm and the probe pulse at 1.06 μm for a sample temperature of 100 K. As we have already noted, this curve exhibits two distinct features. The first is a narrow spike in probe transmission approximately two picoseconds wide and centered about zero delay. The second is a gradual rise and fall in probe transmission lasting hundreds of picoseconds. In this section, we review early attempts to interpret this data.

Parametric Scattering

The narrow spike in probe transmission (shown on an expanded scale in Fig. 9) was first observed by Kennedy *et al.* (1974) and was attributed by them to a saturation and relaxation of the direct absorption. Subsequently, Shank and Auston (1975) observed, in addition to the narrow spike near zero delay, the slower structure at longer delays. In light of this additional structure, they reinterpreted the narrow spike in probe transmission near zero delay as a parametric coupling between the excitation and probe beams caused by an index grating produced by the interference of these two beams in the germanium sample. In this section, we review the manner in which parametric scattering can account for the spike in the probe transmission.

In the picosecond "excite-and-probe" studies described previously, the excitation and probe pulses are derived from a single pulse by means of a beam splitter. Consequently, the probe pulse is simply an attenuated version of the excitation pulse. Near zero delay, the excitation and probe pulses are both spatially and temporally overlapped. As a result, the interference of the two beams will produce a modulation of the optically-created carrier density

TABLE I. Fundamental Processes

Process	Function	Characteristic Constants
a. Direct Inter-band absorption	generates carriers with excess energy, $\Delta\varepsilon \simeq 0.5$ eV	$\alpha_0 = 1.4 \times 10^4 cm^{-1}$
b. Phonon-assisted indirect absorption	increases carrier density	$\alpha_{ID} \simeq 3.5 \times 10^2 cm^{-1}$
c. Phonon-assisted intervalley scattering	populates side valleys	scattering rate/state $\sim 10^{14} sec^{-1}$
d. Carrier-carrier scattering	thermalizes carriers	scattering rate† $\gtrsim 10^{14} sec^{-1}$
e. Phonon-assisted intravalley relaxation	cools carriers, creates "hot" phonon distribution	phonon emission rate/carrier $\sim 10^{12} sec^{-1}$
f. Free-carrier absorption	heats carriers	$\alpha^{\dagger}_{FCA} \simeq 3 \times 10^2 - 3 \times 10^3 cm^{-1}$
g. Radiative recombination	reduces carrier density	recombination rate $\sim 10^9 sec^{-1}$
h. Auger recombination*	reduces carrier density, heats carrier	recombination rate† $< 10^{11} sec^{-1}$

*Important only at high carrier densities
†Strongly dependent on carrier concentration

TABLE I. Fundamental Processes (cont.)

Process	Function	Characteristic Constants
avalanche process	increases carrier density	pair creation rate# $<10^{8}sec^{-1}$
i. Plasmon-assisted recombination*	reduces carrier number, heats carriers	recombination rate† $<10^{13}sec^{-1}$
j. Direct inter-valence band absorption*	heats carriers	$\alpha_{IB}^{\dagger\#} \sim 10^{3}cm^{-1}$
k. Coulomb-assisted indirect absorption*	increases carrier number	$\alpha_{CID}^{\dagger} \sim 10^{3} - 10^{4}cm^{-1}$
Diffusion	decreases carrier density	diffusion rate ~1 μm/100psec
Band-gap narrowing	decreases band gap	<32 meV†

All values are estimated for carrier densities of $\sim10^{20}cm^{-3}$ and carrier temperatures of 1800 K

#Strongly dependent on carrier temperature

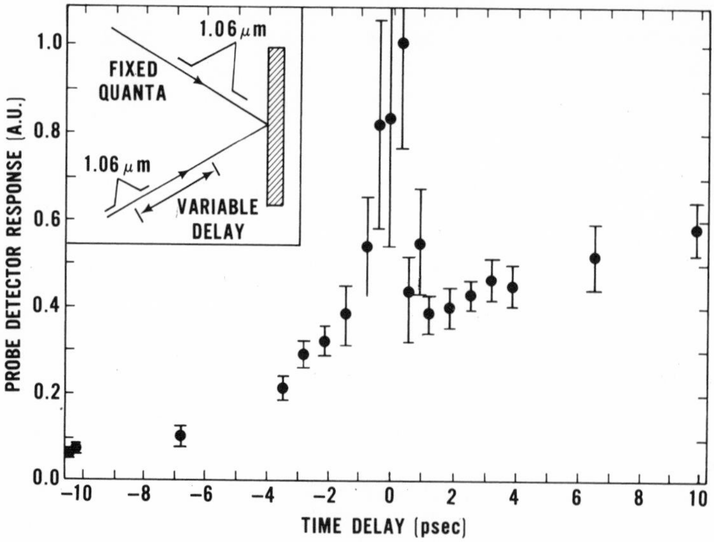

Fig. 9. Normalized response of the probe detector, in arbitrary units, vs. time delay between the excite pulse at 1.06 μm and the probe pulse at 1.06 μm.

to form a grating with spacing $d = \lambda/(2 \sin \theta)$, where λ is the optical wavelength and θ is the angle between each beam and the sample normal as shown in Fig. 10. The grating is formed rapidly and will diffract both excitation and probe pulses as shown in Fig. 10. The first order diffracted beams for both excitation and probe are shown. Notice that one of the first order diffracted beams from the excitation pulse will be scattered into the direction of the probe pulse detector. Also, one of the first order diffracted beams from the probe pulse will be scattered into the direction of the excitation pulse detector. Since the probe pulse energy is only a small fraction of the excitation pulse energy, the amount of light diffracted from the probe pulse into the excitation pulse detector is insignificant. On the other hand, a small fraction of the exciting beam scattered in the direction of the probe detector can produce a signal on the probe detector larger than that produced by the transmitted probe pulse.

The sharp increase (spike) in the signal observed on the probe detector as the two pulses are delayed with respect to one another can then be understood in terms of this parametric scattering in the following manner. An increase in probe detector signal will be observed so long as a grating is produced. Such a grating will be

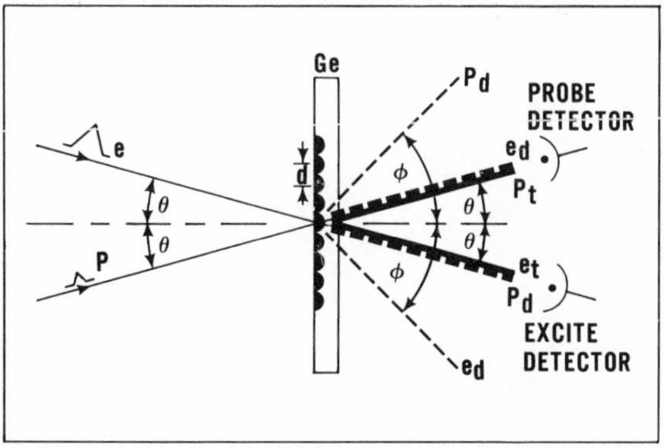

Fig. 10. Geometry for the diffraction of excite and probe beams by a laser-induced grating, where e denotes the incident excite pulse, p the incident probe pulse, e_t the transmitted excite pulse, p_t the transmitted probe pulse, e_d a first order diffracted excite beam, p_d a first order diffracted probe beam, and $\sin\phi = 3\sin\theta$. Solid lines represent transmitted beams and broken lines diffracted beams.

formed only if the delay between excitation and probe pulses does not exceed the coherence length of the two pulses. It is well known that pulses produced by mode-locking glass lasers are usually correlated over a length that is less than their optical pulse width because of various nonlinear processes involved in pulse generation. Consequently, parametric scattering results in an increase in probe detector signal for time delays less than the optical pulse width. Thus, the narrow spike in probe transmission is *not* an increase in sample transmission at all but a scattering of the excitation pulse into the probe pulse. As such, the spike is merely a coherent coupling artifact of the measurement technique.

While recognizing that some parametric scattering is bound to occur during such measurements, Ferry (1978) has recently presented numerical studies that account for the spike in germanium transmission in terms of state filling modulated by band-gap narrowing. If, indeed, these processes were responsible for the narrow rise and fall in probe transmission, a careful study of this structure would yield information concerning carrier scattering rates from the optically coupled states. Subsequently, however, Lindle et al. (1979) have presented the results of measurements that indicate tha parametric scattering fully accounts for the observed spike. As a result, we ignore this spike in probe detector response in the remainder of our discussion. We caution the reader at this point, however, that band-gap narrowing and state filling have been observ in other semiconductor experiments involving optical excitation, an they must certainly be occurring to some degree here as well. They simply do not contribute to the spike in a measureable way.

Hot Electron Relaxation Model

Recently, Elci et al. (1977) have presented an initial first principles theoretical treatment that attempts to account for both the generation and the subsequent transient behavior of the electro hole plasmas created in germanium by the absorption of intense pico second optical pulses. They considered direct band-to-band absorption, free-carrier absorption, phonon-assisted intervalley scatteri phonon-assisted carrier relaxation, carrier-carrier collisions, and nonradiative recombination. In these calculations, rate equations were obtained for the parameters characterizing the electron-hole distributions (electron number, temperature, Fermi energies), and the rates for the individual processes were computed from perturbation theory and Fermi's "Golden Rule" to provide a quantitative description of the transient optical properties of germanium. Briefly, this model (hereafter referred to as the ESSM model) accounts for the transmission of a single optical pulse through a thin germanium sample as a function of incident pulse energy (Fig. and the transmission of a weak probe pulse as a function of time delay after an energetic pulse (Fig. 5) in terms of these processes in the following manner. When an excitation pulse is incident on the germanium sample, the unreflected portion of the pulse enters the sample where most of it is absorbed by direct transitions, creating a large density of electrons (holes) in the central valley of the conduction (valence) band. The electrons are rapdily ($\lesssim 10^{-14}$sec) scattered to the conduction-band side valleys by long-wave-vector phonons. Carrier-carrier scattering events, which occu at a rate comparable to the direct absorption rate, ensure that the carrier distributions are Fermi-like and that both electron and hol distributions have the same temperature, which can be different fro the lattice temperature. Since the photon energy $h\nu$ is greater tha either the direct energy gap E_o or the indirect gap E_L, such a dire absorption event followed by phonon-assisted scattering of an electron to the side valleys results in the photon giving an excess energy of $h\nu$ - E_L to the phonons. This excess energy results in a initial distribution temperature (approximately 1800 K for a lattice temperature of 300 K) due to direct absorption that is greater than the lattice temperature. Thus, the single-pulse transmission (Fig. 4) would begin at its Beer's-law value and increase as a function of incident optical pulse energy because of the partial filling (depletion) of the optically coupled states in the conduction (valence) band as a result of band filling caused by direct absorption. Other processes such as free-carrier absorption and nonradiative recombination events (i.e., Auger and plasmon-assisted recombination) can further raise the carrier temperature during the passage of the excite pulse, while phonon-assisted intravalley relaxation processes can reduce the carrier temperature.

After the passage of the excitation pulse, the interaction region of the sample contains a large number of carriers (10^{19} - 10^{20} cm^{-3}) with a high distribution temperature. The final temperature is determined by the number of quanta in the excitation pulse and the relative strengths of the nonradiative recombination and the phonon-assisted relaxation rates as discussed by Latham et al. (1978). As time progresses, the distribution will continue to cool by phonon-assisted intravalley relaxation. Experimentally, the probe pulse interrogates the evolution of the distribution after the passage of the excitation pulse and is a sensitive measure of whether the optically coupled states are available for absorption or are occupied. The probe pulse transmission versus time delay (Fig. 5) can be understood in the following way. Immediately after the passage of the excitation pulse, the probe transmission is small since the electrons (holes) are located high (low) in the conduction (valence) bands because of the high distribution temperature, leaving the states that are optically coupled available for direct absorption (Fig. 11). Later, as the distribution temperature cools and carriers fill the states needed for absorption, the transmission increases. In short, the ESSM model attributes the slow rise in probe transmission with delay to a cooling of the hot carrier distribution created by the absorption of the excite pulse. The subsequent slow fall in probe transmission at much longer delays is attributed to carrier recombination, which reduces the carrier density and once again frees the optically coupled states for absorption, and to diffusion.

The theoretical fits from Elci et al. (1977) to the single pulse transmission data and probe pulse data of Smirl et al. (1976) are shown as solid lines in Fig. 12 and Fig. 13. Given the complexity of the problem, the overall fit can be regarded as satisfactory. Nonlinear transmission measurements in which the energy band gap of the germanium sample was tuned by hydrostatic pressure (Van Driel et al., 1977) have been accounted for by this model as well.

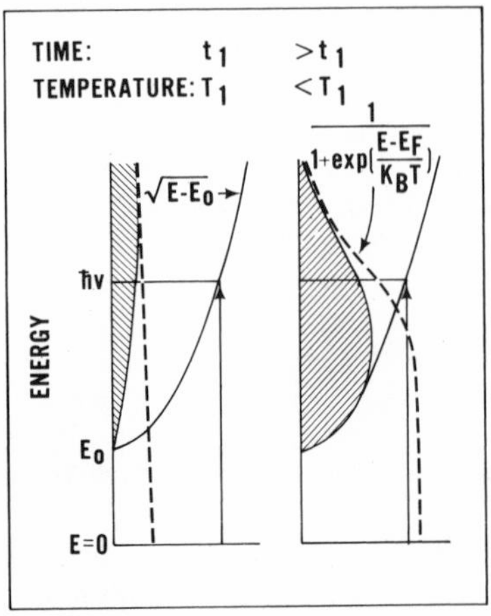

Fig. 11. Schematic diagram for the temporal evolution (cooling) of the carrier distribution created by the absorption of the excite pulse. The solid curve represents the density of states at an energy E, and the broken curve the distribution function. The height of the cross hatched region (density of states times distribution function) is proportional to the number of carriers between E and E + ΔE. The area under the cross hatched curve is proportional to the carrier number (approximately constant here). T_1 and t_1 represent the distribution temperature and time immediately after excitation.

Despite the apparent successes of this model, some basic questions remain concerning the roles of the various physical processes in determining the saturation and temporal evolution of the optical transmission of thin germanium samples under intense optical excitations. Elci et al. (1977) noted that their calculations contained serious assumptions that warranted further theoretical and experimental investigation. The major assumptions were the following: (i) The carrier-carrier collision rate was assumed to be high enough to justify taking the carrier distributions to be Fermi-Dirac. Ferry (1978) has recently reexamined this approximation by calculating the time and energy dependence of the distribution function at the high carrier photogeneration rates encountered here. He concludes that on a time scale of tens of picoseconds the distribution function does indeed approximate a Fermi distribution;

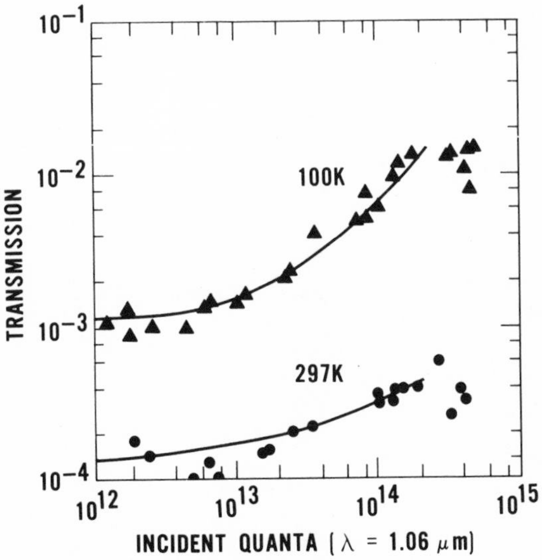

Fig. 12. Transmission of a 5.2-μm-thick germanium sample as a function of incident excite pulse energy at 1.06 μm for sample temperatures of 100 and 297 K. The solid lines are theoretical curves from Elci et al. (1977) and the data are from Smirl et al. (1976). Note that the focused spot size for the optical beam was roughly a factor of 10 smaller than for measurements depicted in Fig. 4.

however, on shorter time scales it contains a δ-function-like spike located at the optically coupled states. Thus, for purposes of calculating the probe-pulse transmission, one may reasonably assume the distribution is Fermi-like, (ii) Carrier Fermi energies and temperatures were taken to depend only on time, rather than on both space and time, thus ignoring the pulse-propagation and carrier-diffusion problems within the optical interaction region of the sample. Therefore, parameters describing the electron-hole plasma, such as the electron number, must be viewed as spatial averages throughout the sample volume. (iii) To simplify the calculations, the actual germanium energy band structure was replaced with a highly idealized parabolic band structure having two degenerate valence bands and a conduction band with a direct valley and 10 equivalent side valleys. The split-off band was totally ignored.

Elci et al. (1977) also noted at the outset that their work contained only a few of the many possible electronic processes.

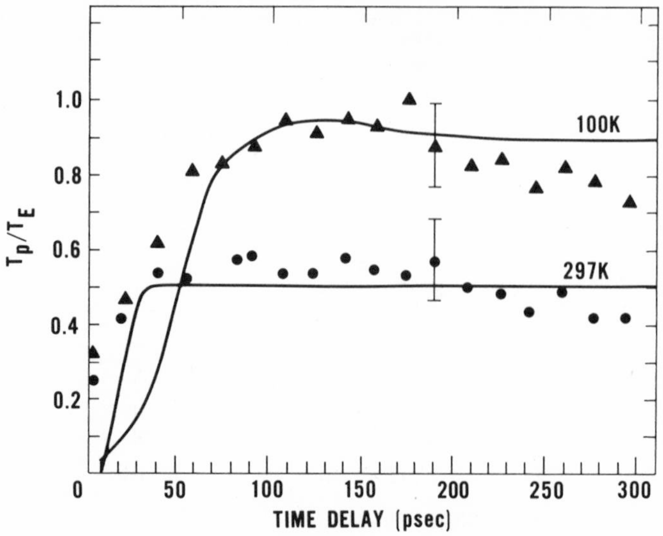

Fig. 13. Probe pulse transmission vs. delay between the excite pulse at 1.06 μm and the probe pulse at 1.06 μm for sample temperatures of 100 and 297 K. The data are plotted as the normalized ratio of probe pulse transmission to excite pulse transmission, T_p/T_E, in arbitrary units. The solid lines are theoretical curves from Elci _et al_. (1977) and the data are from Smirl _et al_. (1976).

Recent studies [Ferry (1978), Leung (1978), Auston _et al_. (1978), Auston _et al_. (1975)] indicate that processes other than those named above may be important. Most of these effects, such as band-gap narrowing (Ferry, 1978), intervalence-band absorption [Leung (1978), Auston _et al_. (1978)], Auger recombination (Auston _et al_. 1975) and Coulomb-assisted indirect absorption (Auston _et al_. 1978) are only observed at large carrier densities. The possible importance of including these processes in any interpretation of the rise in probe transmission will be examined in a seminar to follow.

In the previous two paragraphs, we have outlined the assumptions and omissions of the initial hot-electron model; however, there is another problem associated with the original calculations that is of importance to the present work. The physical constants for germanium, specifically the electron-phonon coupling constants, are not sufficiently well-known to allow a precise calculation of the energy relaxation rate. Latham _et al_. (1978) have previously discussed this point in detail. For the theoretical fits shown in Fig. 13, the electron-phonon coupling constants are chosen as 6×10^{-4} erg/cm at a lattice temperature of 297 K and 2×10^{-4} erg/cm at 100 K. These values are within the range of the accepted theoretically and experimentally determined values listed by Latham et al. (1978); however, they are much lower than the mean value of 1×10^{-3} erg cm^{-1} as obtained from an average of the eight values listed. Since the carrier cooling rate is proportional to the square of the electron-phonon coupling constant, the fitted values result in carrier cooling rates that are 3 and 25 times slower than that obtained by using the average value.

The ESSM model was the first comprehensive theoretical model to attempt to account for the ultrafast response of optically-excited semiconductors. Because of the large number of processes that actually occur, the approximations taken to simplify the mathematics, and the uncertainties in the magnitudes of certain physical constants, the theory is, unavoidably, rather incomplete as it was first presented and as we have reviewed it here. We have chosen to present the model in this early form primarily for tutorial purposes. In spite of its limitations, the theory does represent a first step toward an understanding of a very complicated problem. It provides both a historical perspective and a solid base for further developments. In addition, it is a model that is still evolving. In fact, in Seminar S4, we shall review experimental studies that provide evidence that processes other than those originally included in the model are important and that indicate that certain approximations of the original ESSM model must be removed. We shall then describe recent attempts to modify the model to accomodate these findings. We shall find that these refinements significantly alter our interpretation of the slow rise in probe transmission as shown in Figs. 5 and 13. In this seminar, we shall also discuss alternative models. The interpretation of the rise in probe transmission is still a matter of active debate.

CONCLUSION

In this lecture, we have discussed an early excite and probe experiment in germanium in an attempt to provide the reader with an introduction to the physics of ultrafast carrier relaxation processes in semiconductors. We have enumerated the processes that could occur during such studies, and presented an early interpretation of these experiments. Throughout, we have tried to emphasize the ultrashort time scales and high carrier densities involved in these recent studies. It is evident from our discussions here that these picosecond excite and probe studies can yield *direct* measurements of ultrafast carrier relaxation processes that were heretofore inaccessable. It is equally clear, however, that as experimenters we must be more clever in designing our experiment if we hope to unambiguously extract these rates. We must choose our techniques so as to isolate the effect of a single process. In the experiments that we have described here, almost every imaginable process was active. The large number of these active processes makes these experiments attractive for tutorial purposes, but it ensures that the interpretation of the data will be a nightmare.

This work was supported by the Office of Naval Research and the North Texas State University Faculty Research Fund.

BIBLIOGRAPHY

Auston, D. H., McAfee, S., Shank, C. V., Ippen, E. P., and Teschke, O., 1978, Sol.-State Electr. 21:147.
Auston, D. H., and Shank, C. V., 1974, Phys. Rev. 32:1120.
Auston, D. H., Shank, C. V., and Lefur, P., 1975, Phys. Rev. Letter 35:1022.
Bauer, G., 1974, in "Springer Tracts in Modern Physics", Vol. 74, Springer-Verlag, Berlin.
Bauer, G., 1976, in "Linear and Nonlinear Electron Transport in Solids," Ed. by J. T. DeVreese and H. van Doren, Plenum, New York.
Cardona, M., 1963, J. Phys. Chem. Sol. 24:1543.
Cardona, M., and Pollak, F. H., 1966, Phys. Rev. 142:530.
Conradt, R., and Aengenheister, J., 1972, Sol. State Commun. 10:321
Conwell, E. M., 1967, "High Field Transport in Semiconductors", Academic Press, New York.
Costato, M., Fontanesi, S., and Reggiani, L., 1973, J. Phys. Chem. Sol. 34:547.
Dash, W. C., and Newman, R., 1955, Phys. Rev. 99:1151.
de Veer, S. M., and Meyer, H. J. G., 1962, in "Proc. 6th Intern. Conf. Phys. Semiconductors, Exeter" Inst. of Phys., London.
Elci, A., Scully, M. O., Smirl, A. L., and Matter, J. C., 1977, Phys. Rev. B 16:191.
Fawcett, W., 1965, Proc. Phys. Soc. 85:931.
Fawcett, W., and Paige, E. G. S., 1971, J. Phys. C. 4:1801.
Ferry, D. K., 1978, Phys. Rev. B 18:7033.
Haas, C., 1962, Phys. Rev. 125:1965.
Herbert, D. C., Fawcett, W., Lettington, A. H., and Jones, D., 1972 in "Proc. 11th Intern. Conf. Phys. Semiconductors, Warsaw."
Ito, R., Kawamura, H., and Fukai, M., 1964, Phys. Letters 13:26.
Jorgensen, M. H., 1967, Phys. Rev. 156:834.
Jorgenson, M. H., Meyer, N. I., and Schmidt-Teidemann, K. J., 1964, in "Proc. 7th Intern. Conf. Phys. Semiconductors, Paris", Academic Press, New York.
Kennedy, C. J., Matter, J. C., Smirl, A. L., Weichel, H., Hopf, F. A., and Pappu, S. V., 1974, Phys. Rev. Letters 32:419.

Latham, W. P., Jr., Smirl, A. L., Elci, A., and Bessey, J.S., 1978, Sol.-State Electr. 21:159.
Leite, R. C. C., 1978, Sol.-State Electr. 21:177.
Leung, T. C. Y., 1978, Dissertation, unpublished.
Lindle, J. R., Moss, S. C., and Smirl, A. L., 1979, to be published.
Meyer, H. J. G., 1958, Phys. Rev. 112:298.
Reik, H. S., and Risken, H., 1962, Phys. Rev. 126:1737.
Safran, S., and Lax, B., 1975, J. Phys. Chem. Sol. 36:753.
Seeger, K., 1973, "Semiconductor Physics" Springer-Verlag, Berlin.
Shah, J., 1978, Sol. State Electr. 21:43.
Shank, C. V., and Auston, D. H., 1975, Phys. Rev. Letters 34:479.
Shank, C. V., Ippen, E. P., and Shapiro, S. L., 1978, "Picosecond Phenomena," Springer-Verlag, Berlin.
Shapiro, S. L., 1977, "Ultrashort Light Pulses: Picosecond Techniques and Applications," Springer-Verlag, Berlin.
Shelton, J. W., and Armstrong, J. A., 1967, IEEE J. Quantum Electr. QE-3:696.
Smirl, A. L., Matter, J. C., Elci, A., and Scully, M. O., 1976, Optics Commun. 16:118.
Ulbrich, R. G., 1973, Phys. Rev. B8:5719.
Ulbrich, R. G., 1978, Sol.-State Electr. 21:51.
van Borzeszkowski, J., 1974, Phys. Stat. Sol. (b) 61:607.
van Driel, H. M., Bessey, J. S., and Hansen, R. C., 1977, Optics Commun. 22:346.
van Driel, H. M., Elci, A., Bessey, J. S., and Scully, M. O., 1976, Optics Commun. 20:837.
Weisbuch, C., 1978, Sol.-State Electr. 21:179.

NONEQUILIBRIUM PHONON PROCESSES

P. Kocevar

Universität Graz
Graz, Austria

INTRODUCTION

The coupled systems of electrons ("e") and phonons ("ph") in a conducting solid will be shifted from their equilibrium distribution under the influence of an external electric field $\vec{F}$. Obviously, the momentum and energy imparted to the carriers by the field cannot be dissipated through "normal" e-e, e-ph and ph-ph-collisions, which conserve the total momentum and energy of the e-ph-system. A steady state can only be established by momentum- and energy-loss mechanisms such as "Umklapp"-processes, in which momentum is transferred to the crystal as a whole, or by scattering at the crystal-boundaries, where momentum and energy, predominantly of the phonons, are dissipated into a surrounding "heat bath". Consequently, the nonelectronic dissipation mechanisms for nonequilibrium phonons determine the possibility of time dependent, nonstationary phenomena and can also influence the details of the usually realized steady-state transport (Klemens, 1951; Kaveh and Wiser, 1972). Quite generally one has to expect effects of nonequilibrium of both the e- and ph-subsystem due to the mutual "dragging along" through collisions in any conventional transport experiment at sufficiently low temperatures in sufficiently pure and perfect materials: "phonon drag" by electrons in the lattice thermal conductivity, "phonon drag" of electrons in the electrical conductivity (Sommerfeld and Bethe, 1933; Ziman, 1960; Bailen, 1958; Kaveh and Wiser, 1974; Parrott, 1957) and both drag effects in the thermopower (Ziman, 1960; Geballe and Hull, 1954; Herring, 1958). Moreover, in semiconductors, under nonohmic conditions, we expect effects of "phonon-heating" by hot carriers.

The occurrence of non-thermal phonon distributions in semiconductors has been studied since the fifties in thermoelectric, acoustoelectric (Seeger, 1973) and hot-electron phenomena (Conwell, 1967). Especially the discovery of acoustoelectric current-instabilities (Smith, 1962) and of the Gunn-effect (Gunn, 1963) initiated a great number of investigations through the hope to find cheap and effective ultrasonic and microwave amplifiers or generators by using electric-field-induced phonon instabilities in semiconductors. These processes are characterized by a strong amplification of certain lattice vibrations. Under special circumstances the correspondingly increased mean occupation-number of phonons in these modes can be assumed to be of Planck-type,

$$N_{\vec{q}} = [\exp(\hbar\omega_{\vec{q}}/k_B T_{ph})-1]^{-1}, \tag{1}$$

with some "phonon-temperature" T_{ph} which is larger than the lattice temperature T_L. In analogy to the usual hot-electron terminology one generalizes this rather exceptional situation to speak of "hot phonons" in all cases of phonon-amplification through the influence of external forces, including the cases of optical and magnetic excitation.

There are two distinct approaches to the description of the coupled nonequilibrium electron-lattice system in solids. The classical approach accounts for ultrasonic attenuation or amplification as well as for acoustoelectric phenomena (typical in the MHz-frequency-regime) by treating the charge carriers as an ensemble, which tends to adapt itself instantaneously to the electric field or the deformation-potential induced by an acoustic wave. The presence of an external electric field will produce a driving action together with a bunching of the carriers into the potential troughs of the acoustic wave. This leads to the possibility of transfer of energy from the carrier-system to the acoustic wave with a resulting current instability, if the mean carrier drift velocity exceeds the velocity of sound. This "hydrodynamical regime" is characterized by the conditions $g\ell_e \ll 1$ and $\omega_{\vec{q}}\theta_e \ll 1$, which guarantee that the mean-free-path ℓ_e and the lifetime θ_e of the wave packet of each individual carrier in a trough are much smaller than the spatial extension and the oscillation period of the troughs (White, 1962).

The second, quantum-mechanical, approach to nonequilibrium e-ph systems is characterized by $g\ell_e > 1$ and $\omega_{\vec{q}}\theta_e > 1$. In this case the interaction of the carriers with the lattice-modes can be described by individual scattering events between electrons and the phonons, both of which are treated as ballistic particles. A corresponding Boltzmann equation can be used with quantum-mechanical transition amplitudes for the e-ph-collisions. Since those phonons which dominate the lattice scattering of carriers, for the usual spatially homogeneous case, have frequencies well within this high-frequency regime, our discussion below of the coupled e- and ph-Boltzmann-equations ("EBE" and "PBE") will be restricted to bulk-effects of nonequilibrium phonons.

However, a unified approach, treating both the wave-aspect ($g\ell_e \ll 1$) and quasi-particle aspect ($g\ell_e > 1$) of phonons on the same footing (using quantum-transport techniques) and covering the whole $g\ell_e$-range, can be formulated and has been applied to certain complicated nonlinear acoustoelectric phenomena such as current saturation and domain formation (Yamashita and Nakamura, 1969).

PHONON INSTABILITIES

When current instabilities, accompanied by microwave emission (in the low GHz-regime), were discovered by Gunn (1963) in n-type Gallium Arsenide under hot-electron conditions, the first explanation was in terms of stimulated emission of certain phonons by the carriers. This effect and a great variety of similar emissions in various other semiconductors turned out to be of different origin, or at least to be caused by the interplay of the e-ph-interaction with various other mechanisms such as electron-transfer effects in special bandstructures, plasma- or recombination-instabilities, acoustoelectric phenomena, domain formation, etc. Moreover, up to now most experimental attempts to use the build-up of certain ph-modes by the e-ph-interaction as (through external fields) easily tunable microwave generators have failed to produce devices of sufficient efficiency. Accordingly, there is a fundamental interest in achieving a better understanding of the physics involved.

The most general and instructive approach is the analysis of the time-evolution of the coupled e-ph-system after the application of an external electric field in terms of the coupled time-dependent Boltzmann equations for the distribution functions $f_{\vec{k}}$ for carriers and $N_{\vec{q}}$ for phonons.

The EBE differs from its well-known form for ph-equilibrium only by the replacement of the Planck-distributions $N_{\vec{q}}^{th}$ by the time-dependent $N_{\vec{q}}$ in all e-ph scattering terms. The PBE for ph-branch ν and coupling-mechanism κ reads

$$\partial N_{\vec{q}}^{(\nu)}/\partial t = (\partial N_{\vec{q}}/\partial t)_{ph-e}^{(\nu,\kappa)} + (\partial N_{\vec{q}}^{(\nu)}/\partial t)_{non-el}, \qquad (2)$$

with

$$(\partial N_{\vec{q}}/\partial t)^{(\nu,\kappa)}_{ph-e} = \frac{2\pi}{\hbar}\sum_{\vec{k}}\sum_{+,-}[\pm|\langle\vec{k}\pm\vec{q},N^{(\nu)}_{\vec{q}}\pm1|H^{(\nu,\kappa)}_{ph-e}|\vec{k},N^{(\nu)}_{\vec{q}}\rangle|^2 \times$$

$$\times\ \delta(E_{\vec{k}} - E_{\vec{k}\pm\vec{q}} \pm \hbar\omega^{(\nu)}_{\vec{q}})\, f_{\vec{k}}(1-f_{\vec{k}\pm\vec{q}})], \quad (3)$$

in the usual notation; the non-electronic losses are given by

$$(\partial N^{(\nu)}_{\vec{q}}/\partial t)_{non-el} = (\partial N^{(\nu)}_{\vec{q}}/\partial t)^{(3)}_{ph} + (\partial N^{(\nu)}_{\vec{q}}/\partial t)_b + \ldots, \quad (4)$$

where only the cubic anharmonic and boundary-scattering terms are r tained as the dominant dissipation mechanisms.

Due to the great complexity of ph-ph-interactions, an evaluati of the ph-ph-collision integrals, which involve cubic terms in the unknown ph-distributions, has never been achieved. But fortunately the ph-emission by drifting carriers is very similar to ultrasonic attenuation, where externally induced "hyperthermal" phonons decay into the thermal phonon bath of the host-lattice. In this case a "single-mode" relaxation time $\tau_{ph}(\vec{q},T_L)$ can be defined (Klemens, 1951b). An analogous assumption can be made in our case of interes as long as the number of excited phonons is negligible compared to the thermal phonon background. This "heat-bath"-condition was firs formulated by Gurevitch and Gasymov (1967a), for the case of hot-electron-phonon systems, as $T_e < k_B T_L^2/8ms^2$ (T_e = electron-temperatu m = effective mass, s = velocity of sound). We shall see that typi- cal ph-avalanches occur only in relatively narrow frequency-bands s that the commonly accepted use of single-mode relaxation times migh be questionable only in the later stages of such ph-bursts. In the cases, nonlinear relaxation processes will soon dominate the instab lity and strongly counteract the ph-amplification by the carriers (Murayama, 1967).

Since energy and momentum dissipation by the phonons at the crystal-boundaries can be described well by relaxation times, total non-electronic relaxation rates can now be used in (4), reducing th PBE to a tractable form. The difficult problem of determining $\tau_{ph}(\vec{q},T)$ for the different phonon-branches and in different materia will be discussed in a later section.

Even after the above simplification, the direct numerical solu- tion of the two coupled integro-differential equations is still completely unfeasible. Any further progress is only possible throug certain assumptions about the functional form of the electronic distribution function.

In 1974, Perrin and Budd (1974) analyzed the case of hot elec- trons in non-polar, nondegenerate semiconductors within the diffusic approximation: $f_{\vec{k}} = f_{\vec{k}}^{(o)} + (\vec{k} \cdot \vec{F}/F) f_{\vec{k}}^{(1)}$ and $N_{\vec{q}} = N_{\vec{q}}^{(o)} + (\vec{q} \cdot \vec{F}/F) N_{\vec{q}}^{(1)}$, taking $f_{\vec{k}}^{(o)}$ as a drifted Maxwellian with a time-dependent electron temperature T_e. Assuming a quasi-stationary e-distribution which adapted itself to the instantaneous value of the ph-population (justified by the very fast electron-momentum relaxation), they integrated the PBE in small time-steps, obtaining T_e, the mobility μ, and $N_{\vec{q}}$ as functions of time by solving the electronic energy-balance equation for each time-step. For a simple isotropic-band model of Germanium, acoustic-deformation potential (DA) and ionized-impurity (ii) scattering of the carriers were taken into account. Due to the well-known difficulties of standard transport theory with i-i-scattering at low temperatures, its contribution to the mobility had to be fitted with an adjustable parameter. Noticeable effects of LA-phonon-disturbances in the mobility were found at $T_L = 4.2°K$, assuming boundary losses for the nonequilibrium phonons and an electron-concentration $n_e = 5.10^{14}$ cm^{-3} (Table 1), in qualitative agreement with the experimental results of Baumann, et al., (1968).

Table 1: n-Germanium, $T_L = 4.2°K$, $n_e = 5.10^{14}$ cm^{-3}, F = 10 V/cm: Time-dependence of mobility μ, electron temperature T_e and phonon distributions for τ_b (energy) = 5 μsec and τ_b(momentum) = 0.2 μsec (after Perrin and Budd, 1974).

t [μsec]	μ [$10^4 \frac{cm^2}{V_{sec}}$]	T_e [°K]	q [10^6 cm^{-1}]	$N^{(o)}(q)$			$q \cdot N^{(1)}(q)$		
			=	1	2	4	1	2	4
0	19	53		0.6	0.2	0	0	0	0
0.04	18	51					0.1	0.1	0.1
0.4	12	45					0.8	0.9	0.2
0.6	11	44		4.2	2.7	0.8			
0.7	10	44					1.2	1.2	0.3
1.2	8.5	45		6.2	3.5	1.3			
2.4	6.5	47		7.7	4.3	1.7			
4.0	6	48					1.6	1.3	0.3

Much more drastic effects had been found in a similar analysis by Kocevar (1972,1973) for high mobility polar semiconductors, in particular n-type Indium Antimonide, where both the acoustic (LA an TA) and optic (LO) modes can be disturbed by the DA, PA (piezo-electric) and PO (polar-optic) e-ph-coupling. Since, in contrast to nonpolar materials, the carrier drift velocity can be of the ord of the carrier mean thermal velocity, a displaced and heated Maxwellian $f_{HDM} \propto \exp\{-\hbar^2(\vec{k}-\vec{k}_o)^2/2mkT_e\}$ was used for the electrons. Again, some phenomenological treatment was necessary for i-i-scattering, but this HDM-approach has the advantage of yielding a closed expression for N_q:

$$N_{\vec{q}}(t) = \exp(-\int_o^t \frac{ds}{\tau(s)})\{\int_o^t ds\, \frac{\overline{N}(s)}{\tau(s)} \exp(\int_o^s \frac{d\sigma}{\tau(s)}) + N_{\vec{q}}^{th}\}, \qquad (5)$$

where the formal asymptotic solution $\overline{N}$ and the total ph-rate $\tau^{-1} = \tau_{el}^{-1} + \tau_{non-el}^{-1}$ are specified functions of the time-dependent parameters k_o and T_e, which have to be evaluated, at each step of the time integration of (5), by solving the electronic energy- and momentum-balance for the instantaneous ph-distribution. The explic form of τ_{el}^{-1} shows that the electronic contribution τ_{el}^{-1} to τ^{-1} becom negative when the mean drift-velocity v_o $(= \hbar\vec{k}_o/m)$ fulfills the wel known acoustoelectric Cerenkov-condition $v_o \cos(\beta) > s$, where $\cos(\beta) = \vec{v}_o \cdot \vec{q}/(v_o q)$ for acoustic phonons and $v_o \cos\beta > \hbar\omega_{ph}/q$ for optical phonons. If the corresponding amplification $|\tau_{el}^{-1}|$ exceeds the nonelectronic ph-losses τ_{non-el}^{-1} for a certain group of modes, the resulting negative τ^{-1} acts as an amplification coefficient for these phonons. In this case a steady state can only be reached if k_o and T_e eventually change in such a way as to make $|\tau_{el}^{-1}| \leq \tau_{non-e}^{-1}$ Such a situation is illustrated in Table 2, where all LA-forward-modes burst and eventually act back on the electron parameters x_o ($\propto k_o$) and T_e shifting them from the initial resonance-regime to their steady-state values.

The second line of Table 2 shows some small shifts of x_o and T_e which correspond to a small LO-ph-disturbance. The smallness of these shifts and the short time-scale to reach an intermediate quas steady state, with respect to the LO-phonons, is due to the very short LO-ph-lifetimes (in all materials of order 10^{-11} sec), becaus quite generally the time-scales for ph-built-up must be of the orde of $\tau \approx |\tau_{el}| \approx \tau_{non-el}$.

As far as experimental evidence for ph-bursts, as bulk effects is concerned, the time scale for the LA-burst in Table 2 is (becaus of the very long τ_{LA} at these low temperatures) much larger than th duration of a typical voltage-pulse in low-temperature hot-electron experiments, which is usually of the order of several microseconds to prevent lattice heating. Thermal breakdown through sample heati

Table 2: n-Indium-Antimonide, $T_L = 4.2°K$, $n_e = 1.25 \cdot 10^{13}$ cm^{-3}, $n_i = 1.1 \cdot 10^{14}$ cm^{-3}, F = 10 V/cm: Time Dependence of the Phonon Numbers in the Most Strongly Coupling Acoustic Modes, Expressed by $\Delta_j(z) = N_j(q, \cos\beta = 0.861;t)/N_j^L(q)$, j = LA, TA. Electron Drift-Momentum (x_0) and Phonon Momenta (z) Numerically Given in Units $(2mk)^{\frac{1}{2}}$ (after Kocevar, 1973).

t	x_0	T_e	$\Delta_{TA}(z)$		$\Delta_{LA}(z)$			
			z=0.139	0.712	0.139	0.712	2.858	6.861
0	1.2362	41.8	1	1	1	1	1	1
$8.3 \cdot 10^{-10}$	1.2531	42.3	1	1	1	1	1	1
$6.3 \cdot 10^{-6}$	1.2530	42.3	1.08	1.27	1.06	1.29	1.16	(1.21)
$2.3 \cdot 10^{-5}$	1.2527	42.3	1.19	1.85	1.18	2.12	1.59	(1.70)
$1.2 \cdot 10^{-4}$	1.2525	42.3	1.68	3.43	1.61	10.5	3.67	(3.83)
$3.0 \cdot 10^{-4}$	1.2523	42.3	1.91	3.79	1.79	60.5	5.8	(5.45)
$4.4 \cdot 10^{-4}$	1.2522	42.3	1.95	3.79	1.81	200	6.7	(5.93)
$6.4 \cdot 10^{-4}$	1.2521	42.3	1.95	3.79	1.81	1050	7.3	6.25
$9.2 \cdot 10^{-4}$	1.2506	42.2	1.95	3.79	1.81	10300	7.65	6.25
$1.2 \cdot 10^{-3}$	1.2367	42.2	1.95	3.72	1.81	98500	7.65	6.25
$1.5 \cdot 10^{-3}$	1.1079	41.6	1.84	3.28	1.76	850000	6.95	5.6
$1.9 \cdot 10^{-3}$	0.8654	40.3	1.62	2.53	1.59	3330000	4.51	3.86
$2.2 \cdot 10^{-3}$	0.8173	40.0	1.57	2.41	1.55	4050000	3.83	3.40

for moderate F is known to occur after times of the order of one second, so that the results of Table 2 might possibly describe its first stage, and a later stepwise decay of the enormously amplified modes into the thermal-ph reservoir would probably result in strong heating of the sample. Bursts at lattice temperatures above 20°K are very unlikely due to the very fast ph-ph-relaxation.

The investigation of ballistic-phonon amplification in non-ohmic transport started with the famous Huebner-Shockley (1960) experiment; a time-ordered list of other typical experimental work can be found in Ascalelli (1960), Zylberstejn (1967), Bodo, et al. (1968), Bauman, et al. (1968), Goreli, et al. (1968), Arizumi, et al. (1968), McGroddy and Christensen (1973), Jack, et al. (1974), Reupert, et al. (1975), Kichigan, et al. (1976), and Narayanamurti, et al. (1978).

The first detailed theoretical investigations of time-dependent effects due to ph-excitation by electric fields were published by Yamashita and Nakamura (1965) and by V. V. and B. V. Paranjape (1968) using a linearized displaced Maxwellian and a HDM respectively.

PHONON LIFETIMES

The rather subtle interplay of electronic amplification and non-electronic dissipation of phonons during these instabilities demonstrates the urgent need for experimental determinations of $\tau_{non-el}(\vec{q},T)$. Table 2 was calculated by using a value $\tau_{LA}(\vec{q},T)$ fitted to thermopower-measurements in magnetic fields (Puri, 1965; Gadzhialiev, 1977), but with theoretical expressions for τ_{TA} and τ_{LO}. Due to the theoretical uncertainties it could easily be that the actual τ_{TA}-values are much larger than the theoretical ones and also that TA-bursts might be possible in indium antimonide.

The main reason for our bad knowledge of the lifetimes of acoustic phonons is the fact, that those phonons which are the most efficient scatterers of charge-carriers have wavevectors of the order of the mean thermal wave-vector of the carriers ($\simeq \sqrt{2mk_BT_e}/\hbar$). The corresponding frequencies (typical in the 100 GHz-regime) are therefore much larger than the highest ultrasonic-frequencies in the usual attenuation experiments. Some theoretical understanding has been achieved for the ultrasonic regime $\omega < k_BT/\hbar$ (Landau and Rumer, 1937; Mario, 1964; Simons, 1967; Gurevich and Shklovskii, 1967), at least for the frequently found proportionality of τ_{TA}^{-1} and τ_{LA}^{-1} to ωT^4. For frequencies $\omega > k_BT/\hbar$, Klemens (1967) estimated $\tau_{LA}^{-1} \sim \omega^4 T$, whereas nothing is known about frequencies $\omega \gtrsim k_BT/\hbar$ which are already well within the regime of electric-field induced ph-disturbances (e.g. for $T_L = 4°K$, $T_e = 40°K$ in n-Germanium).

Experimental methods for generating and detecting high-frequency acoustic phonons and measuring their lifetimes cover a wide range of frequencies, temperatures and materials: standard ultrasonic-attenuation techniques with input and output quartz transducers (Polerantz, 1965; Woodruff and Ehrenreich, 1961; Simpson, 1975), piezoelectric excitation of THz-phonons by IR-laser pulses (Grill and Weis, 1975), generation and detection of high-frequency phonons with superconducting tunnel-junctions (Eisenmenger and Dayem, 1967; Dynes and Narayanamurti, 1972), pulse-echo techniques (Dutoit, 1971) heat-pulses from Joule-heated metal-films into a substrate (Maris, 1972; Huet, et al., 1976), detection of high-frequency phonons by vibronic side-band spectrometers (Bron and Grill, 1977), excitation by stimulated Brillouin-scattering and probing by delayed light-pulses (Pohl, et al., 1968; Winterling and Heinicke, 1968), direct (via Brillouin-scattering)(Ishida and Inuishi, 1968; Palik and Bray, 1971; Sussman and Ridley, 1974) or indirect (via time-resolved I-V-characteristics) (Mosekilde, private commun.) determination of decay-rates of acoustoelectrically amplified phonons, and, most recently, stimulated ph-emission from laser-excited impurities (Bron and Grill, 1978). Unfortunately at most qualitative results have been obtained for the standard semiconductor-materials, especially in the high-frequency regime $\hbar\omega \gtrsim kT$.

The understanding of optical-phonon lifetimes is more satisfactory because of their rather universal magnitudes. Theoretical estimates of Klemens (1966) are in good agreement with Raman-linewidths in Silicon (Hart, *et al.*, 1970) and in qualitative agreement with neutron-scattering data in Germanium (Nilsson and Nelin, 1971). Spectroscopic methods for demonstrating nonequilibrium optical-phonon-distributions include: "Stokes" versus "Anti-Stokes"-luminescence from ph-assisted exciton-recombination at acceptor-impurities (Shah, *et al.*, 1974), excitation and probing (again through Stokes-versus Antistokes intensities) with laser light (Mattos and Leite, 1973), or the analysis of photo-luminescence spectral shapes (Shah, 1974; Motsircke, *et al.*, 1975; Meneses, *et al.*, 1975). Moreover, direct determination of optic-ph-lifetimes was also possible by time-resolved delayed-probe techniques in Raman spectroscopy (Shah, *et al.*, 1970; Alfaro and Shapiro, 1971) with results in good agreement with Raman linewidths.

STEADY-STATE EFFECTS OF NONTHERMAL PHONONS

The possible influence of mutual drag and heating in a steady-state e-ph-system on various transport-coefficients can be studied by analyzing the coupled, stationary EBE and PBE. In contrast to the time-dependent case of section 2, simple numerical solutions can be found for the electrical conduction within the diffusion-approximation (Conwell, 1964a) and even simple algebraic expressions for k_0 and T_e as functions of F for a linearized DHM (Kocevar, 1977; Kocevar and Fitz, 1978). Again only few quantitative results were found due to the badly known ph-relaxation-times. Generalizing these cases (of n-Germanium and n-Gallium-Arsenide), one can expect noticeable drag-effects of 10% or more for moderate doping and for lattice temperatures around 20 K in the ohmic mobility and much larger effects under nonohmic conditions. Because of carrier freezeout at low temperatures and rapidly increasing 3-ph-relaxation processes at higher temperatures the general range for effects is given roughly by $T_e > 15$ K, $T_L < 25$ K. Similar effects were predicted for degenerate semiconductors (Sharma and Aggarwal, 1975; Samoilovich and Buda, 1976). Further simplified approaches have either neglected carrier-heating (Spector, 1962) or the drift term in the e-ph-collision integrals (Conwell, 1964b) or used very crude ph-relaxation-models for acoustic (Paranjape, 1973; Keyes, 1974) or optic-phonons (Beattie, 1978).

Analogous studies of the coupled EPE and PBE for the magnetoconductivity (Yamashita, 1965; Gurevich and Gasymov, 1967b, 1969; Lifshitz, et al., 1972; Lang and Pavlov, 1973; Arora and Miller, 1974; Pomortsev and Kharus, 1968; Cassiday and Spector, 1974) and thermoelectric (Gurevich and Gasymov, 1968; Samoilovich and Buda, 1976; Jay-Gerin, 1974, 1975) or photoelectric (Yassievich and Schliefstein, 1970, 1972) effects showed that any detailed analysis of low-temperature transport-measurements should allow for possible effects of mutual e-ph drag and heating.

SOME FURTHER ASPECTS OF PH-DISTURBANCES IN SOLIDS

Almost identical to our discussion in section 2 is the detailed analysis of the spectral distribution of phonons radiated from the metal-film into the substrate in heat-pulse experiments (Perrin and Budd, 1972; Perrin 1975). Also very closely related are the theoretical models for the "phonon-bottleneck effect" in paramagnetic-spin relaxation usually discussed in terms of coupled time-dependent rate equations for occupation-numbers of phonon-modes and spin-level (Bruya, 1971; Leonardi and Peisico, 1973).

As a rather remote, but very spectacular example of nonthermal acoustic emission we finally mention the well-known austenite-martensite phase transition in ferrous alloys, which can be looked upon as localized mode-softening (Kayser, 1972). In this connection we should emphasize, that the interesting phenomenon of TO (transverse optic)-mode softening in narrow-gap ferroelectrics such as Lead-Germanium-Telluride, which within the so-called vibronic theory is supposed to be induced by interband e-ph-interaction (Konstin and Kristoffel, 1978), cannot be considered as non-thermal phonon effect the sudden increase in the TO-phonon-number at or near the structural phase-transition is in complete agreement with the thermal Planck distribution.

BIBLIOGRAPHY

Alfaro, R. R., and Shapiro, S. L., 1971, Phys. Rev. Lett. 26:1247.

Arizumi, T., Aoki, T., and Hayakawa, K., 1968, J. Phys. Soc. Jpn. 25:1361.

Arora, V. K., and Miller, S. C., 1974, Phys. Rev. B10:688.

Ascarelli, G., 1960, Phys. Rev. Lett. 5:367.

Bailyn, M., 1958, Phys. Rev. 112:1587.

Bauman, K., Kocevar, P., Kriechbaum, M., Kahlert, H., and Seeger, K. 1968, Proc. 9th Intern. Conf. Phys. Semicond., Moscow, p. 757.

Beattie, A. R., 1978, Proc. 14th Intern. Conf. Phys. Semicond., Inst. of Phys., London, p. 343.

Bodo, Z., Roesner, R., and Sebestyn, T., 1968, Proc. Intern. Conf. Phys. Semicond., Moscow, p. 686.
Bron, W. E., and Grill, W., 1977, Phys. Rev. B16:5303,5315.
Bron, W. E., and Grill. W., 1978, Phys. Rev. Lett. 40:1459.
Bruya, W., 1971, Phys. Rev. B3:635.
Cassiday, D. R., and Spector, H. N., 1974, J. Phys. Chem. Sol. 36:957.
Conwell, E. M., 1964a, J. Phys. Chem. Sol. 25:593.
Conwell, E. M., 1964b, Phys. Rev. 135:A814.
Conwell, E. M., 1967, "High Field Transport in Semiconductors", Academic Press, New York.
Dynes, R. C., and Narayanamurti, V., 1972, Phys. Rev. B6:143.
Dutoit, M., 1971, Phys. Rev. B3:453.
Eisenmenger, W., and Dayem, A. H., 1967, Phys. Rev. Lett. 18:125
Gadzhialiev, M. M., 1977, Sov. Phys. Semicond. 19:711.
Geballe, T. H., and Hull, G. W., 1954, Phys. Rev. 94:1134.
Gorelik, J., Fisher, B., Pratt, B., Luz, Z., and Many, A., 1968, Phys. Lett. 28A:485.
Grill, W., and Weis, O., 1975, Phys. Rev. Lett. 35:588.
Gunn, J. B., 1963, Sol. State Commun. 1:88.
Gurevich, L. E., and Gasymov, T. M., 1967a, Sov. Phys. Sol. State 9:78.
Gurevich, L. E., and Gasymov, T. M., 1967b, Sov. Phys. Semicond. 1:640.
Gurevich, L. E., and Gasymov, T. M., 1968, Sov. Phys. Sol. State 9:2752.
Gurevich, L. E., and Gasymov, T. M., 1969, Sov. Phys. Sol. State 10:2557.
Gurevich, L. E., and Shklovskii, B. I., 1967, Sov. Phys. Sol. State 9:401.
Hart, T. R., Aggarwal, R. L., and Lax, B., 1970, Phys. Rev. B1:638.
Herring, C., 1958, in "Semiconductors and Phosphors," Ed. by M. Schoen and H. Welker, Vieweg, Berlin, p. 184.
Huebner, K., and Shockley, W., 1960, Phys. Rev. Lett. 4:504.
Huet, D., Pannetier, B., Ladan, F. R., and Maneval, J. P., 1976, J. de Phys. 37:521.
Ishida, A., and Inuishi, Y., 1968, Phys. Lett. 27A:443.
Jack, J. W., Barron, H. W. T., and Smith, T., 1973, Phys. Rev. C7:2961.
Jay-Gerin, J. G., 1974, J. Phys. Chem. Sol. 35:81.
Jay-Gerin, J. G., 1975, Phys. Rev. B12:1418.
Kaveh, M., and Wiser, N., 1972, Phys. Rev. Lett. 29:1374.
Kaveh, M., and Wiser, N., 1974, Phys. Lett. 49A:47.
Kayser, U., 1972, J. Phys. F. 2:L60.
Keyes, R. W., 1974, Commun. Sol. State Phys. 6:45.
Kichigan, D. A., Lobachyov, V. P., Mironov, O. A., and Kalitenko, N. G., 1976, Sol. State Commun. 20:645.
Klemens, P. G., 1951a, Proc. Phys. Soc. A64:1030.
Klemens, P. G., 1951b, Proc. Roy. Soc. A208:108.

Klemens, P. G., 1966, Phys. Rev. 148:845.
Klemens, P. G., 1967, J. Appl. Phys. 38:4573.
Kocevar, P., 1972, J. Phys. C 5:3349.
Kocevar, P., 1973, Acta Phys. Austriaca 37:259,270.
Kocevar, P., 1977, Phys. Stat. Sol. (b) 84:681.
Kocevar, P., and Fitz, E., 1978, Phys. Stat. Sol. (b) 89:225.
Konsin, P., and Kristoffel, N., 1978, "Proc. XIV Intern. Conf. Phys Semicond.", Inst. of Phys., London, p. 453.
Landau, L., and Rumer, G., 1937, Phys. Z. Sov. U. 11:18.
Lang, I. G., and Pavlov, S. T., 1973, Sov. Phys. JETP 36:793.
Leonardi, C., and Persico, F., 1973, Phys. Rev. B3:635.
Lifshitz, T. M., Oleinikov, A. Ya., Romanovtsev, V. V., and Shulman, A. Yu., 1972, "Proc. XI Intern. Conf. Phys. Semicond.", Warsaw, p. 608.
Maris, H. J., 1972, J. de Phys. 33:C403.
Maris, H. J., 1964, Phil. Mag. 9:901.
Mattos, J. C. V., and Leite, R. C. C., 1973, Sol. State Commun. 12:465.
McGroddy, J. C., and Christensen, O., 1973, Phys. Rev. B8:5582.
Meneses, E. A., Januzzi, N., Ramos, J. G. P., Luzzi, R., and Leite, R. C. C., 1975, Phys. Rev. B11:2213.
Motsiuke, P., Argüello, C. A., and Leite, R. C. C., 1975, Sol. State Commun. 16:763.
Murayama, Y., 1967, J. Phys. Soc. Jpn. 23:802.
Narayanamurti, V., Chin, M. A., Logan, R. A., and Lax, M., 1978, "Proc. XIV Intern. Conf. Phys. Semicond.", Inst. of Phys., London, p. 215.
Nilsson, G., and Nelin, G., 1971, Phys. Rev. B3:364.
Palik, E. D., and Bray, R., 1971, Phys. Rev. B3:3302.
Paranjape, B. V., in "Cooperative Phenomena," Ed. by H. Haken and M. Wagner, Springer, Berlin.
Paranjape, V. V., and Paranjape, B. V., 1968, Phys. Rev. 166:757.
Parrott, J. E., 1957, Proc. Phys. Soc. 70:590.
Perrin, N., 1975, Sol. State Commun. 17:131.
Perrin, N., and Budd, H., 1972, Phys. Rev. Lett. 28:1701.
Perrin, N., and Budd, H., 1974, Phys. Rev. B9:3454.
Pohl, D., Maier, M., and Kaiser, W., 1968, Phys. Rev. Lett. 20:366.
Polerantz, M., 1965, Phys. Rev. 139:A501.
Pomortsev, R. V., and Kharus, G. I., 1968, Sov. Phys. Sol. State 9:2256.
Puri, S. M., 1965, Phys. Rev. 139:A995.
Reupert, W., Lassman, K., and de Groot, P., 1975, "Proc. Intern. Conf. on Phonon Scattering," Inst. of Phys., Nottingham, p. 315
Samoilovich, A. G., and Buda, I. S., 1976, Sov. Phys. Semicond. 9:977.
Seeger, K., 1973, "Semiconductor Physics," Springer, New York.
Shah, J., 1974, Phys. Rev. B10:3697.
Shah, J., Leheny, R. F., and Dayem, A. H., 1974, Phys. Rev. Lett. 33:818.

Shah, J., Leite, R. C. C., and Scott, J. F., 1970, Sol. State Commun. 8:1089.
Sharma, S. K., and Aggarwal, D. B., 1975, Phys. Stat. Sol. (b) 69:169.
Simons, S., 1967, Proc. Phys. Soc. 83:748.
Simpson, I. C., 1975, J. Phys. C 8:399.
Smith, R. W., 1962, Phys. Rev. Lett. 9:87.
Sommerfeld, A., and Bethe, H., 1933, Handbook der Physik 24:333.
Spector, H. N., 1962, Phys. Rev. 127:1084.
Sussmann, R. S., and Ridley, B. K., 1974, J. Phys. C 7:3941.
White, D. L., 1962, J. Appl. Phys. 33:2547.
Winterling, G., and Heinicke, W., 1968, Phys. Lett. 27A:443.
Woodruff, T. O., and Ehrenreich, H., 1961, Phys. Rev. 123:1553.
Yamashita, J., 1965, Progr. Theor. Phys. 33:343.
Yamashita, J., and Nakamura, N., 1965, Progr. Theor. Phys. 33:1022.
Yamashita, J., and Nakamura, N., 1969, Progr. Theor. Phys. 41:1123.
Yassievich, I. N., and Schleifstein, M., 1970, Phys. Stat. Sol. (b) 42:415.
Yassievich, I. N., and Schleifstein, M., 1972, Phys. Stat. Sol. (b) 51:161.
Ziman, J. M., 1960, "Electrons and Phonons," Oxford Univ. Press, Oxford.
Zylberstejn, A., 1967, Phys. Rev. Lett. 19:838.

NOISE AND DIFFUSION OF HOT CARRIERS

J. P. Nougier

Université des Sciences et Techniques du Languedoc
Centre d'Etudes d'Electronique des Solides
Montpellier, France

INTRODUCTION

Every electrical system biased by a dc voltage or a dc current exhibits fluctuations of the voltage and/or of the current at its terminals. These fluctuations are called the noise. The bias point is determined as the intersection of the characteristics of the system under consideration and of the output circuit. Therefore, the noise depends partly on the inner properties of the system. In the same way, the noise of the system is also governed by its inner properties.

By studying the noise, one gets information on the physical mechanisms responsible for the electrical behaviour. In addition to this fundamental aspect, noise appears as a limiting factor on the sensitivity of devices; hence studying the physical sources of noise allows one to know whether it can be reduced by technological improvements.

In this section, we are interested in studying the noise of semiconductors in the hot carrier regime. Generally, this noise is due to fluctuations of the velocity of the carriers, which governs their diffusion properties. Therefore, noise and diffusion of hot carriers are closely related properties. The purpose of this paper is to explore the connection between these transport coefficients, then to describe briefly the experimental and theoretical techniques allowing one to measure them. Finally, some examples of noise sources are given, to illustrate examples of calculating the noise of a device once the noise sources are known.

FLUCTUATIONS AND NOISE: GENERAL CONSIDERATIONS

Noise Voltage and Noise Current Spectral Densities

Let us consider a one port network biased by a dc voltage V_o with a dc current I_o passing through it. A small sinusoidal voltage of complex value δV, at frequency ν, superimposed on V_o, results in an additional complex current δI. The differenti impedance $Z(V_o,\nu)$ and admittance $Y(V_o,\nu)$ are

$$\delta V = Z(V_o,\nu)\,\delta I \tag{1}$$

$$\delta I = Y(V_o,\nu)\,\delta V \tag{2}$$

$$Y(V_o,\nu) = I/Z(V_o,\nu) \tag{3}$$

If the network is linear, Z and Y do not depend on V_o. This is not the case for example in semiconductor samples or devices operating in the hot carrier range, where Z and Y actually depend on V_o.

Further, if no sinusoidal voltage is superimposed to the dc bias, the voltage at the terminals of the network is indeed not constant, but fluctuates around its average value V_o. It may be written as

$$V(t) = V_o + \Delta V(t), \tag{4}$$

where $\Delta V(t)$ is a small fluctuating voltage of zero average value: $\overline{\Delta V(t)} = 0$. The correlation function of ΔV is given by

$$\Gamma_{\Delta V}(V_o,\tau) = \overline{\Delta V(t) \cdot \Delta V^*(t - \tau)}, \tag{5}$$

where the star denotes the complex conjugate. The associated noise spectral density $\gamma_{\Delta v}$ is the Fourier transform of $\Gamma_{\Delta V}$:

$$\gamma_{\Delta V}(V_o,\nu) = \int_{-\infty}^{+\infty} \Gamma_{\Delta V}(V_o,\tau)\, e^{-2i\pi\nu\tau} d\tau. \tag{6}$$

Further, the noise voltage spectral density $S_V(V_o,f)$ is defined at real frequency $f(f > 0)$ by

$$S_V(V_o f) = \gamma_{\Delta V}(\nu = f) + \gamma_{\Delta V}(\nu = - f). \tag{7}$$

Now, since $\Delta V(t)$ is a real quantity, $\Gamma_{\Delta V}$ is real, and $\gamma_{\Delta V}(\nu) = \gamma_{\Delta V}(-\nu$ so that:

$$S_{\Delta V}(V_o,f) = 4\int_0^{\infty} \Gamma_{\Delta V}(V_o,\tau)\cos 2\pi f\tau . d\tau \tag{8}$$

$$\Gamma_{\Delta V}(V_o,\tau) = \int_0^{\infty} S_{\Delta V}(V_o,f)\cos 2\pi f\tau . df . \tag{9}$$

If one now puts $\tau = 0$ in (5) and (8), one gets

$$\overline{|\Delta V(t)|^2} = \int_0^{\infty} S_{\Delta V}(V_o,f)\, df \tag{10}$$

Since $\overline{|\Delta V(t)|^2}$ is proportional to the average electric power, it follows from (10) that the average power is the sum of the powers $S_{\Delta V}(V_o,f)$ df inside the bandwidth df of the device spectrum. Therefore $S_{\Delta V}(V_o,f)$ is proportional to the average power per unit bandwidth. All of the above discussion for the fluctuation voltage $\Delta V(t)$ can be extended to the fluctuation current $\Delta I(t)$, which gives the noise current spectral density $S_{\Delta I}(V_o,f)$.

A noise is said to be white when the corresponding spectral density does not depend on f, but is constant throughout the bandwidth. The Wiener-Kintchine theorem (Wiener, 1930; Kintchine, 1934) states that, if the Fourier components of two variables are related through an impedance Z, their noise spectral densities are related through $|Z|^2$. Hence, (1) to (3) give

$$S_{\Delta V}(V_o,f) = |Z(V_o,f)|^2\, S_{\Delta I}(V_o,f) \tag{11}$$

$$S_{\Delta I}(V_o,f) = |Y(V_o,f)|^2\, S_{\Delta V}(V_o,f) \tag{12}$$

The Langevin Method

The Langevin (1908) method consists of replacing the real system, having noise, by an ideal system without noise, in series with its noise source (or in parallel with it, if the Norton description is used instead of the Thevenin one). Let X(t) be the quantity to be studied (current, voltage, electric field, carrier's density, etc. . . .). X(t) is the solution of the equation

$$g[\langle X(t)\rangle_{cond}] = 0, \tag{13}$$

where g is some operator. The notation used in (13) means that, in fact, dynamic equations such as (13) involve macroscopic quantities, which are average values (labelled by the brackets) submitted to some initial conditions (indicated by the subscript "cond"). For example, in a circuit composed of a self-inductance L, in parallel with a resistance R, X(t) is the instantaneous current I(t), and (13) becomes

$$R\langle I(t)\rangle_{cond} + L\,\frac{d\langle I(t)\rangle_{cond}}{dt} = 0; \tag{14}$$

the initial condition being for example $\langle i(t = 0)\rangle = i_o$.

If we let X_o be the steady state solution, and $\Delta X(t)$ a small perturbation around X_o, we then have

$$X(t) = X_o + \Delta X(t). \tag{15}$$

Replacing X(t) in (13) by (15), and keeping only the first order terms, leads to an equation of the type

$$\hat{L}\,\langle \Delta X(t)\rangle_{cond} = 0, \tag{16}$$

where $\hat{L}$ is a linear operator.

Langevin said that, in fact, the parameter $\Delta X(t)$ fluctuates around its average value $\langle \Delta X(t)\rangle$, and that this fluctuation is produced by some random noise source $\Delta\Xi(t)$. Therefore, the fluctuating quantity $\Delta X(t)$ is the solution of the equation

$$\hat{L}\,\Delta X(t) = \Delta\Xi(t)\,. \tag{17}$$

Indeed, it can be proved (Van Vliet, 1971) that (16) and (17) are identical provided that $\hat{L}$ is a linear operator, that $\langle \Delta\Xi(t)\rangle = 0$, and that $\Delta\Xi(t)$ is a white noise source. Therefore, the Langevin description does not predict the noise, but relates the fluctuation of the parameter X to the noise source, assumed to be white, responsible of this fluctuation. In order to get the expression for the noise source itself, one must generally consider some microscopic description of the system, and solve a master equation (van der Ziel, 1970).

Instead of considering the time dependence formalism (17), it is easier to get the frequency response. Let us take the Fourier transform of (17). Let $\Delta x(\nu)$ and $\Delta\xi(\nu)$ be the Fourier components of $\Delta X(t)$ and $\Delta\Xi(t)$, respectively, at frequency ν. Since $\exp(2i\pi\nu t)$ is an eigenvector of every linear operator $\hat{L}$, the eigenvalue of whic will be labelled $Z(\nu)$, the Fourier transform of (17) leads to

$$Z(f)\,\Delta x(f) = \Delta\xi(f), \tag{18}$$

and the Wiener-Kintchine theorem then gives

$$|Z(f)|^2 S_{\Delta x}(f) = S_{\Delta\xi}, \tag{19}$$

where $S_{\Delta\xi}$ is frequency independent.

Ohmic Case: The Nyquist and Einstein Relations:

We shall illustrate the above considerations by deriving the Nyquist (1928) relation. Consider an R-L circuit. According to the Langevin theory, the usual equation (14) of the circuit becomes (note that L and $\hat{L}$ are quite different quantities!)

$$R\ \Delta I(t) + L\ \frac{d\ \Delta I(t)}{dt} = \Delta V(t), \tag{20}$$

where $\Delta V(t)$ is the noise source. The Fourier transform of (20) gives, for (18):

$$(R + i\ \omega L)\ \Delta I(\nu) = \Delta V(\nu), \tag{21}$$

and

$$S_{\Delta I}(f) = S_{\Delta V}/(R^2 + L^2\omega^2). \tag{22}$$

By integration of (22) over the frequency range $[0, \infty]$, by (10) the left hand side gives the quadratic current $I^2(t)$. With the Langevin assumption that $S_{\Delta V}$ is a white noise source, the right hand side is readily integrated, and gives

$$\overline{I^2(t)} = S_{\Delta V}/4\ RL. \tag{23}$$

Now, $\overline{I^2(t)}$ is obtained by noting that at thermal equilibrium the average energy of the circuit $\frac{1}{2} L\,\overline{I^2(t)}$ is equal to $\frac{1}{2}k_B T$, where k_B is the Boltzmann constant, which yields the Nyquist relation

$$S_{\Delta V} = 4\ k_B\ R\ T \tag{24}$$

In a series of papers on Brownian motion, Einstein (1905, 1906) established the wellknown relation between the diffusion coefficient D_o, the temperature, and the ohmic mobility μ_o

$$\frac{kT}{q} = \frac{D_o}{\mu_o} \tag{25}$$

This relation holds in thermal equilibrium, and was later shown to be a particular case of the fluctuation-dissipation theorem (Callen and Welton, 1951; Kubo, 1957).

In a recent paper, Van Vliet and Van der Ziel (1977) established a more general relation, also valid for degenerate semiconductors and metals. This equation is

$$q\, D_o \frac{\partial \ln n}{\partial E_F} = \mu_o, \tag{26}$$

where E_F is the Fermi energy.

Although the definitions given above were quite general, the Nyquist and Einstein relations are valid only in thermal equilibrium that is, in the ohmic case. We shall now proceed to a study of the hot carrier regime. Then these relations are no longer valid, since the fluctuation-dissipation theorem fails for non-linear systems (Van Kampen, 1965).

NOISE TEMPERATURES OF HOT CARRIERS

Definition

Let us consider once more a one-port network (say a semiconductor bar), biased by a dc voltage V_o with a dc current I_o passing through it. We first define, around the bias point, the a.c. impedance $Z(V_o,f)$, and the noise voltage and noise current spectral densities $S_{\Delta V}(V_o,f)$ and $S_{\Delta I}(V_o,f)$. Then, in analogy with the Nyquist relation (24), we define (Nougier, 1973; Nougier and Rolland, 1973) the noise temperature $T_n(V_o,f)$ of the network as

$$S_{\Delta V}(V_o,f) = 4\, k_B\, T_n(V_o,f)\ \mathrm{Re}\ \{Z(V_o,f)\}, \tag{27}$$

where Re $\{Z(V_o,f)\}$ is the real part of the impedance $Z(V_o,f)$. According to (11), an equivalent definition of T_n is

$$S_{\Delta I}(V_o,f) = 4\, k_B\, T_n(V_o,f)\ \mathrm{Re}\{Y(V_o,f)\}. \tag{28}$$

It must be noted that: (i) T_n has nothing to do with the electron temperature (at the present state, T_n is merely the definition of some electrical parameter), (ii) All these quantities (impedances and fluctuations) are a.c. values which must be defined around the bias point (V_o, I_o), and once biased, the network "ignores" its situation at zero bias. Therefore, all quantities, other than the differential ones around the bias point, have no physical meaning except for linear networks. In particular the noise temperature, defined using in (27) V_o/I_o instead of $Z(V_o,f) = \delta V/\delta I$, is a parameter which has no physical interest and which cannot be measured for non-linear networks.

It follows from the above definitions that, concerning the noise properties in the bandwidth Δf, any one-port network can be represented by its differential impedance $Z(V_o,f)$ in series with a voltage generator $[S_V(V_o,f)\ \Delta f]^{\frac{1}{2}}$ or in parallel with a current generator $[S_I(V_o,f)\ \Delta f]^{\frac{1}{2}}$, where S_V and S_I can be as well characterized by $T_n(V_o,f)$.

Physical Meaning

When the one-port network of impedance Z is loaded by an output impedance $Z' = R' + i\,X'$, the noise power ΔP dissipated in Z' within the bandwidth Δf around f is

$$\Delta P = R'\ S_{\Delta I}(V_o,f)\ \Delta f.$$

The fluctuating current through Z' is produced by a fluctuating voltage $S_{\Delta V}(V_o,f)$ at the terminals of the network. Therefore, because of (19) and (27),

$$\Delta P = 4\ k_B\ T_n\ \frac{R'\ \mathrm{Re}\ \{Z\}}{|Z + Z'|^2}\ \Delta f.$$

This quantity is maximum when $Z' = Z^* = \mathrm{Re}\{Z\} - i\ \mathrm{Im}\ \{Z\}$, that is when the impedances are matched, and takes the value

$$\Delta P_{max} = k_B\ T_n\ \Delta f, \tag{29}$$

Therefore T_n has a physical meaning and can be measured, since $k_BT_n(V_o,f)$ is the maximum noise power at frequency f per unit bandwidth which can be displayed by the network in an output circuit.

All the above definitions are quite general and apply to every one-port network, whether linear or non-linear, which is biased by a d.c. voltage V_o (which can be of course set equal to zero). In the following we shall mainly restrict ourselves to homogeneous samples of a semiconductor.

Longitudinal and Transverse Noise Temperatures

Let us consider a bar of homogeneous semiconductor, with ohmic contacts at the ends A and B, in a constant electric field E (Fig. 1).

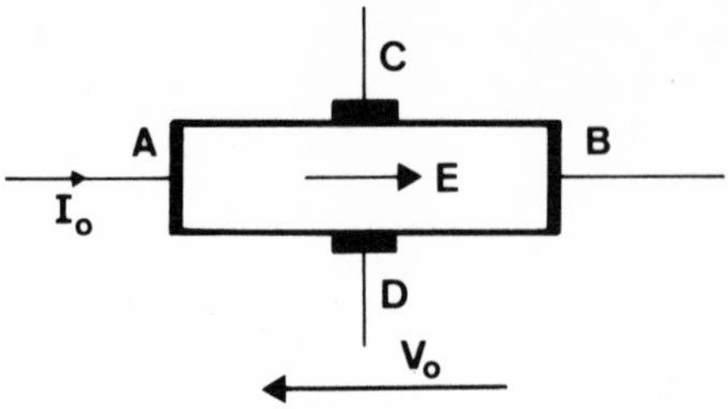

Fig. 1. Geometry of the samples for studying longitudinal and transverse noise temperatures.

The bias V, applied to the arms A and B, fluctuates around its average value V_0. Along the direction A-B, parallel to the electric field, one may define the differential impedance $Z_{\|}(E,f)$, the noise spectral densities $S_{\Delta V\|}(E,f)$ and $S_{\Delta I\|}(E,f)$, and the associate longitudinal noise temperature $T_{n\|}(E,f)$. But, one way also conside transverse quantities, measured between the arms C and D in a direction transverse to the applied electric field. A small transverse sinusoidal voltage $\delta V_{\perp}$ applied to the ends C and D, and superimposed on the longitudinal voltage V_0 along A-B, results in an additional transverse current $\delta I_{\perp}$ flowing from C to D superimposed on the d.c. current I_0 along A-B. This allows one to measure the transverse differential impedance $Z_{\perp}(E,f) = \delta V_{\perp}/\delta I_{\perp}$. The voltage between C and D fluctuates around zero, which allows us to define (and measure a transverse noise voltage or current spectral density $S_{\Delta V\perp}(E,f)$ or $S_{\Delta I\perp}(E,f)$ and the associated transverse noise temperature $T_{n\perp}(E,f) = S_{\Delta V\perp}(E,f)/[4\ k_B\ \mathrm{Re}\ \{Z_{\perp}(E,f)\}]$.

DIFFUSION OF HOT CARRIERS

In this section, we shall first give the definitions of the various diffusion coefficients involved, then establish the link between these coefficients.

Noise Diffusion Coefficients

Let $\vec{v}_i(t)$ be the velocity of the carrier number i at time t, and $\vec{v}_d(\vec{E})$ the average velocity of the carriers, at a point where the electric field is $\vec{E}$. Then

$$\vec{v}_i(t) = \vec{v}_d(\vec{E}) + \Delta\vec{v}_i(t). \tag{30}$$

In analogy with Price (1958), the noise diffusion coefficient $D_{n\alpha\beta}(E)$ may be defined as

$$D_{n\alpha\beta}(E) = \frac{1}{2}\int_{-\infty}^{\infty} \overline{[v_{i\alpha}(t) - \overline{v_{i\alpha}}^{t}][v_{i\beta}(t-\theta) - \overline{v_{i\beta}}^{t}]}^{t} e^{-2i\pi\nu\theta} d\theta, \qquad (31)$$

where the subscripts α and β mean projections of vectors along the directions α and β. The symbol $\overline{(\quad)}^{t}$ means averaging over the time t. Using the ergodic theorem, averaging over the time for a given carrier can be replaced by an average over all the carriers at a given instant t using the distribution function, provided stationarity holds. Therefore $\overline{\vec{v}_i(t)}^{t} = \vec{v}_d$, so that according to (30), (31) becomes

$$D_{n\alpha\beta}(E) = \frac{1}{2}\int_{-\infty}^{+\infty} \langle \Delta v_{i\alpha}(t)\cdot\Delta v_{i\beta}(t-\theta)\rangle\, e^{-2i\pi\nu\theta} d\theta, \qquad (32)$$

where the brackets mean ensemble averages. Therefore, the function under the integral does not depend on the time t. Equivalence between (31) and (32) supposes that the distance travelled by a given carrier between two collisions is much shorter than the distance over which the electric field (or more precisely the drift velocity) varies significantly, so that $\overline{v_{i\alpha}}$ and the correlation functions can be defined everywhere in a unique manner for a given carrier. Of course $D_{n\alpha\beta}(E)$ does not depend on the time, but depends both on the frequency and on the electric field E, since $\vec{v}_d$ depends on $\vec{E}$ in the hot carrier range.

Relation Between the Noise Diffusion and the Noise Temperature

One is now able to establish the link between the noise diffusion coefficients and the noise temperatures. Let us consider a slice of semiconductor, thin enough so that the electric field $\vec{E}$ is uniform. The current produced at time t in the direction α by all of the N carriers inside this slice of thickness L is

$$I_\alpha(t) = \frac{q}{L}\sum_i v_{i\alpha}(t). \qquad (33)$$

The correlation function of the fluctuations of I(t) along the directions α and β is then, in analogy with (5),

$$\Gamma_{\Delta I\alpha\beta}(\theta) = \frac{q^2}{L^2}\sum_i\sum_j \overline{[v_{i\alpha}(t) - \overline{v_{i\alpha}(t)}^{t}][v_{j\beta}(t-\theta) - \overline{v_{j\beta}(t)}^{t}]}^{t}. \qquad (34)$$

Let us now suppose that two different carriers are uncorrelated. This occurs if the carrier density is not too high, that is in the usual operating conditions of solid-state devices. Then

$$\overline{[v_{i\alpha}(t) - \overline{v_{i\alpha}(t)}^{t}][v_{j\beta}(t-\theta) - \overline{v_{j\beta}(t)}^{t}]}^{t}$$

$$= \overline{[v_{i\alpha}(t) - \overline{v_{i\alpha}(t)}^{t}][v_{i\beta}(t-\theta) - \overline{v_{i\beta}(t)}^{t}]}^{t}\delta_{ij}, \qquad (35)$$

where δ_{ij} is the Kronecker Delta. Equation (34) then reduces to

$$\Gamma_{\Delta I\alpha\beta}(\theta) = \frac{q^2}{L^2} N\overline{[v_{i\alpha}(t) - \overline{v_{i\alpha}(t)}^{t}][v_{i\beta}(t-\theta) - \overline{v_{i\beta}(t)}^{t}]}^{t} \qquad (36)$$

The Fourier transform of (36) gives $\gamma_{\Delta I\alpha\beta}(\nu)$, which by comparison with (31) gives

$$\gamma_{\Delta I\alpha\beta}(\nu) = 2\,\frac{q^2 N}{L^2}\,D_{n\alpha\beta}(E).$$

The noise current spectral density is then

$$S_{\Delta I\alpha\beta}(f) = \frac{4\,q^2 n(E)\,A}{L}\,D_{n\alpha\beta}(E), \qquad (37)$$

where A is the cross-sectional area of the sample and n(E) is the carrier density. We now use the definition (28) of the noise temperature to get

$$S_{\Delta I\alpha\beta}(f) = 4\,k_B\,T_{n\alpha\beta}(E,f)\,\mathrm{Re}\{Y_{\alpha\beta}(E,f)\}, \qquad (38)$$

where $Y_{\alpha\beta}$ is the admittance tensor and is related to the mobility tensor $\mu'_{\alpha\beta}$ by

$$Y_{\alpha\beta} = \frac{q\,n(E)\,A}{L}\,\mu'_{\alpha\beta}(E). \qquad (39)$$

Equations (38) and (39) carried in (37) give

$$\frac{k_B\, T_{n\alpha\beta}(E)}{q} = \frac{D_{n\alpha\beta}(E)}{\mathrm{Re}\{\mu'_{\alpha\beta}(E)\}} \; . \tag{40}$$

It is necessary to insist once more on the fact that Y is a differential admittance and therefore $\mu'_{\alpha\beta}$ is the differential mobility tensor defined in the following way: Suppose that a small sinusoidal perturbating field δE_β at frequency f is superimposed along the direction β, on the d.c. field E (the orientation of which is not necessarily along β). Let δv_α be the component along the direction α of the sinusoidal drift velocity $\delta\vec{v}$ produced by δE_β. Then,

$$\mu'_{\alpha\beta} = \delta v_\alpha / \delta E_\beta . \tag{41}$$

At usual frequencies (lower than 10 GHz), $\mu'_{\alpha\beta}$ is real. When α and β are parallel to the d.c. field, one gets the longitudinal quantities

$$\frac{k_B\, T_{n||}(E)}{q} = \frac{D_{n||}(E)}{dv_d/dE} \; . \tag{42}$$

When α and β are perpendicular to the dc field, one gets the transverse quantities

$$\frac{k_B\, T_{n\perp}(E)}{q} = \frac{D_{n\perp}(E)}{dv_\perp/dE_\perp} \; . \tag{43}$$

One may note that for isotropic semiconductors, that is for semiconductors such that $\vec{v}_d(\vec{E})$ is always co-linear to $\vec{E}$, $dv_\perp/dE_\perp = v_d/E$ (Nougier, 1973b), but this is not so for non-isotropic semiconductors, as was experimentally shown (Nougier and Gasquet, 1978).

Another demonstration of (37), from which (40) follows, can be given in order to illustrate the Langevin theory. One first notices that (32) written according to (5) - (7) gives

$$S_{\Delta vi\alpha\beta}(f) = 4\, D_{n\alpha\beta}(E), \tag{44}$$

where $S_{\Delta vi}(f)$ is the noise spectral density of the velocity of a single carrier. Following Langevin, we write the current

$$I = q\, n\, v\, A = \frac{q\, N\, v}{L} \; . \tag{45}$$

Therefore, if the fluctuations of I are due to fluctuations of velocity (diffusion noise),

$$\Delta I = \frac{q\,N}{L}\,\Delta v, \tag{46}$$

where ΔI and Δv can be taken either parallel or perpendicular to E. Then, according to (18) and (19),

$$S_{\Delta I} = \frac{q^2N^2}{L^2}\,S_{\Delta v}. \tag{47}$$

Here, $S_{\Delta v}$ is the fluctuation of the average value of v. It is well-known in statistics that the fluctuation of the average value over N trials is N times smaller than the fluctuation of a single random event, thus

$$S_{\Delta v} = \frac{1}{N}\,S_{\Delta v_i}\,,$$

which gives

$$S_{\Delta I} = \frac{q^2N}{L^2}\,S_{\Delta v_i} = \frac{4\,q^2\,N}{L^2}\,D_n = \frac{4\,q^2\,n\,A}{L}\,D_n.$$

This is the result (37).

Equation (40) looks like the Einstein relation (15). In fact, it is different since the ohmic value D_o is also the diffusion coefficient related to the diffusion current, although at the present stage $D_n(E)$ is related only to the fluctuations of the velocity. Then (40) is a direct consequence of the definitions of T_n and D_n. Equation (40) was established twenty years ago (Price, 1958, 1965) in a sophisticated and rather unclear way, but was recently demonstrated (Nougier, 1973) in a much simpler way using the impedance field method (Shockley, et al., 1966). The two demonstrations give in the present section are quite straightforward, the first having the advantage of showing the microscopic significance of the noise.

Spreading Diffusion Coefficients

When a packet of carriers is injected into the semiconductor, it is drifted by the electric field and spreads because of collisio Let $\vec{r}_i(t)$ be the position of carrier number i at time t, and $\vec{r}_d = \vec{v}_d t$ be the position of the center of the packet:

$$\vec{r}_i(t) = \vec{r}_d + \Delta\vec{r}_i(t). \tag{48}$$

When t is large enough, the variance of $r_i(t)$ becomes proportional to t, the proportionality constant being twice the spreading diffusion coefficient $D_{S\alpha\beta}$. More precisely, $D_{S\alpha\beta}$ is defined as

$$\overline{\Delta r_{i\alpha}(t)\cdot\Delta r_{i\beta}(t)}^{t} = 2\, D_{S\alpha\beta} t. \tag{49}$$

Because of ergodicity, this equation can be written as

$$\langle\Delta r_{i\alpha}(t)\cdot\Delta r_{i\beta}(t)\rangle = 2\, D_{S\alpha\beta} t \tag{50}$$

Current Diffusion Coefficients:

The diffusion current $\vec{j}_D$ is given by

$$\vec{j}_D = -q\,\overleftrightarrow{D}_c\vec{\nabla}n. \tag{51}$$

This defines the current diffusion tensor $\overleftrightarrow{D}_c$. It is a macroscopic and phenomenological expression, which could not have been derived using microscopic considerations. However, by assuming its validity, one can prove (Fawcett, 1973) that $\overleftrightarrow{D}_c$ is identical with $\overleftrightarrow{D}_S$ defined above:

$$\overleftrightarrow{D}_c \equiv \overleftrightarrow{D}_S. \tag{52}$$

To demonstrate this, we start from the kinetic equations

$$\begin{aligned} \vec{j} &= q\,n\,\vec{v}_d - q\,D_c\vec{\nabla}n, \\ \vec{\nabla}\cdot\vec{j} &+ q\frac{\partial n}{\partial t} = 0. \end{aligned} \tag{53}$$

Eliminating $\vec{j}$ between these equations and taking $\vec{E}$ parallel to the z-axis, we obtain

$$\frac{\partial n}{\partial t} = D_{c\perp}\left(\frac{\partial^2 n}{\partial x^2} + \frac{\partial^2 n}{\partial y^2}\right) + D_{c\parallel}\frac{\partial^2 n}{\partial z^2} - v_d\frac{\partial n}{\partial z}, \tag{54}$$

where $D_{c\perp}$ and $D_{c\parallel}$ are the transverse and longitudinal components of $\overleftrightarrow{D}_c$. If the packet of carriers is initially located at $\vec{r} = 0$,

$$n(\vec{r}, t{=}0) = n_0\delta(\vec{r}),$$

and the solution of (54) is

$$n(\vec{r},t) = \frac{n_o}{8(\pi t)^{3/2} D_{c\perp} \sqrt{D_{c\parallel}}} \exp\{-[\frac{x^2+y^2}{4 D_{c\perp} t} + \frac{(z - v_d t)}{4 D_{c\parallel} t}]\}. \quad (55)$$

This shows that $<x^2> = 2 D_{c\perp} t$ and $<(z - v_d t)^2> = 2 D_{c\parallel} t$, which demonstrates (52). Besides the phenomenological aspect of (53), this result neglects the space charge repulsion and supposes that the field is constant everywhere, including the space-charge region

Identity Between Spreading and Noise Diffusion Coefficients

The identity between D_S and D_n for hot carriers was demonstrat very recently (Nougier, 1978). This identity arises from the fact that, neglecting space-charge effects, both spreading and noise result from scattering of the carriers: if there is no scattering, n spreading and no noise occur.

Let us then consider (32), which gives the definition of the noise diffusion coefficient $D_{n\alpha\beta}$. The average value under the integral is the carrier velocity-autocorrelation function, which is non-zero only within a time interval $-\tau \leq \theta \leq \tau$, where τ is the average time between two collisions (average relaxation time). Therefore, the integral actually extends only from $-\tau$ to $+\tau$. Suppose that the frequency ν at which $D_{n\alpha\beta}$ is measured is much lowe than the cutoff frequency corresponding to the different scattering mechanisms involved, that is

$$2\pi\nu\tau << 1, \quad (56)$$

then (32) can then be written

$$D_{n\alpha\beta} = \frac{1}{2} \int_{-\infty}^{+\infty} <[v_\alpha(t) - <v_\alpha>][v_\beta(t-\theta) - <v_\beta>]>d\theta. \quad (57)$$

Equation (57) is the expression of $D_{n\alpha\beta}$ in its white noise spectral range.

In order to identify $D_{n\alpha\beta}$, given by (57), with $D_{S\alpha\beta}$, given by (50), let us consider a carrier at $\vec{r} = 0$ at time $t = 0$. Its departure from its average path at time t, along the direction α, is given by

$$\Delta r_\alpha(t) = \int_0^t [v_\alpha(u) - <v_\alpha>]du, \quad (58)$$

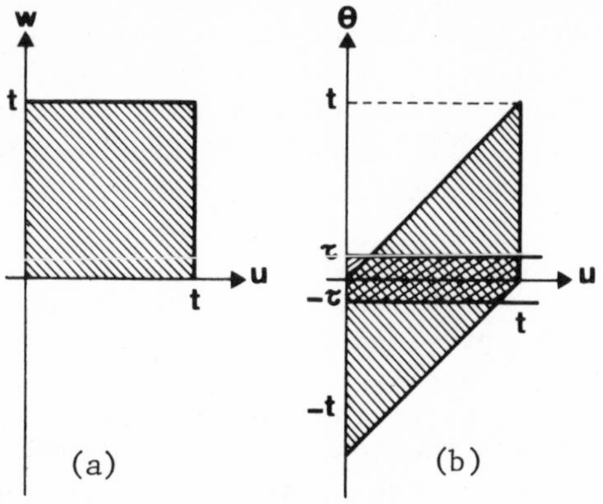

Fig. 2. Integration domain for (59) (a) and (60) (b).

which, upon multiplying by Δr_β (t) and averaging, gives

$$<\Delta r_\alpha(t)\Delta r_\beta(t)> = <\int_0^t \int_0^t [v_\alpha(u) - <v_\alpha>][v_\beta(w) - <v_\beta>]du\ dw>. \quad (59)$$

By changing variables, and substituting $w = u - \theta$, one obtains

$$<\Delta r_\alpha(t)\Delta r_\beta(t)> = \int_0^t du \int_{u-t}^{u} <[v_\alpha(u) - <v_\alpha>][v_\beta(u-\theta) - <v_\beta>]d\theta> \quad (60)$$

The velocity-autocorrelation function appears in (60), and as we said above, the first integral vanishes outside of the interval $[-\tau < \theta < \tau]$. Therefore (see Fig. 2), if the spreading is measured during a time $t >> \tau$, the first integration in (60) can be extended from $-\infty$ to $+\infty$, which by comparison with (57), gives

$$<\Delta r_\alpha(t)\Delta r_\beta(t)> = 2\ D_{n\alpha\beta} t. \quad (61)$$

Comparison of Eqs. (61) and (50) shows that

$$\overleftrightarrow{D}_S = \overleftrightarrow{D}_n. \quad (62)$$

The correct expression for (62), taking into account the above demonstration, should be:

$$\overleftrightarrow{D}_s(\vec{E}, t >> \tau) = \overleftrightarrow{D}_h(\vec{E}, \omega << \tau^{-1}) \quad (62b)$$

Therefore, noise and spreading diffusion coefficients are identical under the three following hypothesis: (i) velocity correlations and spectral densities exist; (ii) the noise diffusion coefficients are measured at a frequency much lower than the cutoff frequency 1/τ (white spectrum); (iii) the spreading diffusion coefficients are measured within a time t much greater than τ.

Note that this demonstration does not involve any assumption concerning either the scattering mechanisms or the form of the distribution function, provided that the three conditions listed above are fulfilled. Therefore, since $\overleftrightarrow{D}_S = \overleftrightarrow{D}_n = \overleftrightarrow{D}_c$, (40) now appears as being analogous to the Einstein relation. Indeed, when the electric field tends towards zero, it follows from the definition (27) that T_n tends towards the thermodynamic temperature T, so that the Einstein relation (25) appears as a particular case, valid in the ohmic regime, of the general relation (40).

The proof that $\overleftrightarrow{D}_n = \overleftrightarrow{D}_S$ allows one to extend the notion of diffusion to trapping mechanisms. In this mechanism, the instantaneous carrier velocity fluctuates from 0 (when the carrier is trapped) to $\vec{v}(t)$ (when it is released). This phenomenon is not essentially different from intravalley optical phonon scattering of carriers with energy equal to the optical phonon energy, and gives the justification for defining the trapping diffusion coefficient used in some noise theories (Rigaud, et al., 1973). Several consequences follow from the demonstration given in this section:

a) Two different techniques are now available to determine the same diffusion coefficient (see section 5 below). One can measure D_S using a time-of-flight technique, and D_n using noise measurements and applying (40).

b) The equivalence between D_S and D_n allow an interpretation, not available in the past, from noise measurements which appears thus as being a powerful tool for both (i) engineering, since it allows characterizing and modeling the noise of devices operating at high electric fields, and (ii) fundamental physics, since the determination of the diffusion coefficients provides new information which can be used in studying the scattering mechanisms and the coupling constants in semiconductors.

c) The noise temperature concept is not only an electrical quantity, but a physical quantity related to the microscopic behaviour of the system. However, T_n is generally different from the electron temperature T_e, defined through the average energy $\langle\varepsilon\rangle = 3/2\ k_B\ T_e$, as was shown from Monte Carlo calculations (Canali et al., 1979), except when the distribution function is a (displaced) Maxwellian (Nougier and Rolland, 1977), which does not occur for most cases of practical interest.

Other Definitions for the Diffusion Coefficients

Of course, these definitions of the various diffusion coefficients are not the only ones available. We may define a generalized diffusion noise coefficient $D_{n\alpha\beta}$ as

$$D_{n\alpha\beta}(E,f) = \frac{1}{N} \sum_i \sum_j \int_0^\infty <\Delta v_{i\alpha}(t)\Delta v_{j\beta}(t-\theta)> e^{-2if\theta\pi} d\theta. \tag{63}$$

This diffusion coefficient, contrary to the previous ones mentioned in this section, is complex and takes into account the cross-correlations between the carriers. The comparison of (63) with the Fourier transform of (34) from $-\infty$ to $+\infty$ shows that

$$S_{\Delta I\alpha\beta}(f) = 4 \frac{q^2 n(E)A}{L} \text{Re}\{D_{n\alpha\beta}(E,f)\}, \tag{64}$$

which is valid even when carrier-carrier scattering is important, contrary to (37). When two different carriers are not correlated, (63) becomes:

$$D_{n\alpha\beta}(E,f) = \hat{D}_{n\alpha\beta}(E,f) = <\int_0^\infty \Delta v_\alpha(t)\Delta v_\alpha(t-\theta)e^{-2i\pi f\theta} d\theta>, \tag{65}$$

where $\hat{D}_{n\alpha\beta}$ is the noise diffusion coefficient defined by Van Vliet and Van der Ziel (1977).

It follows from (5) that $\Gamma(\tau) = \Gamma^*(-\tau)$. Since Δv is real, by changing θ into $-\theta$, and then taking the complex conjugate in (65), one readily gets

$$D_{n\alpha\beta} = \text{Re}\{\hat{D}_{n\alpha\beta}\}, \tag{66}$$

where $D_{n\alpha\beta}$ is the coefficient used throughout this paper.

For modeling devices, several authors use the following expression of the diffusion current D'_c instead of (51)

$$\vec{j}_D = q\, \vec{\nabla}(D'_c\, n). \tag{67}$$

This coefficient D'_c cannot be identified with $\overleftrightarrow{D}_S$ and $\overleftrightarrow{D}_n$, and therefore cannot be measured. Furthermore, since D'_c is a scalar, contrarily to the other ones, (67) predicts that a spherical packet would result in the same current parallel and transverse to the electric field, which is obviously in error since $D_\parallel$ is different from $D_\perp$, as predicted theoretically and verified experimentally

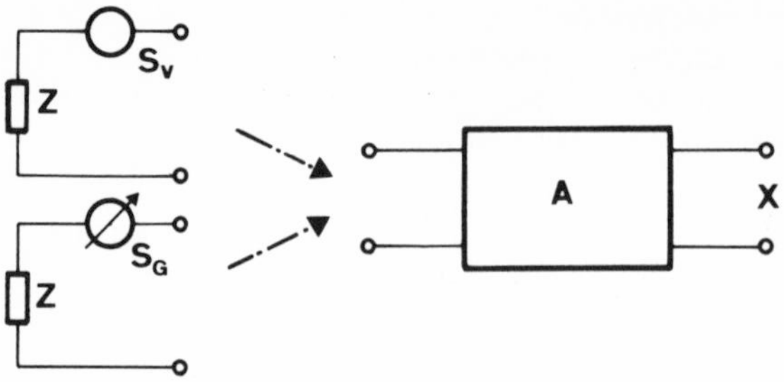

Fig. 3. Ideal technique for noise measurements at high frequencies

(Nougier and Rolland, 1977). Therefore D'_c, as defined by (67), ha no physical meaning and therefore has to be dropped. (See Note 1).

EXPERIMENTAL TECHNIQUES

One of the most important aspects of experiments is the fact that, because of the high electric fields involved, all experiments must be performed in pulses to avoid thermal heating of the devices For example, in semiconductors of usual doping 10^{15} to 10^{16} cm^{-3}, for electric fields higher than 3 kV cm^{-1}, a significant heating occurs during pulses as short as 20 μs and one must use pulses shorter than 5 μs at a repetition rate of one pulse every 10 sec., to avoid heating during one pulse as well as cumulative heating from one pulse to another. Two basic techniques are used: noise measurements, and time-of-flight measurements.

Noise Temperatures Measurements

The measurements must be performed on homogeneous bars with ohmic contacts so that the field remains homogeneous. Because of the short pulses, the noise analysis must be performed at high frequency (say greater than approximately 100 MHz). The usual techniques for noise measurements are then no longer available. For transverse measurements, a cross shaped geometry such as described in Fig. 1 is required. The ideal technique is described in Fig. 3. The device to be studied, with differential impedance Z, is at the input of the apparatus A, the output signal is X. The device is then replaced by a standard noise generator in series with an impedance equal to Z. When the output is X, the noise voltage spectr density S_G of the generator is equal to that S_V of the device. It is absolutely necessary to have the same impedance Z in both

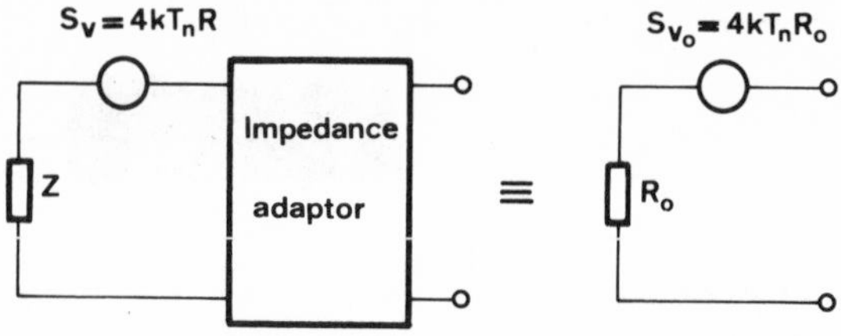

Fig. 4. Transformation of the noise of a one port network through an impedance adaptor.

experiments, since the amplification and the noise of the apparatus strongly depend on the input impedance.

In practice this technique cannot be used since, at high frequencies, the impedance Z of the sample cannot be reproduced and be put in series with a standard noise generator of zero internal impedance. It is then necessary to work at the characteristic impedance R_o (generally 50Ω) of the apparatus, although the impedance of the sample varies according to the dc field applied. Two solutions are then available: the first one is to insert a passive loss-free impedance adaptor between the sample and the apparatus so that the noise power $k_B T_n$ delivered by the sample remains then unchanged (Fig. 4). The second solution consists of inserting a directional coupler at the input of the apparatus, so that the input impedance is always 50Ω. Then, the line between the sample and the coupler presents a reflection coefficient $\rho(\nu)$ and therefore an attentuation $a(\nu)$ $(a > 1)$. The noise temperature T_n of the sample, and $T_{n,app}$ at the input of the apparatus, are related through (Nougier, et al., 1974; Nougier, 1972)

$$T_{n,app} - T = (T_{n,sample} - T)/a(\nu) = (1 - |\rho|^2)(T_{n,sample} - T) \quad (68)$$

To get the desired noise $T_{n,sample}$ from the measured $T_{n,app}$, one needs to measure the losses, that is, the reflection coefficient. Note that the first solution is the particular case where $\rho = 0$.

The apparatus uses a filter, a square-law detector, and an integrator, followed by an analog-digital convertor (counter) to increase the accuracy. The signals after these different steps are shown in Fig. 5. Since the duration t_o of the pulse is short, the output signal X undergoes a statistical fluctuation. It can be shown that the average value $\langle X \rangle$ and the root mean square deviation σ_X of X are given by

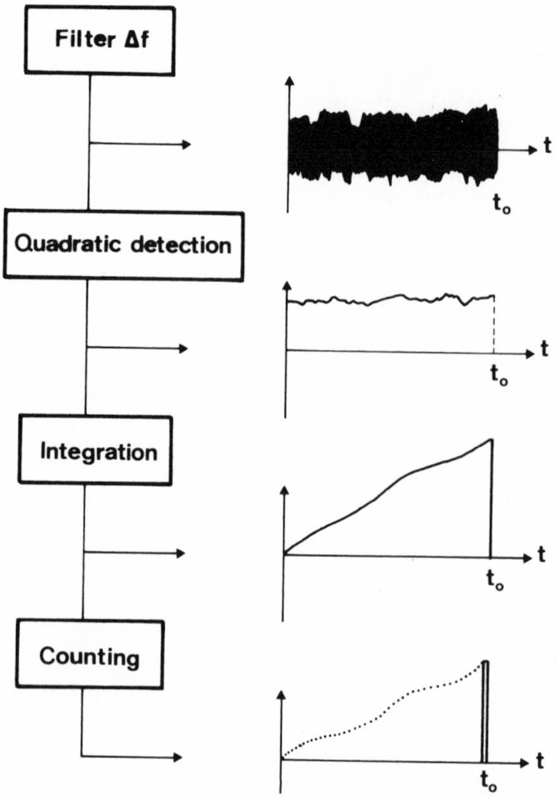

Fig. 5. Signals obtained at the different stages of the apparatus.

$$\langle X \rangle = \lambda(T_A + T_n), \tag{69}$$

$$\sigma_X / \langle X \rangle = (2/N t_o \Delta f)^{\frac{1}{2}}, \tag{70}$$

where λ and T_A are the amplification and the input noise temperature of the apparatus, N is the number of pulses, and Δf is the bandwidth of the filter. Two experiments measuring $\langle X_o \rangle$ and $\langle X_G \rangle$, using the standard noise generator with noise temperatures T and T_G, allow determining λ and T_A, which gives

$$(\langle X \rangle - \langle X_o \rangle)/(\langle X_G \rangle - \langle X_o \rangle) = (T_n - T)/(T_G - T) \tag{71}$$

The measurements performed at the ends A and B (Fig. 1) give $T_{n\|}$, those performed at the ends C and D give $T_{n\perp}$. Subsequent measurements of $\mu'_\|$ and $\mu'_\perp$ lead to $D_\|$ and $D_\perp$. It is necessary to analyze the noise at high frequencies in order to minimize the uncertainty

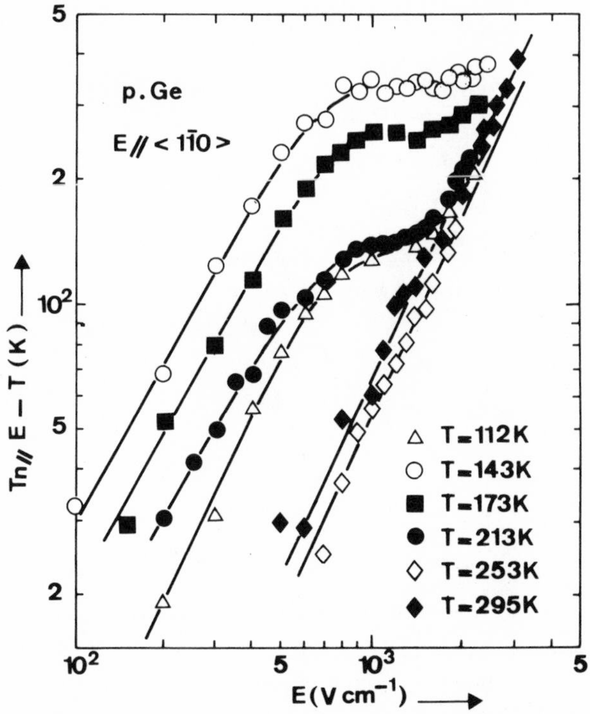

Fig. 6. Experimental excess longitudinal noise temperatures $T_{n\parallel}$ (E) - T versus E in p-Ge 40 cm, from Nougier and Rolland (1973).

and to avoid parasitic effects such as 1/f contact noise, which may occur at quite high frequencies (it has been observed up to 1 GHz).

The first measurements of noise temperatures were performed by Erlbach and Gunn (1962a, 1962b). The subsequent ones (Nougier and Rolland, 1973; Erlbach, 1963; Bareikis, et al., 1968; Pozhela, et al., 1968, Hart, 1970, Nougier and Rolland, 1978) exhibit important general features illustrated in Figs. 6-13. As shown on Fig. 6, the excess noise temperature varies strongly with the lattice temperature. The transverse excess noise temperature is lower than the longitudinal one (see Fig. 7), and both vary in the same way with respect to the electric field. As a consequence, the longitudinal and transverse diffusion coefficients are generally different (Fig. 8). The noise increases with increasing impurity concentration as is clearly shown in Figs. 9 and 10, and for a given doping the noise depends on the orientation of the electric field (Fig. 11). As a consequence, the diffusion coefficients depend both on the

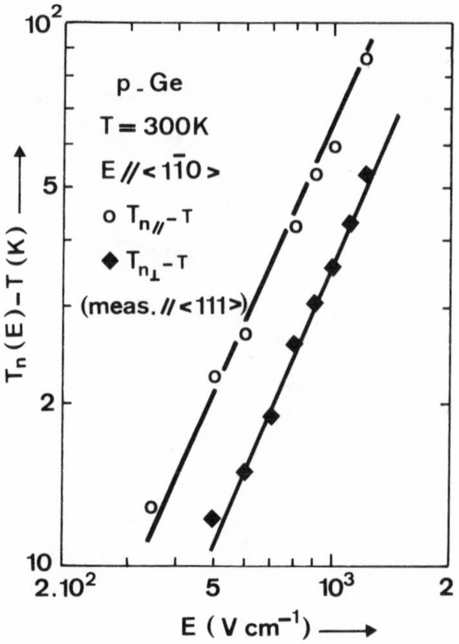

Fig. 7. Comparison between experimental longitudinal and transverse excess noise temperatures in p-Ge, 1Ωcm, from Nougier and Rolland (1973).

impurity concentration (Fig. 12) and on the orientation (Fig. 13). Another way to determine D(E) by noise techniques is to perform measurements using actual devices.

The Time-of-Flight (TOF) Technique

In order to measure $D_{||}$ (E), a packet of carriers is injected at one electrode of a semiconductor sample, using for example a laser, or an electron gun, etc. . . . One type of carrier recombines at the electrode, the other type drifts towards the opposite electrode and the packet spreads because of collisions, as it moves through the sample. The time between the injection and the collection of the pulse gives the drift velocity, and the shape of the current collected is related to the diffusion coefficient (Fig. 14).

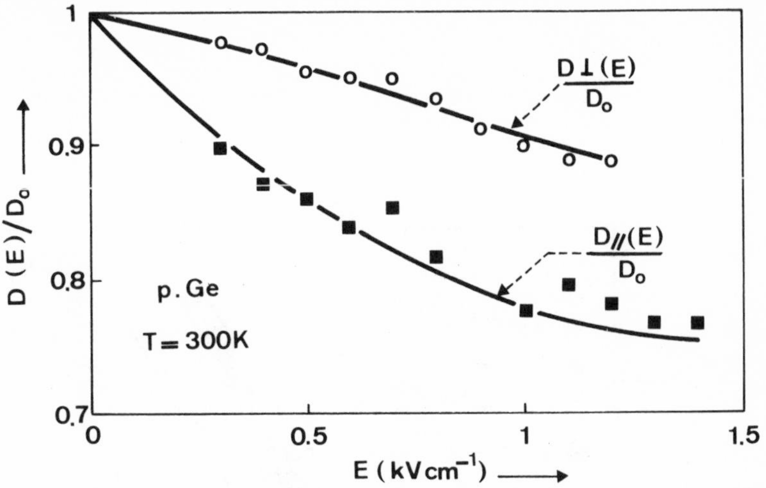

Fig. 8. Comparison between experimental $D_{||}$ (E) and $D_{\perp}$(E) in p-Ge, 40Ωcm, from Nougier and Rolland (1973). $\vec{E}$//<110>. Transverse measurement parallel to <111>.

The drift velocity v_d is obtained by $v_d = L/t$ where L is the length of the sample and t the transit time (duration of the pulse at half height). The longitudinal diffusion coefficient is given by (Ruch and Kinto, 1968; Sigmon and Gibbons, 1969)

$$D_{||}\,(E) = (\tau_F^2 - \tau_R^2)\; v_d^3/21.6\; L, \tag{72}$$

where τ_F and τ_R are the fall time and the rise time of the current, defined at 95% of its maximum amplitude, and the factor 21.6 results from the Maxwellian shape of the packet of carriers.

Because the number of injected carriers is relatively small, one must use very high purity samples, with one rectifying contact at a sufficiently high bias so as to create a depleted zone in the whole sample in order to get a constant electric field. Furthermore, very fast electronic circuits must be used (Canali, et al., 1975a). In order to measure $D_{\perp}$(E), the injection is localized in a very small area of about 1 μm diameter (Fig. 15). $D_{\perp}$(E) is then deduced (Bartelink and Persky, 1970; Persky and Bartelink, 1971) from the comparison of the radius r_o of the injected packet with

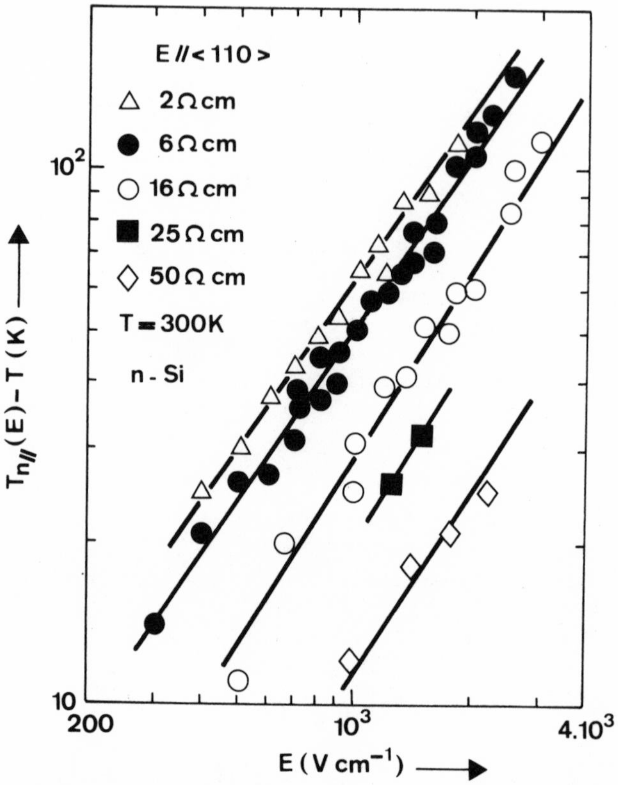

Fig. 9. Effect of the impurity concentration on the longitudinal excess noise temperature in n-Si, after Nougier and Rolland (1976).

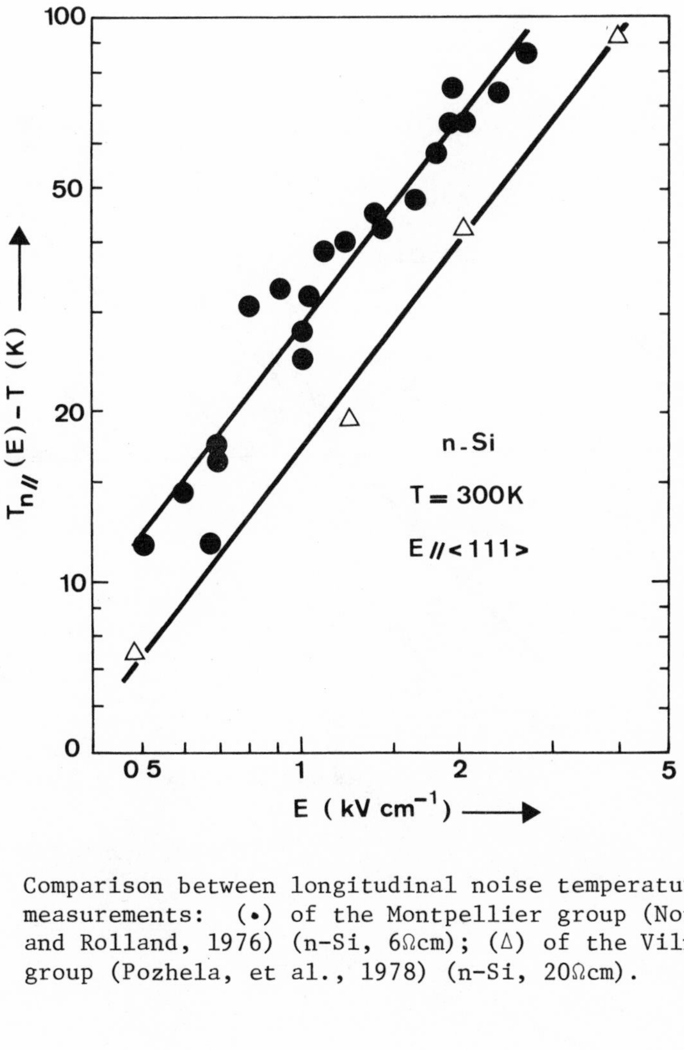

Fig. 10. Comparison between longitudinal noise temperature measurements: (•) of the Montpellier group (Nougier and Rolland, 1976) (n-Si, 6Ωcm); (Δ) of the Vilnius group (Pozhela, et al., 1978) (n-Si, 20Ωcm).

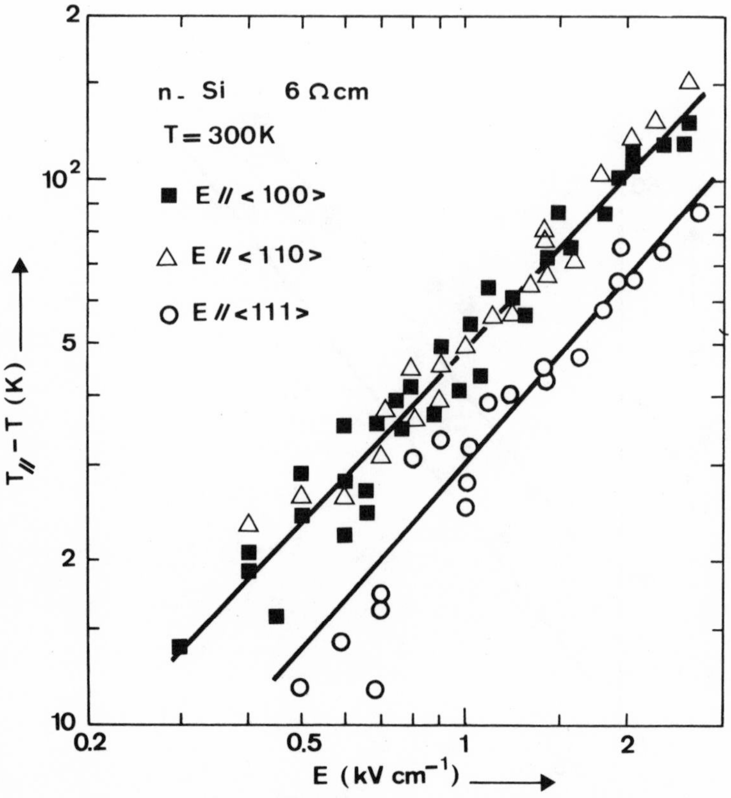

Fig. 11. Anisotropy of the excess longitudinal noise temperature in n-Si, after Nougier and Rolland (1976).

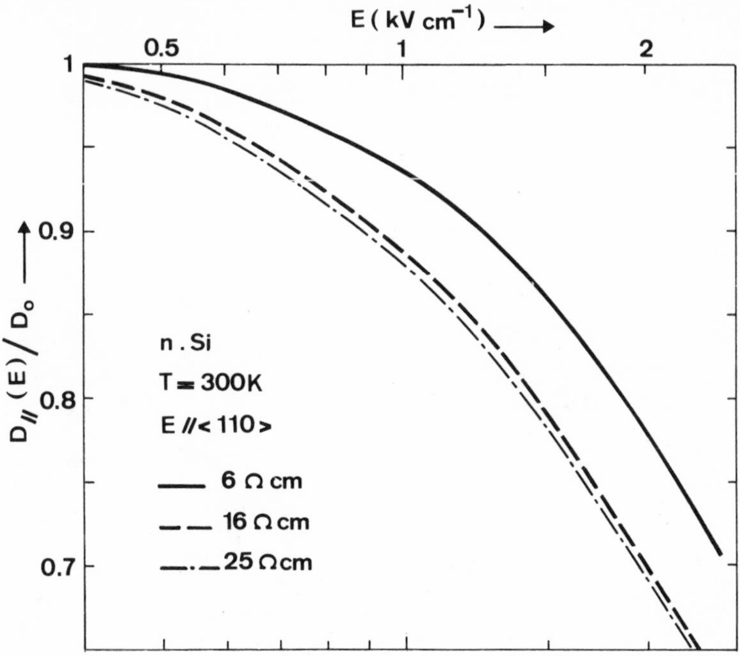

Fig. 12. Effect of the impurity concentration on the longitudinal diffusion coefficient in n-Si, after Nougier and Rolland (1976).

the quantity $r^2{}_f = w^2/2 \ \ell n2$ where w is the full-width at the half-maximum of the current, determined by a probe:

$$D_\perp(E) = (r^2{}_f - r_o{}^2)v_d/4\ L.$$

The accuracy of transverse diffusion measurements using the TOF technique is not good, and only the two experiments mentioned above have been reported in the literature. Conversely, longitudinal measurements have received increasing importance because of the reliability of the experiments of the Modena group (Canali, et al., 1975a, 1975b, 1975c, 1976). Some illustrative results are given in Figs. 16 to 19. The important result outlined by these experiments is that, generally $D_{||}$ (E) first decreases with increasing E, and then reaches a plateau at high fields. This is evidenced in n-Si (Fig. 16) and in p-Ge (Fig. 17). This also holds for p-Si (see Fig. 19, taken from recent results of the Montpellier and Modena groups (Reggiani, et al., 1979) in contradiction with previous

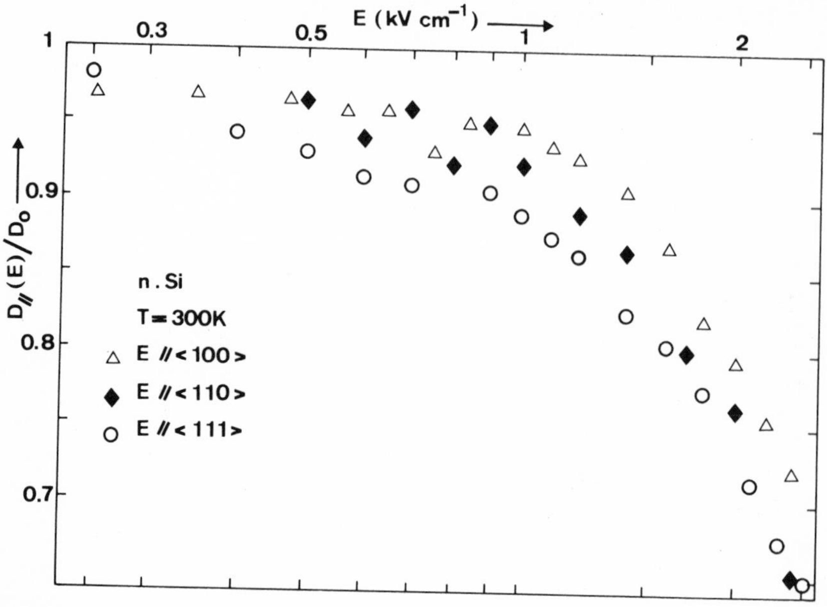

Fig. 13. Anisotropy of the longitudinal diffusion coefficient in n-Si, after Nougier and Rolland (1976).

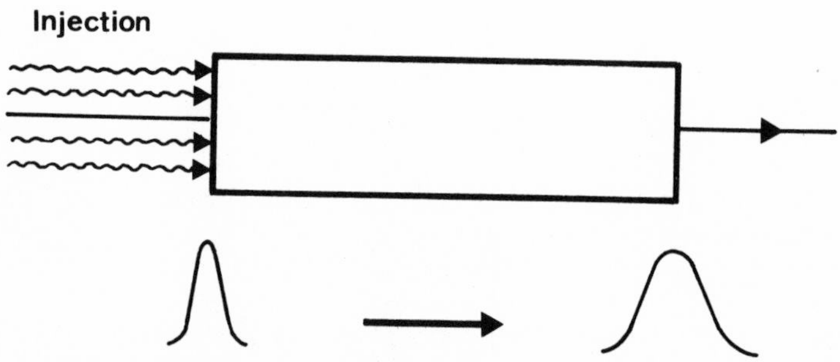

Fig. 14. Principle of the time of flight technique for longitudinal measurement, showing the flight of a packet of carriers through a sample.

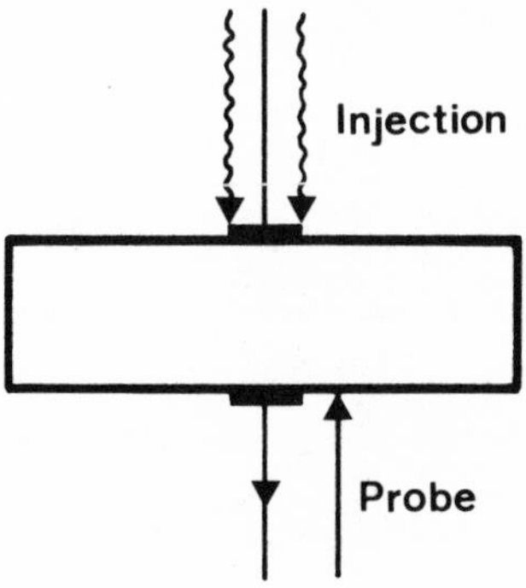

Fig. 15. Time of flight technique for transverse diffusion measurements.

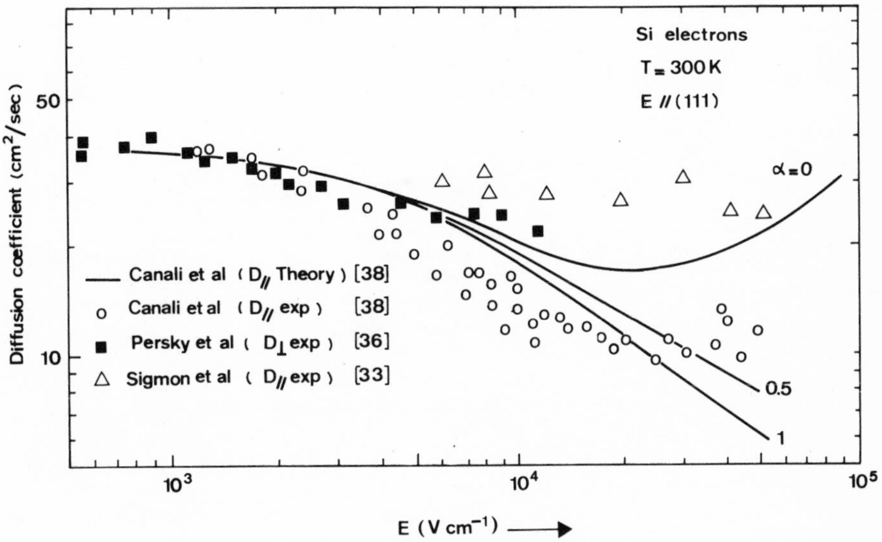

Fig. 16. Comparison between experimental and theoretical determinations of longitudinal diffusion coefficients in n-Si, after Canali, et al. (1975).

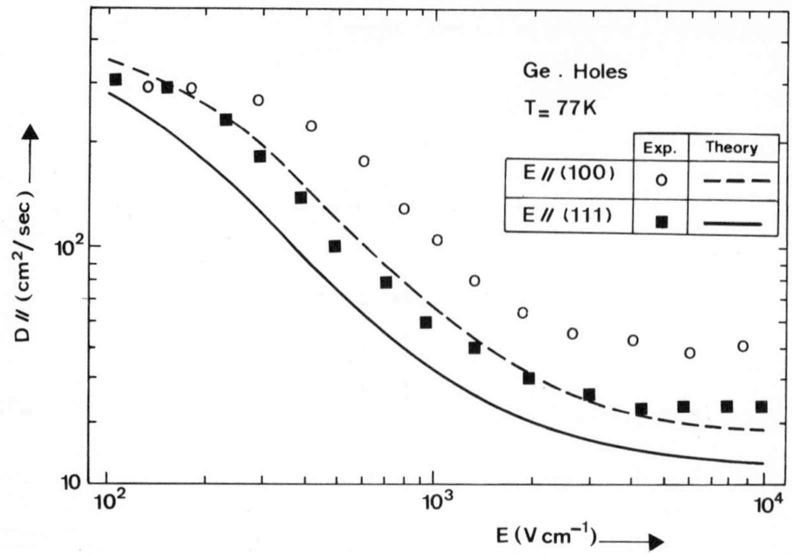

Fig. 17. Theoretical and experimental longitudinal diffusion coefficients in p-Ge, after Canali, et al. (1976).

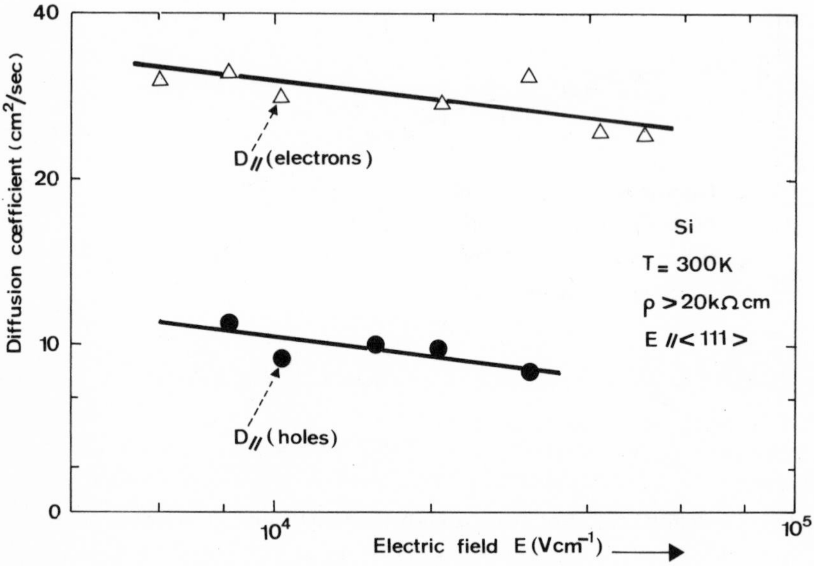

Fig. 18. Experimental $D_{||}$ (E) of electrons and holes in Si, after Sigmon and Gibbons (1969).

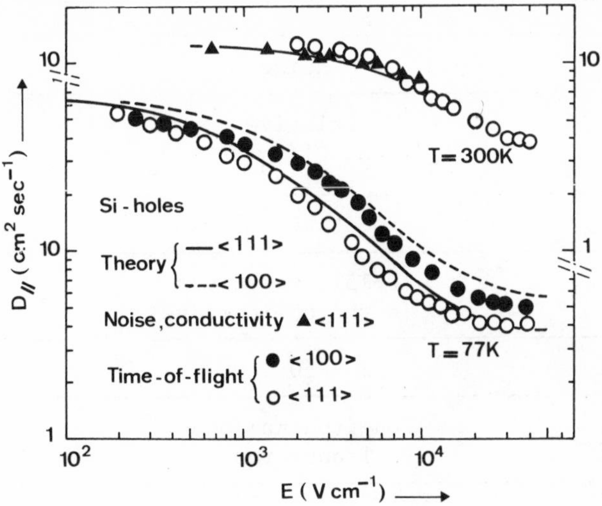

Fig. 19. Theoretical and experimental longitudinal diffusion coefficients of hot holes in silicon (after Reggiani, Canali, Nava and Vaissière, Gasquet, Nougier, unpublished work).

results obtained in earlier experiments (see Fig. 18). GaAs exhibits a particular behaviour: $D_{||}$ (E) increases and reaches a maximum value around the field at which oscillations may occur, and then decreases (Fig. 20). Such a situation also occurs in CdTe. One should note that the anisotropy appears in all semiconductors at high enough fields.

Comparison Between Noise and Time of Flight Techniques:

The main characteristics of these two techniques, at this time, are summarized in Table 1, where the values indicated must not be considered as precise values, but rather as order-of-magnitude, which may be modified according to the evolution of the techniques. Table 1 clearly shows that these two techniques are quite complementary, so that it is difficult to compare experimentally D_s and D_n. However, in the small range where the experimental data overlap (Nougier and Rolland, 1976, Canali et al., 1975a), the agreement is quite satisfactory, as is shown in Fig. 19.

Table 1. Comparison Between Noise and TOF Techniques

	Noise	Time of flight
Type of measurement	Relative T_n/T and $\mu'/\mu_o \leftrightarrow D(E)/D_o$	Absolute $v_d(E)$, $D(E)$
Range of E	0 - 10 kV cm^{-1}	0 - 50 kV cm^{-1}
Accuracy	10-15%, very good in the ohmic range	20 - 25%
Resistivities	2 - 50 Ωcm	> 50 Ωcm
Direction of measurement	Longitudinal and Transverse	Longitudinal

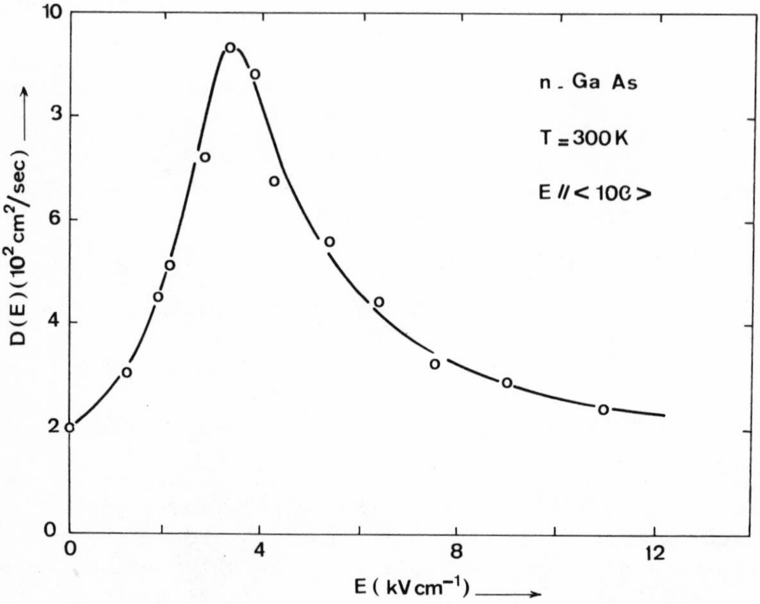

Fig. 20. Longitudinal diffusion coefficient in GaAs, after Ruch and Kino (1968).

THEORETICAL DETERMINATIONS

The Monte Carlo Technique

In the Monte Carlo technique [see for example the review papers by Fawcett (1973) and Alberigi-Quaranta, et al., (1971)], the random motion of one carrier is described in $\vec{k}$-space by a great number of sequences, each of them including events simulated by random numbers: one free flight, one collision and one transition due to the collisions. The wave vector $\vec{k}$ of the carrier is then known at every instant of time, and therefore the position is obtained by simple integration. The distance L_d travelled during the time t_d corresponding to many sequences (5000 to 50000) gives the drift velocity $v_d = L_d/t_d$. One gets then the average position and the deviation with respect to this position, which gives the various transport coefficients, among them the spreading diffusion coefficients through the variance $<\Delta r_\alpha(t)\cdot\Delta r_\beta(t)>$. The only fitting parameters are the coupling constants involved in the transition rates associated with the scattering mechanisms. Figures 16, 17 and 19 show how good is the fit between theory and experiment. Since $\vec{k}$, hence $\vec{v}$, is known at every instant, it is possible to calculate the velocity correlation function and to derive the noise diffusion coefficient D^n (Zimmerman, et al., 1978) which is in good agreement with D_S, according to the theory developed by Nougier (1978). (See Note 2).

Iterative Methods

Contrary to the Monte Carlo method which simulates the flight of one carrier, thus reconstructing the distribution function after a great number of simulations, the iterative methods consist in a numerical solution of the entire distribution function. For this purpose (Fawcett, 1973) the integro-differential Boltzmann equation is often transformed into an integral equation which is then solved using an iterative method. The Boltzmann equation is

$$\frac{\partial f(\vec{k},\vec{E},t)}{\partial t} + \frac{e}{\hbar}\vec{E}\cdot\vec{\nabla}_k f(\vec{k},\vec{E},t) = \int f(\vec{k}',\vec{E},t)P(\vec{k}',\vec{k})d^3k' - \frac{f(\vec{k},\vec{E},t)}{\tau(\vec{k})}, \qquad (73)$$

where $P(\vec{k}',\vec{k})$ is the transition rate corresponding to the scattering mechanisms and $\tau(k)$ is the usual relaxation time defined as

$$1/\tau(\vec{k}) = \int P(\vec{k},\vec{k}')\, d^3k'. \qquad (74)$$

Obtaining the diffusion coefficients, using the distribution function, is not so straightforward as using the Monte Carlo technique. It is necessary to define a differential relaxation time $\tau(\vec{k},\vec{E})$ as (Nougier and Rolland, 1973, 1977)

$$\lim_{\delta\vec{E}\to 0}\left[\frac{f(\vec{k},\vec{E}+\delta\vec{E},t)}{\delta t}\right]_{t=0} = \lim_{\delta\vec{E}\to 0}\left[\frac{f(\vec{k},\vec{E}) - f(\vec{k},\vec{E}+\delta\vec{E})}{\tau(\vec{k},\vec{E})}\right], \quad (75)$$

where $\delta\vec{E}$ is a perturbating field superimposed on $\vec{E}$, $f(\vec{k},\vec{E})$ and $f(\vec{k},\vec{E}+\delta\vec{E})$ being the stationary distribution functions in the fields $\vec{E}$ and $\vec{E}+\delta\vec{E}$. It can be shown that $\tau(\vec{k},\vec{E})$ defined using (75) depends on the direction of the vanishing $\delta\vec{E}$ with respect to $\vec{E}$, and this gives $\tau_{||}(\vec{k},\vec{E})$ when $\delta\vec{E}//\vec{E}$ and $\tau_{\perp}(\vec{k},\vec{E})$ when $\delta\vec{E}\perp\vec{E}$. Indeed $\tau_{||}(\vec{k},\vec{E})$ and $\tau_{\perp}(\vec{k},\vec{E})$ are not identical although being of the same order of magnitude. One then defines

$$D_{||}(E) = \langle\tau_{||}(\vec{k},\vec{E})[v_{||} - v_d(\vec{E})]^2\rangle , \quad (76)$$

$$D_{\perp}(E) = \langle\tau_{\perp}(\vec{k},\vec{E})\, v_{\perp}^2\rangle , \quad (77)$$

where the brackets mean ensemble averages taken over the stationary distribution function $f(\vec{k},\vec{E})$, $v_{||}$ and $v_{\perp}$ are the components, along directions parallel and perpendicular to $\vec{E}$, of the velocity associated with the $\vec{k}$ vector of a carrier. Figure 21, taken from Nougier and Rolland (1977), shows a comparison between the experimental $D_{||}(E)$ and the theoretical ones computed using eq. (76), where τ was taken equal to $\tau(\vec{k})$, $\tau_{||}(\vec{k},\vec{E})$, and $\tau_{\perp}(\vec{k},\vec{E})$; Fig. 21 obviously shows that the agreement is quite good using $\tau_{||}(\vec{k},\vec{E})$. Analogous results were obtained for $D_{\perp}(E)$ which showed the necessity to use there $\tau_{\perp}(\vec{k},\vec{E})$ as indicated in (77).

It must be noted that (76) and (77) have not yet been demonstrated in a rigorous way, so that the agreement which occurred with the experiments performed does not ensure that these are applicable in every case. Besides being necessary for computing $D_{||}$ and $D_{\perp}$, the definitions of $\tau_{||}(\vec{k},\vec{E})$ and $\tau_{\perp}(\vec{k},\vec{E})$ are interesting in themselve since they allow one to demonstrate that the distribution function may be reduced to a displaced Maxwellian only when $\tau(\vec{k})$ defined in (74) does not depend on k, which occurs if the scattering mechanism involve carrier-carrier and/or neutral impurity interactions. In that case, using (76), (77), and (40) lead to the result that $T_n = T_e$.

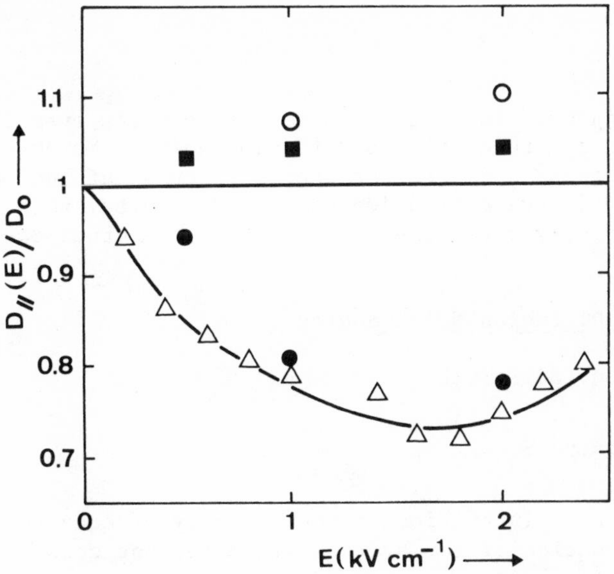

Fig. 21. Comparison between experimental and theoretical longitudinal diffusion coefficients in p-Ge at 300 K, after Nougier and Rolland (1977): Δ experiment ($\vec{E}$//<111>, $N_A - N_D = 7.5 \times 10^{13}\ cm^{-3}$); o theory using $\tau(\vec{k})$ given by eq. (6.2); ● theory using $\tau_{||}(\vec{k},\vec{E})$, see eq. (6.4); ■ theory using $\tau_{\perp}(\vec{k},\vec{E})$ instead of $\tau_{||}(\vec{k},\vec{E})$ in eq. (76).

NOISE SOURCES

One generally supposes that there is no spatial correlation between the microscopic noise sources. Let $S_{\Delta\vec{j}}(f)$ be the local fluctuation of the current density $\vec{j}$. If $\Delta\vec{j}$ is the sinusoidal component of $\vec{j}$, one takes

$$S_{\Delta j}(\vec{r},\vec{r}',f)\Delta f = \langle \Delta\vec{j}(\vec{r}) \cdot \Delta\vec{j}(\vec{r}')\rangle \Delta f. \tag{78}$$

The noise source $K(\vec{r},f)$ is defined at point $\vec{r}$ as (Van Vliet, et al., 1975a, 1975b)

$$S_{\Delta j}(\vec{r},\vec{r}'f)\Delta f = K(\vec{r})\delta(\vec{r}-\vec{r}')\Delta f. \tag{79}$$

The expression for the drift current

$$\vec{j}_d = q\, n\, \vec{v} \tag{80}$$

shows that $\vec{j}_d$ may fluctuate because $\vec{v}$ fluctuates even if n remains constant and one then obtains diffusion noise. Moreover, $\vec{j}_d$ may fluctuate because n fluctuates when $\vec{v}$ is constant and one then obtains generation-recombination noise. Of course, both are actually present, but very often one noise cancels the other one.

Hot Carrier Diffusion Noise Source

Equation (80) gives

$$S_{\Delta j} = q^2\, n^2\, S_{\Delta v}. \tag{81}$$

Considering the fluctuation of the velocity of each carrier, instead of the fluctuation of the average velocity, one gets

$$S_{\Delta j} = q^2\, n^2\, S_{\Delta v_i}/n = 4\, q^2\, n\, D.$$

Therefore, according to (79), the hot carrier diffusion noise source is

$$K(\vec{r}) = 4\, q^2\, n(\vec{r})\, D(\vec{r}), \tag{82}$$

where $D(\vec{r})$ is determined either experimentally, or theoretically using microscopic equations described above.

Hot Carrier Generation-Recombination (G-R) Noise Source

Now, according to (80), we have

$$S_{\Delta j} = q^2\, v^2\, S_{\Delta n},$$

which gives

$$K(\vec{r}) = q^2\, v^2\, S_{\Delta n}\, (\vec{r}), \tag{83}$$

or, according to (80),

$$K(\vec{r}) = \frac{j^2(\vec{r})}{n^2(\vec{r})} S_{\Delta n}(\vec{r}). \tag{84}$$

This expression gives hot carrier G-R noise since the various quantities involved in (84) depend on the electric field at high bias. Anyway, the noise spectral density for G-R noise is always proportional to the square of the current flowing through the sample.

Once more, $S_{\Delta n}$ must be deduced from microscopic considerations. For example, if the fluctuation of n is due to trapping on a single energy level, with a lifetime τ, one gets (Van der Ziel, 1970) for the fluctuation of the total number inside the sample of valume AL

$$N = AL\, n, \tag{85}$$

and

$$S_{\Delta N} = A^2L^2\, S_{\Delta n} = 4\langle\Delta N^2\rangle \frac{\tau}{1+\omega^2\tau^2} = 4\alpha N \frac{\tau}{(1+\omega^2\tau^2)},$$

where α is related to the generation and recombination rates. This gives

$$S_{\Delta n} = \frac{4\alpha n}{AL} \frac{\tau}{(1+\omega^2\tau^2)}, \tag{86}$$

so that

$$K(\vec{r}) = 4 \frac{j^2(\vec{r})}{A\ L\ n(\vec{r})} \frac{\alpha\tau}{(1+\omega^2\tau^2)}. \tag{87}$$

Both coefficients α and τ depend on the electric field, as was experimentally evidenced (Nougier, et al., 1978). An example of such a situation occurs in n-type silicon at 77 K: at zero bias, only part of the donors are thermally ionized (Nash and Holm-Kennedy, 1974, 1977). The electric field produces ionization of the neutral impurities through two mechanisms: impact ionization due to free-carriers with high energy, and the lowering of the potential barrier due to the electric field dependent Poole-Frenkel effect, which allows a supplementary thermal ionization. The electric field induced impurity ionization causes n to then fluctuate between the conduction band and the donor energy level, thus producing hot carrier G-R noise which competes with, and probably overcomes, the hot carrier diffusion noise. In fact the Poole-Frenkel effect alone could perhaps explain the high noise level observed (Van der Ziel, et al., 1979).

When several levels are involved, in the G-R process, the expression is much more complicated, because many time constants occur, and because the process is then non-Markovian. In particular the frequency dependence is not as simple. The case of 1/f noise, occuring either in the interfaces or the bulk, may appear as a particular case of (84) where $S_{\Delta n}$ takes the form

$$S_{\Delta n} = \frac{\alpha' \, n}{f} , \tag{88}$$

which gives for (84) (Zijlstra, 1978)

$$K_{1/f}(\vec{r}) = \frac{\alpha' \, j^2(\vec{r})}{f \; n(\vec{r})} . \tag{89}$$

However the 1/f dependence of this noise has not yet been explained in a satisfactory way.

NOISE OF HOT CARRIERS IN DEVICES

The most efficient method for computing the noise in devices is the transfer impedance method recently established by Van Vliet, et al., (1975a, 1975b). One writes the transport equations and then eliminates the auxiliary variables, so as to obtain a relation between the electric field $\vec{E}(\vec{r})$ and the current density $\vec{j}(\vec{r})$ at point $\vec{r}$ inside the device. One then applies a perturbation $\delta\vec{E}\, e^{i\omega t}$, from which results $\delta\vec{j}\, e^{i\omega t}$. The zero-order terms give the static distribution, while the first-order terms give a linear equation of the Langevin type

$$\hat{L} \; \delta\vec{E}(\vec{r},\omega) = \vec{\delta}j(\vec{r},\omega) . \tag{90}$$

By inversion of the operator $\hat{L}$, one obtains the solution of this equation, which can be written as

$$\vec{\delta}E(\vec{r},\omega) = \iiint \vec{\vec{z}}(\vec{r},\vec{r}\,',\omega) \; \delta\vec{j}(r',\omega) \; d^3r' . \tag{91}$$

The kernel $\vec{\vec{z}}(\vec{r},\vec{r}\,',\omega)$ of this integral is a propagator, which describes the influence at point $\vec{r}$ of the current density at point $\vec{r}\,'$: it is the Green's function of the operator $\hat{L}$ and is called the transfer impedance operator. Integration of (91) from one electrode to the other gives the voltage perturbation as a function of the current perturbation, that is the differential impedance around the bias point. The impedance field $\nabla\vec{Z}(\vec{r}\,',\omega)$ is given by

$$\nabla\vec{Z}(\vec{r}',\omega) = -\int_0^L \vec{dr}^{\dagger}\vec{\vec{z}}(\vec{r},\vec{r}',\omega), \tag{92}$$

where $\vec{dr}^{\dagger} = \{dx, dy, dz\}$ is the hermitian conjugate of the column vector $\vec{dr}$. The differential impedance is then

$$Z(\omega) = \int_0^L \int_0^L \vec{dr}^{\dagger}\vec{\vec{z}}(\vec{r},\vec{r}',\omega)\vec{dr}'. \tag{93}$$

The noise voltage spectral density S_V between these two electrodes is then given by:

$$S_V(\omega) = \int_0^L \int_0^L \vec{dr} \iiint d^3r''\vec{\vec{z}}(\vec{r},\vec{r}'',\omega)\ K(\vec{r}'',\omega)\vec{\vec{z}}^{\dagger}(\vec{r}',\vec{r}'',\omega)\vec{dr}', \tag{94}$$

where $K(\vec{r}'',\omega)$ is the local noise source defined by (78) and (79).

For one-dimensional devices, and for diffusion noise, taking into account (82), (94) reduces to

$$S_V(\omega) = 4\ q^2\ A\int_0^L n(x)\ D(x)\left|\frac{\partial Z(x)}{dx}\right|^2\ dx. \tag{95}$$

This is the expression of $S_V(\omega)$ given by the impedance field method of Shockley, et al., (1966). The transfer impedance method has been applied for calculating the noise of various devices (Van Vliet, et al., 1975b), and is also applicable for devices working in hot carrier regime (Van Vliet, et al., 1975b; Nougier, et al., 1978). One notes that this method takes into account the influence of each slice on the other slices of the device. In contrast, in the "salami" method, one considers that the slices are completely independent. If $\Delta V_i(t)$ is the voltage variation across the ith slice of a one-dimensional sample, the total voltage variation of the sample is $\Delta V(t) = \sum_i \Delta V_i(t)$. If one takes the correlation function of ΔV, then its Fourier transform, assuming that two different slices are uncorrelated, yields

$$S_{\Delta V} = \int_0^L dS_{\Delta V}, \tag{96}$$

where $S_{\Delta V}$ is the noise voltage spectral density of the sample, $dS_{\Delta V}$ being that of the slice of thickness dx at abscissa x. Thus, using the definitions of the local noise temperature T_n and of the total noise temperature T_{Σ} according to (30), one readily gets (Nougier, 1973)

$$T_\Sigma = \frac{1}{R}\int_0^L T_n \, dR, \tag{97}$$

where dR and R are the real parts of the impedance of the slice dx and of the entire sample

$$dR = - dx \; \mathrm{Re} \; \{\frac{\delta E(x,I)}{\delta I}\}. \tag{98}$$

Because the mutual influence between slices is neglected, (97) is obviously wrong and must be considered as an approximation. Indeed, it can be shown that (98) is valid provided that both the cross section A and the carrier density n at point x do not depend on the local electric field E(x), which does not occur in most devices. However, such a simple model has been applied (Nougier, et al., 1978) to field-effect transistors at 77 K and reduced the discrepancy between experiment and theory to a factor 2, instead of a factor of 40 to 100. More accurate noise models are now being applied for such computations (Fig. 22, taken from Sodini, 1979).

Finally, it ought to be mentioned that if the knowledge of the local noise sources allow determination of the noise of the sample, then also one has that the measurement of the noise of the sample gives information on the noise sources. This can be considered as a method for studying noise sources. It has been applied for studying D(E) from the measurement of the noise of a n^+-n-n^+ semiconductor sample (Zimmermann, et al., 1978) assuming a variation for D(E). Applying (95) gives S_V which is compared with the experimental value. If D(E) is the right one, the two results agree. This is in fact a self-consistent method. Provided that the comparison is performed at numerous values of the biases, beginning with the ohmic ones, it gives the correct variation of D(E).

SOME QUANTUM EFFECTS

The Quantum Correction Factor:

In the quantum limit, hf (where h is Planck's constant) is no longer large compared with k_BT, so that the average energy $<\frac{1}{2} mv^2>$ has to be replaced by that of a harmonic oscillator. The Nyquist relation (24) becomes

$$S_{\Delta V} = 4 \; R\left[\frac{h\,f}{2} + \frac{h\,f}{\exp\,(h\,f/k_BT) - 1}\right] \tag{99}$$

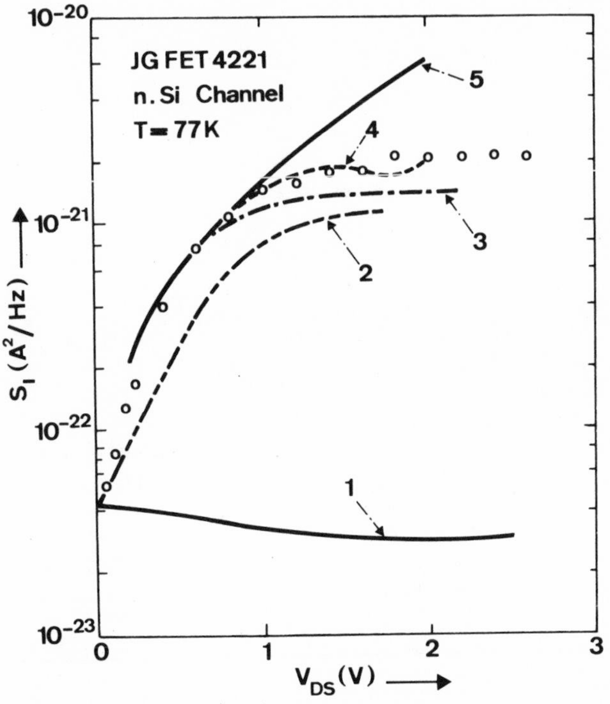

Fig. 22. Noise current spectral density of the channel of JG.FET (2N4221 n-Si channel) at 77 K, after Nougier, et al. (1978b) and Sodini (1979). o Experiment (Sodini, 1979). 1) Theory neglecting hot carriers. 2 to 5, theoreis including hot carriers: 2) Salami method; 3) Active line model; 4) Deduced from Van der Ziel theory; 5) Impedance field method.

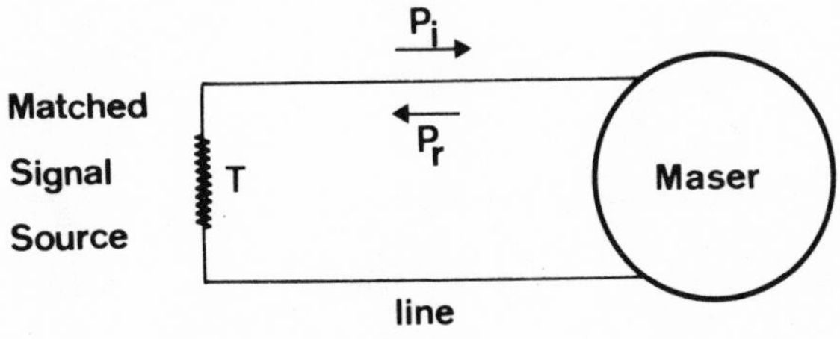

Fig. 23. Basic maser configuration.

In his original paper, Nyquist (1928) gave (99) without the zero point energy hf/2, as a consequence of the old quantum theory. Equation (99) was extensively discussed by Robinson (1974).

Noise in Masers

Studying hot carriers shows that the noise temperature is superior to the thermodynamic temperature of the samples. Masers show another example of noise temperatures T_n different from the temperature of the source T, T_n possibly being lower than T in that case. One may consider the basic system of Fig. 23. The signal source has an internal resistance at temperature T, matched to a line coupled at its other end to a maser. The power incident on the active element is reflected with an increased amplitude (gain G >> 1), and the reflected wave is the output of the device. Let n_+ and n_- be the populations of the energy levels separated by hf. The reflected signal power P_{sr} corresponding to the incident signal power P_{si} is, since $P_{sr} >> P_{si}$,

$$P_{rs} \simeq A\ hf(n_+ - n_-)\ P_{si}, \tag{100}$$

where A is a constant at a given frequency. The power gain is then

$$G = A\ hf\ (n_+ - n_-). \tag{101}$$

For computing the noise, one remarks that the incident signal is now the noise produced by the resistance in the bandwidth df, according to (99), as

$$d\ P_{ni} = \left[\frac{1}{2} h\ f + \frac{h\ f}{\exp(h\ f/k_B T) - 1}\right] df. \tag{102}$$

The reflected power noise dP_{nr} is related to the rate r_u of upward transitions from the lower state to the higher state, and to the rate r_d of downward transitions, as

$$dP_{nr} = h\ f(r_d - r_u). \tag{103}$$

The output noise is then given by

$$dP_{no} = dP_{ni} + dP_{nr}/G. \tag{104}$$

Now we need to evaluate r_u and r_d. Obviously

$$r_u = A\, n_- \left(dP_{ni} - \frac{1}{2} h\, f\, d\, f\right). \tag{105}$$

In analogy, r_d is given by

$$r_d = An_+ \left(dP_{ni} + B\right),$$

and the constant B is determined by the conditions that, when the maser and the source are in thermal equilibrium, $n_+ = n_- \exp(-hf/k_B T)$, and $r_d = r_u$, which gives

$$r_d = An_+ \left(dP_{ni} + \frac{1}{2} h\, f\, d\, f\right). \tag{106}$$

Combining the above equations, we then find

$$\frac{dP_{nr}}{G} = \frac{1}{2} h\, f \frac{n_+ + n_-}{n_+ - n_-} d\, f. \tag{107}$$

When the populations are inverted, one defines the negative temperature $-T_m$ as

$$n_+/n_- = \exp\left(hf/k_B\, T_m\right), \tag{108}$$

so that

$$d\, P_{nr}/G = \left[\frac{1}{2} h\, f + \frac{h\, f}{\exp\left(hf/k_B\, T_m\right) - 1}\right] df. \tag{109}$$

Using (108) and (109) in (104) gives

$$dP_{no} = \left[h\, f + \frac{h\, f}{\exp\left(hf/k_B T\right) - 1} + \frac{h\, f}{\exp\left(hf/k_B T_m\right) - 1}\right] df. \tag{110}$$

Thus the minimum detectable noise power is hf and not $\frac{1}{2}$ h f. In usual conditions, h f is negligible compared with $k_B T$ or $k_B T_m$ (λ = 3 mm at T = 4K), so that

$$dP_{no} \simeq k_B \left(T + T_m\right) df, \tag{111}$$

which corresponds to a noise temperature of the maser $T_n = T_m$, which can be lower than T.

Photon Noise

The noise of a photocurrent is the superposition of two noises The first one is the shot noise, due to the Poisson-distributed fluctuations in the random arrival of the photons. The second one is caused by the fluctuations of the intensity of the associated electromagnetic field, called the "excess photon noise". This excess photon noise can be measured through the spectrum $S_{\Delta i}(f)$ of a photodetector current and is directly related to the line shape of the radiation (Van Vliet, 1976). For example a Gaussian line shape $(\sigma\sqrt{2\pi})^{-1} \exp[-(\nu-\nu_o)^2/2\sigma^2]$ gives $S_{\Delta i} = (2/\sigma\sqrt{2\pi}) \times \exp(-f^2/2\sigma^2)$.

Noise in Josephson Junctions

When particles of charge q are randomly emitted, according to a Poisson law, at a rate of n particles per unit time, the associat current fluctuates around the average vlaue

$$I = q\, n. \tag{112}$$

This fluctuation is called shot noise, and the corresponding spectr density can be shown to be

$$S_{\Delta I} = 2\, q\, I. \tag{113}$$

The average current through a junction biased by a voltage V is

$$I = I_S \left(\exp \frac{qV}{k_B T} - 1\right), \tag{114}$$

and this current is the sum of the forward one $I_S \exp(qV/k_BT)$ and o the reverse one $-I_S$. Since these two currents are independent, the fluctuations add, each of them producing shot noise, and one has

$$S_{\Delta I} = 2\, q\, I_S \exp(q\, V/k_B T) + 2\, q\, I_S,$$

which can be written

$$S_{\Delta I} = 2\, q\, I_S [\exp(q\, V/k_B T) + 1],$$

that is, according to (114)

$$S_{\Delta I} = 2\, q\, I \coth(q\, V/2\, k_B T). \tag{115}$$

A Josephson junction consists of two superconducting materials separated by a thin oxide layer. The current I is

$$I = I_n + I_p, \tag{116}$$

where I_n is due to single electrons and I_p is due to Cooper pairs. The noise of such a device (Stephen, 1969) can be calculated in the following way (Van der Ziel, 1976). Since I_n and I_p are uncorrelated,

$$S_{\Delta I} = S_{\Delta I_n} + S_{\Delta I_p}, \tag{117}$$

with I_n and I_p being of the form (114). Their spectral densities are given by (115), where $q = e$ for the single electrons and $q = 2\,e$ for the Cooper pairs. Hence

$$S_{\Delta I} = 2\,e\,I_n \coth(eV/2\,k_B T) + 4\,e\,I_p \coth(eV/k_B T). \tag{118}$$

At high bias, such that $eV/k_B T >> 1$, (118) reduces to

$$S_{\Delta I} = 2\,e\,I_n + 4\,e\,I_p$$

and is therefore the sum of the shot noises of I_n and I_p. Indeed the operating voltage is usually a few microvolts. Since around 4 K, $k_B\,T/e$ is 300 μV, $eV/k_B T << 1$. Furthermore, $\coth x \simeq 1/x$ for small x, and (118) then reduces to

$$S_{\Delta I} \simeq \frac{4\,k_B\,T}{V}(I_n + I_p) = 4\,k_B\,T\,\frac{I}{V}\,. \tag{119}$$

At such low bias, (114) gives

$$I = (I_{ns} + 2\,I_{ps})\,eV/k_B T.$$

Therefore $dI/dV = I/V$ and (119) shows that the noise temperature is equal to T. This is no longer true at higher bias.

Throughout this section it has been assumed that the noise in quantum systems can be treated as if the system had a classical behaviour. This point was discussed by Robinson (1974).

CONCLUSION

As a conclusion, one must mention that the determination of the noise temperatures and of the diffusion coefficients of hot carriers is of great interest for both applied and fundamental physics.

Applied Physics

A great variety of devices operate in the high field region. In some of them, especially those working at high frequency and usi transit time effects, the diffusion current is not negligible. It then necessary to know the variation of D(E) for modeling the I-V characteristic of these devices.

The variation of D(E), or $T_n(E)$, is also necessary to predict the noise of devices operating in high electric fields. This can be done using the impedance field method or the transfer impedance method. This latter method, applied in $n^+n\ n^+$ devices (Rolland, 1975), in SCLC diodes (Nougier, et al., 1978a), and in FETs at 77 K (Nougier, et al., 1978b), clearly shows the necessity of taking int account hot carrier effects.

Fundamental Physics

The coupling constants of the scattering mechanisms are determined so as to fit transport coefficients. These are generall the ohmic ones, or the variation of the drift velocity versus the electric field. However "second-order coefficients" such as D(E) and $T_n(E)$ are often much more sensitive to the scattering mechanism than first-order coefficients. It may be expected that fitting the coupling constants using D(E) or $T_n(E)$ would result sometimes in a better accuracy.

Figure 24 shows the variation of the longitudinal excess noise temperature $T_n(E) - T$ versus the lattice temperature T in n-type silicon at E = 1 kV/cm: the kink at about 90 K clearly shows a change in the predominant scattering mechanisms. We think that at temperatures lower than 90 K the observed noise is due to hot carri G-R noise, the impurity ionization hypothesis being substantiated by conductivity measurements (Rolland, 1975). Figure 16, taken fro Canali, et al., (1975c), clearly shows that the non-parabolicity factors have a great influence on the calculated D(E) and that comparison between experimental and theoretical D(E) can bring importa information about the energy bandshapes.

Finally, one should mention that the demonstration that $D_n(E) = D_S(E)$ allows interpretations, not available in the past, of noise measurements which appear to be a powerful tool of investigation, as well as from diffusion measurements using the time of flight technique.

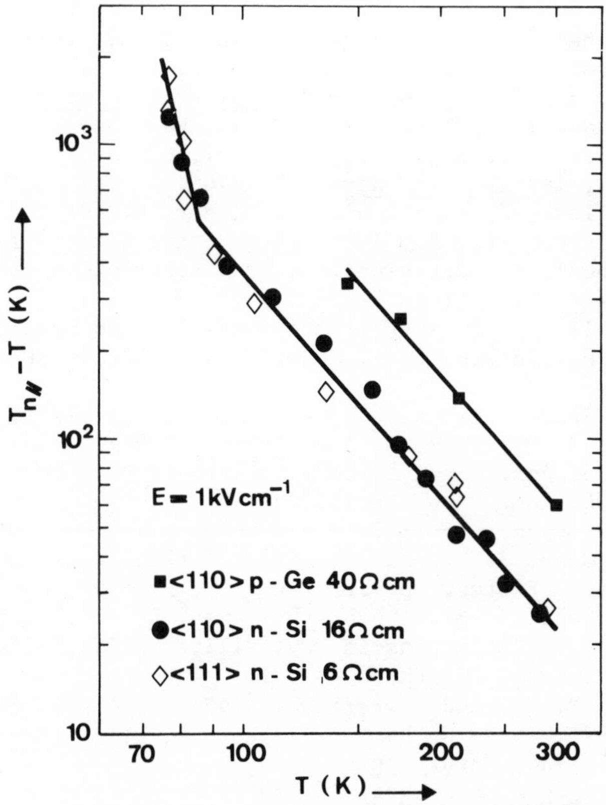

Fig. 24. Experimental variation of the longitudinal excess noise temperature $T_n - T$ versus the lattice temperature T at $E = 1$ kV cm^{-1} in n-Si and p-Ge, after Rolland (1975).

BIBLIOGRAPHY

Alberigi-Quaranta, A., Borsari, V., Jacoboni, C., and Zanarani, G., 1973, Appl. Phys. Letters 22:103.
Alberigi-Quaranta, A., Jacoboni, C., and Ottaviani, G., 1971, Rev. Nouvo Cim. 1:445.
Bareikis, V., Pozhela, Yu. K., and Matulenene, I. B., 1968, in "Proc. Intern. Conf. Phys. Semiconductors, Moscow" p. 760.
Bartelink, D. J., and Persky, G., 1970, Appl. Phys. Letters 16:191.
Callen, H. B., and Welton, T. A., 1951, Phys. Rev. 83:34.
Canali, C., Jacoboni, C., Nava, F., Ottaviani, G., and Alberigi-Quaranta, A., 1975a, Phys. Rev. B 12:2265.
Canali, C., Catellani, F., Jacoboni, C., Minder, R., Ottaviani, G., and Alberigi-Quaranta, A., 1975b, Sol. State Commun. 17:1443.
Canali, C., Jacoboni, C., Ottaviani, G., and Alberigi-Quaranta, A., 1975a, Appl. Phys. Letters 27:278.
Canali, C., Gavioli, G., Nava, F., Ottaviani, G., and Reggiani, L., 1976, in "Proc. Intern. Conf. Phys. Semiconductors, Rome", p. 1231.
Canali, C., Nava, F., Reggiani, L., Gasquet, D., Vaissiere, J. C., and Nougier, J. P., 1979, J. Appl. Phys. 50:922.
Einstein, A., 1905, Ann. Phys. 17:549.
Einstein, A., 1906, Ann. Phys. 19:289,371.
Erlbach, E., 1963, Phys. Rev. 132:1976.
Erlbach, E., and Gunn, J. B., 1962a, "Proc. Intern. Conf. Phys. Semiconductors, Exeter," p. 128.
Erlbach, E., and Gunn, J. B., 1962b, Phys. Rev. Letters 8:280.
Fawcett, W., 1973, in "Electrons in Crystalline Solids," IAEA, Vien
Hart, L. G., 1970, Canad. J. Phys. 48:531.
Hill, G., Robson, P. N., and Fawcett, W., 1979, J. Appl. Phys. 50:3
Kaszynski, A., 1979, Thèse de Docteur Ingenieur, Lille (France).
Kintchine, A. I., 1934, Math. Ann. 109:604.
Kubo, R., 1957, J. Phys. Soc. Jpn. 12:570.
Langevin, M. P., 1908, Comp. Rend. Acad. Sci., Paris 146:530.
Nash, J. G., and Holm-Kennedy, J. W., 1974, Appl. Phys. Letters 24:139.
Nash, J. G., and Holm-Kennedy, J. W., 1977, Phys. Rev. B 16:2834.
Nougier, J. P., 1972, Thesis, unpublished.
Nougier, J. P., 1973a, Phys. State Sol. (b) 55:K43.
Nougier, J. P., 1973b, Physica 64:209.
Nougier, J. P., 1978, Appl. Phys. Letters 37:671.
Nougier, J. P., and Rolland, M., 1973, Phys. Rev. 138:5728.
Nougier, J. P., and Rolland, M., 1976, in "Proc. 13 Intern. Conf. Phys. Semiconductors, Rome", p. 1227.
Nougier, J. P., and Rolland, M., 1977, J. Appl. Phys. 48:1683.
Nougier, J. P., Comallonga, J., and Rolland, M., 1974, J. Phys. E 7:287.
Nougier, J. P., Gasquet, D., and Vaissiere, J. C., 1978a, Rev. Phys. Appl. 13:715.

Nougier, J. P., Gasquet, D., Vaissiere, J. C., and Bilger, H. R., 1978b, in "Proc. 5th Intern. Conf. on Noise in Phys. Systems," Springer-Verlag, p. 110.
Nougier, J. P., Gasquet, D., 1978, Appl. Phys. Letters 33:89.
Nougier, J. P., Sodini, D., Rolland, M., Gasquet, D., and Lecoy, G., 1978c, Sol.-State Electr. 21:133.
Nyquist, H., 1928, Phys. Rev. 32:110.
Persky, G., and Bartelink, D. J., 1971, J. Appl. Phys. 42:4414.
Price, P. J., 1958, IBM J. Res. Develop. 3:191.
Price, P. J., 1965, in "Fluctuation Phenomena in Solids," Ed. by Burgess, R. E., Academic Press, New York.
Pozhela, Yu. K., Bareikis, V. A., and Matulenene, I. B., 1968, Sov. Phys.-Semicond. 2:503.
Reggiani, L., Canali, C., Nava, F., and Gagliani, G., 1979, unpublished.
Rigaud, A., Nicolet, M. A., and Savelli, M., 1973, Phys. Stat. Sol.(a) 18:531.
Robinson, F. N. H., 1974, "Noise and Fluctuations in Electronic Devices and Circuits," Clarendon Press, Oxford.
Rolland, M., 1975, Thesis, unpublished.
Ruch, J. G., and Kino, G. S., 1968, Phys. Rev. 174:921.
Shockley, W., Copeland, J. A., and James, R. P., 1966, in "Quantum Theory of Atoms, Molecules, and the Solid State," P. Löwdin, Ed., Academic Press, New York.
Sigmon, T. W., and Gibbons, J. F., 1969, Appl. Phys. Letters 15:320.
Sodini, D., 1979, Thesis, unpublished.
Stephen, S., 1969, Phys. Rev. 182:531.
Van der Ziel, A., 1970, "Noise: Sources, Characterization, Measurement," Prentice-Hall, Englewood Cliffs, New Jersey.
Van der Ziel, A., 1976, "Proc. 4th Intern. Conf. Physical Aspects of Noise in Sol.-State Devices," Physica 83B:41.
Van der Ziel, A., Jindal, R., Kim, S. K., Park, H., and Nougier, J. P., 1979, Sol.-State Electr. 22:177.
Van Kampen, N. G., 1965, in "Fluctuation Phenomena in Solids," Ed. by Burgess, R. E., Academic Press, New York.
Van Vliet, K. M., 1971, J. Math. Phys. 12:1981.
Van Vliet, K. M., 1976, in "Proc. 4th Intern. Conf. on Physical Aspects of Noise in Sol.-State Devices," Physica 83B:52.
Van Vliet, K. M., and Van der Ziel, A., 1977, Sol.-State Electr. 20:231.
Van Vliet, K. M., Friedman, A., Zijlstra, R. J. J., Gisolf, A., and Van der Ziel, A., 1975a, J. Appl. Phys. 46:1804.
Van Vliet, K. M., Friedman, A., Zijlstra, R. J. J., Gisolf, A., and Van der Ziel, A., 1975b, J. Appl. Phys. 46:1814.
Wiener, N., 1930, Acta Math. Stockholm 55:117.
Zijlstra, R. J. J., 1978, in "Proc. 5th Intern. Conf. on Noise in Physical Systems," Springer-Verlag, Springer Series in Electro-Physics 2, p. 90.

Zimmerman, J., Leroy, Y., Kaszynski, A., and Carnez, B., 1978, in "Proc. 5th Intern. Conf. on Noise in Physical Systems," Springer-Verlag, Springer Series in Electrophysics 2, p. 101.

NOTES ADDED IN PROOF

Note 1

It has been shown (see (61)) that, for long times, the variance of the position becomes proportional to the time. This is no more true for short time investigations. One may then define a transient spreading diffusion coefficient $D_{S\alpha\beta}(E,t)$ as:

$$< \Delta r_\alpha(t) \cdot \Delta r_\beta(t) > = 2\ t\ D_{S\alpha\beta}(E,t) \qquad (67.a)$$

It has been recently shown (Hill et al. 1979) that $D_{S\alpha\beta}(E,t)$ could be related to the frequency dependent $D_{n\alpha\beta}(E,f)$ (see (31) or (32)) through the relation:

$$D_{S\alpha\beta}(E,t) = 2\ t \int_0^\infty \left(\frac{\sin \pi f\ t}{\pi\ f\ t}\right)^2 D_{n\alpha\beta}(E,f)\ df$$

Another quantity, the transient differential spreading diffusion coefficient $D'_{S\alpha\beta}(E,t)$, may be defined as (Alberigi-Quaranta et al., 1973)

$$\frac{1}{2}\frac{d}{dt} < \Delta r_\alpha(t) \cdot \Delta r_\alpha(t) > = D'_{S\alpha\beta}(E,t) \qquad (67.b)$$

The link between the two transient spreading diffusion coefficients is obviously:

$$D'_{S\alpha\beta}(E,t) = \frac{\partial}{\partial t}[t\ D_{S\alpha\beta}(E,t)]$$

of course both tend towards the steady state spreading diffusion coefficient $D_{S\alpha\beta}(E)$ as t tends towards infinity (see (61)):

$$\lim_{t\to\infty} D_{S\alpha\beta}(E,t) = \lim_{t\to\infty} D'_{S\alpha\beta}(E,t) = D_{S\alpha\beta}(E)$$

Therefore, in principle, both can be used as estimators for evaluating $D_{S\alpha\beta}(E)$.

Note 2

$<\Delta r_\alpha(t)\ \Delta r_\beta(t)>$ as a function of t, is a curve the slope of which tends towards $D_{S\alpha\beta}(E)$. This slope (see (67.b)) is identical with $D'_{S\alpha\beta}(E,t)$. As a consequence, $D'_{S\alpha\beta}(E,t)$ given by (67.b) is a much better estimate of $D_{S\alpha\beta}(E)$ than $\tilde{D}_{S\alpha\beta}(E,t)$ given by (67.a). Of course this study can be performed as a function of the frequency, and for the first time very interesting results have been obtained concerning $D_n(E,f)$ in GaAs.

HIGH-FIELD TRANSPORT OF HOLES IN ELEMENTAL SEMICONDUCTORS

Lino Reggiani

Gruppo Nazionale di Struttura della Materia
Istituto di Fisica, Università di Modena
41100 Modena, Italy

INTRODUCTION

This paper reports a brief review on recent results of hot-hole transport in the elemental semiconductors diamond, silicon and germanium. The macroscopic quantities of interest are the drift velocity v_d and the longitudinal diffusion coefficient D_ℓ, and in the present study we shall be concerned with their dependence upon electric field strength, crystallographic directions and temperature. Comparison between theory and experiments enables the values of the deformation potential parameters, which describe the hole-lattice scattering, to be evaluated. Furthermore, correlations between peculiarities of the valence band structure, such as warping and nonparabolicity, and macroscopic quantities are analyzed in detail.

THEORETICAL MODEL

The valence band of elemental semiconductors consists of three subbands (see Fig. 1). Two of them are degenerate at k = 0 and the third is split in energy by an amount Δ (spin-orbit energy). They are usually called: heavy-hole band, light-hole band and spin-orbit-band, respectively. The energy-wavevector relationship $\varepsilon(\vec{k})$ of each band was calculated by Kane (1956) with a $\vec{k}\cdot\vec{p}$ approach and, to a good approximation (the better the larger the direct energy gap), it can be described in terms of four parameters as:

$$\varepsilon = \varepsilon(A, B, C, \Delta) \quad (1)$$

where A, B and C are the inverse valence band parameters, usually determined by cyclotron resonance measurements.

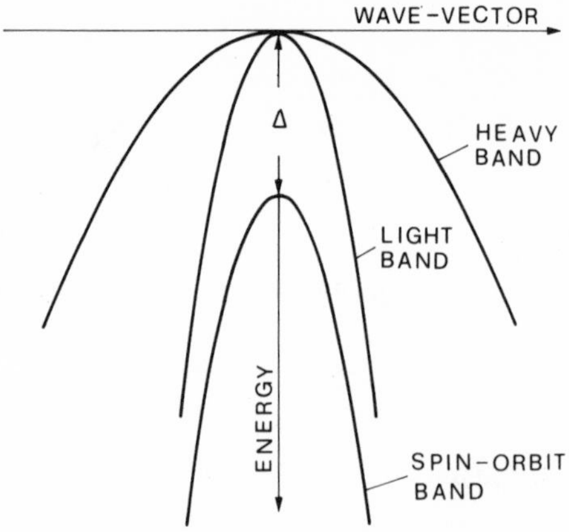

Fig. 1. Valence band model of elemental semiconductors.

Owing to its separation in energy and small density-of-states, the third band gives negligible contribution to the transport properties reported here, thus it will be neglected. As a consequence of degeneracy, the surfaces of constant energy of the heavy and light-band are characterized by a warped shape. Away from k = 0, the three hole bands interact and the $\varepsilon(\vec{k})$ dependence is no longer parabolic. The effects of nonparabolicity are strongest in the region of energy centered around $\Delta/3$. Figure 2 exemplifies these features for the case of Si. It is worth mentioning that warping

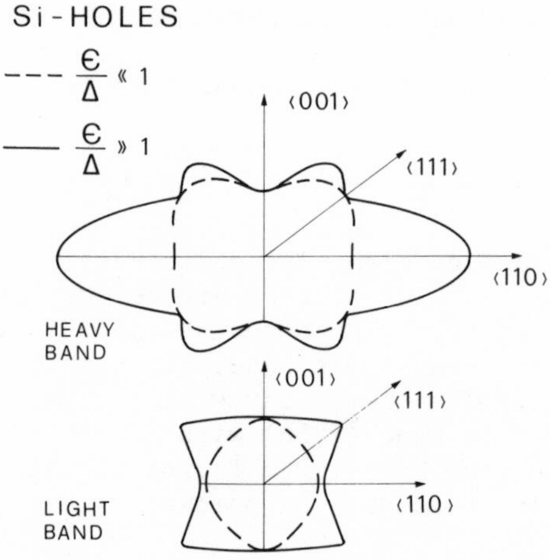

Fig. 2. Warped shape of the equienergetic surfaces of the heavy and light band in Si. Broken curves refer to the case $\varepsilon/\Delta \ll 1$ full curves to the case $\varepsilon/\Delta \gg 1$.

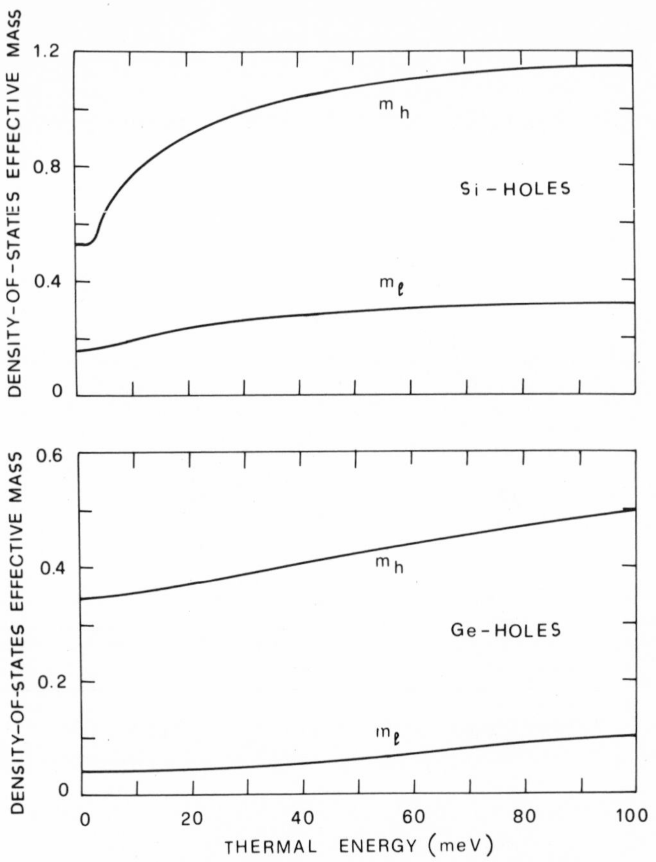

Fig. 3. Density-of-states effective mass as a function of thermal energy for the case of Si and Ge.

leads to a hole effective-mass which is a function of the wavevector direction $\vec{k}$, and that nonparabolicity leads to an effective-mass which is a function of energy, as illustrated in Fig. 3 for the cases of Si and Ge.

Since high-purity crystals are considered here, transport properties are controlled by lattice scattering only. Thus, interactions with acoustic and nonpolar optical phonons are taken into account. The former includes overlap effects due to the p-like symmetry of the hole wavefunctions [Costato and Reggiani (1973)] and acoustic dissipation through an exact Bose-Einstein phonon distribution function.

The Boltzmann equation is solved through a Monte Carlo technique in the usual way [see, e.g., Reggiani, et al., (1977)]. Uncertainty

Table 1. Constants used in calculations.*

	Diamond[a]	Silicon[b,c]	Germanium[c,d]	Units
$\|A\|$		4.22	13.38	- - -
$\|B\|$		0.78	8.48	- - -
$\|C\|$		4.80	13.14	- - -
Δ	0.006	0.044	0.295	eV
$m_h(\varepsilon/\Delta << 1)$		0.53	0.346	- - -
$m_h(\varepsilon/\Delta >> 1)$	1.1	1.26	0.73	- - -
$m_\ell(\varepsilon/\Delta << 1)$		0.155	0.042	- - -
$m_\ell(\varepsilon/\Delta >> 1)$	0.3	0.36	0.25	- - -
ρ	3.51	2.329	5.32	gr/cm^3
s	14.27×10^5	6.57×10^5	3.93×10^5	cm/sec
θ_{op}	1938	735	430	K
E_1^o	5.5	4.0	4.6	eV
d_o	61.2	26.6	40.3	eV

*Notation: m_h, m_ℓ are the density-of-states effective mass of heavy and light hole; ρ is the crystal density; s is an average sound velocity; θ_{op} is the equivalent temperature of the optical phonon; E_1^o is an average acoustic deformation potential; d_o is the optical deformation potential.

[a]Reggiani, et al., (1979).
[b]Jacoboni, et al., (1978).
[c]Gagliani and Reggiani (1975).
[d]Reggiani, et al., (1977).

of theoretical results is estimated within 5% for the drift velocity and 10% for the diffusion coefficient, respectively. The constants used in calculations are reported in Table 1.

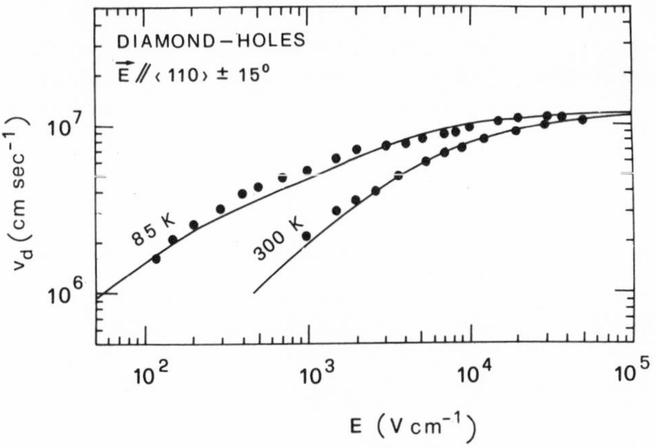

Fig. 4. Drift velocity as a function of electric field for holes in diamond at different temperatures. Full points represent experimental data; the lines report theoretical results.

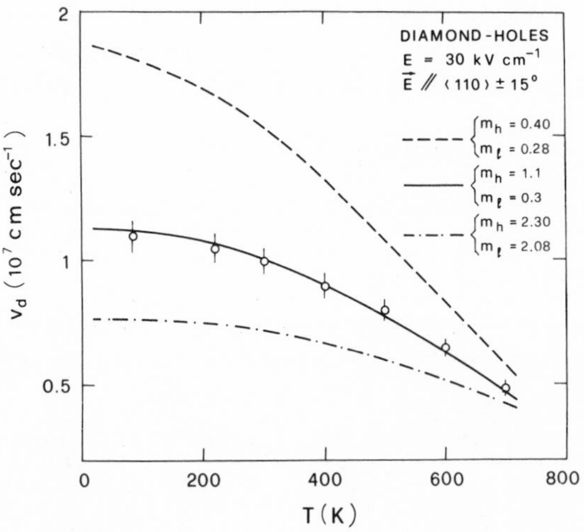

Fig. 5. Experimental (open circles) and theoretical (lines) hole drift velocity in diamond at fixed field strength as a function of temperature [see Reggiani, et al., (1979)]. Dot-dashed curve refers to calculations performed with the values of the hole effective masses deduced from Rauch (1962); dashed curve refers to those deduced from Lawaetz (1971); continuous curve refers to present results (see Table 1).

RESULTS AND DISCUSSION

Measurements have been performed with a time-of-flight technique making use of the electron gun at the Modena University and with a noise-conductivity technique (for more details the reader is referred to a previous chapter). Uncertainty of experimental data is estimated within 5% and 25% for v_d and D_ℓ, respectively.

Diamond

The v_d versus E dependence at 85 and 300 K is shown in Fig. 4. The high-field (E = 30 kV/cm) drift velocity as a function of temperature is shown in Fig. 5. Here data suggest a highest value of the saturated drift velocity $v_{s<110>} = 1.1 \times 10^7$ cm/sec.

The theoretical model makes use of a spherical and parabolic two-band model. The comparison between theory and experiment, within the values of the deformation potential parameters, suggests values of the effective masses [Reggiani, et al., (1979)].

Silicon

The v_d versus E dependence at temperatures between 6 and 430 K and for the two crystallographic directions <100> and <111> is shown in Fig. 6. The high-field drift velocity as a function of temperature is given in Fig. 7. Here data suggest a highest value of the saturated drift velocity $v_{s<100>} = 1.0 \times 10^7$ cm/sec and $v_{s<111>} = 0.9 \times 10^7$ cm/sec. The D_ℓ versus E dependence at 300 K is shown in Fig. 8.

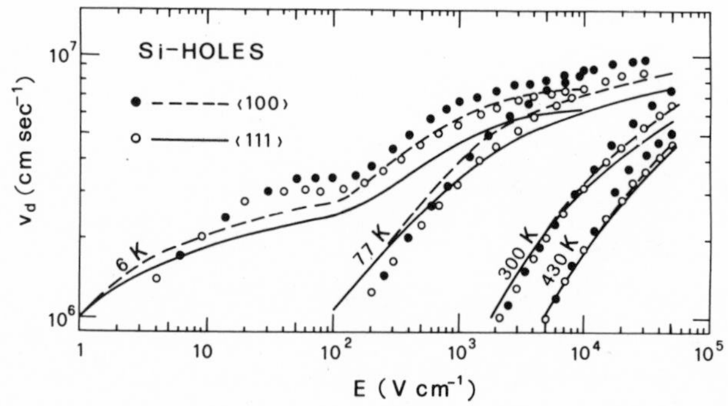

Fig. 6. Drift velocity as a function of electric field for holes in Si at different temperatures. Full and open points represent experimental data from Ottaviani, et al., (1975); the broken and full lines report theoretical results.

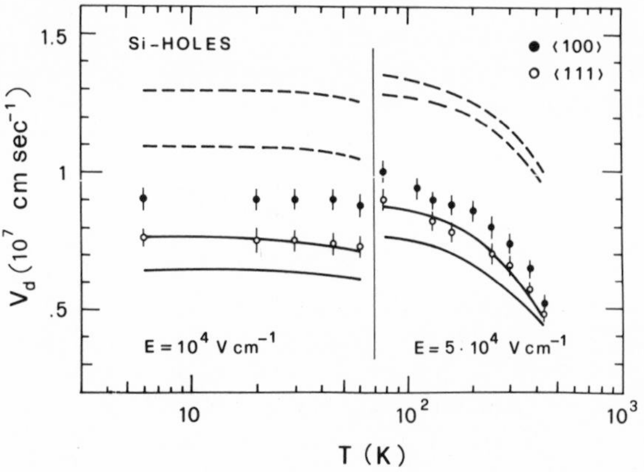

Fig. 7. Drift velocity as a function of temperature at fixed electric-field strengths for holes in Si. Full and open points represent experimental data from Ottaviani, et al., (1975); broken and full lines report theoretical results for a parabolic and nonparabolic band model, respectively.

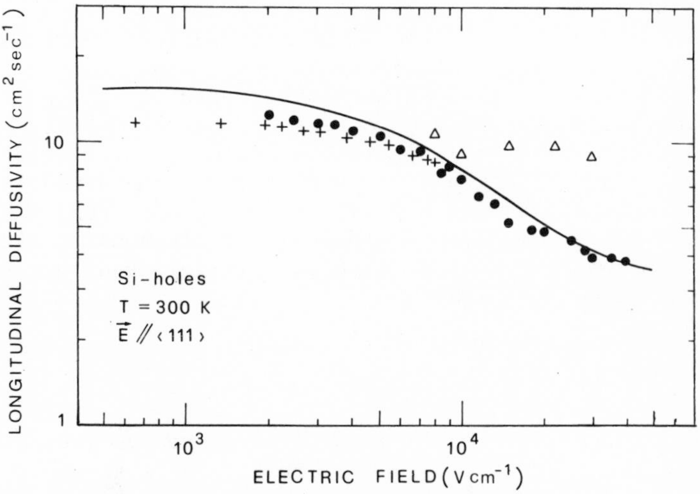

Fig. 8. Longitudinal diffusion coefficient of holes in Si at 300 K as a function of field oriented along a <111> direction. Different points refer to experimental results; full line refers to theoretical calculations [see Nava, et al., (1979)].

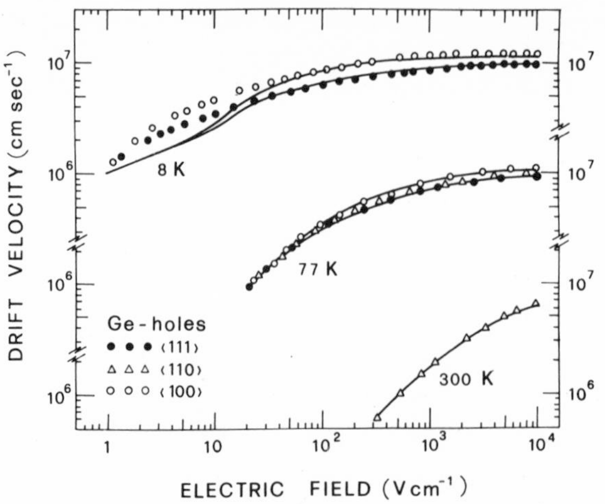

Fig. 9. Drift velocity as a function of electric field for holes in Ge at different temperatures. Different points represent experimental data; full lines reports theoretical results [Jacoboni and Reggiani (1979)].

The theoretical model makes use of a single warped and nonparabolic band (the heavy-hole band). The comparison between theory and experiments leads to the determination of the deformation potential parameters (see Table 1). Anisotropy effects of v_d (see Fig. 6) are found to be correlated to the warping of the heavy band. The value of the saturated drift velocity is found to be correlated to the effects of nonparabolicity. To support this conclusion in Fig. 7 calculations performed for a parabolic (broken curves) and nonparabolic (full curves) model are reported. It is seen that the agreement between theory and experiment is well-improved when nonparabolicity is taken into account.

Germanium

The v_d versus E dependence at temperatures between 8 and 300 K and for different crystallographic directions is shown in Fig. 9. The high-field drift velocity as a function of temperature is given in Fig. 10. Here data suggest a highest value of the saturated drift velocity $v_{s<100>} = 1.2 \times 10^7$ cm/sec and $v_{s<111>} = 1.0 \times 10^7$ cm/sec. The D_ℓ versus E dependence at temperatures between 112 and 295 K is shown in Fig. 11. The D_ℓ versus E at 77 K for the two crystallographic directions <100> and <111> is given in Fig. 12.

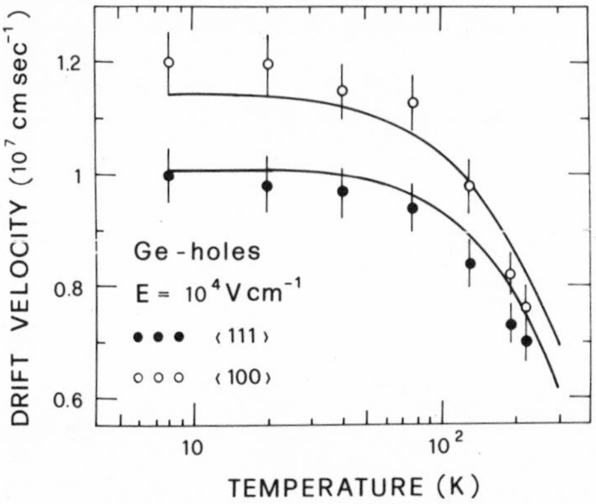

Fig. 10. Hole drift velocity in Ge as a function of temperature for an electric field $E = 10^4$ V/cm. Closed and open circles refer to experimental data; full curves indicate theoretical results.

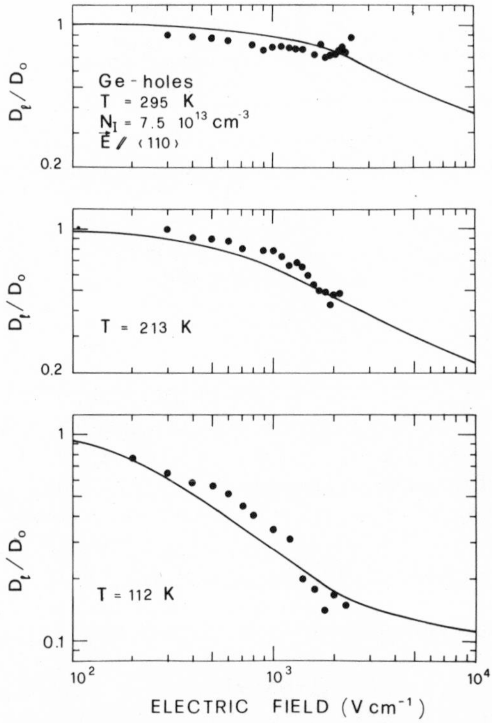

Fig. 11. Longitudinal diffusion coefficient of holes in Ge as a function of field strength at different temperatures. Closed circles show experimental data, the continuous curves indicate theoretical results of Reggiani, et al., (1978).

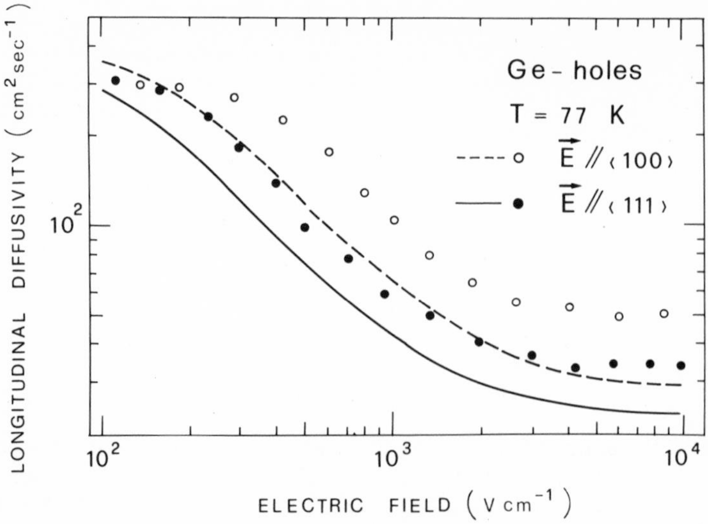

Fig. 12. Longitudinal diffusion coefficient of holes in Ge as a function of field at 77 K. Closed and open circles indicate the experimental data. The full and broken lines indicate theoretical results of Reggiani, et al., (1978).

As for the case of Si the theoretical model makes use of a single warped and nonparabolic band. Even for the case of Ge, anisotropy of v_d and D_ℓ and saturation of v_d are found to be mainly correlated to warping and nonparabolicity of the heavy-band.

CONCLUSIONS

Drift velocity and longitudinal diffusion coefficient of holes in elemental semiconductors have been found to exhibit similar behaviour at increasing field strength when they are controlled by lattice scattering only. The general trends and expectations under nonlinear conditions can be summarized as follows:

i) v_d exhibits a sublinear dependence;

ii) D_ℓ decreases from its thermal equilibrium value;

iii) Both v_d and D_ℓ give evidence of anisotropic effects with respect to crystallographic axes;

iv) A saturation region of v_d is attained at the highest fields.

The microscopic interpretation explains points i) and ii) on the basis of a scattering rate which increases with carrier energy (typical of acoustic and nonpolar optical mechanisms). Point iii) is found to be correlated to the warped shape of the heavy hole equienergetic surfaces. Finally, point iv) is associated with the strong nonpolar optical interaction and with the nonparabolic behaviour of the energy wavevector relationship.

BIBLIOGRAPHY

Costato, M., and Reggiani, L., 1973, Phys. Stat. Sol. (b) 58:47,461.

Gagliani, G., and Reggiani, L., 1975, Nuovo Cimento 30B:207.

Jacoboni, C., Gagliani, G., Reggiani, L., and Turci, O., 1978, Sol.-State Electr. 21:315.

Jacoboni, C., and Reggiani, L., 1979, Adv. Phys., in press.

Kane, E. O., 1956, J. Phys. Chem. Sol. 1:82.

Lawaetz, P., 1971, Phys. Rev. 4B:3460; in this paper the value of V_2 should be corrected to 0.42.

Nava, F., Canali, C., Reggiani, L., Gasquet, D., Vaissiere, J. C., and Nougier, J. P., 1979, J. Appl. Phys., 50:922.

Ottaviani, G., Reggiani, L., Canali, C., Nava, F., and Alberigi-Quaranta, A., 1975, Phys. Rev. 12B:3318.

Rauch, J. C., 1962, in "Proc. 6th Intern. Conf. Phys. Semicond.," Institute of Physics, London, p. 276.

Reggiani, L., Canali, C., Nava, F., and Ottaviani, G., 1977, Phys. Rev. 16B:2781, and references therein.

Reggiani, L., Canali, C., Nava, F., and Alberigi-Quaranta, A., 1978, J. Appl. Phys. 49:4446.

Reggiani, L., Bosi, S., Canali, C., Nava, F., and Kozlov, S. F., 1979, Solid-State Commun., 30:333.

NONLINEAR TRANSPORT IN QUASI-ONE-DIMENSIONAL CONDUCTORS

H. Kahlert+)

Ludwig Boltzmann Institut für Festkörperphysik and
Institut für Festkörperphysik der Universität Wien
A-1090 Vienna, Austria

INTRODUCTION

Quasi-one-dimensional (1d) solids have recently attracted considerable interest among solid-state physicists, both experimental and theoretical, because of their unusual physical properties [for a recent review, see Schuster (1975), Keller (1975) Päl, et al. (1977), Keller (1977), Miller and Epstein (1978)]. Some of their remarkable features are a high, sometimes even metallic conductivity at room temperature, metallic optical properties, the occurrence of a Peierls' distortion of the linear chains in some cases, and the high anisotropy of most of their properties. Chemically and structurally very different groups of materials are presently considered to belong to the family of quasi-1D solids. Organic charge-transfer complexes (TTF-TCNQ and related materials), linear-chain transition-metal compounds (KCP and its analogs), and trichalcogenides such as $NbSe_3$ are only the most prominent members of this large and rapidly growing class of solids.

From the very beginning of the widespread interest in these substances, which was sparked by the discovery of a large conductivity peak in TTF-TCNQ at 58 K by Coleman et al. (1973), the efforts of experimentalists and theoreticians were aimed at the understanding of their unusual electronic transport properties. At low temperatures, many quasi-1D conductors are found to undergo periodic lattice distortions leading to superlattice incommensurate with the primitive lattice and to the formation of charge-density waves coupled to the

+)Work supported by the Fonds zur Förderung der Wissenschaftlichen Forschung in Österreich and by the Ludwig-Boltzmann-Gesellschaft.

periodic distortion (Friend and Jerome, 1979). Such an incommensurate periodic lattice distortion coupled to a charge density wave has no preferred position in the lattice, being therefore free to move and able to carry current. This is the collective mode proposed by Fröhlich (1955) as a mechanism for superconductivity. However, besides commensurability with the lattice, impurities and defects can pin the CDW to the lattice (Lee et al. 1974). A phenomenological theory of the electrodynamics of the pinned CDW has been developed by Rice et al. (1975). Up to now, a contribution of the collective Fröhlich mode to the dc conductivity in materials like TTF-TCNQ and KCP, where a periodic lattice distortion has been experimentally found by diffraction studies with X-rays and neutrons, has not been unambiguously identified. However, in these materials, and also recently in the quasi-1D system $NbSe_3$, an electric-field dependent increase of the conductivity has been observed, which is interpreted in terms of unpinning of a pinned CDW by the applied field.

The highly conducting polymers polysulfurnitride, $(SN)_x$, and doped polyacetylene, $(CH)_x$, are sometimes also considered as belonging to the group of quasi-1D solids. For the case of $(SN)_x$, the electronic anisotropy is much smaller than e.g. in TTF-TCNQ or KCP, and no transition to a Peierls distorted state has been observed, whereas the intrinsic anisotropy of $(CH)_x$ is not known, since single crystals of this polymer are not yet available.

Nevertheless, there exists experimental evidence for a nonlinear behavior of the conductivity in these materials, obtained from harmonic mixing experiments. An explanation of these phenomena is based on a theory of fluctuation-induced tunneling conduction recently developed by Sheng (unpublished) for materials, where large conducting regions are separated by small insulating barriers.

NONLINEAR TRANSPORT IN ORGANIC CHARGE TRANSFER SALTS: TTF-TCNQ AND RELATED COMPOUNDS

The first report on the non-Ohmic behavior of TTF-TCNQ (tetrathiofulvalene-tetracyanoquinodimethane) at 4.2 K, where it is in its low-temperature low-conductivity phase, was given by Kahlert (1975). Two distinct features were observed, one being drastic deviations from Ohm's law at applied electric fields of about 60 V/cm at power levels in the sample as low as 1 μW, as displayed in Fig. 1. The second observation was a breakdown-type transition to a well-conducting state occurring at electric fields in the range between 300 to 600 V/cm, dependent on the sample investigated. The measurements were done with a continuously applied dc electric field, since the high sample resistance of a few 100 MΩ, in conjunction with the high value of the dielectric constant, prevents the use of reasonably short pulses. The breakdown effect was soon identified as being of

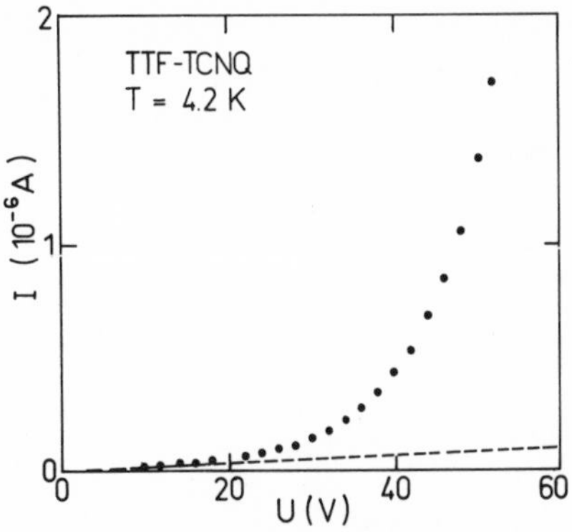

Fig. 1. Current-voltage characteristic of a sample of TTF-TCNQ at 4.2 K. The dashed line represents the zero-field resistance (Kahlert, 1975).

purely thermal origin, both by comparing it to the well-known characteristic of thermistors (Seeger, 1976) and by direct experimental evidence coming from an experiment in which a TTF-TCNQ sample and an n-type Ge sample were brought into close thermal contact in order to monitor the temperature increase associated with the TTF-TCNQ breakdown by an increase of the Ge-sample conductivity (Kahlert, unpublished). However, it was noted that the nonlinearity at low electric fields could not consistently be explained by the thermistor behavior at high electric fields and appeared to be an independent phenomenon. An explanation was given by Rice et al. (1975) who predicted a hopping type motion of the pinned CDW condensate and a resulting field-dependent conductivity. The non-Ohmic properties were further investigated by Cohen et al. (1976) and Cohen and Heeger (1977), who extended the measurements to temperatures below 4.2 K and supplemented their dc data below current densities of 5×10^{-2} A/cm^2 by pulsed data at higher current densities as shown in Fig. 2. It is interesting to note that the deviations from Ohm's law at 4.2 K become observable at appreciably lower electric fields than reported earlier (Kahlert, 1975). If the data are analyzed by fitting a relation of the form $\sigma = \sigma_0(1 + \beta E^2)$, values for the coefficient β of the order of 10^{-4} cm^2/V^2 for the data of Kahlert (1975) are obtained, and β-values in the range of 0.5 to 1×10^{-2} cm^2/V^2 for the later data are found. It should be mentioned that this simple relation describes this later data to the highest values of electric field with a remarkable precision. One possible reason for the large variation of β might be a different content of impurities or defects in the various samples. Cohen and Heeger (1977) have shown that the irradiation of TTF-TCNQ samples with deuterons increases the critical field strength for the onset of deviations from Ohm's law at 4.2 K.

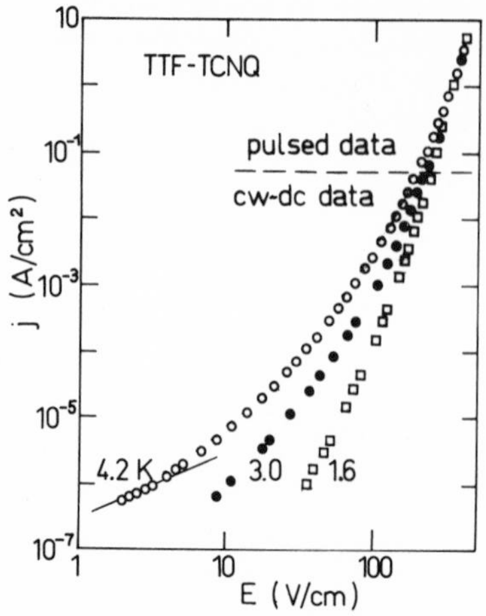

Fig. 2. Current density vs. field strength for TTF-TCNQ at three different temperatures. "Pulsed data" were obtained by a superposition of a d.c. bias and a pulse (Cohen et al., 1976, Cohen and Heeger, 1977).

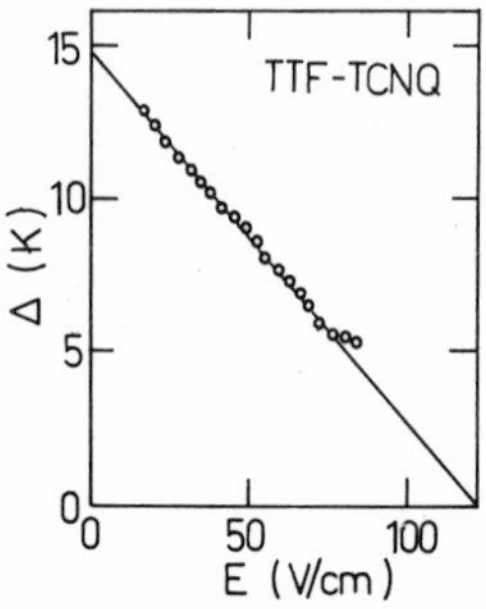

Fig. 3. Field-dependent activation energy of the low-temperature conductivity of TTF-TCNQ (Cohen et al., 1976, Cohen and Heeger, 1977).

These authors have described their data by the expression $j = \sigma_0 E \exp[-(\Delta/T)(1 - E/E_0)]$. The characteristic field obtained from several samples was 150 ± 50 V/cm and the conductivity factor σ_0 had a typical value of $10^{-4}(\Omega cm)^{-1}$. The activation energy Δ is approximately 14 K for E = 0 and decreases linearly with increasing field, extrapolating to zero at a field of about 125 V/cm (see Fig. 3). Their interpretation resorts to a new type of nonlinear current carrying elementary excitation in a weakly pinned Fröhlich CDW condensate theoretically postulated by Rice et al. (1976) as a solution of a nonlinear wave equation for the local phase $\phi(x,t)$ of the CDW. These elementary excitations correspond to localized compressions (called ϕ-particles by the authors) or rarefactions (called anti-ϕ-particles) in the local condensed electron density. However, the nonlinear transport in TTF-TCNQ was at the time the only, and indirect, evidence for these ϕ-particles or phase-kink solitons in the pinned CDW and further experiments are required which are more directly sensitive to the presence of such excitations in this material.

An alternative theoretical approach was presented by Wonneberger and Gleisberg (1977) who have treated the nonlinear transport properties of a pinned CDW by considering the Brownian motion of a charged particle in a periodic potential. The resulting j-E characteristics were in qualitative agreement with the experiments in TTF-TCNQ even for a simple sinusoidal pinning potential, but implied values for the effective charge and the pinning frequency are different from those currently assumed. For a quantitative agreement, a highly anharmonic form of the pinning potential was necessary (Wonnenberger, 1979).

Similar effects as those observed in TTF-TCNQ have been found in tetraselenofulvalenium-TCNQ (TSeF-TCNQ), which is isostructural to TTF-TCNQ, but where the four sulfur atoms in the TTF-molecule are replaced with selenium. Although the ohmic conductivity at 1.5 K is more than seven orders of magnitude greater than that of TTF-TCNQ, deviations from linear behavior show up at approximately the same electric fields (Cohen et al., 1976, Cohen and Heeger, 1977). The data are displayed in Fig. 4.

A truly unambiguous proof for the intrinsic nature of the nonlinear transport in TTF-TCNQ and TSeF-TCNQ came from microwave harmonic mixing (MHM) experiments reported by Seeger and Maurer (1978) and Maurer (1978). The result for the amplitude ΔU and the phase of the MHM signal is shown in Fig. 5 for TTF-TCNQ and in Fig. 6 for TSeF-TCNQ. In both cases the amplitude increases rapidly by warming the sample from 4.2 K to higher temperatures, exhibiting a maximum at 20 K for TTF-TCNQ and at 10 K for TSeF-TCNQ and decreasing with further increasing temperature. One should note that in both cases, one finds at the temperatures of the phase transition, known from scattering experiments and from resistivity data, shoulders or even extrema of the MHM signal are clearly visible. Unfortunately, the amplitude data have a deficiency, in that the amount of nonlinearity cannot easily be extracted from them, since the microwave electric field strength, which enters the expression for ΔU and which depends strongly upon the sample resistivity, could not be determined in the experimental setup used. However, the phase-data do not depend on the magnitude of the microwave field. They show a very different behavior for the two materials. Whereas the phase changes by 45°

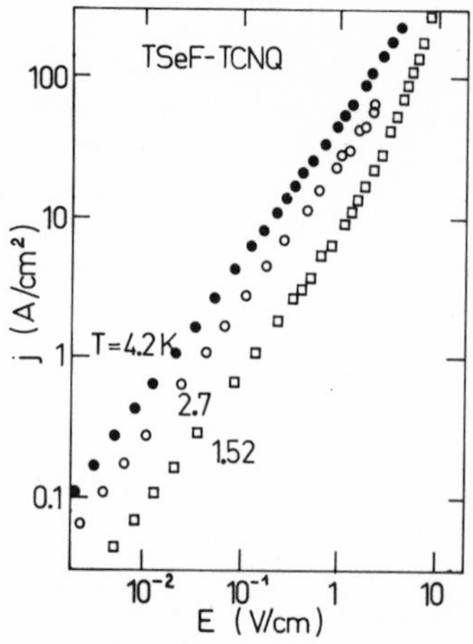

Fig. 4. Current density-field strength characteristic for TSeF-TCNQ at three different temperatures (Cohen et al., 1976, Cohen and Heeger, 1977).

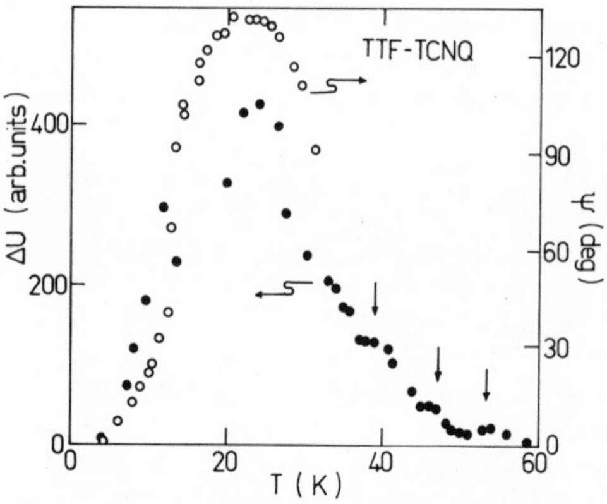

Fig. 5. Amplitude ΔU and relative phase ψ of the MHM signal vs. temperature for TTF-TCNQ. The vertical arrows mark the temperatures, where phase transitions occur (Seeger and Maurer, 1978, and Maurer, 1978).

between 4.2 K and 12 K in TTF-TCNQ and then shows an additional chan of 90° in the temperature interval between 12 and 20 K, there is almost no dependence of the phase on temperature in TSeF-TCNQ. It is important to note that the absolute value of ψ cannot be determined and the data reported here are known within an arbitrary, but temperature-independent additive constant. An explanation of the

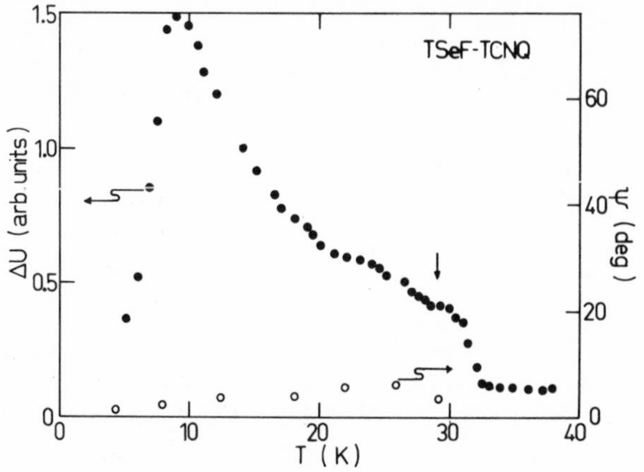

Fig. 6. Amplitude ΔU and relative phase ψ of the MHM signal vs. temperature for TSeF-TCNQ (Seeger and Maurer, 1978, and Maurer, 1978).

amplitude change of the MHM-signal in TTF-TCNQ between 4.2 K and 12 K was recently presented by Wonneberger (1979) within his theory of the Brownian motion of a charged heavy particle in a 1d-periodic potential. A phase shift ψ, however, is not produced by this theory. In contrast, a consideration of oscillations of the carrier concentration driven by the microwave field, where the term "carrier" may include single particles as well as CDWs, was recently proposed by Seeger (1979) and supplies an explanation for the rapid phase change by 90° between 12 and 20 K in TTF-TCNQ.

It should be mentioned that some basic transport data like the carrier concentration and the mobility have not yet been determined for the semiconducting phase of TTF-TCNQ. Information from Hall-effect measurements, both in the Ohmic and non-Ohmic regime, though extremely difficult to obtain experimentally, would be highly desirable for a thorough understanding of the transport properties.

Non-linear transport has also been observed in quinolinium-$(TCNQ)_2$ ($(Qn(TCNQ)_2)$) by G. Mihaly et al. (1979). They have applied pulses with a duration in the range of 10 μsec to 10 msec to two-probe and four-probe samples. The normalized conductivity is plotted versus the applied field strength for three different temperatures in Fig. 7. Up to 40 K, the deviations from Ohm's law increase with increasing temperature; above 40 K they decrease again as shown in Fig. 8, where the conductivity at 100 V/cm, normalized by the zero field conductivity, is plotted versus temperature. This maximum in the nonlinear behavior at 40 K is only an apparent one: it corresponds to the maximum visibility of the non-Ohmic component

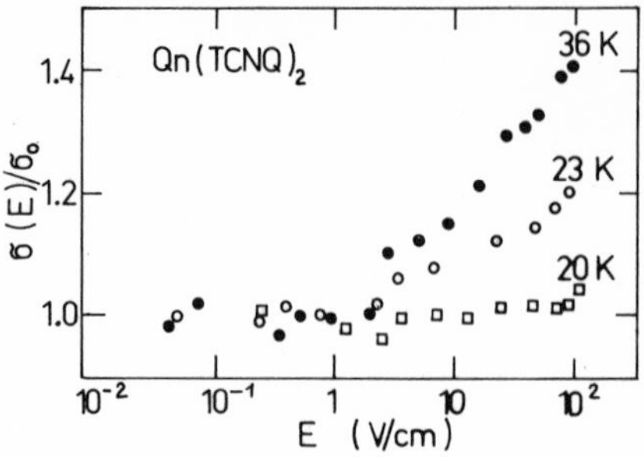

Fig. 7. Normalized conductivity vs. electric field in $Qn(TCNQ)_2$ at various temperatures (Mihaly et al., to be published).

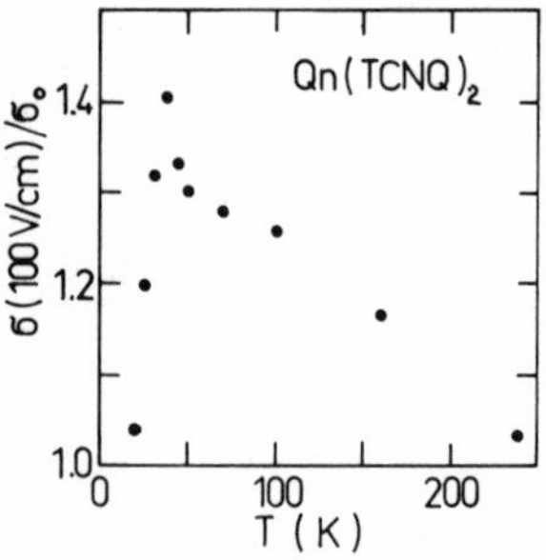

Fig. 8. Normalized conductivity at an electric field of 100 V/cm vs. temperature for $Qn(TCNQ)_2$ (Mihaly et al., to be published).

σ_A with a different temperature dependence. An analysis of the conductivity according to $\sigma(E) = \sigma_A + \sigma_B^0 \exp(-\Delta/T - E\ E)$ yields values of $\sigma_B^0 \sim 40\ (\Omega cm)^{-1}$, $\Delta \sim 20$ meV, and $E_0 = 10$ V/cm independent of the temperature. The authors also report that samples containing microscopic breaks exhibit strongly enhanced nonlinearities thereby demonstrating that spurious effects in this and similar materials must be carefully studied, before the results are discussed as bulk properties.

NON-OHMIC EFFECTS IN KCP

In many respects, $K_2[Pt(CN)_4]Br_{0.3} \cdot 3H_2O$ (KCP) was a unique model system for the early studies of 1D-effects (Zeller, 1975). The occurrence of a Peierls instability was experimentally confirmed by X-ray and neutron diffraction studies (Renker and Comes, 1975).

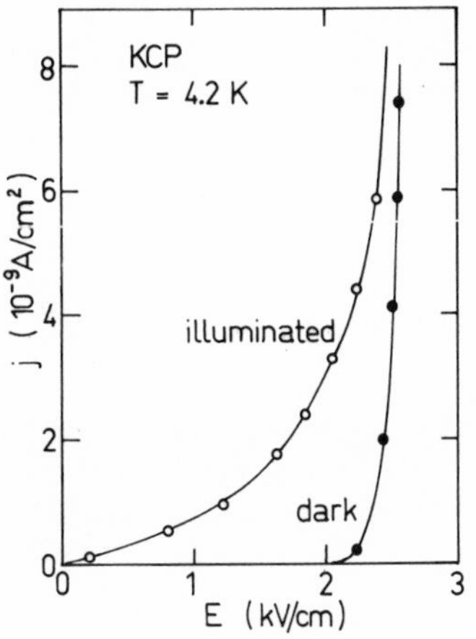

Fig. 9. DC current density-field strength characteristic for KCP at 4.2 K (Zeller, 1975).

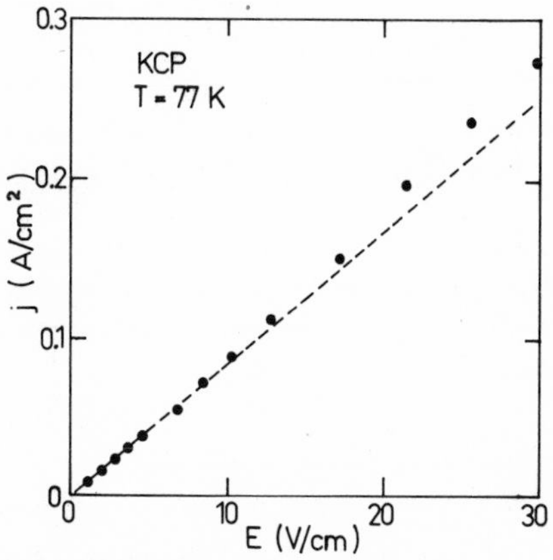

Fig. 10. Pulsed current density-field strength characteristic for KCP at 77 K (Phillipp, 1979).

In his review article, Zeller (1975) reported data for the electric field dependence of the current density in KCP at 4.2 K, which are reproduced in Fig. 9. His tentative explanation was a field-induced unpinning of a pinned charge density wave. However, he did not comment on the difference between the two characteristics obtained under dark and illuminated conditions and a detailed description of the experiment has never been published. At these low temperatures the sample resistance is tremendously high and prevents the use of pulsed electric fields. At 77 K, however, good samples exhibit

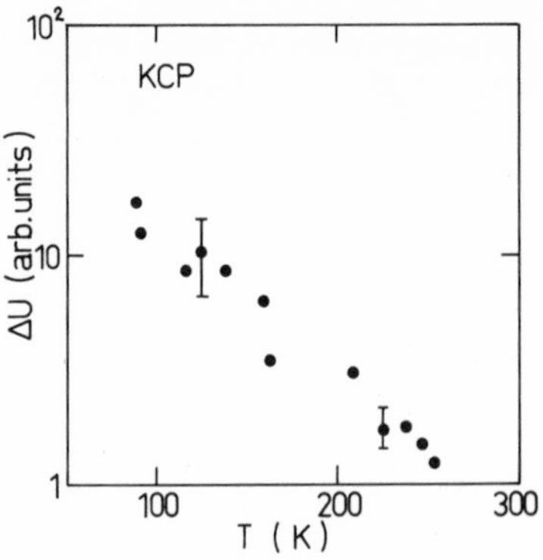

Fig. 11. MHM signal ΔU vs. temperature for KCP (Mayr et al., unpublished).

resistance values of a few kΩ only, and Phillip (1979) recently succeeded to perform measurements of the j-E characteristic of KCP as shown in Fig. 10 using pulses of 4 μsec duration. A super-linea behavior shows up at field strength as low as 10 V/cm with a value of β about 1 x 10^{-4} cm^2/V^2. In addition, we have recently obtained MHM data on samples of KCP as a function of temperature (Mayr et a unpublished). The amplitude of the signal, plotted in Fig. 11 vs. temperature shows a monotonic decrease between 77 K and 250 K, wher it is finally no longer distinguishable from noise, which is a part cular problem in such measurements on KCP. Comparing the data to t results in TTF-TCNQ, one notes that the critical electric field strengths for super-linear deviations from Ohm's law are almost ide tical in both cases and one is tempted to attribute them to the sam physical origin (For a review of the electrical properties of KCP, see Underhill and Wood, 1978).

NON-OHMIC CONDUCTIVITY OF QUASI-1D TRICHALCOGENIDES: $NbSe_3$

$NbSe_3$ crystallizes in the form of fibrous strands and exhibits anomalous transport properties (Haen et al., 1975). Upon cooling, the resistivity drops approximately linearly with temperature but shows two anomalies at T_1 = 144 K and T_2 = 59 K as displayed in Fig. 12, which is taken from a publication by Ong and Monceau (1977 At microwave frequencies of 9.3 GHz, the anomaly below T_2 is completely wiped out and that below T_1 is nearly so. It was found tha the anomalies in the dc conductivity can be suppressed by the appli cation of electric fields (Monceau et al., 1976). This suppressio is shown in Fig. 13 for the T_1 = 144 K anomaly and in Fig. 14 for the T_2 = 59 K anomaly, in both cases for three different values of the current density. The nonohmic conductivity can be expressed in

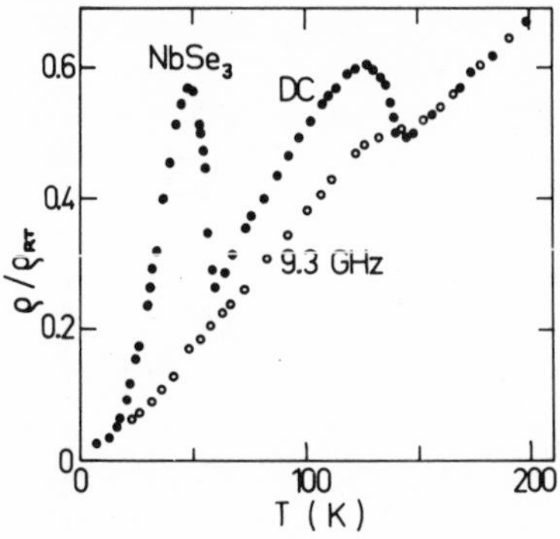

Fig. 12. Normalized resistivity as a function of temperature for $NbSe_3$. Full circles: dc-data; open circles: microwave data at 9.3 GHz (Ong and Monceau, 1977).

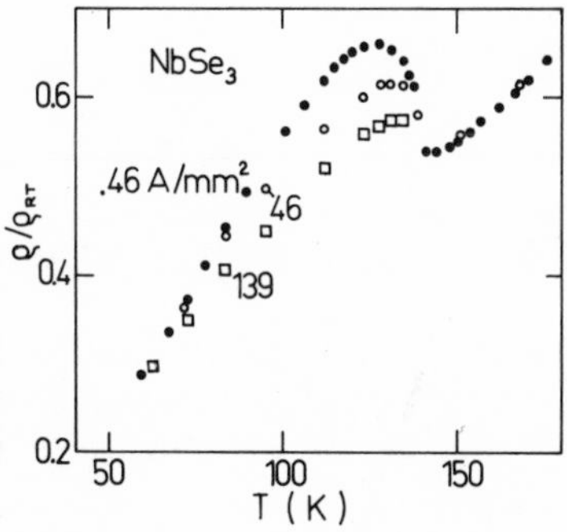

Fig. 13. Normalized resistivity as a function of temperature at three current densities near the T = 59 K resistivity anomaly in $NbSe_3$ (Monceau et al., 1976).

the form $\sigma = \sigma_a + \sigma_b \exp(-E_o/E)$ where σ_a is the low-field conductivity (Ong, 1978). The critical field E_o strongly depends on temperature and has a minimum of about 1.5 V/cm below T_1 and 0.1 V/cm below T_2. X-ray scattering studies by Fleming et al. (1978) have proven the formation of two independent, incommensurate CDW's at T_1 and T_2. These X-ray studies were also performed in the presence of electric fields large enough to wipe out at least half of the resistive anomalies and gave evidence for an undiminished presence of the CDW's. Measurements on samples with different resistance ratios revealed that the characteristic field E_o scales with the number of defects: the samples having a lower resistance ratio due to a higher impurity or defect concentration exhibit a higher critical electric field E_o.

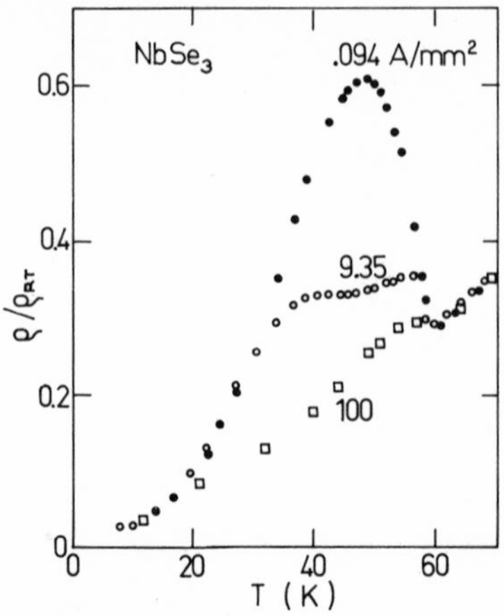

Fig. 14. Normalized resistivity as a function of temperature at three current densities near the T = 59 K resistivity anomaly in $NbSe_3$ (Monceau et al., 1976).

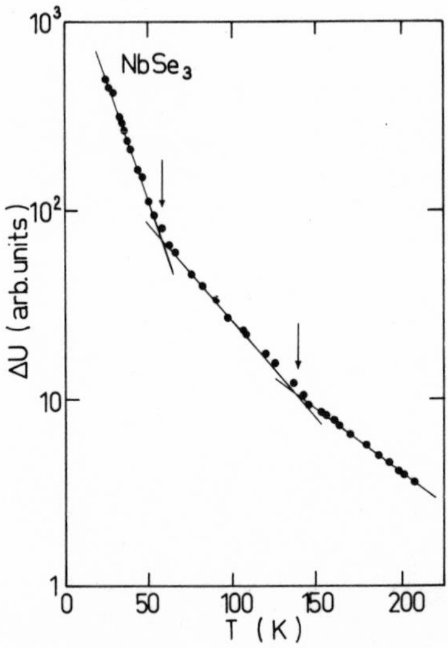

Fig. 15. MHM mixing signal ΔU vs. temperature for $NbSe_3$. The arrow show the temperatures of the phase transitions at T = 59 T = 144 K (Mayr et al., 1979).

These authors therefore conclude "that $NbSe_3$ is the first example of a Fröhlich 'sliding mode' conductor with CDW's pinned at impurity or defect sites".

In view of the disappearance of the resistance anomalies at microwave frequencies it seemed to be interesting to perform a MHM experiment on samples of $NbSe_3$. The result of such a measurement is shown in Fig. 15 (Mayr et al., 1979). Surprisingly, a harmonic mixing signal, which increases with decreasing temperature is also observed at temperatures higher than 144 K. At the temperature of the superlattice formation, breaks show up in the slope of the logarithm of the harmonic mixing signal vs. temperature indicating an abrupt change of the parameter β characterizing the magnitude of the nonlinearity. Although Tsutsumi et al. (1977) have reported 1D-precursor scattering in $NbSe_3$, observable in their electron diffraction measurements above T_1, which seems to be evidence for fluctuations to a CDW-distorted state and an associated fluctuating nonlinear conductivity, no such precursor effects were found by Fleming et al. (1978), and the role of fluctuations in $NbSe_3$ remains controversial.

The theoretical approaches to nonlinear transport in $NbSe_3$ generally accept the CDW formation as the basic physical phenomenon involved. Maki (1977, 1978) considered tunneling of phase solitons in an external electric field. His theory accounts for the experiments only at very low temperatures and there is no agreement between the proposed value of the characteristic field and the experimental results. G. Mihály and L. Mihály (1979) have extended Maki's work taking into account the effect of bound soliton-antisoliton pairs. Using reasonable microscopic parameters, they find non-Ohmic conductivity setting in at low electric fields and persisting to high temperatures. Alternative theories were developed by Bardeen (1979), who considers Zener-type tunneling of the CDW across a gap at the Fermi surface that is determined by the pinning frequency, and by Lee and Rice (1979) who treat the pinning of a CDW by impurities in systems that exhibit at least short range order in three dimensions.

NONLINEAR TRANSPORT IN HIGHLY CONDUCTING POLYMERS

The highly conducting polymer $(SN)_x$ is different from the materials discussed in the previous sections insofar as its electronic anisotropy is much lower and a periodic lattice distortion associated with a CDW has never been observed (for a review, see Greene and Street, 1977). A measurement of the MHM signal by Maurer (1978) is presented in Fig. 16 for the temperature range between 4.2 and 50 K. Apparently, there exists a nonlinear contribution to the conductivity, which strongly increases with decreasing temperature. At this point, the importance of lattice defects for the Ohmic transport properties

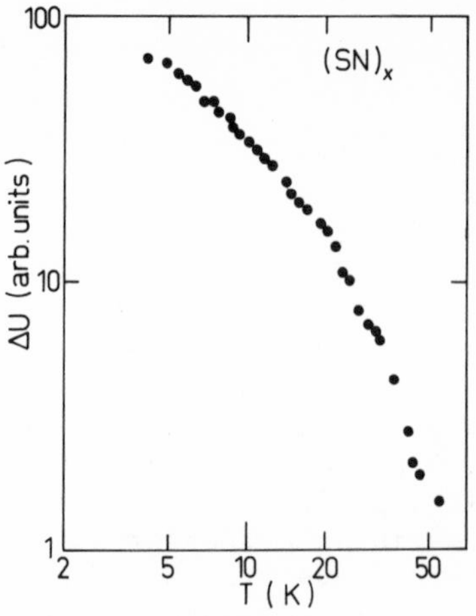

Fig. 16. MHM signal ΔU vs. temperature for $(SN)_x$ (Seeger and Maur[e] 1978, and Maurer, 1978).

of $(SN)_x$ should be emphasized, which was reported by Kahlert and Seeger (1976).

Doped polyacetylene $(CH)_x$ is the second highly conducting poly-mer, which has been extensively studied during the last two years (for a review of the electrical and optical properties, see Seeger and Gill, 1979). Films doped with iodine or AsF_5 exhibit a metalli[c] behavior of the thermoelectric power and optical properties typical for a metal. The temperature dependence of the resistivity, howeve[r] is anomalous for a metal: the resistivity increases with decreasin[g] temperature. This anomalous behavior was recently explained by Sheng (1979), who developed a theory for fluctuation-induced tunnel-ing conduction in disordered materials. In this mechanism, thermal[ly] activated voltage fluctuations across insulating gaps play an important role in determining the temperature and field dependences o[f] the conductivity. The theory is ideally suited to be applied to conduction in doped $(CH)_x$, which is not a single-crystal material, but consists mostly of fibrils about 200-300 Å in width and of extreme length, which are entangled in all directions. Measurements of the amplitude and relative phase of the MHM signal are plotted for $(CH)_x$ doped with iodine in Fig. 17 (Fig. 18 shows the correspon[d]ing temperature variation of the sample resistance) and for $(CH)_x$ doped with AsF_5 in Fig. 19 (Fig. 20 displays the temperature variation of the sample resistance)(Kahlert et al., 1979). In both cas[es] there is clear evidence for a nonlinear behavior of the conductivit[y]. We believe that these results are strong additional evidence for th[e] adequateness of Sheng's theory to the electronic transport in doped $(CH)_x$, since it predicts a nonlinear conductivity, where the magnit

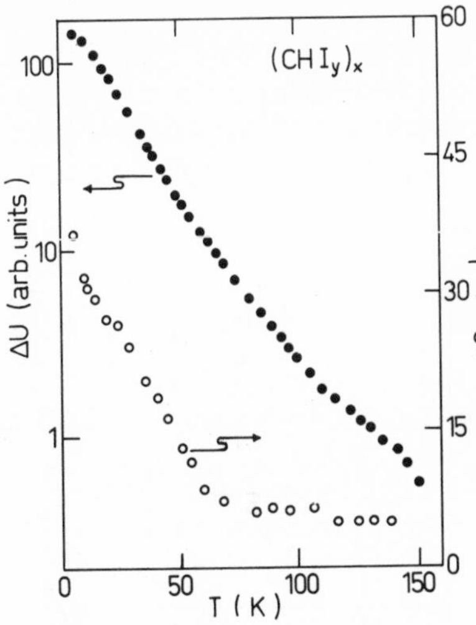

Fig. 17. Amplitude ΔU and relative phase ψ of the MHM signal for $(CH)_x$ doped with iodine (Kahlert et al., to be published).

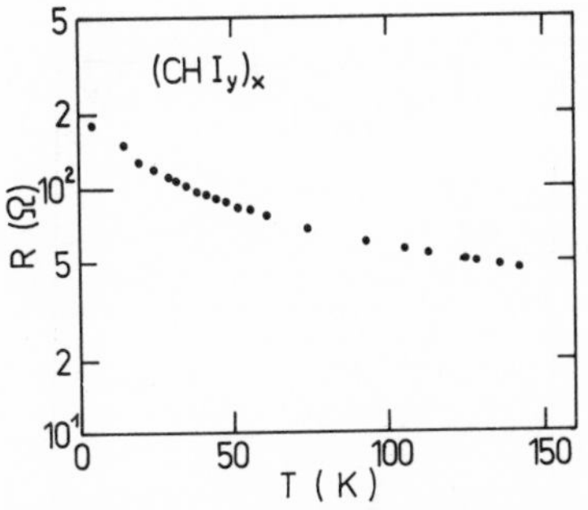

Fig. 18. Resistance of the $(CHI_y)_x$ sample of Fig. 16 vs. temperature (Kahlert et al., to be published).

of the nonlinearity should decrease with increasing temperature, as observed in our experiment. A detailed comparison of the data with the theory is presently in progress.

SUMMARY

The occurrence of nonlinear transport effects appears to be a common phenomenon in many quasi-1D conductors studied so far. In contrast to the field dependent conductivity in semiconductors

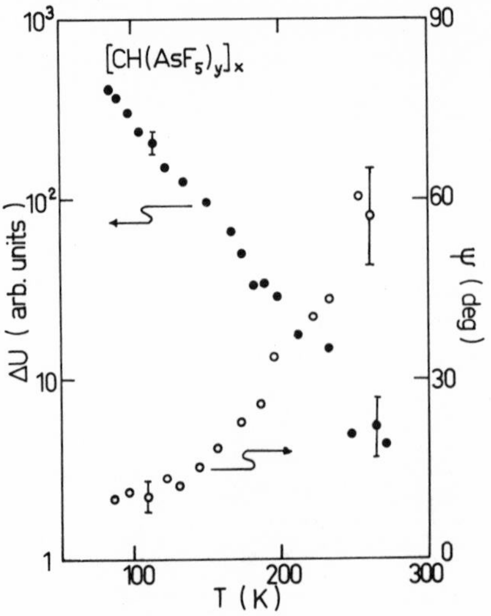

Fig. 19. Amplitude ΔU and relative phase ψ of the MHM signal for $(CH)_x$ doped with AsF (Kahlert et al., to be published).

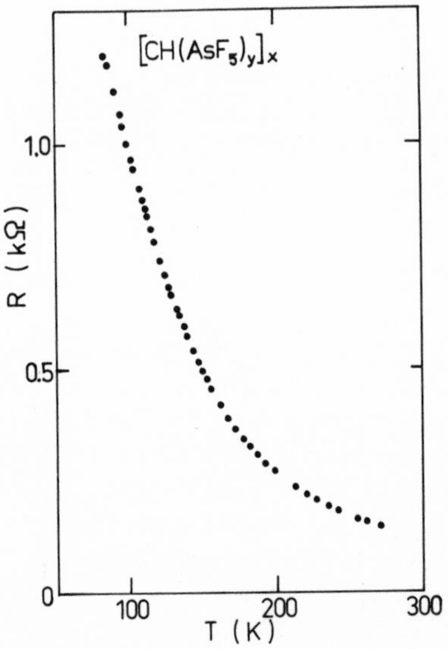

Fig. 20. Resistance of the $[CH(AsF_5)_y]_x$ sample of Fig. 18 vs. temperature (Kahlert et al., to be published).

which either increases or decreases with electric field, depending on the operating mechanisms like impact ionization, carrier repopulation or energy dependence of the scattering rates, the conductivity in quasi-1D conductors is invariably found to increase with applied electric field. For the case of $NbSe_3$, there is general agreement that the formation of CDWs is the physical basis responsible for the nonlinear transport properties. Because of the lack of very fundamental experimental information, such as Hall-effect measurements over a wide range of temperatures and electric fields, the question of the origin of the nonlinearity in the charge transfer salts like TTF-TCNQ and in KCP is not yet satisfactorily resolved. The non-Ohmic phenomena in highly conducting polymers, where only a few experiments have been performed and more information is certainly required for a deeper understanding, seem to be related to the particular non-single-crystalline morphology of the macroscopic samples, where barriers between conducting regions or strands play a dominant role.

ACKNOWLEDGEMENTS

I am grateful to Prof. K. Seeger, who first gave me the impulse to study nonlinear transport phenomena in quasi-1D solids, for his continuous encouragement and support. I thank my colleagues W. Maurer, W. Mayr, B. Kundu, and A. Philipp for the permission to include unpublished data.

BIBLIOGRAPHY

Bardeen, J., 1979, submitted for publication.
Cohen, M. J., and Heeger, A. J., 1979, Phys. Rev. B 16:688.
Cohen, M. J., Newman, P. R., and Heeger, A. J., 1976, Phys. Rev. Letters 37:1500.
Coleman, L. B., Cohen, M. J., Sandman, D. J., Yamagishi, F. G., Garito, A. F., and Heeger, A. J., 1973, Sol. State Commun. 12:1125.
Fleming, R. M., Moncton, D. E., and McWhan, D. B., 1978, Phys. Rev. B 18:5560.
Friend, R. H., and Jerome, D., 1979, J. Phys. C 12:1448.
Fröhlich, H., 1955, Proc. Roy. Soc. A223:296.
Greene, R. L., and Street, G. B., 1977, in "Chemistry and Physics of 1D Metals," H. J. Keller, Ed., Plenum Press, New York, p. 167.
Haen, P., Monceau, P., Tissier, B., Waysand, G., Meerschaut, A., Molinie, P., and Rouxel, J., 1975, in "Proc. 14th Intern. Conf. on Low Temp. Phys.", M. Krusius and M. Vuorio, Eds., North Holland, Amsterdam, 5:445.
Kahlert, H., 1975, Sol. State Commun. 17:1161.
Kahlert, H., unpublished.

Kahlert, H., Mayr, W., Kundu, B., and Seeger, K., 1979, to be published.

Kahlert, H., and Seeger, K., 1976, in "Proc. 13th Intern. Conf. on Phys. Semiconductors", F. G. Fumi, Ed., Tipografia Marves, Rome, p. 353.

Keller, H. J., Ed., 1975, "Low-dimensional Cooperative Phenomena," Plenum Press, New York.

Keller, H. J., Ed., 1977, "Chemistry and Physics of 1D Metals", Plenum Press, New York.

Lee, P. A., and Rice, T. M., 1979, Phys. Rev. B, in press.

Lee, P. A., Rice, T. M., and Anderson, P. W., 1974, Sol. State Commun. 14:703.

Maki, K., 1977, Phys. Rev. Letters 39:46.

Maki, K., 1978, Phys. Rev. B 18:1641.

Maurer, W., 1978, thesis, unpublished.

Mayr, W., Kahlert, H., Monceau, P., and Seeger, K., 1979, to be published.

Mayr, Z., Kahlert, H., and Philipp, A., unpublished.

Mihaly, G., Jánossy, A., Kurti, J., Forro, L., and Gruner, G., 1979 Phys. Stat. Sol. (b), in press.

Mihaly, G., and Mihaly, L., 1979, Sol. State Commun., in press.

Miller, J. S., and Epstein, A. J., Eds., 1978, "Synthesis and Properties of Low-dimensional Materials," New York Acad. Sci., New York.

Monceau, P., Ong, N. P., Portis, A. M., Meerschaut, A., and Rouxel, J., 1976, Phys. Rev. Letters 37:602.

Ong, N. P., 1978, Phys. Rev. B 17:3243.

Ong, N. P., and Monceau, P., 1977, Phys. Rev. B 16:3443.

Päl, L., Gruner, G., Jánossy, A., and Solyom, J., Eds., 1977, "Organic Conductors and Semiconductors," Springer-Verlag, New York.

Philipp, A., 1979, to be published.

Renker, B., and Comés, R., 1975, in "Low Dimensional Cooperative Phenomena," H. R. Zeller, Ed., Plenum Press, New York, p. 235.

Rice, M. J., Bishop, A. R., Krumhansl, J. A., and Trullinger, S. E., 1976, Phys. Rev. Letters 36:432.

Rice, J. M., Strässler, S., and Schneider, W. R., 1975, in "One Dimensional Conductors," H. G. Schuster, Ed., Springer-Verlag, New York, p. 151.

Schuster, H. G., Ed., 1975, "One Dimensional Conductors," Springer-Verlag, New York.

Seeger, K., 1976, Sol. State Commun. 19:245.

Seeger, K., 1979, submitted for publication.

Seeger, K., and Gill, W. D., 1979, Colloid and Polymer Science, in press.

Seeger, K., and Maurer, W., 1978, Sol. State Commun. 27:603.

Sheng, P., 1979, to be published.

Tsutsumi, K., Takagaki, T., Yamamoto, M., Shiozaki, Y., Ido, M., Shambongi, T., Yamaya, K., and Abe, Y., 1977, Phys. Rev. Letter 39:1675.

Underhill, A. E., and Wood, D. J., 1978, in "Synthesis and Properties of Low Dimensional Materials," J. S. Miller and A. J. Epstein, Eds., New York Acad. Sci., New York, 313:516.
Wonneberger, W., 1979, in "Proc. Intern. Conf. on Quasi-1D Conductors, Dubrovnik, 1978," Springer-Verlag, Berlin.
Wonneberger, W., 1979, Sol. State Commun., in press.
Wonneberger, W., and Gleisberg, F., 1977, Sol. State Commun. 23:665.
Zeller, H. R., 1975, in "Low Dimensional Cooperative Phenomena," H. J. Keller, Ed., Plenum Press, New York, p. 215.

OPTICAL ABSORPTION OF SOLIDS UNDER LASER IRRADIATION*

L. D. Laude and M. Wautelet

Faculté des Sciences,
Université de l'Etat
Mons, Belgium

INTRODUCTION

Little attention has been paid so far on the effects produced on the electronic structure of solids by intense electromagnetic fields. Since the latter is of fundamental interest in interpreting a number of physical properties of solids, it would seem necessary to delineate and evaluate properly such effects both theoretically and experimentally. It is the object of this paper to specifically examine this problem. In particular, it will be shown that band-gaps are opened in $\vec{k}$-space regions where direct transitions occur. The resulting perturbations of the optical spectra of the laser irradiated solids will be discussed, comparison with existing data will be drawn, and meaningful experiments which have yet to be performed will be outlined.

BAND STRUCTURE EFFECTS

Theory

In the one-electron approximation, and in the limit of optical wavelengths much greater than the lattice parameter, the Hamiltonian of the electron-phonon-photon system is given by

$$H(r,t) = H_{el}(r,t) + H_1(r,t) \ , \qquad (1)$$

*Work supported by project IRIS of the Belgian Ministry for Science Policy.

where

$$H_{el}(r,t) = \frac{1}{2m}[\vec{p} + \alpha\,\vec{A}(r,t)]^2 + V(r) \qquad (2)$$

is the electron-photon Hamiltonian in which p is the momentum operator of the electron, A(r,t) is the vector potential of the electromagnetic wave V(r) is the periodic potential seen by the electron, and α is a coupling factor. In addition,

$$H_1(r,t) = -\,(NMC)^{-\frac{1}{2}} \sum_{q,\lambda} Q_{q,\lambda}(t)e_{q,\lambda}(a)\Delta V_a(r)\exp(iq\cdot R_a) \qquad (3)$$

is the electron-phonon Hamiltonian (Ziman, 1967), a labels the atom position, $V_a(r)$ is the potential associated with the atom in A, Q_q, are the normal coordinates of atomic displacements at frequency ω, $\vec{q}$ is the wave vector, and λ denotes the phonon branch.

One has to solve the time-dependent Shrödinger equation

$$i\hbar\,\frac{\delta|\psi(r,t)>}{\delta t} = H(r,t)|\psi(r,t)>, \qquad (4)$$

where $|\psi(r,t{=}0)>$ are wave functions of the system when $H_1(r,t)=0$ and without laser irradiation.

In the following, the procedure given by Tzoar and Gersten (1975), based upon that of Shirley (1965) is adopted. First, it is necessary to transform the system Hamiltonian (1) into a new Hamiltonian via a cononical transformation. For this purpose we set

$$\psi = \phi\cdot\exp(\theta), \qquad (5)$$

where

$$\theta = -i\hbar \int dt[\frac{\alpha}{m}\,Ap + \frac{\alpha^2}{2m}\,A^2 + H_1(r,t)] \qquad (6)$$

It is found that

$$[\frac{p^2}{2m} + U(r,t) - \frac{i}{\hbar}\frac{\partial}{\partial t}]\,\phi = 0, \qquad (7)$$

with

$$U(r,t) = V(r) + 2\vec{p}\theta\cdot\vec{p} + p^2\theta + (p\theta)^2 + \vec{A}(r,t)\cdot\vec{p}\,\theta\,. \qquad (8)$$

U(r,t) is spatially and temporally periodic, but with two time periods: one corresponding to the photons, the other to the lattice vibrations. It then appears useful to Fourier analyze U(r,t), i.e.

$$U(r,t) = \sum_{G,q,n} U_{G,q,n} \exp\left[i((\vec{G}+\vec{q})\cdot\vec{r}-n\cdot\hbar\omega t)\right] \quad (9)$$

and

$$U_{G,q,n} = \int dr \int dt \; U(r,t) \exp\left[-i((\vec{G}+\vec{q})\cdot\vec{r}-n\cdot\hbar\omega t)\right] \quad , \quad (10)$$

where G is a reciprocal lattice vector, n and ω are each defined from a set of two values: integers n_1 and n_2, and frequencies ω_1 and ω_2, respectively, where the subscripts refer to photons "1" or phonons "2". After relevant approximations, i.e., neglecting the photon-photon and photon-phonon interactions, contained in $\vec{A}.\vec{p}\,\Theta$, as well as the quadratic term $(p\Theta)^2$, (8) reduces to

$$U(r,t) + V(r) = 2p\Theta\, p + p^2\Theta. \quad (11)$$

Let

$$\phi(r,t) = \exp\; i(\vec{k}\cdot\vec{r}-\frac{\varepsilon t}{\hbar}) \quad u(r,t) \quad , \quad (12)$$

in which

$$u(r,t) = \sum_{G,q,n} u_{G,q,n} \exp\; i((\vec{G}+\vec{q})\cdot\vec{r}-n\cdot\hbar\omega t) \quad . \quad (13)$$

Inserting (9), (10, (12), (13) into (7), leads to

$$\left[\frac{\hbar^2}{2m}(k+G+q)^2 - \varepsilon - n\cdot\hbar\omega\right] u_{G,q,n} + \sum_{G',q',n'} U_{G-G',q-q',n-n'} \; x$$

$$x \; G',q',n' = 0. \quad (14)$$

The eigenvalues ε of (14) are found by zeroing the Hill determinant:

$$\Delta(k,\varepsilon) = \det\left[\delta_{G,G',qq',nn'} + \frac{U_{G-G',q-q',n-n'}}{(\hbar^2/2m)(k+G+q)^2-\varepsilon-n\cdot\hbar\omega}\right]. \quad (15)$$

One may note that the actual solutions are the solutions of a steady-state problem (Tzoar and Eersten, 1975). In addition, eigenfunctions at times t and t+τ depend on each other following the relation

$$\phi(r,t+\tau) = \exp(-\frac{i\varepsilon t}{\hbar})\,\phi(r,t).$$

Moreover,

$$\Delta(k,\varepsilon+n\cdot\hbar\omega) = \Delta(k,\varepsilon), \quad (17)$$

i.e., all the electronic strucutre is contained in an energy domain which extends from zero to $\hbar(\omega_1+\omega_2)$. This range is variable, since ω_2 is very small compared to ω_1, particularly in the visible and ultraviolet part of the electromagentic spectrum. As a result, the conclusion of Tzoar and Gersten (1975) that the laser-perturbed band structure of the solid is contained in a fundamental hyper-rectangle in (ε,k) space is still aproximatively valid when electron-phonon interactions are taken into consideration. The latter would appear to simply broaden the electron bands of the phonon-free system.

When phonons are not included in the system, the term $U_{G-G', q-q' n-n'}$ in (15) reduces to $U_{G-G', n-n'}$ of the direct transition problem (Tzoar and Gersten, 1975), i.e.,

$$U_{G,n_1} = V_G J_{n_1}\left(\frac{\alpha}{\omega_1}\vec{G}\cdot\vec{A}\right), \tag{18}$$

where $V_G = \int dr\,[V(r)\exp(-i\vec{G}\cdot\vec{r})]$, and J_{n_1} is the Bessel function of order n_1. When phonons are introduced in the system, solutions of (15) are different but close to the ones of the phonon-free case. Since the electron-phonon transition matrix elements are proportional to the number of phonons, which increases with increasing temperature, effects in the band structure are expected to be more important at high temperature.

Discussion

As already noted for atomic systems, displacements of the energy levels arise from static effects of the electromagnetic field, and characterize the Lamb effect observed in atomic hydrogen (Shirley, 1965, Bartolo, 1963). A similar effect may be predicted in the case of semiconductors. For simplicity, consider the Kröning-Penney model, for which rigorous solutions of E(k) may be calculated. In the case of a δ-potential, E is related to k by (Haug, 1972)

$$\cos(2\eta k) = \frac{\eta y}{2}\sin(2\eta\sqrt{E}) + \cos(2\eta\sqrt{E}), \tag{19}$$

in which n=m=e=1 and y measures the depth of the potential well. The two first bands, calculated for y=0.3, are shown in Fig. 1(a). The corresponding density-of-states appears in Fig. 1(b). In both of these figures, the effect of a laser field is indicated at a given laser photon energy, as calculated from (16). When an optical transition occurs away from the Brillouin zone boundaries, a gap appears in the structure. For comparison, a transition occuring between valence and conduction band edges produces opposite shifts of these edges, in a somewhat similar fashion as observed for atoms (Mollow, 1969, Schuda, et al., 1974).

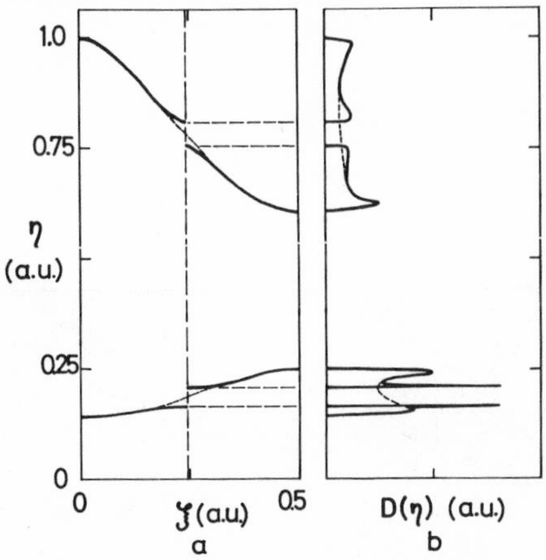

Fig. 1. a) Band structure η (ε) of the Krönig-Penney model (equation 19), without laser field perturbation (dashed line) and with intense laser field (solid line).

b) Density-of-states corresponding to a dashed line without laser. Continuous line in the presence of an intense laser field.

Constant energy surfaces in three-dimensional k-space

Let us now consider a three-dimensional solid with spherical constant energy surfaces. Effects of laser photon abosrption are: a) to bend these surfaces and b) to create stateless (forbidden) pockets within $\vec{k}$-space. When direct transitions alone are considered, a small volume in $\vec{k}$-space is perturbed (Fig. 2). With electron-phonon interactions taken into account from (15), a much greater number of $\vec{k}$-directions are sensitized to laser abosrption. In addition, indirect (phonon-assisted) optical transitions may initiate within these perturbed $\vec{k}$-space regions. Consider two of these regions centered along a particular axis (say in $\vec{k}_1$ and $\vec{k}_2$, such that $\vec{k}_2=\vec{k}_1+G$). An indirect transition initiating along $\vec{k}$, near $\vec{k}_1$ and toward the vicinity of $\vec{k}_2$, and terminating along a different axis, $\vec{k}'$, would have the highest probability if the state in $\vec{k}'$ is nearly degenerate with the state about $\vec{k}_2$.

Therefore, practically all indirect transitions proceeding from states about $\vec{k}_1$ towards states about $\vec{k}_2$ would eventually terminate on states confined to a narrow energy range comprising states about $\vec{k}_2$, i.e., those which are effectively perturbed by laser radiation. That energy range may cover a large number of k-directions. As a result, the occurrence of indirect transitions is seen to spread the laser-induced perturbation of the band structure over large regions of $\vec{k}$-space.

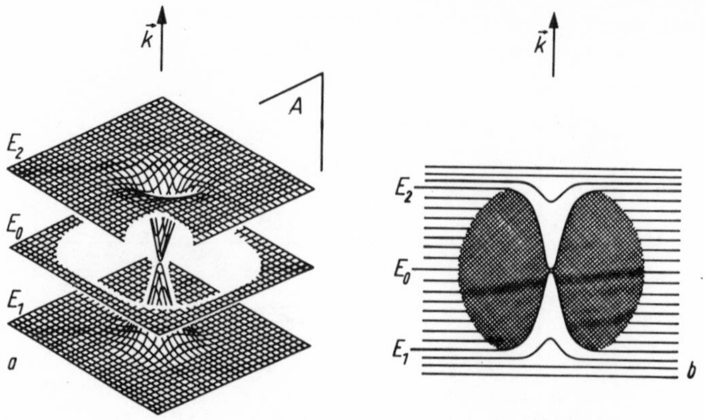

Fig. 2. a) Equal energy surface in k-space, in the presence of an intense laser field; E_o refers to energy level when the direct transition takes place; E_1 and E_2 are the split levels.

b) Section of surfaces described in a) in the place A.

Altogether, the introduction of electron-phonon interactions within the three-dimensional model appears i) to produce band-splitting comparable to the phonon-free case, and ii) to extend the notion of such energy band splitting to the occurrence of forbidden band-gaps in laser irradiated solids in the same sense as discussed by Cronstrom and Noga (1977) and Noga (1977).

OPTICAL PROPERTIES

Band splittings of the kind derived in the previous section need now to be sought experimentally. Obviously, it is the response of the irradiated solid to a second, tunable and weak (non-perturbing) photon beam which will allow us to probe the expected perturbation of the band structure resulting from the absorption of the laser photons. However, before devising an experiment along this line, one has to calculate a two photon abosrption coefficient which would account for the absorption of a non-perturbing photon-flux within a (laser) perturbed band structure.

Theory

In the one-electron approximation, the non-relativistic time-dependent Schrödinger equation describing the behavior of an electron in a solid, which interacts with two electromagnetic fields, is

$$i\hbar \frac{\partial}{\partial t} \mid \Psi (\underline{k},r,t)> = H(r,t) \mid \Psi (k,r,t)> \quad , \tag{20}$$

with

$$H(r,t)=(1/2m) \ \{p+\frac{e}{c} \ [A_1(r,t)+A_2(r,t)]\}^2+V(r), \tag{21}$$

where p is the mementum operator of the electron, V(r) is the cyrstal potential seen by the electron,

$$\vec{A}_n(r,t)=A_n\vec{e}_n\exp[i\vec{\eta}_n\cdot\vec{r}-\omega_n t)] \tag{22}$$

is the vector potential of the radiation field associated with the beam n, e_n and η_n are the polarization and wave vectors respectively, and ω_n is the frequency of the electromagnetic wave. In the absence of the electromagnetic field, the wave functions of the system are written:

$$\Psi_j(k,r,t) = \exp[-i(E_j(k)/n)t] \ \exp(i\vec{k}\cdot\vec{r})u_j(kr,) \quad , \tag{23}$$

where u (kr,) have the translational periodicity of the lattice. In the presence of photons, the system Hamiltonian (Dimmock, 1967, Girlanda, 1977, Bassani, 1966) transforms into a new Hamiltonian via a canonical transformation. One then obtains

$$\Psi = \phi \exp (\Theta), \tag{24}$$

with

$$\Theta = i\hbar \int dt \ [\frac{e}{mc}(\vec{A}_1+\vec{A}_2)\cdot\vec{p}+(\frac{e}{mc})^2 \ (A_1+A_2)^2] \ . \tag{25}$$

the Schrödinger equation becomes

$$[(p^2/2m)+U(r,t)-i\hbar(\partial/\partial t)] \ \phi + 0 \tag{26}$$

and

$$U(r,t)=V(r)+V_1 \ [\Theta,p,A_n(r,t)] \ . \tag{27}$$

When one source of photons (of energy $\hbar\omega_1$) is present, some approximations are made on the new system Hamiltonian. If the corresponding radiation field is weak, V_1 may be assumed to be zero and the ϕ are taken to be the wavefunctions of the system without photons. Θ is also small so that exp(Θ) may be approximated by 1+Θ. In the limit of large time, this gives the classical optical absorption theory.

It is generally considered that V_1 can be neglected in the pres ence of laser fields, but exp(Θ) must be calculated explicitly. How ever, the ϕ are usually taken to be the unperturbed Ψ_j given by (23). In particular, when $n\omega_1$ is such that $\hbar\omega_1 < E_g < 2\hbar\omega_1$, where E_g is the fundamental energy gap of a semiconductor, this approximation leads to the normal multiphonon absorption theory (Girlanda, 1977). With intense laser fields and $\hbar\omega_1 \geq E_g$, these approximations are no longer valid since the ϕ are different from the Ψ_j and V_1 differs from zero as shown by Tzoar and Gertsen (1975). For instance, band gaps as large as 10^{-4}eV are expected to appear in Ge for photon energies of $\hbar\omega_1$=2eV and laser pulse power of about 10kW/cm^2, for a pulse duration of 10^{-6}s.

When a second, but weak, beam (photon energy $\hbar\omega_2$) is simultaneously focused onto the solid, the perturbation of the electronic structure introduced by the corresponding field is negligible, so that the first approximation remains valid for such a beam. However the modifications of the eigenstates (24) and V_1 (27) still have to be considered explicitly.

In the following, the effect of the modification of ϕ on the absorption coefficient, measured when varying ω_2 with a fixed value of ω_1, will be considered in the semi-classical approximation, i.e. assuming $\exp\Theta \simeq 1+\Theta$ and $V_1=0$. In a second step, the influence of V_1 will be discussed in what will be called the semi-realistic approach.

The semi-classical approximation

In the limit of large t and for direct transtitions, the transition probability per unit time for photons of beam 2 (energy $\hbar\omega_2$) is given by (Bassani, 1966)

$$d\omega_2/dt-2\eta\ \hbar(e/mc)^2\ A_2^2\ |\langle\Psi_i(k,r,0)|e_2\cdot p\ |\ \Psi_f(k,r,0)\rangle|^2$$

$$(E_f-E_i-\hbar\omega_2) \quad , \tag{28}$$

where f and i refer to final (conduction) and initial states, respectively. In the limit where the two electromagnetic fields are weak,

$$\Psi_j(k,r,0) = \Psi_V(k,r,0) \quad , \tag{29}$$

where V refers to valence band one-electron Bloch states. If beam 1 produces a strong (laser) field, the initial state may no longer be considered as a pure valence state. Therefore, using the approximation $\exp(\Theta)\simeq 1+\Theta$, (Landau and Lifschitz, 1967),

$$\Psi_j(k,r,t) = (1+\alpha^2)^{-\frac{1}{2}}[\Psi_V(k,r,t) + \alpha\,\Psi_L(k,r,t)] \tag{29}$$

where the coefficient α, for direct transitions, is proportional to the transition matrix element which couples states $\Psi_V(k,r,t)$ and $\Psi_L(k,r,t)$

$$\alpha(t) = -i\frac{e}{mc}\int_o^t dt'\ \exp\{i[E_L(k)-E_V(k)-\hbar\omega_1]t'/h\}$$

$$x <\Psi_V(k,r,0)|\ \vec{A}_1(r,0)\cdot\vec{p}\,|\Psi_L(k,r,0)> . \tag{30}$$

For large t and direct transitions, the transition probability for photons in beam 2 is now given by inserting (10) into (28)

$$\frac{d\omega_2}{dt} = 2\pi\hbar(\frac{e}{mc})^2 \cdot \frac{A_2^2}{1+\alpha^2}\,[|<\Psi_V|\vec{e}_2\cdot\vec{p}\,|\Psi_{f1}>|^2\delta(E_{f_1}-E_V-\hbar\omega_2)$$

$$+\ \alpha^2|<\Psi_L|\vec{e}_2\cdot\vec{p}\,|\Psi_{f_2}>|^2\delta(E_{f_2}-E_L-h\omega_2) \quad , \tag{31}$$

with

$$\alpha^2 = 2\,\hbar\eta(\frac{e}{mc})^2\ A_1^2\ |<\Psi_V|\vec{e}_1\cdot\vec{p}\,|\Psi_L'\ >|^2\ \delta(E_L-E_V-\hbar\omega_1) . \tag{32}$$

The total transition probability per unit time and per unit volume, as calculated at $\omega=\omega_2$, is obtained by integrating (31) over the entire Brillouin zone, as

$$W(\hbar\omega_2) = 4\pi\hbar(\frac{e}{mc})^2 A_2^2 \int_{BZ} \frac{d^3k}{8\pi^3(1+\alpha^2)}\ [|<\Psi_V|\vec{e}_2\cdot\vec{p}\,|\Psi_{f_2}>|^2\delta$$

$$+\ (E_{f_1}-E_V-h\omega_2) + |\alpha|^2|<\Psi_L|\vec{e}_2\cdot\vec{p}\,|\Psi_{f_2}>|^2\delta(E_{f_2}-E_L-h\omega_2)] \tag{32}$$

This equation may be easily extended to the case of non-$\vec{k}$-conserving transitions. As long as ω_2 remains distinct from ω_1 (within experimental resolution), (33) is unambiguous. When ω_2 equals ω_1, $W(\omega_2{=}\omega_1)$ differs obviously from (33).

The absorption coefficient, K_2, for photons of energy $\hbar\omega_2$ in the presence of an intense laser beam of photon energy $\hbar\omega_1$ (different from $\hbar\omega_2$) is easily calculated from (33), and

$$K_2 = \frac{W(\hbar\omega_2)}{\text{flux}} = K_2^{\circ}+\Delta K_2(\omega_1,\omega_2) \ , \tag{34}$$

where K° is the absorption coefficient in the absence of beam 1. Using the properties of the δ-function (Bassani, 1966), we obtain

$$\Delta K_2(\omega_1,\omega_2) = c \int_L \frac{dL}{8\pi^3} \frac{1}{1+c^2|<|_V|A_1\cdot p|\Psi_L>|^2} \times \frac{|\Phi\Psi_V|\vec{A}_1\cdot\vec{p}|\Psi_L>|^2}{\nabla_k(E_L-E_V)_{E_L-E_V=\hbar\omega_1}}$$

$$\times \left(- \frac{|<\Psi_V|\vec{c}_2\cdot\vec{p}|\Psi_{f_1}>|^2}{|\nabla_K(E_{f_1}-E_V)|_{E_{f_1}-E_V=\hbar\omega_2}} + \frac{|<\Psi_L|\vec{e}_2\cdot\vec{p}|\Psi_{f_2}>|^2}{|\nabla_K(E_{f_2}-E_L)|_{E_{f_2}-E_V=\hbar\omega_2}}\right. . \tag{35}$$

The line integration runs along the intersection of equal-energy surfaces $E_L-E_V=\hbar\omega_1$ and $E_{f_i}-E_i=\hbar\omega_2$. Singularities occur when $\nabla_k(E_L-E_V)=0$ $\nabla_k(E_{f_1}-E_V)=0$ and/or $\nabla_k(E_{f_2}-E_L)=0$.

The semi-realistic approach for one-photon processes

In the limit of rather large laser fields, (33) and (35) are no longer valid and (24)-(27) have to be solved exactly in order to take into account indirect as well as multi-photon processes. Before discussing these last processes, the influence of the band structure modification on the one-photon absorption coefficient K_1 of the laser photon will first be considered in what we call the semi-realistic approach.

In this approximation, one assumes that (21) remains valid but that the effect of the laser field can be described by considering only the actual electronic structure of the laser-solid system. Then the modifications of K_1 can be considered as being due to the increase in volume, caused by the energy resolution E of the laser beam, of the Brillouin zone involved in the integration probabilities. For low laser power, the integration has to be performed in a broad zone corresponding to the laser energy resolution. By increasing the laser power, band gaps open and the integration zone increases, i.e., the absorption coefficient K_1 increases like $K_1+A_1xK_1$ +..., where K_1 is the classical optical absorption coefficient for photons $\hbar\omega_1$. As noted previously, the band splitting induced by laser fields, to a first approximation, are proportional to the one-photon transition matrix elements, i.e., proportional to A_1. In the limit of a linear electron dispersion relation E(k), the number of states involved in direct laser transitions is given by $(n \ldots A_1 n)^3$, where n^3 is the number of states to be considered at weak radiation field, and n is a proportionality factor depending on the slope of the E(k) curve at the band states under investigation. At a given A_1 value, this relation might differ between a flat-band portion and a light-electron one.

If the band splitting is of the same order of magnitude as the laser resolution, a flat band will give $(n_0+A_1n)^3 \simeq A_1^3n^3$, while the light-electron one will be like n_0^3, i.e., it will not be affected by the intense field. More generally, we obtain

$$K_1 = \frac{W_1}{\text{flux}} = \frac{[(W_1^\circ)^{1/3}+C''A_1]^3}{A_1^2(1+C''A_1^2)} = \frac{[(K_1^\circ)^3+C''A_1^{1/3}]^3}{(1+C''A_1^2)} \tag{36}$$

in which C" may be calculated from a knowledge of all points contributing to the optical absorption. The term $(1+C''A_1^2)$ arises from (29) and is due to saturation effects in the relative populations of initial and final states.

The next step consists of including the multi-photon and indirect processes in the model. This means that $|\Psi_i(k,r,t)\rangle$ would be written as

$$|\Psi_i(k,r,t)\rangle = (1+\sum_g \alpha_g^2)^{-1/2}[|\Psi_v(k,r,t)\rangle+\sum_g \alpha_q(t)|\Psi_g(k',r,t)\rangle], \tag{37}$$

in which $\alpha_g(t)$ are the transition matrix elements coupling the states $|\Psi_v(k,r,t)\rangle$ and $|\Psi_g(k',r,t)\rangle$. Calculation of the absorption coefficient could then proceed as in the previous section. At this point, it should be noted that the effect of indirect transitions could be more important than might be expected at the actual sample temperature. After say 10^{-13}s, electrons decay via electron-phonon collisions and thus create phonons, i.e., "heat" the sample. Since indirect transition probabilities are proportional to the number of phonons in the system, their contribution to K_1 would increase by a factor depending on pulse duration and laser intensity.

Semi-realistic approach to K_2

It is now possible to calculate more accurately the variation of K_2 induced by the laser irradiation at $\hbar\omega_1$. There are two cases to be considered concerning the energy resolution ΔE_2 of the weak (classical) beam 2. If ΔE_2 is of the same order of magnitude as the band splitting, modification of the integration zone has to be taken explicitly into account. However, this would seldom be the case since, due to detection problems, ΔE_2 far exceeds the energy splittings. This will be taken into account in the following.

In this case, the number of states coupled by photons $\hbar\omega_2$ would not vary significantly under laser irradiation. Then (33) and (35) remain valid once the new electronic structure is taken into consideration, i.e., replacing $W_1^\circ=C'|\langle\Psi_v|A_1\cdot p|\Psi_L\rangle|^2$ by

$$\Delta K_2(\omega_1,\omega_2)=C\int_L \frac{dl}{8\pi^3} \times \frac{1}{1+C^2|\langle\Psi_v|\vec{A}_1\cdot\vec{p}|\Psi_L\rangle|^2} \times \frac{(|\langle\Psi_v|\vec{A}_1\cdot\vec{p}\Psi_L\rangle|^{2/3}+C'A_1)^3}{|\nabla_k(E_L-E_v)}_{E_L-E_v=\hbar\omega_1}$$

$$\times \left(-\frac{|\langle\Psi_v|\vec{e}_2\cdot\vec{p}|\Psi_{f_1}\rangle|^2}{|\nabla_k(E_{f_1}-E_v)|}_{E_{f_1}-E_v=\hbar\omega_2} + \frac{|\langle\Psi_L|\vec{e}_2\cdot\vec{p}|\Psi_{f_2}\rangle|^2}{|\nabla_k(E_{f_2}-E_L)}_{E_{f_2}-E_L=\hbar\omega_2}\right) . \quad (38)$$

This equation shows that ΔK_2 may be very small in some cases, due to the opposite signs of the last two terms. Also, should optical excitations (via photons of energy $\hbar\omega_2$) be allowed both from $|\Psi_v\rangle$ and $|\Psi_L\rangle$, four levels would be coupled at the same $\vec{k}$-point in the reduced band scheme: levels $|\Psi_v\rangle$ and $|\Psi_L\rangle$ via $\hbar\omega_1$, and $|\Psi_v\rangle$ $|\Psi_{f_1}\rangle$ as well as $|\Psi_L\rangle$ and $|\Psi_{f_2}\rangle$ via $\hbar\omega_2$. The introduction of indirect and multi-photon absorption processes is easily performed by using the same procedure.

Discussion

Effects of the laser beam intensity. As discussed previously, the validity of the different approaches depends on the magnitude of the laser intensity, L, and on the respective energy resolutions of the two beams. In the case of a large energy resolution, ΔE_2, (38) or its generalization including other types of optical absorption is valid. As a result, for a given structural feature, ΔK_2 is proportional to the laser transition probability, W_1 , i.e.,

$$\Delta K_2 \sim W_1 = W_1^\circ \, [1+(I/I_A)^{1/2}]^3 \;\; /(1+I/I_s) \quad , \qquad (39)$$

where I_s measures the saturation intensity, and I_A is the intensity above which laser-field deformations become important.

The influence of I_A would be detectable for laser powers of the order of 1 kW/cm^{-2}. A simple way to evaluate the saturation intensity I_s is to assume $|\langle\Psi_v|\vec{e}_1\cdot\vec{p}|\Psi_L\rangle^2$ equal to unity. Further, the volume of the Brillouin zone which is explored by the laser beam at I_s is given by the product of the surface of energy $E_L-E_v = \hbar\omega_1$ with the resolution of the laser energy, ΔE_1. Approximating the equal-energy surface by the Fermi surface, S_F, the number of excited electrons at saturation is then

$$n = (NS_F/V_B)\Delta E \ ,$$

where N is the total number of valence electrons and V_B is the occupied volume of the Brillouin zone. In the case of Ge, at $\hbar\omega_1$=2eV and ΔE_1-10^{-5}eV, we obtain n ≈ $10^{21}cm^{-3}$ over 10^{-13}s, the electron lifetime before electron-phonon scattering occurs. This corresponds to a power of about 20 MW/cm^2, in agreement with experiment (Gibson, et al., 1972, Keilmann, 1976).

In the same approximation, it should be noted that I_s would be reduced by several orders of magnitude in direct-band-gap semiconductors, when $\hbar\omega_1$ is equal to the gap. In this case, the lifetime of electrons is the readiative one, i.e., about 10^{-9}s. This reduces I_s to about 2kW/cm^2. I_s might be reduced more drastically if impurity states lying within the gap are populated. From the above discussion, it would appear that the influence of I_A would be best seen in the kW/cm^2 range of laser power.

At laser powers below I_A, the absorption coefficient K_1 in (36) is directly related to $\overset{\circ}{K}_1$. Near I_A, the effect of V_1 would become detectable, since laser-induced band splittings are then greater than ΔE_1. At this point K_1 is described by the semi-realistic model. When I increases further, other facts have to be considered: (i) E(k) relations are not linear so that, $C''A_1^{1/3}$ in (36) has to be replaced by a function F(A1) which describes the shape of E(k) near the optical transition region; (ii) the splitting is not linear, but follows a law described by a function G(A1) (Tzoar and Gertsen, 1975, Quang, 1978). Together these lead to the replacement of $C''A_1^{1/3}$ in (36) by $F[G(A_1)]$ which depends explicitly on the non-perturbed one-electron energy structure, e.g. given the analytic expressions of $G(A_1)$, the effect of V_1 would saturate for a value of I which would depend on the band structure.

When I increases up to I_s, saturation effects become predominant (Gibson, et al., 1972, Keilmann, 1976). However, (36) predicts that K_1 at saturation is higher than the value $\overset{\circ}{K}_1(1+I/I_s)^{-1}$ given by the classical theory.

To our knowledge, experimental data were nearly all collected in the MW range, i.e., in actual saturation conditions. In contrast, I_A effects would be demonstrated at much lower powers (kW).

Structural features in K_2. As an example, let us consider three bands (labelled 1 to 3) at k points belonging to the same $\vec{k}$-direction; two of these bands are taken to be structureless, the third one shows for instance a minimum, as shown schematically in Fig. 3(a). Structure appears at $\Delta K_2(\overset{\circ}{\omega}_1,\overset{\circ}{\omega}_2)$. When $\hbar\omega_1$ is decreased, the contribution to ΔK_2 goes to zero, since no coupling exists between states in bands 1 and 2. If $\hbar\omega_1$ is increased by a small quantity, coupling between bands 1 and 2 may exist at two different $\vec{k}$-points (one-dimensional case) and that coupling may initiate additional coupling between bands 2 and 3 for two different values of

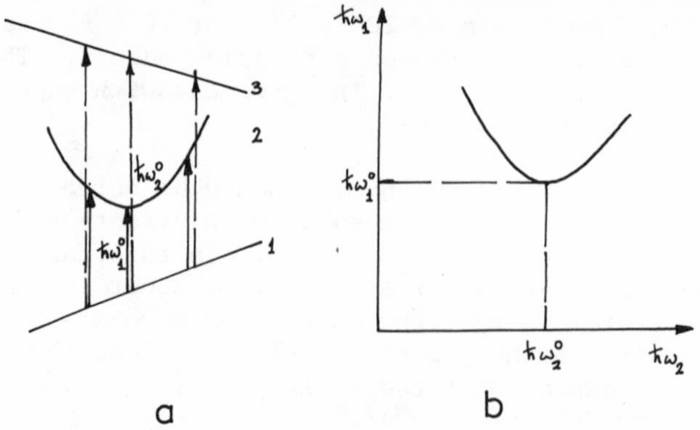

Fig. 3. a) Schematic band structure discussed in the text; curve 1 is the (E,k) profile of a valence band whilst curves 2 and 3 represent conduction bands.

b) Evolution of the iso-absorption curves ΔK_2, with ($\hbar\omega_1$, $\hbar\omega_2$), as discussed in the text, from band structure show n in a); when $\hbar\omega_1$ increases above $\hbar\omega_1^\circ$, K_2 appears at values of $\hbar\omega_2$ shown in the figure.

$\hbar\omega_2$, i.e., $\hbar\omega_2^\circ + \varepsilon_2$ and $\hbar\omega_2^\circ + \varepsilon_2'$. The same thing occurs when $\hbar\omega_1$ increases continuosly, leading to iso-abosrption curves as shown schematically in Fig. 3(b), which can be asymmetric about the $\hbar\omega_2 = \hbar\omega_2^\circ$ line. This asymmetry may give intersting information about the profile of the bands in the vicinity of $\vec{k}$-points where $\hbar\omega_1^\circ$ and $\hbar\omega_2^\circ$ transitions occur. In a general way, ΔK_2 may be evaluated in the framework of the Krönig-Penney model, assuming to a first approximation that the transition matric elements are constant. Following Fig. 1, optical absorption between the two bands is calculated and displayed in Fig. 4(a) without the laser [K_2° in (34)]-solid line) and with the laser (K_2 - dashed line). The difference between these two curves [ΔK_2 of (34)] is shown in Fig. 4(b). In relation with the above model, the band splitting of Fig. 1(a) is directly related to the width of the two peaks lying on the high and low energy sides of the ΔK_2 curve. Further, the energy separation between these two peaks equals the energy of the laser photons, $\hbar\omega_L$.

In addition, transition matrix element effects may annihilate ΔK_2. As shown in (38), ΔK_2 might be equal to zero if the two last terms on the right are equal in absolute values. Detection of ΔK_2 is possible when

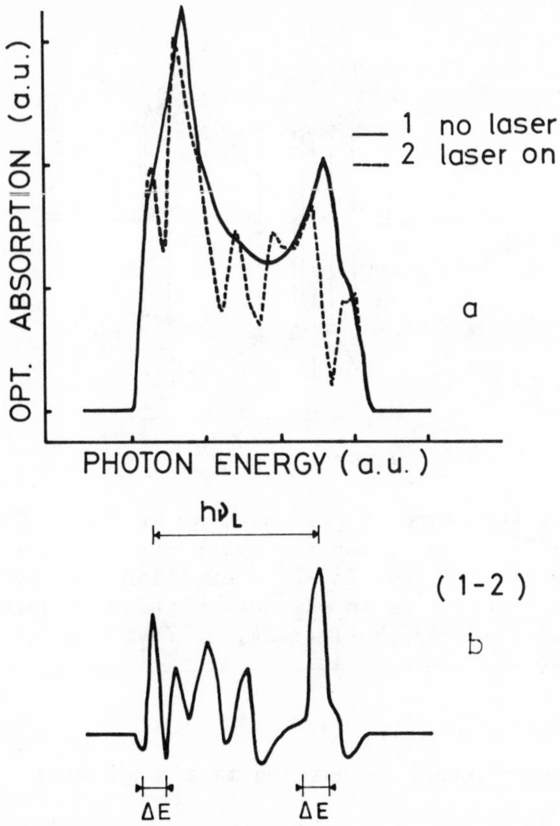

Fig. 4. a) Optical absorption coefficient, K_2, calculated within the Krönig-Penney model, without (solid line) or with (dashed line) laser.
b) Difference between the two preceding curves (ΔK_2); band splitting (ΔE) and laser photon energy ($H\nu_L$) are indicated.

$$\left|\nabla_k(E_{f_1} - E_v)\right|_{E_{f_1} - E_v = \hbar\omega_2} = 0$$

or

$$\left|\nabla_k(E_{f_2} - E_L)\right|_{E_{f_2} - E_L = \hbar\omega_2} = 0,$$

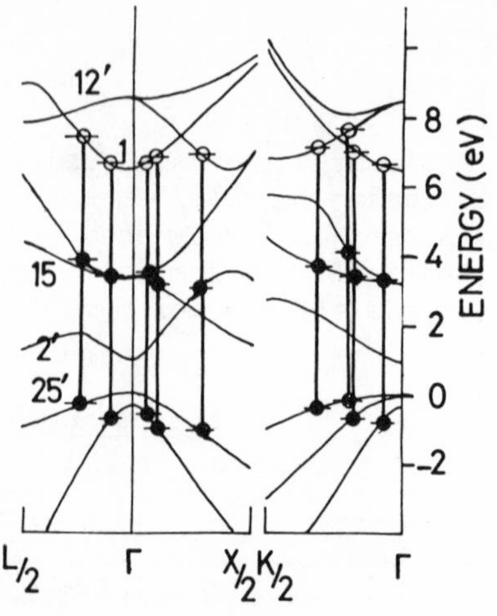

Fig. 5. Band structure of germanium about Γ from Grobman et al. (1975). From group theiry, $\Gamma_{25}'-\Gamma_{15}$ and $\Gamma_{15}-\Gamma_1$ transitions are allowed, but $\Gamma_{25}'-\Gamma_1$ transitions are forbidden. Therefore, structure in ΔK_2 due to these transitions might appear at $(\hbar\omega_1,\hbar\omega_2)=(3.24,\ 3.27\text{eV})$ but not at (6.50, 3.27eV).

for approximately equal transition matrix elements, and/or when

$$|\langle\Psi_v|\vec{e}_2\cdot\vec{p}|\Psi_{f_1}\rangle|^2 \text{ and } |\langle\Psi_L|\vec{e}_2\cdot\vec{p}|\Psi_{f_2}\rangle|^2$$

are very different. One example of the latter possibility occurs near the Γ-point in the band structure of germanium, as seen in Fig. 5 (Grobman, et al., 1975). Group theory indicates that $\Gamma_{25}'-\Gamma_{15}$ and $\Gamma_{15}-\Gamma_1$ transitions are allowed, but the $\Gamma_{25}'-\Gamma_1$ transition is forbidden; i.e., $|\langle\Psi_v|\vec{e}_2\cdot\vec{p}|\Psi_{f1}\rangle|^2 = 0$. This means that this structural feature should be detected at a ΔK_2 of 3.23 and 3.27 eV, but not at a ΔK_2 of 6.50 and 3.27 eV.

Effects of beam polarization. Another aspect of (33) refers to the polarization dependence of ΔK_2(Belousov and Oleinik, 1979). In classical (one-photon) absorption, the effects of beam polarization are broadly averaged due to the many $\vec{k}$-directions contributing to the absorption coefficient. In "double'beam" absorption, assuming that ΔK_2 can be experimentally separated from K_2, $\vec{k}$-directions are severely restricted to a few $\vec{k}$-points in the Brillouin zone, so that a more precise determination of the band structure is available. Beam polarization is introduced in the model via the matrix elements products of (33), but not via $(\vec{e}_1 \times \vec{e}_2)$. This means that if the two beams are perpendicularly polarized, ΔK_2 might be non-zero, if $\vec{e}_j \cdot \vec{k}$ differs from zero. For a single crystal, the result is that it would be possible to determine experimentally the $\vec{k}$-directions along which specific "double absorption" occurs. This may be performed in several different ways, for instance: (i) one of the beams (say $\hbar\omega_1$) is polarized, the other being left unpolarized; (ii) the two beams have parallel polarizations; (iii) the beam polarizations made an angle with each other. In the two first cases, ΔK_2 would be at a maximum when beam 1 is polarized parallel to $\vec{k}$, since $|\vec{e}_1 \cdot \vec{k}|$ would then be maximum.

<u>Energy distribution of excited electrons</u>. Another interesting consequence of (38) is that electrons which are co-operatively excited by the two beams would be distributed in two different energy regions at given photon energies. Indeed electrons would populate level E_f by absorbing either ($\hbar\omega_2$) or ($\hbar\omega_1 + \hbar\omega_2$). This effect has already been observed in the double-beam photoemission on semiconductors (Wautelet and Laude, 1977, Laude et. al., 1977). When a band gap E_g exists, excited electrons are distributed: (i) between ($\hbar\omega_2 - \hbar\omega_1 + E_g$) and $\hbar\omega_2$; and (ii) between ($\hbar\omega_2 + E_g$) and ($\hbar\omega_2 + \hbar\omega_1$), energy values being referred to the top of the valence band. As previously noted, these two energy distribution curves would have the same profile only if the actual transition matrix elements contributing to (19) were equal.

<u>Experimental considerations</u>. In one-photon classical absorption experiments, transitions occur in the vicinity of equal-difference energy surfaces in the reciprocal lattice, while ΔK_2 results from transitions along lines intersecting two equal-energy surfaces. Therefore, ΔK_2 is expected to be weak compared to K_2^o.

In order to increase the intensity of the ΔK_2 spectral features, the line integral in (33) must be transformed into a volume integral by allowing the respective photon energies to vary about their nominal values, over a range set for instance by the actual optical resolution fo the beams, which we write as ΔE_1 and ΔE_2, for beams 1 and 2 respectively. The larger $\Delta E_{1,2}$, the larger would be the volume within which optically excited $\vec{k}$-points would have to be searched for. However, increasing the laser width, ΔE_1, decreases experimentally the intensity of the central line, so that the coupling of $|\Psi_v(k)\rangle$ and $|\Psi_L(k)\rangle$ might become less efficient.

CONCLUSIONS

In this paper, a theoretical approach to the optical absorption of solids under laser- and double-beam irradiation is developed and shown to be a powerful method for investigating the band structure of solids. The model presented here predicts that an enhancement of the laser optical absorption coefficient would stem from the modification of the solid electronic structure under the laser field. The evolution of the "double-beam" absorption coefficient with ener and polarization of the beams is shown to allow the experimental determination of the transition matrix elements.

BIBLIOGRAPHY

Bartolo, B., 1963, "Optical Interactions in Solids," John Wiley, New York.

Bassani, G.F., 1966, in "Optical Properties of Solids," Ed. by J. Tauc, Academic, New York.

Belousov, I.V., and Oleinik, V.P., 1979, J. Phys. C 12: 655.

Cronstrom, C., and Noga, M., 1977, Phys. Letters A 60:137.

Dimmock, J.O., 1967, in "Semiconductors and Semimetals," Ed. by R.K. Willardson and A.C. Beer, Academic, New York, 3: 296.

Gibson, A.F., Rosito, C.A., Raffo, C.A., and Kimmit, M.F., 1972, Appl. Phys. Letters 21: 356.

Girlanda, R., 1977, Nuovo Cim. 32b: 593.

Grobman, W.D., Eastman, D.E., and Freeouf, J.L., 1975, Phys. Rev. B 12: 4405.

Hassan, A.R., 1970, Nuovo Cim. 70B: 21.

Haug, A., 1972, "Theoretical Solid State Physics," Vol. I, Pergamon, Oxford.

Keilmann, F., 1976, IEEE J. Quant. Electr. 12: 592.

Landua, L., and Lifschitz, E., 1967, "Quantum Mechanics," MIR, Moscow.

Lande, L.D., Lovato, M., Martin, M.L., and Wautelet, M., 1977, Phys. Rev. Letters 39: 1565.

Mollow, B.R., 1969, Phys. Rev. 188: 1969.

Nickle, H.H., 1967, Phys. Rev. 160: 553.

Noga, M., 1977, Phys. Letters A 62: 102.

Quang, N.V., 1978, Phys. Stat. Sol. (b) 90: 597.

Schuda, F., Stroud, C.R., Jr., and Hercher, M., 1974, J. Phys. B 7: L198.

Shirley, J.H., 1965, Phys. Rev. 138: B979.

Tzoar, N., and Gersten, J.I., 1975, Phys. Rev. B 12: 1132.

Wautelet, M., and Laude, L.D., 1977, Phys. Rev. Letters 39: 1565.

Ziman, J.M., 1967, "Electrons and Phonons," Clarendon Press, Oxford.

HIGH INTENSITY PICOSECOND PHOTOEXITATION OF SEMICONDUCTORS

Arthur L. Smirl

North Texas State University
Denton, Texas 76203

INTRODUCTION

In the past half-decade, studies of the optical properties of high-density electron-hole plasmas, generated in undoped semiconductors by the direct absorption of intense, ultrashort pulses from mode-locked lasers, have provided direct information concerning ultrafast electronic processes (the reader is referred to a previous chapter by Smirl, which reviews these studies, and to the Bibliography accompanying this paper). Generally, early experimental studies in this area employed mode-locked pulses, obtained from a Nd-glass laser, as an excitation source to generate the electron-hole plasma. This source produces optical pulses that are approximately 10 psec in duration and that often have peak powers in excess of 10^8 watts at a wavelength of 1.06 μm. These pulses when focused on the surface of a thin semiconductor sample can produce a measured irradiance of $10^{-2}J/cm^2$. Direct absorption of such an optical pulse can create carier densities exceeding $10^{20}cm^{-3}$. Germanium was chosen as a candidate for study, in many of these early investigations, primarily because it is a readily-available, well-characterized semiconductor whose bandgap energy is comparable to but less than the energy of a photon at a wavelength of 1.06 μm (1.17 eV).

As a rule, in such studies, investigators have used a variation of the "excite-and-probe" technique. Here, the semiconductor sample is first irradiated with an intense optical pulse ("excite" pulse) that causes a change in the transmission or reflection properties of the germanium. This initial pulse is followed, at some later time, by a weak pulse ("probe" pulse) that monitors the change in transmission or reflectivity of the germanium as it returns to its equilibrium condition. (A more detailed description of this measure ment technique is contained in a previous chapter.) There are a number of variations on this technique, some of which will be discussed in this paper.

This "excite-and-probe" technique is embarrassingly simple in its concept. In practice, however, quite the opposite often is true Some of the experimental difficulties that arise can be attributed t the statistical, nonlinear evolution of the optical pulse within the Nd-glass laser cavity. Typically, this laser produces a mode-locked train of 10 - 50 optical pulses in a single firing; however, the pulses vary in energy and in duration from the first pulse to the last. Moreover, the pulse train envelope usually varies from one laser firing to the next. This irreproducible and random nature of the pulse that evolves within the laser cavity precludes the selection of identical excitation pulses on subsequent firings. In addition to the uncertainty in the energy and width of the pulse from shot to shot, the transverse spatial mode structure of the laser is also of questionable quality. Deviation of the transverse mode structure from the TEM_{00} mode leads to "hot" spots on the surface of the semiconductor when the pulse is focused. Variations in the posi tion of these "hot" spots, caused by irreproducible day-to-day align ment or random changes in laser mode structure, will result in varia tions in the degree and quality of spatial overlap between the excitation and probe beams. One must keep in mind as well that one is trying to maintain exact spatial overlap of the excitation and probe pulses on the sample surface (each focused to a diameter of one milli meter or less) and that one is often irradiating the sample surface with optical intensities close to the damage threshold. Another limitation is the low repetition rate of Nd-glass laser systems; typically less than 10 firings per minute. Furthermore, if a probe wavelength different from 1.06 μm is desired, then it must be generated by some nonlinear process such as frequency doubling, tripling, or stimulated-Raman scattering. Thus, data acquisition can be a tedious and exasperating procedure.

Because of the huge carrier densities generated and the complex nature of the germanium band structure, interpretation of these experiments has also been difficult. By its very nature, the proble is a complex many-body problem involving the simultaneous interactio of many processes. As a result of these experimental and theoretica difficulties (as well as others not discussed), progress in this field has been painstakingly slow and is tedious to achieve. None-the-less, progress has slowly and steadily been achieved.

In an earlier chapter we provided the reader with an introduction to the physics of ultrafast relaxation processes in semiconductors. There, we discussed one of the early "excite-and-probe" experiments in germanium, enumerated and discussed the important physical processes that could occur during such studies on picosecond time scales and at high carrier densities, and presented an early interpretation of this experiment. We assume here that the reader is familiar with that material.

In this paper, we provide a semi-chronological account of our progress in understanding the temporal evolution of photogenerated electron-hole plasmas in germanium on a picosecond time scale. We believe such a review will be useful in providing insight into why certain investigations were undertaken and in providing a perspective of our progress in this area. It also allows us, in light of more recent studies, to make a few arbitrary comments concerning some of the earlier work. Throughout this current work, we shall again (as in our previous lecture) emphasize the recent "excite-and-probe" studies in germanium. The remainder of this seminar is organized as follows. In the next section, we review experiments in which picosecond optical pulses are used to measure the saturation and the decay of the optical absorption of germanium at 1.06 μm. We then describe investigations that attempt to isolate and measure the effects of diffusion, Auger recombination, free-carrier absorption, intervalence-band absorption, and Coulomb-assisted indirect absorption. We then discuss the possible contributions of hot phonon distributions to the picosecond optical response of germanium, and following that, we outline a recent modification of the original ESSM model (Elci et al., 1977) that includes Auger recombination and inter-valence-band absorption, as well as accounting for spatially inhomogeneous effects such as diffusion. Finally, we make some concluding remarks concerning the present status of our understanding of picosecond optical interactions in germanium and concerning remaining problems.

DYNAMIC SATURATION OF THE OPTICAL ABSORPTION

The direct absorption of a photon of energy greater than the direct-bandgap energy in germanium induces an electron to make a transition from the valence band to a state high in the conduction band, leaving behind a hole in the valence band. If a large enough number of such electron-hole pairs can be created on a short enough time scale, we can partially fill the states that are resonant with the optical transition, and the transmission of the germanium should be enhanced. As we discussed earlier, the narrow set of optically-coupled states that are resonant with the approximately monochromatic light from a mode-locked laser can be partially filled or saturated in two ways. If the optical generation rate into the optically-coupled states exceeds the out-scattering rate by which carriers are scattered out of the state, the states will be partially filled and the transmission of the germanium will increase. This process is known as state-filling. A condition of increased transparency will also be observed if the

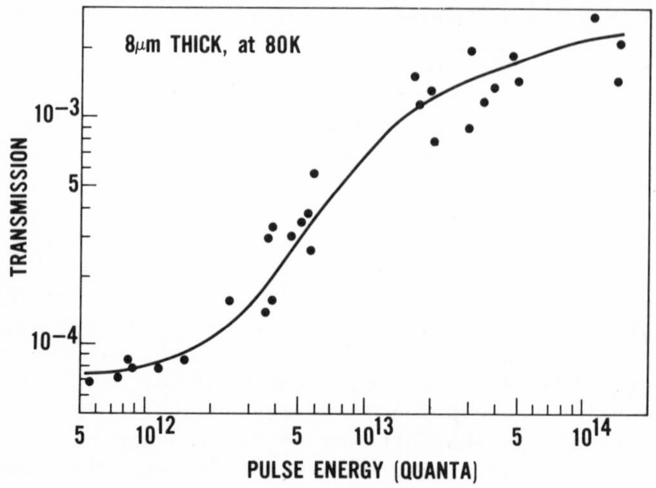

Fig. 1. Nonlinear transmission of an 8-μm-thick germanium wafer versus incident optical pulse energy at a wavelength of 1.06 μm in units of quanta (from Kennedy et al., 1974).

optical pulse generates enough electron-hole pairs to fill all of the states in either the valence or conduction band up to, and including, those required for the direct optical transition--a process called band filling. In this section, we review the technique and results of experiments that measure the degree, and duration, o this bleaching of the optical transmission of germanium on a picosecond time scale.

The first observations of a saturation of the germanium transmission were reported by Kennedy et al. (1974). They observed a decrease in the absorption of picosecond pulses of high optical intensities at 1.06 μm in thin germanium wafers. They performed tw experiments. In the first, they irradiated an 8 μm-thick, single-crystal germanium sample with picosecond pulses of varying intensity at 1.06 μm, and they measured the transmission of each pulse. A plot of germanium transmission versus incident optical pulse ener (Fig. 1) showed that the germanium transmission was bleached, or enhanced, by a factor of approximately 20 over its linear value at low intensities. Second, these authors employed the "excite-and probe" technique in an attempt to measure the decay of this enhance transmission in the following manner. They irradiated the sample

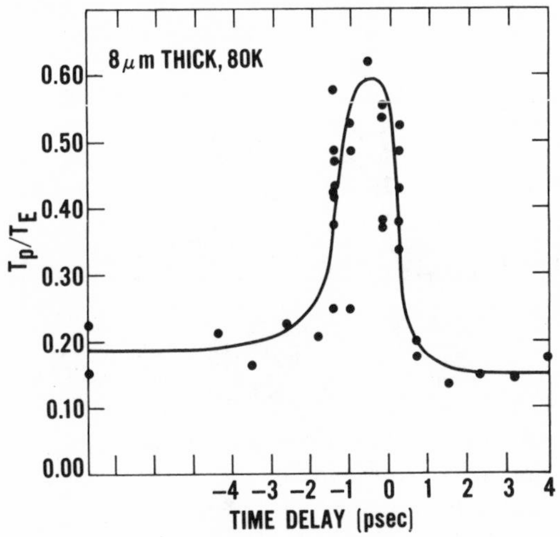

Figure 2. Probe pulse transmission versus time delay between the excite pulse at 1.06 μm and the probe pulse at 1.06 μm for a sample temperature of 80 K. The data are plotted as the normalized ratio or probe pulse transmission to excite pulse transmission T_P/T_E, in arbitrary units (from Kennedy *et al.*, 1974).

with an excitation pulse intense enough to bleach the sample transmission by a factor of approximately 20. This excitation pulse was then followed, at some later time, by a probe pulse that monitored the decay of the enhanced transmission. The authors observed (Fig. 2) in the probe transmission a narrow spike located near zero delay. The width of the spike was approximately 2 psec. No further structure in probe transmission was seen at this time due to problems related to experimental configuration and laser performance. Further structure would later be reported by Shank and Auston (1975) and Smirl *et al.* (1976), as we shall discuss below. In the absence of further structure, however, the authors erroneously interpreted this narrow spike near zero delay as evidence for an intraband relaxation time of hot electrons of less than 5 psec.

Shank and Auston (1975) repeated the 1.06 μm "excite-and-probe" measurements of Kennedy *et al.* (1975). In addition to the narrow spike in the probe transmission near zero delay, the measurements revealed a slower, broader structure in the probe transmission (Fig. 3). This probe transmission exhibited a slow rise lasting

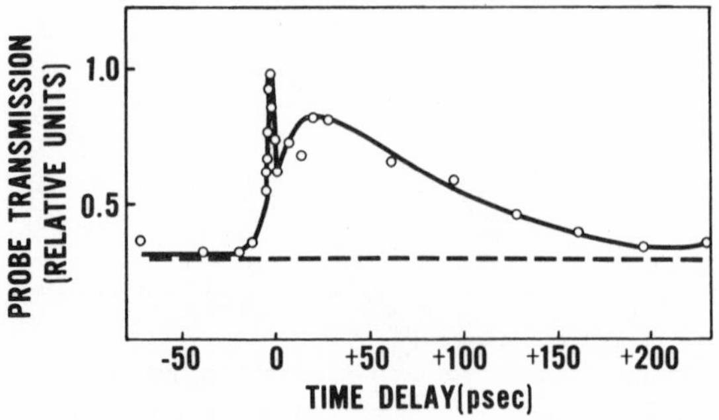

Figure 3. Probe pulse transmission as a function of relative time delay between excite and probe pulse at room temperature (from Shank and Auston, 1975).

approximately 20 to 30 psec followed by a gradual decrease lasting several hundred psec. In view of this additional structure, Shank and Auston reinterpreted the narrow spike in probe transmission nea zero delay as a parametric scattering of the strong excitation beam into the direction of the probe beam by an index grating that was produced by the interference of the two beams in the germanium sampl In addition, they attributed the slower rise in probe transmission band-filling. That is, they attributed it to a filling of conducti (valence) band states by electrons (holes) to the point where the electron (hole) Fermi energy approached the optically coupled state As a result, the buildup of this effect should be proportional to t total number of carriers created, i.e. it should follow the integrated optical pulse energy. Note that this interpretation does no involve hot electron effects. According to this interpretation, th correlation spike and rise in probe transmission contain little new physics. They are merely artifacts of the measurement techniques: one being a correlation between the excitation pulse and probe puls and the other, the integral of the intensity correlation function. These conclusions were based on observations performed only at room temperature.

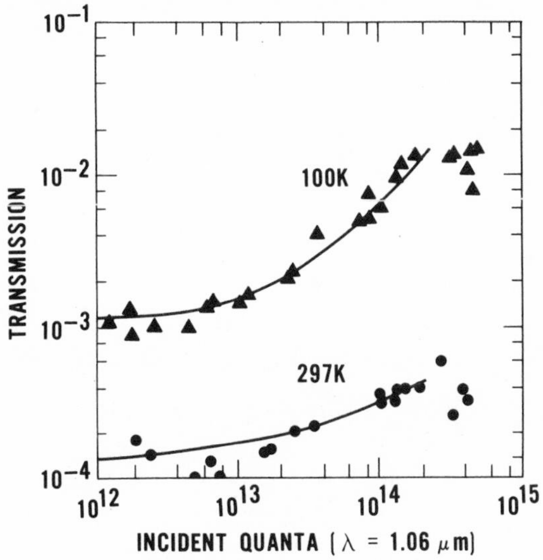

Figure 4. Transmission of a 5.2-μm-thick germanium sample as a function of incident quanta at 1.06 μm for sample temperatures of 100 K and 297 K. The solid lines are theoretical curves from Elci et al. (1977). The data are from Smirl et al. (1976).

Later, Smirl et al. (1976) independently extended the 1.06 μm excite and probe measurements of Kennedy et al. (1974) to include probe structure at longer delays. In addition, they determined the dependence of the "excite-and-probe" measurements on sample temperature and excitation pulse energy levels. Specifically, the nonlinear germanium transmission was measured as a function of the incident optical pulse energy at sample temperatures of 105 K and 297 K (Fig. 4). Further, the normalized transmission of the probe pulse as a function of time delay after an excitation pulse was measured for the same two temperatures (Fig. 5) and for three different excitation pulse energy levels (not shown). The temperature dependence of the probe transmission measurements contained surprising new information: the rise in probe transmission at 100 K was too slow (~100 psec) to be attributed to an integration effect (i.e. it did not appear to follow the integrated optical energy of the excitation pulse). The authors suggested that this slow rise in probe transmission might be attributed to a cooling of the energetic electrons (holes) created in the conduction (valence) band by the direct absorption of the excitation pulse. Thus, the rise in probe transmission was taken to be an indication of the carrier energy relaxation time.

At this point, Elci et al. (1977) presented the first detailed theoretical treatment of these problems. Their model (hereafter referred to as the ESSM model) attempts to account for the nonlinear transmission and the "excite-and-probe" response of germanium

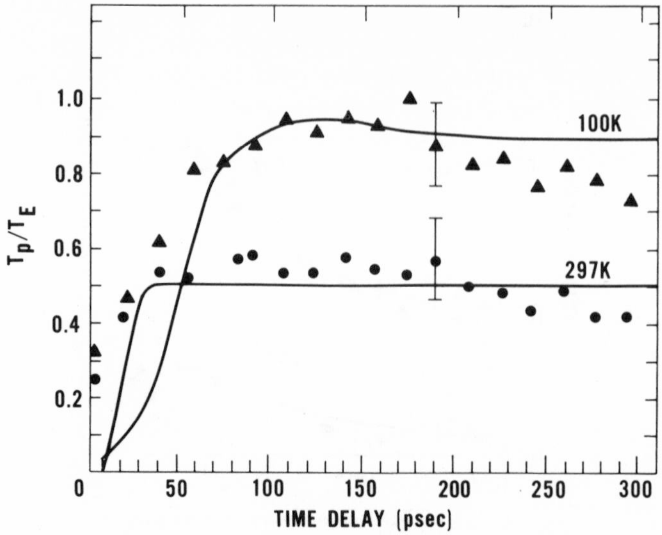

Figure 5. Probe pulse transmission versus delay between the excite pulse at 1.06 μm and the probe pulse at 1.06 μm for samp temperatures of 100 and 297 K. The data are plotted as the normalized ratio of probe pulse transmission to exci pulse transmission, T_P/T_E, in arbitrary units. The soli lines are theoretical curves from Elci et al. (1977). T experimental data are from Smirl et al. (1976).

in terms of: (1) direct band-to-band absorption, (2) free-carrier absorption, (3) long wavevector phonon-assisted intervalley scatter ing, (4) phonon-assisted carrier relaxation, (5) carrier-carrier Coulombic collisions, and (6) plasmon-assisted recombination. We have presented a detailed overview of the ESSM model previously, an we will not repeat those discussions here. However, in short, the authors attributed the rise in the probe transmission with the dela after an intense excitation pulse to a cooling of the hot electron-hole plasma created by the absorption of the excitation pulse. The results of these calculations are presented as the solid lines in Figs. 4 and 5. As we stated in a previous chapter, the theoretical fit to the nonlinear transmission data and the probe transmission data can be regarded as satisfactory, given the complexity of the problem. Subsequently, van Driel et al. (1977) conducted further nonlinear transmission studies, in which the energy-band gap of the germanium sample was tuned by hydrostatic pressure, which seemed to corroborate the proposed model.

One of the interesting features of the ESSM model was that it predicted that the nonlinear transmission of the thin germanium sample should depend on the width of the optical pulses. In fact, Elci et al. (1977) suggested that the transmission of the germanium be measured as a function of incident optical pulse energy for pulses of various widths to further test the model. Initial measurements by Bessey et al. (1978) were found to differ substantially from the predictions of the ESSM model. This was the first disagreement between experiment and a heretofore successful model.

As a result of the disagreement between theory and experiment reported by Bessey et al. (1978), Latham et al. (1978) performed numerical studies to determine whether the differences between model and experiment could be attributed to assumptions made in the calculations, experimental uncertainties in the physical constants used in the calculations, or the limited number of physical processes included in the model. Their results indicated that, due to uncertainties in the optical phonon-electron coupling constant in germanium, the optical pulsewidth experiments did not provide a definitive test of the ESSM model. This investigation of the uncertainties in the known values of the physical coupling constants produced an important and a disquieting result: accepted values of the optical phonon-electron coupling constant ranged from 6.4×10^{-4} erg/cm to 18.5×10^{-4} erg/cm. An average of the eight values listed by Latham et al. (1978) is 1×10^{-3} erg/cm. Elci et al. (1977) had originally used constants of 6×10^{-4} erg/cm at a lattice temperature of 297 K and 2×10^{-4} erg/cm at 100 K. These values result in carrier cooling rates that are 3 and 25 times slower than that obtained by using the average value. In fact, if the average value for the optical-phonon-electron coupling constant is substituted into the ESSM model, the energy relaxation rate for the hot carriers is too rapid to account for the rise in probe transmission. The theoretical probe transmission is plotted as a function of time delay for several values of optical-phonon-electron coupling constant in Fig. 6. Consequently, the rise in probe transmission with delay time after excitation can, or can not, be accounted for by carrier cooling, depending on the value chosen for the coupling constant. As we shall discuss later, the cooling rate for the photogenerated hot carriers can be further complicated by hot phonon effects.

In view of the experimental uncertainties in key physical constants used in the original ESSM calculations (as discussed above), the magnitude of the energy relaxation rate and the origin of the rise of the probe transmission are in doubt. These, however, were not the only indications that the model was incomplete. Other problems were related to certain major assumptions made in performing the calculations and to the limited number of physical processes included. We shall review these complications in the following section.

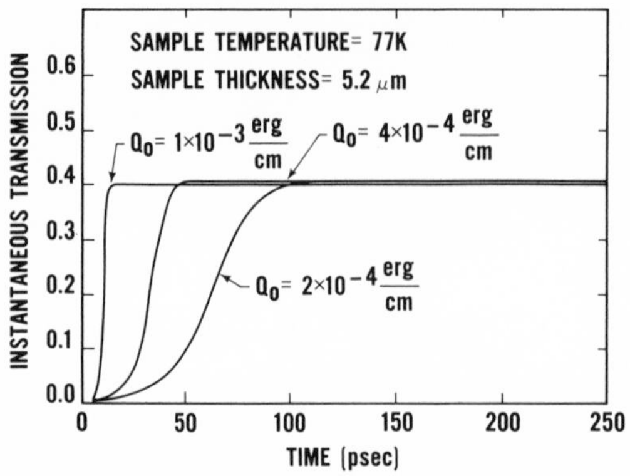

Figure 6. Instantaneous probe transmission as a function of relative time delay between probe and excite pulses for several values of the optical phonon-electron coupling constant, Q_o (from Latham et al., 1978).

HIGH PHOTOGENERATED-CARRIER DENSITIES

Studies, to be discussed in this section, indicate that processes other than those included in the original ESSM model can be important. Most of these effects, such as intervalence-band absorption, Auger recombination, and Coulomb-assisted indirect absorption, are only significant at large carrier densities. Other processes, such as diffusion, are enhanced at high carrier densities. The possible importance of including these processes in any theoretical model is discussed in this section. Previously, most information concerning these high-density phenomena has been obtained from measurements on highly-doped samples, in the presence of large donor and/or acceptor concentrations. One advantage of intense, picosecond excitation is the opportunity to study these processes in the absence of impurity effects.

Diffusion

One of the more drastic assumptions of the ESSM model was that the parameters that characterize the electron and hole distributions (i.e. Fermi energies, temperatures, and carrier densities) were taken to depend only on time, rather than on both space and time. The linear absorption coefficient α_o for germanium at 1.06 μm is approximately 1.4×10^4 cm^{-1}. Consequently, most of the excitation pulse

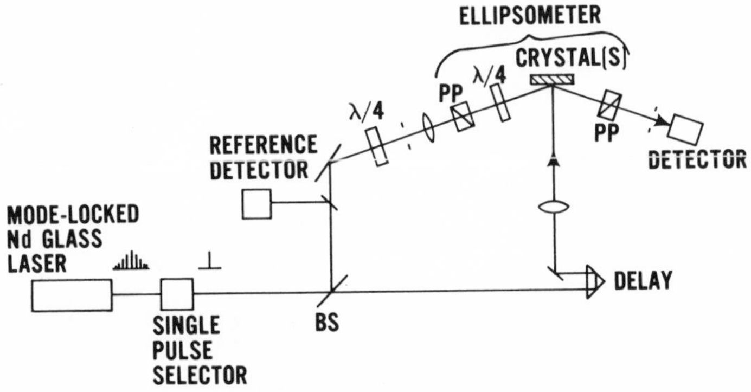

Figure 7. Picosecond ellipsometer used to measure the time evolution of optically generated electron-hole plasmas in germanium, where PP denotes a polarizing prism, $\lambda/4$ a quarter waveplate (from Auston and Shank, 1974).

will be absorbed within a micron ($\sim 1/\alpha_0$) of the sample surface, creating a dense photogenerated electron-hole plasma localized to this region. Thus, neglect of the spatial variation of these parameters is not a reasonable assumption for typical sample thicknesses (~ 5 μm) used in recent "excite-and-probe" experiments. Elci _et al._ (1977) recognized this problem, but, in order to simplify their initial calculations, they chose to view the parameters describing the electron-hole plasma as spatial averages throughout the sample volume.

Recent studies performed by Auston and Shank (1974) indicate that longitudinal diffusion (that is, diffusion along the direction of light propagation) can be important on picosecond time scales. In these experiments, the authors first irradiated a germanium sample near normal incidence with an intense 1.06 μm picosecond pulse that produced a large carrier density near the sample surface. This excitation pulse was then followed by a weak, circularly polarized probe pulse of the same wavelength. The change in polarization of the reflected probe light was monitored by ellipsometric techniques as shown in Fig 7. The transmission of the ellipsometer as a function of time delay between excitation and probe pulses in shown in Fig. 8. The transmission of the ellipsometer is proportional to the square of the fractional change in the index of refraction $|\delta n/n|^2$ induced by the absorption of the excitation pulse, where δn is the change in index and n the index of refraction. The index change $\delta n/n$, in turn, is proportional to the photogenerated carrier density. As a result, we can see from Fig. 8, that the photogenerated carrier

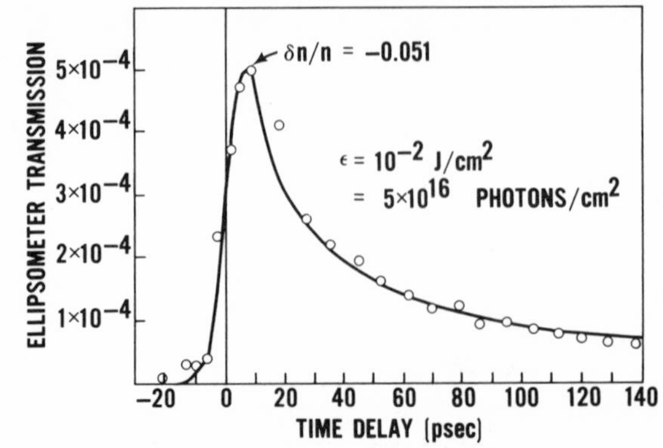

Figure 8. Ellipsometer transmission, $\Delta T \alpha |\delta n/n|^2$, versus time delay between carrier generation by absorption of the excite pulse and probe pulse (from Auston and Shank, 1974).

density at the sample surface is reduced to half its initial value in 30 psec following excitation. Auston and Shank attributed this decay of the surface density to a diffusion of the carriers into the sample bulk, and they deduced a diffusivity of 230 cm^2/sec at estimated surface carrier densities of approximately 10^{20} cm^{-3}. This value is 3.5 times larger than the low density ambipolar diffusion constant in germanium. Using this value, we would expect the initial optically-created carrier layer to double its thickness in 75 to 100 psec. Recombination effects were considered to be negligible during these measurements.

Auger Recombination

In another novel application of the "excite-and-probe" technique Auston et al. (1975) demonstrated the importance of Auger recombination at these densities. They first illuminated a 300 μm-thick slab of germanium with a 1.06 μm excitation pulse, creating a large carrier density. The decay of this carrier density was then probed by a pulse at 1.55 μm, generated by stimulated Raman scattering in benzene, as shown in Fig. 9. A photon of the excitation pulse (1.17 eV) is energetic enough to excite direct band-to-band transitions. The energy of a probe quantum (0.8 eV), on the other hand, is less than the direct band-gap energy but larger than the indirect ga

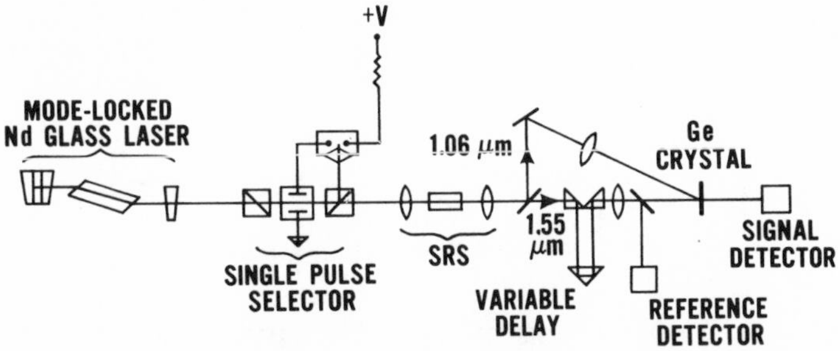

Figure 9. Schematic diagram of experimental apparatus used for the measurement of Auger recombination. The excite pulse was at 1.06 μm and the probe, generated by stimulated Raman scattering in Benzene, was at 1.55 μm (from Auston et al., 1975).

at L. Thus, the probe pulse can be absorbed as a result of both indirect and free-carrier transitions. At the carrier densities encountered in their experiments, these authors judged the indirect absorption coefficient to be negligible compared to the free-carrier coefficient. A plot of the change in absorbance of the 1.55 μm probe pulse as a function of time delay after the 1.06 μm excitation pulse is shown in Fig. 10 for two different optically-created carrier densities. Since the change in absorbance of the probe is taken to be a measure of the change in the free carrier absorbance and since the free carrier absorption coefficient is proportional to carrier number, Fig. 10 directly displays the decay of the carrier density. The probe absorbance decays significantly in the first 100 psec following excitation, indicating the importance of carrier recombination on a picosecond time scale. Auston et al. (1975) attributed this decay to an Auger process and extracted an Auger rate constant of approximately $10^{-31}cm^6\,sec^{-1}$. The reader should note that the carrier recombination rate exhibits a cubic dependence on the carrier density, that is,

$$dn/dt = -\gamma_A n^3 \quad ,$$

where n is the carrier density and γ_A is the rate constant. A sensitive estimate of the rate constant requires a precise knowledge of the carrier density.

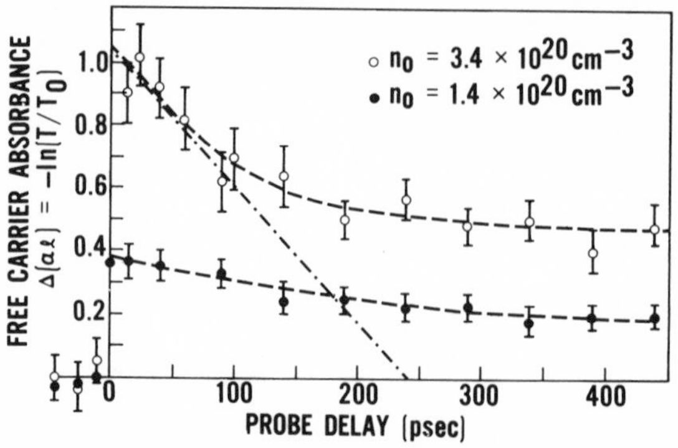

Figure 10. The change in free-carrier absorbance as a function of time delay between the excite pulse at 1.06 μm and the probe pulse at 1.55 μm for two carrier densities n_0. Here, T_0 represents the sample transmission before excitation (from Auston et al., 1975).

Indirect, Intervalence-Band, and Free-Carrier Absorption and Auger Recombination

At this point, perhaps we should pause to summarize the state of our understanding of the origin of the slow rise in probe transmission observed in the early "excite-and-probe" studies discussed above and displayed in Fig. 5. Originally, Shank and Auston (1975) attributed this rise in probe transmission to a saturation of the direct absorption as a result of band filling. We remind the reader that this interpretation was based on measurements performed only at room temperature. Subsequently, Elci et al. (1977) attributed this rise in probe transmission to a cooling of a hot electron-hole plasma created by direct absorption of the excitation pulse. Although the original calculations by Elci et al. are sound, time has shown that the proposed model (ESSM model) has several objectionable features, as detailed in the previous two sections: (1) uncertainties in the optical-phonon-electron coupling constant, (2) neglect of the spatially inhomogeneous nature of the parameters that characterize the carrier distributions, and (3) the omission of important processes such as diffusion and Auger recombination from the model. The authors realized and stated at the outset that their model contained serious assumptions and approximations that warranted further study and that the model contained only a few of the many possible processes. It was, however, hoped that the model would serve as a basis for further study and development.

In sharp contrast to the interpretation by Elci et al. (1977), Auston et al. (1978) and McAfee and Auston (private communication) stated that they expected the energy relaxation time in germanium to be too short to account for the rise in probe transmission shown in Fig. 5. This suggestion is, of course, consistant with the more detailed numerical studies presented by Latham et al. (1978), as discussed earlier and shown in Fig. 6. More importantly, in the spirit of suggesting plausible alternative models for evaluation, Auston et al. (1978) stated that enhanced intervalence-band and Coulomb-assisted indirect absorption effects might be important at the high photogenerated carrier densities encountered in these "excite-and-probe" experiments. Furthermore, they suggested that these processes might introduce a *minimum* in the absorption versus carrier density curve in germanium in the following way: The direct absorption coefficient will remain approximately constant as a function of photogenerated carrier density until the density reaches the point where the electrons (holes) clog the states needed for direct electronic transitions in the conduction (valence) band. At this point, the direct absorption coefficient rapidly decreases. On the other hand, Coulomb-assisted indirect, intervalence-band, and free-carrier absorption coefficients monotonically increase with carrier density. Thus, the absorption coefficient could initially decrease with increasing density, as the direct absorption coefficient saturates, then increase with increasing density as the free-carrier, intervalence-band and indirect absorption coefficients become large enough to dominate. In a private communication, McAfee and Auston further explained how an absorption curve containing a minimum could be combined with Auger recombination to account for the rise in probe transmission of Fig. 5. Briefly, the absorption of the excite pulse creates an initial carrier density greater than n_{min}, where n_{min} denotes the density at which the minimum total absorption coefficient occurs. As the initial photogenerated carrier density is decreased in time by Auger recombination, the absorption coefficient of the germanium will decrease in time until the carrier density reaches n_{min}, then increase. Thus, the probe transmission will increase then decrease, if the initial optically-created density is greater than n_{min}. In direct contrast to the ESSM model, this interpretation does not require hot electron effects. This model does, however, require a minimum in the absorption versus carrier density curve.

Consequently, we summarize and emphasize that there were at this point in time at least three possible explanations for the rise in probe transmission with delay (see Fig. 5): (1) the rise is caused by band-filling and is, as a result, an integration effect that follows the integrated optical energy of our excite pulse, (2) the rise is due to a cooling of a hot carrier distribution created by direct absorption of the excitation pulse, or (3) the rise can be attributed to Auger recombination combined with an absorption versus carrier density relationship containing a minimum.

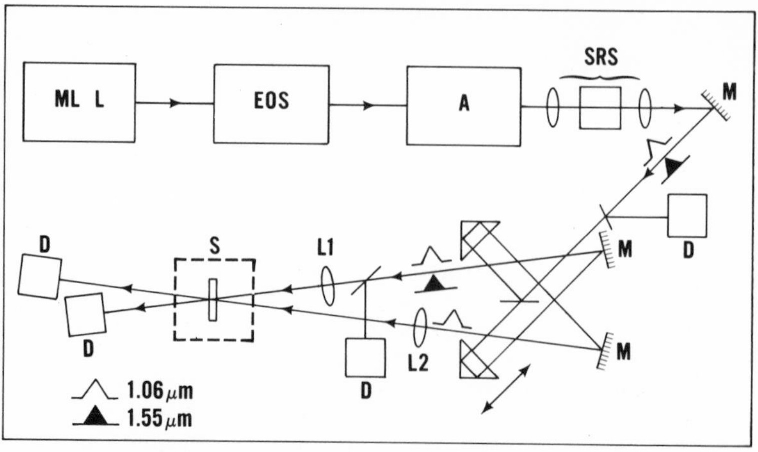

Figure 11. Block diagram of the experimental configuration for excite and probe measurements at 1.06 and 1.55 μm, where MLL denotes the mode-locked laser, EOS the electro optical switch, A the laser amplifier, SRS the stimulate Raman-scattering cell, M a mirror, D a detector, L1 and L2 lens, and S the sample (from Smirl et al. 1978).

In recent work, Smirl et al. (1978) have attempted to test the first and third possibilities listed above and have attempted to ascertain the importance of free-carrier, intervalence-band, and indirect absorption effects in excite and probe experiments at 1.06 μm. The experimental configuration used in these studies is similar to that used by Auston et al. (1975) and is shown in Fig. 11. The excitation pulses used here were approximately 10 psec in duration and had peak powers of approximately 10^8 W at a wavelength of 1.06 μm, and they produced a measured irradiance of approximately 10^{-2}J/cm when focused on the crystal surface. The plasma produced by the absorption of the excitation pulse was probed using weak pulses of two types: one at 1.06 μm had a photon energy greater than the direc band-gap energy for germanium, and the other at 1.55 μm had an energ less than the direct gap but greater than the indirect gap. The latter probe, at a wavelength of 1.55 μm, was generated by stimulate Raman scattering in benzene. We emphasize that the energy of a photon at 1.06 μm (1.17 eV) is sufficient to excite direct band-to-band transitions in germanium as well as free-carrier, intervalence band, and indirect transitions; whereas, the energy of a quantum at 1.55 μm (0.08 eV) falls below the direct gap and is only a measure of the combined free-carrier, intervalence-band, and indirect processes

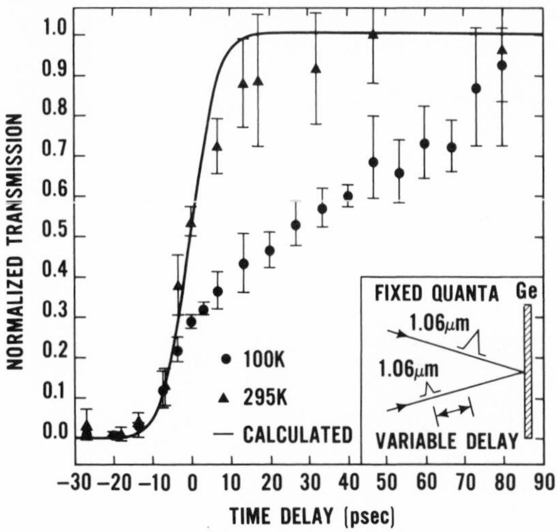

Figure 12. Normalized probe pulse transmission in arbitrary units versus delay between the excite pulse at 1.06 μm and the probe pulse at 1.06 μm for sample temperatures of 100 and 295 K. The solid line represents a theoretical integration curve assuming Gaussian-shaped optical pulses of 10 psec width (from Smirl et al., 1978).

Smirl et al. (1978) performed three separate measurements. In the first of these, they carefully repeated the measurements by Smirl et al. (1976) (Fig. 5) of the transmission of a 1.06 μm probe pulse as a function of time delay after an intense 1.06 μm excite pulse for sample temperatures of 100 K and 295 K. These original measurements were repeated so that the authors could more carefully investigate the possibility that the rise in probe transmission follows the integrated excitation pulse autocorrelation function. The rises in probe transmissions for the two sample temperatures are carefully compared to a calculated integration curve in Fig. 12, assuming an optical pulsewidth of 10 psec. The authors concluded from this comparison that the experimental rise in probe transmission at 295 K was indistinguishable from an integration effect, in agreement with the original interpretation of room temperature data by Shank and Auston (1975). However, the rise at 100 K is much slower than the integration curve or the rise at 295 K and cannot be attributed to such artifacts. For the remainder of this chapter, the rise in probe transmission at 100 K will be the object of our discussion.

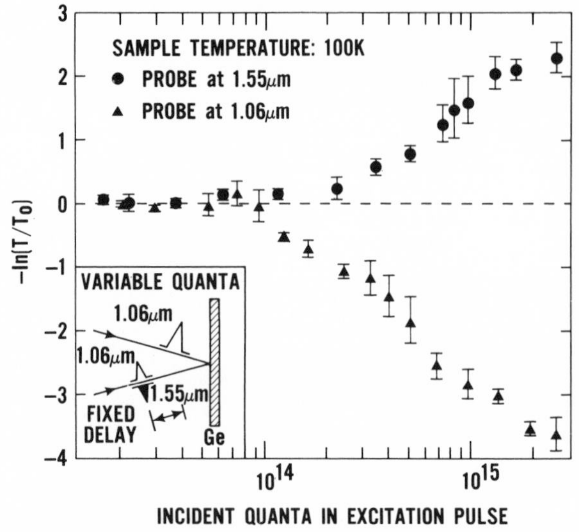

Figure 13. Change in absorbance, $-\ell n\ (T/T_0)$, of a 6-μm-thick germanium sample at 1.06 μm and 1.55 μm as a function of incident excite pulse energy at 1.06 μm, where T_0 is the linear transmission of the sample at the wavelength under consideration. Note that an excite pulse energy of 2×10^{15} quanta corresponds to an incident energy density of approximately $10^{-2} J/cm^2$ (from Smirl et al., 1978).

The authors next measured the transmission of a thin germanium sample at 1.55 and 1.06 μm as a function of optically-created carrier densities as shown in Fig. 13. The data were obtained in the following manner. The crystal was illuminated by variable energy pulses with a wavelength of 1.06 μm. Each pulse at 1.06 μm was followed immediately at a fixed delay by pulses that monitored the absorbance of the crystal at wavelengths of 1.55 μm and 1.06 μm. The optical absorbance at 1.17 eV is seen to *decrease* by approximately 3.5 as the carrier number increases. By contrast the absorbance at 0.8 eV *increases* roughly by 2.3. Over the range of densities encountered in these experiments, the absorption versus density relationship at 1.17 eV *does not* exhibit a minimum. Thus, a temporal decay of the carrier density alone cannot be combined with this absorption versus density relationship to account for the rise in probe transmission at 1.06 μm exactly as we discussed earlier. In addition, these measurements indicate that the *combined* free-carrier, intervalence-band, and indirect absorbance changes are opposite in sign and smaller in magnitude than changes caused by saturation of the direct absorption. As a result, the authors concluded that the decrease in absorbance at 1.06 μm with increasing carrier number is dominated by a saturation of the direct absorption coefficient; however, the rate of this decrease in absorbance is slowed by the contributions of these "other" processes that are opposite in sign. Note, that when comparing the data discussed here (Fig. 13) with the earlier data by Smirl et al. (1976) (Fig. 4), one must realize that the sample thickness and focused optical spot sizes are not identical.

Finally, Smirl et al. (1978) measured the temporal evolution of the absorbance of a 1.55 μm probe pulse as a function of time delay after an intense excitation pulse at 1.06 μm. In this experiment, the sample was irradiated by an optical pulse at 1.06 μm containing roughly 2 x 10^{15} quanta (corresponding to surface energy density of ~10^{-2}J/cm^2) and was probed by a weak pulse having a wavelength of 1.55 μm (See Fig. 14). The results of these probe measurements are similar to those obtained by Auston et al. (1975). However, these latter authors stated that they performed their measurements at excitation intensities such that the absorption of the excitation pulse was *linear*. These experiments were clearly performed in the *nonlinear* region. In addition, the measurements of Auston et al. (1975) were performed on a 300 μm-thick sample, while our sample was 6 μm thick. The measurements presented in Fig. 14 indicate that free-carrier, intervalence-band, and indirect absorption can be significant at the carrier densities encountered during "excite-and-probe" experiments described here. Recall that Auston et al. (1975) attributed the decrease in probe pulse absorbance at 1.55 μm, shown in Fig. 10, to a decrease in free-carrier absorption caused by a temporal decay in carrier density due to Auger recombination. The experiments that we have just described only allow the measurement of the change in the *combined* free-carrier, intervalence-band and indirect absorbance, and they do not provide for a convenient separation of the individual contributions.

Summarizing the results of the measurements described in the previous three paragraphs, we conclude that the rise in probe transmission during the 1.06 μm excitation and 1.06 μm probe experiments at 100 K is not an integration effect (i.e. not a simple band filling) and that it cannot be attributed to free-carrier, intervalence-band, and Coulomb-assisted transitions combined with Auger recombination. The contributions of these latter processes are significant, however, and they must be accounted for by any successful model. Unfortunately, the measurements described here yielded no *direct* information concerning carrier distribution temperatures or energy relaxation rates, and the question of attributing the rise in 1.06 μm probe transmission to a cooling of a hot carrier plasma created by the excite pulse remains unresolved.

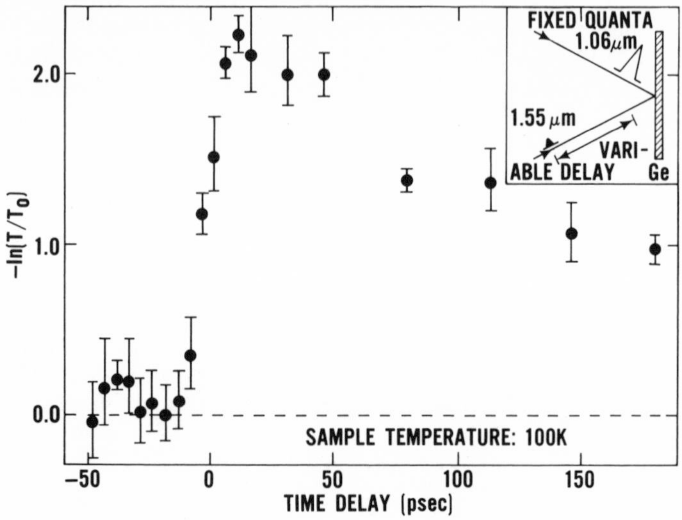

Figure 14. Change in probe pulse absorbance, $-\ell n\ (T/T_0)$, versus delay between the excite pulse at 1.06 μm and the probe pulse at 1.55 μm, where T_0 is the linear transmission of the probe pulse at 1.55 μm (from Smirl et al., 1978).

Having rejected two of the three possible explanations for the probe transmission listed earlier and with the other explanation all but rejected, to what do we attribute this rise in probe transmission? Recent suggestions are reviewed in the next two sections.

HOT PHONONS

As we recall from our discussion of the work of Latham et al. (1978), the physical constants for germanium are not well-enough known to allow a precise calculation of the energy relaxation rate. For theoretical fits to experiment by Elci et al. (1977) (Fig. 5), the optical phonon-electron coupling constants were chosen as 6×10^{-4} erg/cm for a lattice temperature of 297 K and 2×10^{-4} erg/cm at 100 K. These values are within the accepted theoretically and experimentally determined values listed by Latham et al.; however, they are much lower than the mean value of 1×10^{-3} erg/cm as obtained from an average of the eight values listed. In fact, as we have seen (Fig. 6), a repetition of the original calculations substituting the *average* phonon-electron coupling constant shows that carrier cooling is too rapid to account for the rise in probe transmission, in complete agreement with the statements of Auston et al. (197

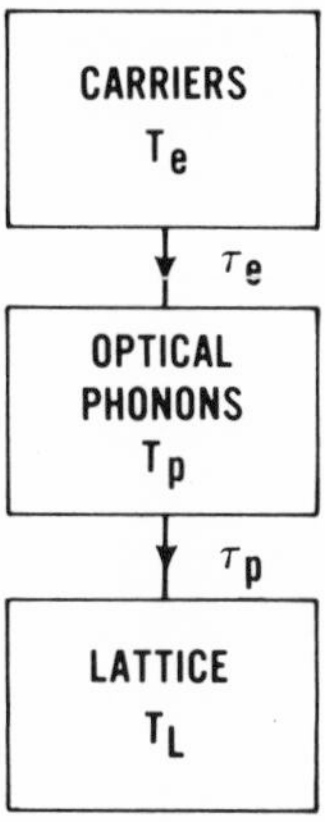

Figure 15. Schematic diagram illustrating the relaxation of hot electrons by the emission of optical phonons and the subsequent decay of the optical phonons into acoustic phonons (from van Driel, 1979).

However, van Driel (1979) has recently calculated the influence of hot phonons on the carrier energy-relaxation rate in these problems. In his calculations, van Driel adopted the ESSM model and extended it to include optical phonon heating effects. Briefly, the modified picture for the probe transmission is as follows. Just as in the ESSM model, the carriers generated by the absorption of the excite pulse cool by emitting optical phonons with a characteristic relaxation time τ_e, where τ_e is determined by using the *average* phonon-electron coupling constant. Since these phonon-assisted electronic transitions are intraband, they occur between states separated by small wavevectors. Consequently, the optical phonons emitted during these transitions also have a short wavevector and are located near the center of the Brillouin zone. Van Driel estimated that approximatley 10^{-2} of the volume of the Brillouin zone is involved in hot carrier relaxation. As 10^{19} carriers/cm^3 relax within the conduction bands (each one emitting approximately 15 optical phonons), an enormous number of these short wavevector optical phonons is created. These short wavevector optical phonons are believed to decay into two long wavevector acoustic phonons in a characteristic time τ_p. This decay brings the optical phonons into equilibrium with the lattice. This simplified picture is illustrated in Fig. 15. The optically-created carriers give their excess energy to the optical phonon reservoir with a characteristic time τ_e; the

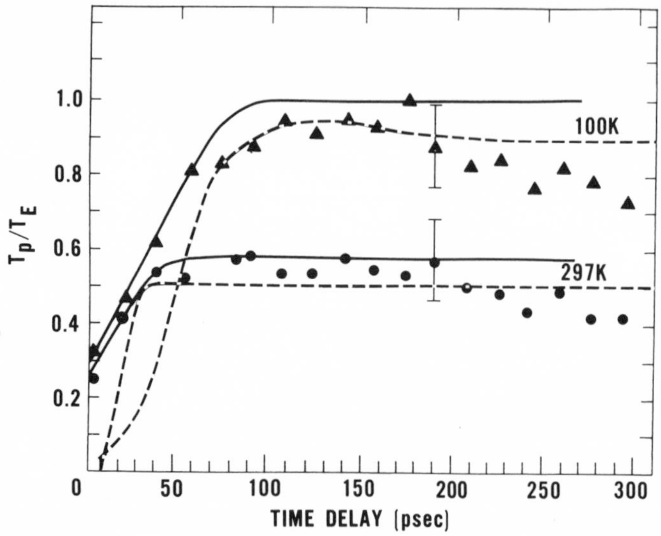

Figure 16. Probe pulse transmission versus delay between the excite pulse at 1.06 μm and the probe pulse at 1.06 μm for sample temperatures of 100 and 297 K. The data are plotted as the normalized ratio of probe pulse transmission to excite pulse transmission, T_P/T_E, in arbitrary units. The dashed lines are theoretical curves from Elci et al. (1977). The solid lines are theoretical curves from van Driel (1979). The experimental data are from Smirl et al. (1976).

optical phonon reservoir, in turn, gives its excess energy to the lattice with a time constant τ_P. As we discussed in a previous chapter, the optical phonon lifetime for germanium at 77 K is 10 psec. This lifetime is relatively long compared to τ_e when a single average temperature-independent optical phonon-electron coupling constant is employed. This results in a relaxation bottleneck for the hot carriers due to the buildup of the optical-phonon population on a picosecond time scale.

The results of these calculations, taking into account optical-phonon heating and using the *average* phonon-electron coupling constant, are shown as solid curves in Fig. 16. Note that the inclusion of hot phonons accounts for one of the major discrepancies between the original theory and experiment. Namely, in contrast to the original theory that predicted a delayed, steep rise (dotted curve, Fig. 16), the present theory shows a steep rise with gradual leveling off in agreement with the data. The solid curves in Fig. 16 were taken from van Driel (1979). The agreement between the modified theory and experiment is remarkable; however, this should be regarded as somewhat fortuitous in view of the simplifications of the model, the limited number of processes included, and the uncertainty in some of the physical constants.

THE RELAXATION-DIFFUSION-RECOMBINATION MODEL

In previous sections, we have reviewed evidence that some of the assumptions included in the early ESSM model are not well justified. For example, the experiments by Auston and Shank (1974) (Fig. 8) clearly indicate that the carrier density is inhomogeneously distributed throughout the interaction volume of the germanium sample and that diffusion is important on picosecond time scales. By contrast, the ESSM model had assumed that all parameters characterizing the electron and hole distributions were functions of time only, independent of spatial coordinates, and all diffusion effects were neglected. In addition, studies by Auston *et al.* (1975) and Smirl, *et al.* (1978) have demonstrated that processes originally omitted from the ESSM model such as Auger recombination and intervalence band absorption are important. And, finally, studies by Latham *et al.* (1978) have shown that the values chosen by Elci *et al.* (1977) for the optical phonon-electron coupling constant were extreme.

Recently, Leung (1978) has modified and extended the original ESSM model to remove most of these objections. In this model, he (1) allowed all parameters characterizing the electron and hole distributions to depend on both spatial coordinates and time, (2) used an optical-phonon-electron coupling constant approximately equal to the mean value determined by averaging the values listed by Latham *et al.* (1978), and (3) included the effects of intervalence-band absorption and Auger recombination. Hot phonon effects were, however, neglected. As we shall now discuss, this model leads to a radically different interpretation for the rise in probe transmission from that proposed by Elci *et al.* (1977). This model is briefly reviewed in the following paragraphs.

Just as in the ESSM model, the direct absorption of the excitation pulse creates a large density of electrons (holes) in the central valley of the conduction (valence) band. The electrons are rapidly scattered to the conduction-band side valleys by long wavevector phonons. Carrier-carrier scattering events, which occur at a rate comparable to the direct absorption rate, ensure that the carrier distributions are Fermi-like. Since the excitation pulse photon energy (1.17 eV) is greater than either the direct-gap energy (0.8 eV) or the indirect gap energy (0.7 eV), such a direct absorption event followed by the scattering of an electron to the side valleys results in the

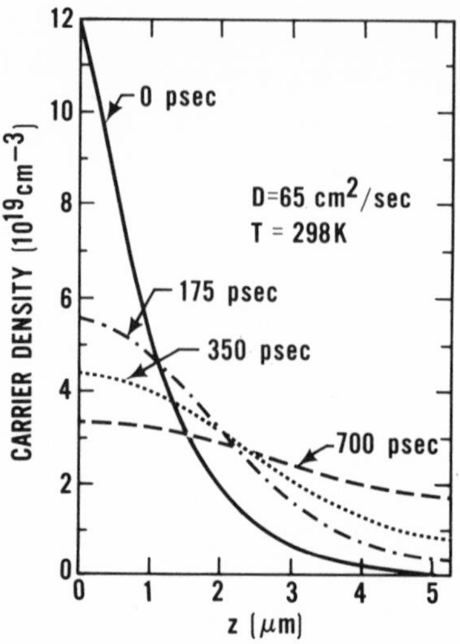

Figure 17. The spatial variation of the carrier density as a function of longitudinal position z and time after excitation in a 5.2 μm-thick germanium sample (from Leung, 197

electron giving an excess energy of approximately 0.5 eV to thermal energy. As a result, absorption of the excitation pulse results in the generation of a huge carrier distribution with an initial distribution temperature greater than the lattice temperature. Other processes such as free-carrier absorption and nonradiative recombinatic can raise the carrier temperature during passage of the excitation pulse, while phonon-assisted carrier relaxation processes can reduce the carrier temperature. So far, the description of the carrier evolution during the period that the excitation pulse is present in the sample is identical to that for the ESSM model. The present model differs in two respects. First, the inclusion of intervalence band absorption results in additional carrier heating effects as electrons are induced to make transitions from the split-off valence band to the light-hole and heavy-hole bands. Second, the carrier density, temperature, and Fermi energies are strongly dependent on longitudinal position within the semiconductor sample. For example, a typical plot of the carrier density as a function of longitudinal position immediately following excitation is shown as a solid line i Fig. 17.

Immediately following the passage of the excitation pulse, the interaction region of the sample contains a large number of carriers with a high distribution temperature. The final number and temperature are complicated functions of position as determined by the

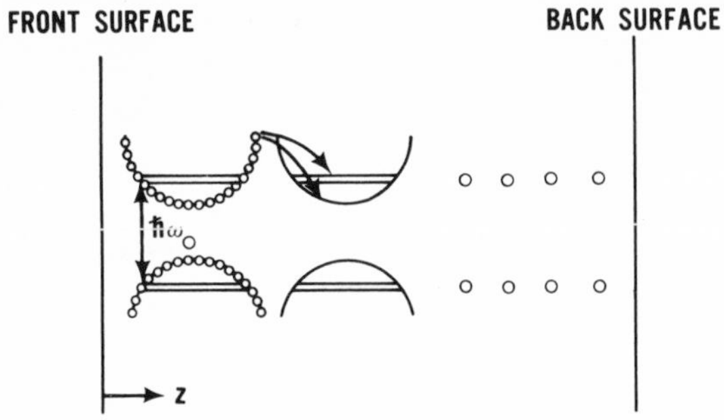

Figure 18. Schematic of the diffusion process (from Elci et al., 1978).

relative strengths of direct absorption, nonradiative recombination, intervalence band absorption and phonon-assisted relaxation rates. Experimentally, the probe pulse interrogates the evolution of the distribution after the passage of the excitation pulse, and its transmission is a sensitive measure of whether the optically-coupled states are available for absorption or are occupied. The probe pulse transmission versus time delay (Fig. 5) can be understood in the following way. Initially, after the passage of the excitation pulse, the probe transmission is small since the electrons (holes) are located high (low) in the conduction (valence) band because of the high distribution temperature, leaving the states that are optically coupled available for direct absorption. As the carrier distribution temperature cools and carriers fill the states needed for absorption, the transmission increases. In contrast to the ESSM model, however, the phonon-assisted relaxation here is extremely rapid. For an optical phonon-electron coupling constant of 10^{-3} erg/cm, the energy relaxation time of the carrier distribution is estimated to be less than 10 psec. Consequently, the electron and hole distributions, while still spatially inhomogeneous, have cooled to lattice temperatures within 5 to 10 psec following excitation. As a result, any initial rise in probe transmission as a result of hot carrier relaxation will be too rapid to account for the protracted rise displayed in Fig. 5. Diffusion is a slow process on a time scale of 10 psec.

For longer delay times (greater than 10 psec), longitudinal carrier diffusion is a dominant process in determining the evolution of the probe pulse transmission. Specifically, for large carrier densities, according to Leung's calculations, longitudinal diffusion can cause a rise in the probe transmission. This may seem surprising at first, but it can be understood by considering a simple schematic of the diffusion process, such as the one shown in Fig. 18. Since

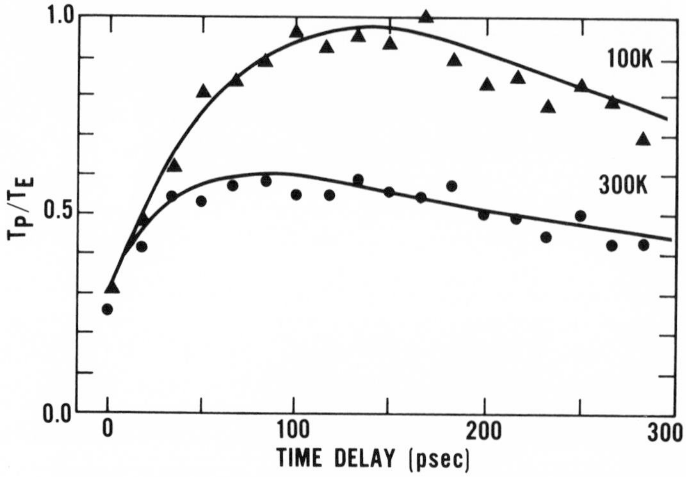

Figure 19. Probe pulse transmission, in arbitrary units, as a function of relative time delay between the excite pulse at 1.06 μm and the probe pulse at 1.06 μm for sample temperatures of 100 K and 298 K. The experimental data ar from Smirl et al. (1976). The solid lines are theoretical curves from Leung (1978).

the carrier number remains constant, the probe pulse "sees" the same total number of carriers regardless of time delay. As time progresses, however, the carriers diffuse into the sample bulk as illustrated in Fig. 17. At first, one might guess that this would reduce the carrier density and, thereby, free the states near the front surface of the sample for direct absorption of the probe. However, recall that the states that are resonant with the probe transmission are localized to narrow regions in the conduction and valence bands As a result, not all carriers are effective in filling these optically coupled states and preventing absorption. The total number of carriers effective in preventing absorption can be altered by diffusion as illustrated in Fig. 18. If the carrier density near the front sample surface is large, as the carriers migrate from this region, they can fill the states needed for absorption away from the front surface without depleting the optically-coupled states near the surface. The number of carriers in the sample effective in preventing absorption increases, and the probe transmission will rise. Note, however, that if the initial density is small in the front region, diffusion will decrease the number of carriers effective in preventing absorption in the front region without significantly increasing the *effective* density at the back; probe transmission will decrease. Thus, depending on the initial carrier density, *longitudinal* (along the direction of light propagation) diffusion can cause a rise or fall in probe transmission. In this model, the slow rise in probe transmission is attributed to a diffusion of the photogenerated carriers from the front sample surface into the sample bulk, in direct contrast to the original interpretation of the ESSM model. Note, however, that the high carrier temperature still plays a key role during the generation of the carrier distribution as a result of the absorption of the excite pulse. In this model the slow fall in probe transmission for much longer delays is attributed to a reduction in carrier density as a result of Auger recombination.

A comparison of the calculations of Leung (1978) to the probe pulse transmission data of Smirl et al. (1976) is shown in Fig. 19. Again, as with the hot phonon model of the last section, the agreement between theory and experiment is excellent. And again, we feel that, given the complexity of the model, the agreement must be considered somewhat fortuitous. Measurements that attempt to determine which, if either, of these two proposed models is correct are in progress.

SUMMARY

In this seminar, we have described experiments that attempt to measure the evolution of electronic processes in germanium with a time resolution approaching 10^{-12} sec. We first surveyed experiments that measured the saturation and relaxation of the germanium transmission at high photo-generated carrier densities. These measurements are important because they have the potential of yielding direct information on the ultrafast relaxation of optically-created hot carriers. However, as we have stressed throughout, investigators have been unable to provide a clear, unique interpretation of these experiments, since so many competing processes are simultaneously active. For example, workers have been unable to unambiguously attribute the rise in probe transmission to a single process.

We then reviewed experiments that provided information on diffusion, nonradiative recombination, and the combined effects of free carrier, intervalence-band, and Coulomb-assisted indirect absorption at high carrier densities. These measurements are interesting in two respects. First, they illustrate that by proper choice of the experimental configuration one can isolate and identify the contributions of single processes, and, second, they provide the opportunity to study these processes on a picosecond time scale and in the absence of impurity effects. The studies reviewed here clearly indicate that picosecond techniques have matured to the point of providing precise quantitative information concerning ultrafast processes in semiconductors.

We then summarized recent attempts to assemble the information provided by the experimental studies into a single theoretical model that describes the generation and evolution of the electron-hole plasma during and following excitation with a single picosecond puls at 1.06 μm. To date, no single, palatable description of the evolution of the carrier distributions has emerged. Further experimental studies are needed to substantiate or reject the two models reviewed here.

In conclusion, the reader should note that we have made no eff to provide a complete review of picosecond studies in semiconductors Further information concerning this subject can be found in a recent review article by von der Linde (1977) and in the Proceedings of the First International Conference on Picosecond Phenomena (Shank et al. 1978). In particular, the reader should be aware of the recent work performed by Shank et al. (1979) and von der Linde and Lambrich (197 in GaAs. These studies provide picosecond time-resolved measurement of hot-carrier relaxation, band-gap narrowing, and screening effects in GaAs, and they represent, in our opinion, some of the best experimental picosecond semiconductor studies to date. Finally, we commen that the recent development of continuous subpicosecond mode-locked dye laser systems has eliminated many of the data acquisition problems detailed in the introduction of this seminar. However, because these systems are relatively new and because high intensity systems are presently expensive to construct, Nd-glass and Nd-YAG systems are still, at this point, the most readily available to workers in the field.

This work was supported by the Office of Naval Research and the North Texas State University Faculty Research Fund.

BIBLIOGRAPHY

Auston, D. H., McAfee, S., Shank, C. V., Ippen, E. P., and Teschke, O., 1978, Sol.-State Electr. 21:147.
Auston, D. H., and Shank, C. V., 1974, Phys. Rev. Letters 32:1120.
Auston, D. H., Shank, C. V., and LeFur, P., 1975, Phys. Rev. Letters 35:1022.
Bessey, J. S., Bosacchi, B., van Driel, H. M., and Smirl, A. L., 1978, Phys. Rev. B 17:2782.
Elci, A., Scully, M. O., Smirl, A. L., and Matter, J. C., 1977, Phys. Rev. B 16:191.
Elci, A., Smirl, A. L., Leung, C. Y., and Scully, M. O., 1978, Sol.-State Electr. 21:151.
Kennedy, C. J., Matter, J. C., Smirl, A. L., Weichel, H., Hopf, F. A Pappu, S. V., and Scully, M. O., 1974, Phys. Rev. Letters 32:41
Latham, W. P., Jr., Smirl, A. L., Elci, A., and Bessey, J. S., 1978, Sol.-State Electr. 21:151.
Leung, T. C. Y., 1978, Dissertation, unpublished.
Shank, C. V., and Auston, D. H., 1975, Phys. Rev. Letters 30:479.

Shank, C. V., Fork, R. L., Leheny, R. F., and Shah, J., 1979, Phys. Rev. Letters 42:112:

Shank, C. V., Ippen, E. P., and Shapiro, S. L., 1978, "Picosecond Phenomena," Springer-Verlag, Berlin.

Smirl, A. L., Lindle, J. R., and Moss, S. C., 1978, Phys. Rev. B 18:5489.

Smirl, A. L., Matter, J. C., Elci, A., and Scully, M. O., 1976, Optics Commun. 16:118.

von der Linde, D., 1977, "Ultrashort Light Pulses: Picosecond Techniques and Applications," Ed. by S. L. Shapiro, Springer-Verlag, Berlin.

von der Linde, D., and Lambrich, R., 1979, Phys. Rev. Letters, 42:1090.

van Driel, H. M., 1979, to be published.

van Driel, H. M., Bessey, J. S., and Hanson, R. C., 1977, Optics Commun. 22:346.

HOT ELECTRON CONTRIBUTIONS IN TWO AND THREE TERMINAL SEMICONDUCTOR DEVICES

H. L. Grubin

United Technologies Research Center
East Hartford, Connecticut 06108

INTRODUCTION

In the previous discussion (Grubin, 1977) we concentrated on transient hot electron effects in semiconductor devices. We noted that these effects may be dominant for spatial dimensions of the order of the carrier mean free path and for doping levels common to currently fabricated field effect transistors. Perhaps one of the most significant consequences of these relaxation effects is that they will ultimately serve to elimate the presence of negative differential mobility from such semiconductors as gallium arsenide and indium phosphide. Thus, there are at least two questions that we must consider: (1) Under what conditions will a semiconductor device possess a region of negative differential mobility, and (2) if it does possess this region, what are its consequences? We will examine these questions in reverse order.

In discussing the consequences of negative differential mobility (NDM) we are at that point in time where most of our device pictures are derived from models in which the equilibrium velocity electric field relation is taken to represent carrier drift in nonequilibrium situations. While the discussion of transient effects points to severe limitations with this model at extremes of size and doping levels, dynamic effects associated with such things as intervally scattering times will limit its usefulness to frequencies at and below the low millimeter-wave scale. Further, the equilibrium velocity electric field curve is dependent on impurity densities and so may vary from one region of the device to another. With these qualifications, it is clear that equilibrium velocity electric field models cannot be used confidentially for providing quantitative determination of device behvaior. Rather, we may expect that in those devices whose dynamic-relaxation contributions still allow a region of NDM to be present, such equilibrium velocity electric fiel models will provide the information necessary to present qualitative explanations of the oscillatory characteristics of NDM semiconductor devices. Indeed this is the very least we can expect of the equilibrium models; and as we will see, they have served us well.

In discussing NDM semiconductor device behavior through the equili brium velocity electric field model we are going to concentrate on: (1) the role of the low resistance contacts on the oscillatory properties of the NDM device, and (2) the influence of the external cir cuit. We will first examine the two-terminal device, then move to the three-terminal device.

TWO-TERMINAL DEVICES

To highlight the importance of the cathode or anode region properti on the oscillatory behavior of two terminal NDM devices, we point ou that while at the early stages of device development most devices were fabricated with ostensibly similar cathode and anode region con tacts, it was the usual case to find that the current voltage charac teristics of the device were not symmetrical with respect to the polarity of the applied bias (Shaw et al., 1969). The importance of the circuit emerges when we realize that all current oscillations in NDM devices are not characterized by isolated propagating domains whose transit time determines the oscillation frequence of the devic (Gunn, 1964). In some cases, the oscillation frequence bears no resemblance to the transit time of the carriers (Thim, 1968).

With regard to the low-resistance contact to materials such as galli arsenide and indium phosphide, there appears to be general agreement that these contacts are highly nonlinear and have a non-negligible effect on device behavior at all except the lowest bias levels. There is not, however, any theory that self-consistently determines the properties of these contacts from first principles and then relates these properties to device behavior. Rather, there are a variety of phenomenological models. In our case, we have adopted th point of view that the alloyed metal/semiconductor contact imitates the properties of the unalloyed metal/semiconductor contact. The important distinction between these being that the alloyed contact has a substantially lower barrier height than the unalloyed contact. In analyzing the role of the contacts to the NDM device, we conceptually divide the device into three sections, as shown in Fig. 1.

The center region in Fig. 1 represents the NDM semiconductor. It i characterized by an equilibrium velocity electric field relation v(E and an equilibrium diffusion electric field relation D(E). The time-

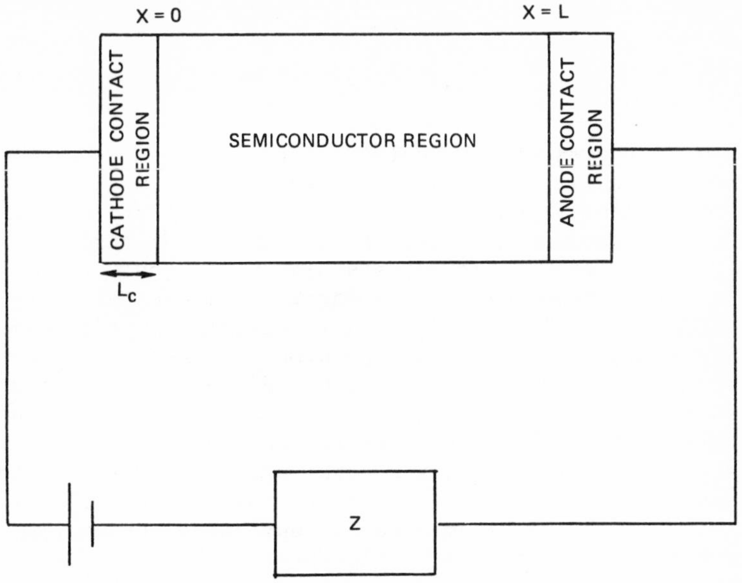

Fig. 1. Device - circuit configuration.

dependent dynamics of the carriers are obtained by solving the semiconductor equations, which for one dimension reduce to that shown below, for the total current density through the NDM semiconductor (see, e.g., Shaw et al., 1979),

$$j_0(t) = v(E) \left[eN_0(x) + \varepsilon \frac{\partial E(x,t)}{\partial x} \right] -$$

$$-D(E) \left[e \frac{\partial^2 E(x,t)}{\partial x^2} + e \frac{\partial N_0(x)}{\partial x} \right] + \varepsilon \frac{\partial E(x,t)}{\partial t} \quad . \qquad (1)$$

The above equation deals with only one carrier specie, whose charge we take conveniently as the magnitude of the electron charge. Here, ε is the permittivity of the semiconductor, $N_0(x)$ is the background doping density, which may be spatially dependent, and $E(x,t)$ is the electric field.

The carrier dynamics in the cathode contact region are taken to governed by the differential equation (Grubin, 1976a)

$$j_0(t) = j_c(E_c) + \lambda\varepsilon E_c/dt, \qquad (2)$$

where $j_c(E_c)$ represents the cathode conduction current density and has the same general properties as the 'control characteristic' dis cussed by Kroemer (1968). E_c is the electric field at the boundary between the cathode region and the semiconductor region. Dependin on the contact, it may assume very high values as well as values ne zero. λ is a dimensionless parameter. As indicated in Fig. 1, we assume that the cathode region, as represented by (2), separates th NDM semiconductor from the cathode proper. The rationale for this separation lies in experiments which indicate that for contacts wit high cathode fields, the transition to the high field values may ta place outside the NDM semiconductor. The cathode contact region is also taken to be electrically neutral and, for the cases considered below, to have a permittivity equal to that of the NDM semiconducto (i.e., $\lambda = 1$). E_c is assumed to be continuous across the interface between the NDM and separating elements.

As indicated above, the principle assumption we make about the alloyed contact is that it imitates the properties of the unalloyed met semiconductor contact. As used here, this implies that

$$j_c(E_c) = -j_r \; \{ \exp(-eV_c/nkT) - \exp\,[-(n^{-1} - 1)eV_c/kT] \} \;, \qquad (3)$$

with $V_c = E_c L_c$. The form of (3) was adpated from studies of the unalloyed metal/semiconductor contact (Rideout, 1975). Its use her presumes a similar description. For the unalloyed contact, 'n' is the 'ideality' factor and describes the contact as dominated by thermionic emission ($n \approx 1$) or by tunneling ($n \gg 1$); j_r is the revers current flux and may be related to the barrier height through the Richardson-Dushmann equation

$$j_r = m^* A T^2 \exp[-(B/kT)] \;, \qquad (4)$$

where B is the barrier height (in ev), A is the Richardson-Dushmann constant (= 120 $A/cm^2/k^2$), and T is the absolute temperature. m^* i the ratio of an appropriate effective mass to that of the free elec tron mass. The results depend significantly on the value of j_r whi we translate into values for B. For this we have arbitrarily assig a value of 0.072 to m^* as representative of the central valley of Ga The relation between j_r and B is displayed in Table 1, where $j_p = N_0 e v_p$, with v_p being the peak drift velocity of the carriers as obtained from the equilibrium velocity electric field relation. Typical barrier heights for our discussion are about 0.2 ev. The

TABLE 1

BARRIER HEIGHT

$-B = kT \log_e(j_r/m^*AT^2)$

j_r/j_p	B
1.00	0.14
.80	0.15
.40	0.16
.20	0.18
.10	0.20
.05	0.22

parameter L_c, associated with (3), is specific to the forumulation we u
and does not appear in the formulation for metal/semiconductor contact
While the parameter L_c is conceptually ambiguous, we regard it as representing the width of the alloyed region; n and j_r are thought as more representative of the properties of the metal/semiconductor interface.

Figure 2 displays the dependence of the form of $jc(E_c)$ on the choice of parameters. Note that significantly different parameters yield similar $jc(E_c)$ curves. We have also included for reference t neutral current density curve $j_c(E) = N_0ev(E)$ which scales the velocity electric field curve for gallium arsenide. The significance o these curves for the space charge distribution within the NDM semiconductor device is considered below for steady-state time independent conditions. For this situation, we return to the total curren density equation and examine it for zero diffusion and for constant doping density. At the cathode boundary,

$$\frac{\partial E}{\partial x} = \frac{j_c(E_c) - j_n(E_c)}{\varepsilon v(E_c)} \qquad (5)$$

whereas within the bulk of the semiconductor, away from either the cathode or anode regions

$$E \approx j/N_0 e\mu_0 \quad , \qquad (6)$$

where μ_0 is the low field mobility of the semiconductor.

In Fig. 3, we sketch the electric field profile for $x \geq 0$ for the situation where the cathode conduction current density curve intersects the neutral current density curve at a point somewhere below the threshold field for negative differential mobility. For succes fully higher current density values, by virtue of current continuit the intersections of the vertical line $j = j_1$ with j_c and j_n determines, respectively, the cathode field E_{c1} and the bulk field E_{b1}. The value of j_n at E_{c1} exceeds j_c at this point, and there is a partially depleted region of charge adjacent to the cathode region boundary. Increasing the current density to the value j_2, where j_c and j_n intersect, yields a situation of uniform fields throughout t cathode plus semiconductor region. For the situation where the curr density is increased to j_3, the field at the cathode is less than t field in the bulk, j_n is less than j_c and we have a region of charg accumulation adjacent to the cathode contact region boundary. A cu rent instability may be expected to occur when the field away from the cathode region enters the region of negative differential mobility.

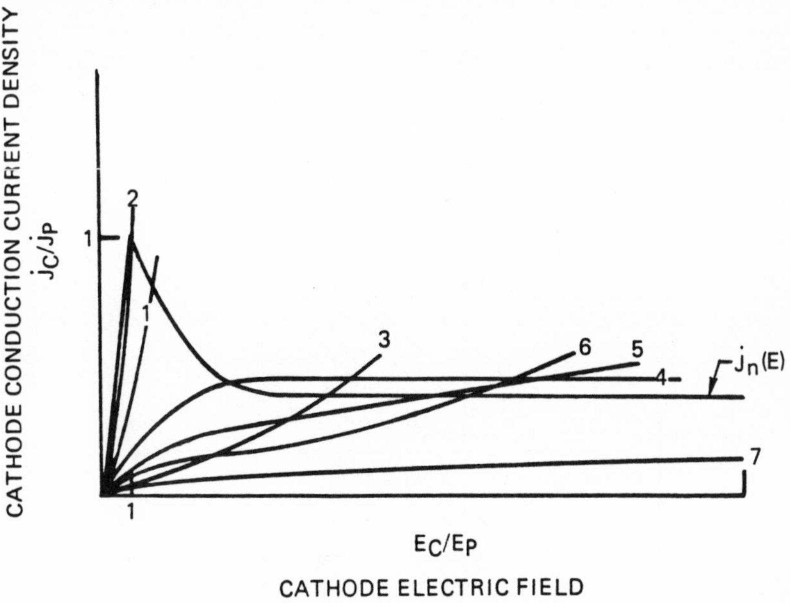

CURVE	j_r/j_P	n	$L_c \times 10^5$
1	0.1	10	1.5
2	0.1	10	3.0
3	0.1	10	0.2
4	0.4	1.0	1.0
5	0.2	1.05	1.0
6	0.1	1.1	1.0
7	0.1	1.1	0.2

Fig. 2. Cathode conduction current density versus electric field for seven of parameters, n, j_r and L_c. Also shown is the neutral conduction current density curve for gallium arsenide, denoted by $J_n(E)$. From Shaw et al. (1979).

The next situation to consider is that shown in Fig. 4, where the cathode current density curve intersects the neutral curve in the region of negative differential mobility. At the current density j_1, we have as in the previous case, a region of charge depletion occurs. adjacent to the cathode region boundary. Increasing the current

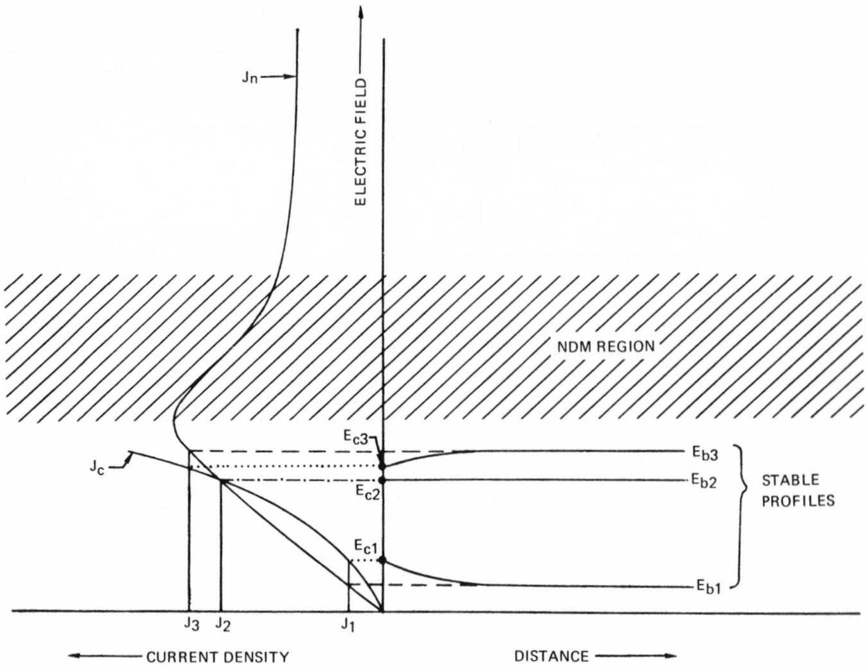

Fig. 3. Electric field versus distance profiles for the situation where $j_c(E_c)$ and $J_n(E_c)$ intersect at values of electric field below the NDM threshold field. The shaded region denotes the region of negative differential mobility.

density to j_2 results in an increase in cathode field to a value so where within the NDM region. The field within the bulk is still within the low field region as indicated. Further increases in current density j_3 result in the intersection of the j_n and j_c curv and charge neutrality at the cathode region boundary. Note that th field within the bulk is still significantly less than the field at the cathode boundary region. The situation at j_3 is unstable becau now further increases in current density result in the formation of an accumulation layer adjacent to the cathode region, followed by a downstream depletion layer. This domain is wholly within the NDM region and may generate a large signal instability. A point worth drawing attention to at this time is made by comparing the current densities and the estimates of the average fields across the semiconductor elements, represented by Figs. 3 and 4. In Fig. 3, the current density at the instability threshold is near the peak value for NDM. Also, the average field just prior to the instability is approximately equal the threshold field for NDM. For the results

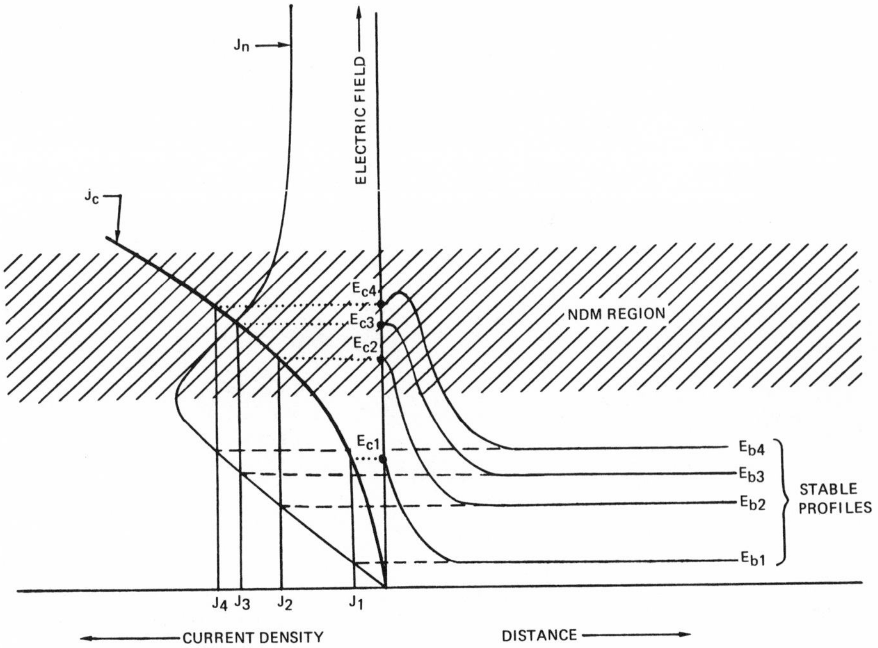

Fig. 4. As in Fig. 3, but $j_c(E_c)$ and $j_n(E_c)$ intersect within the region of negative differential mobility.

of Fig. 4, both the current densities and the average fields prior to the instability are significantly less than the NDM threshold values.

In Fig. 5, we illustrate the situation where the cathode conduction current density intersects the neutral curve somewhere within the region of saturated drift velocity region. For this case, we do not expect any charge accumulation until the value of current density exceeds the current density associated with the saturated drift velocity.

To examine the dynamic properties of these space charge layers, we resort to numerical simulation and solve the semiconductor equations simultaneously with the circuit equations. For this we need the anode contact region conditions. This is an easier contact to specify phenomenologically, for we have found that if the region is not tailored to meet specific and unusual requirements, its effect on the oscillation and instability conditions can be simulated by setting the anode field value to zero. With this in mind, we examine the simulation results.

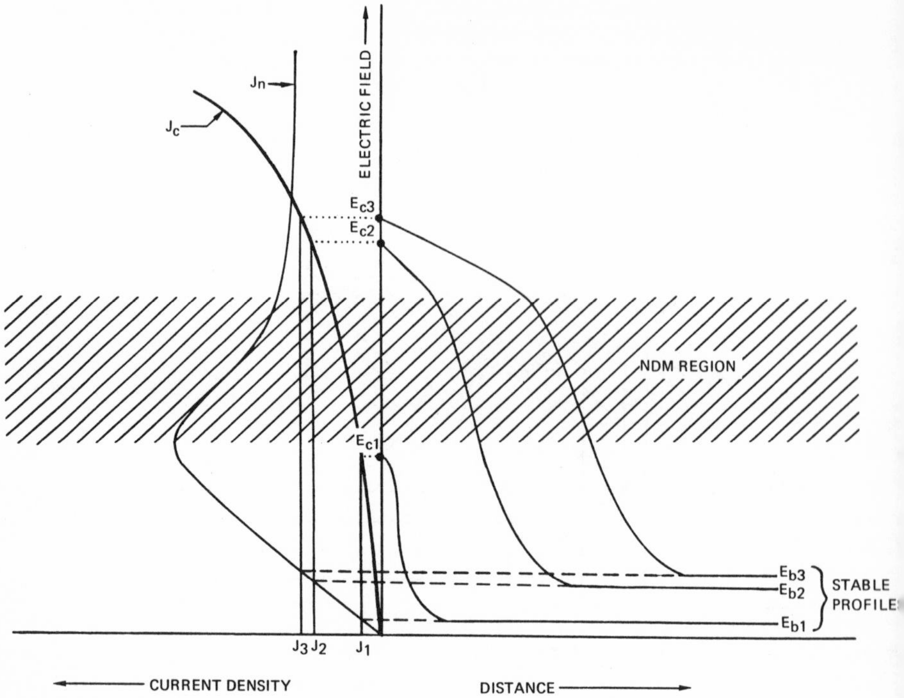

Fig. 5. As in Fig. 3, but $j_c(E_c)$ and $j_n(E_c)$ intersect within the region of saturated drift velocity.

We being with the situation where the cathode conduction curre density curve intersects the neutral curve within the region of neg tive differential mobility. For a resistive circuit, Z = R, we get result in Fig. 6 (Shaw et al., 1979) which shows the prethreshold electric field versus distance profile and the postthreshold electr field versus distance profile (at one instant of time). The nuclea- tion criteria is the same as that discussed in Fig. 4, and the re- sulting domain travels between the contacts. It drains at the anod and is renucleated at the cathode region. The dc current voltage curves (Grubin, 1976b) are shown in the inset. We point out the sublinearity, just prior to the instability (represented by the 'wiggle' in the curve). The sublinearity is due to the enhanced voltage drop in the vicinity of the cathode contact. In a circuit containing reactive components, e.g., serial inductance and paralle package capacitance, the dynamic behavior of an NDM element with a similar contact may be different. This is displayed in Fig. 7a whe we plot current density through a dc load resistor versus an averag electric field, $\overline{E}$ (Grubin, 1978). Time is eliminated between the tw

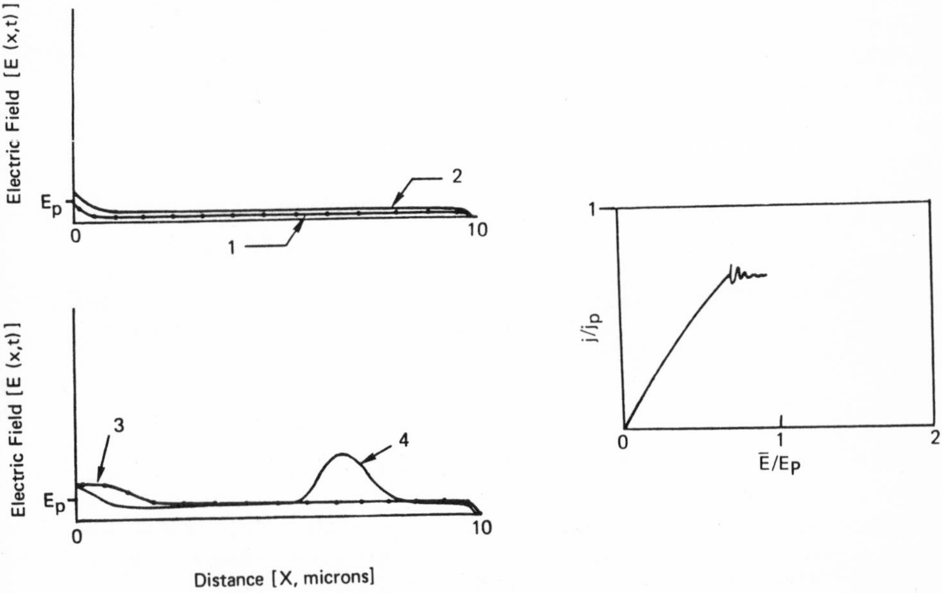

Fig. 6. Resistive circuit calculations for a 10 micron long element with $j_c(E_c)$ and $j_n(E_c)$ intersecting within the region of NDM. Curves 1, 2, and 3 denote pre-instability curves at increasing values of bias. Curve 4 denotes an instability at one instant of time. From Shaw et al. (1979). The current density versus average electric field curve for this element is shown in the graph on the right. From Grubin (1976b).

In Fig. 7b, we show the electric field versus distance profile at four instants of time. Also shown is the cathode conduction current density curve. We note that in the case of Fig. 6, the instability is dominated by the cathode contact and the frequency is determined by the transit time of the domain. In the circuit containing reactive components, the frequency here is determined primarily by the circuit elements (although this is not always the case). We note here that the oscillation appears to consist of a moving oversized dipole layer.

We show a similar set of calculations for a device in which the cathode conduction current density curve intersects the neutral curve further within the region of negative differential mobility, close to the region of velocity saturation. The results show a much weaker

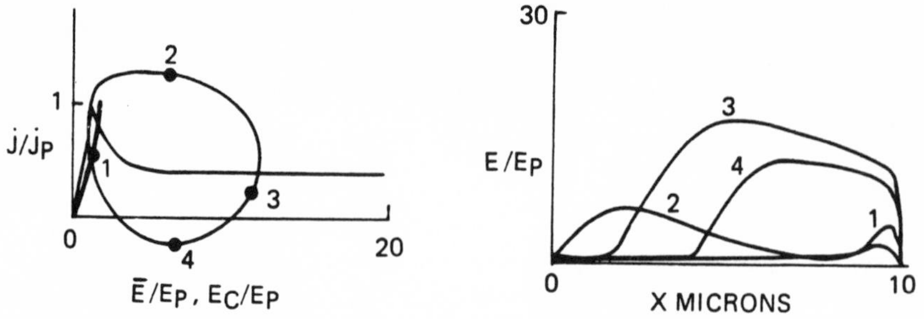

Fig. 7. Reactive circuit computations for a j_c curve that intersects j_n within the NDM region. From Grubin (1976b).

current oscillation. Indeed with cathode curves of this type, as they intersect further out we may expect the oscillatory activity to disappear. The current voltage curves for this device are shown in the inset (Grubin, 1976a) and we see a situation where the sublinearity is extreme enough to approach saturation. This excessive sublinearity is due to the large voltage drop in the vicinity of the cathode contact, as illustrated in Fig. 8.

The space charge profiles discussed above can be intuitively under stood on the basis of some simple analytical concepts associated wit the intersection of the neutral current density curves and the catho conduction current density curve. There is another profile that occurs frequently in device operation that is less amenable to simpl arguments. It occurs most often in devices whose cathode current density curves intersect the neutral curves at cathode field values $E_c \lesssim E_p$, and it manifests itself as a large voltage drop in the vici nity of the anode contact. This domain requires a region of NDM for its existence, but does not require the presence of any special anod property. The presence of these anode domains was first predicted b Shockley (1954). In resistive circuits its presence usually occurs at high bias levels and often accompanies the cessation of oscillations. In circuits containing reactive elements it usually manifest itself as a permanent and residual voltage drop in the vicinity of t anode as illustrated in Fig. 9. (In Fig. 9, rather than deal with t cathode conduction current curves of Fig. 2, we have simply set the cathode field to zero.) We show the lissajous figure for the oscillation as well as a multiple exposure of the electric field versus distance profile. This profile represents an accumulation layer tha propagates toward the anode but never quenches at the anode. Instea there is always a residual space charge layer adjacent to the anode contact.

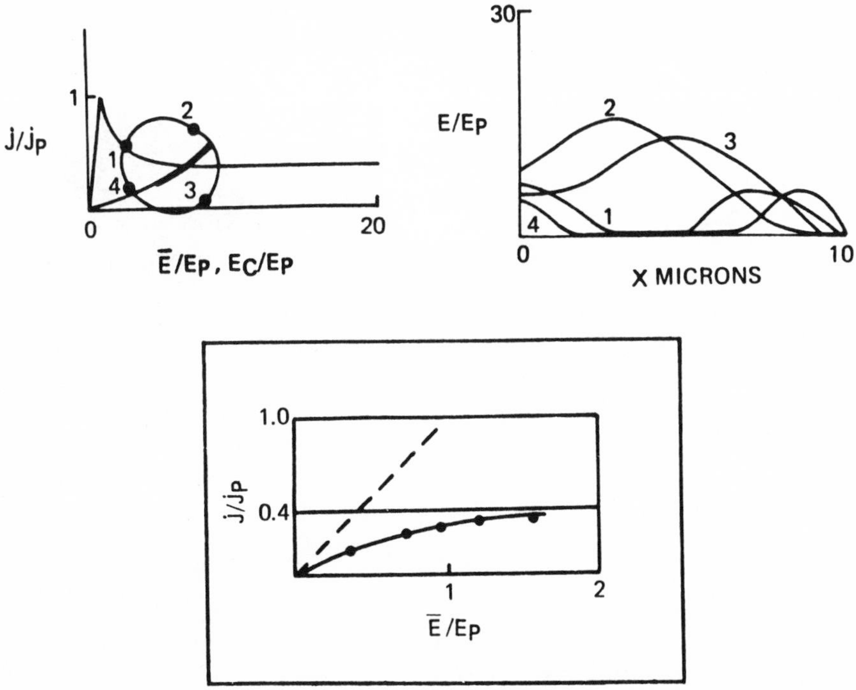

Fig. 8. As in Fig. 7, but for intersections within the region of saturated drfit velocity. From Grubin (1976n). The inset shows the current density versus average electric field. From Shaw et al. (1979).

Each of the above calculations was carried out for a simulated gallium arsenide device that had an active region length of 10 microns and a doping density of $10^{15}/cm^3$. The space charge profiles were developed for a rather sophisticated model of the cathode. We point out, however, that the use of this model is not necessary for simulating the qualitative properties of the instability in GaAs. A far simpler model can be used to generate these results; one in which the cathode field, instead of being dependent on current density, is held fixed (Shaw et al., 1969). We have carried out simulations with this model for 10 micron long devices as well as 100 micron long devices. The cathode dependent results are qualitatively similar. We have also performed experiments and used the fixed cathode field model in their interpretation. The experimental results are shown in Figs. 10 through 12; and the summary of the gallium arsenide results is shown in Fig. 13.

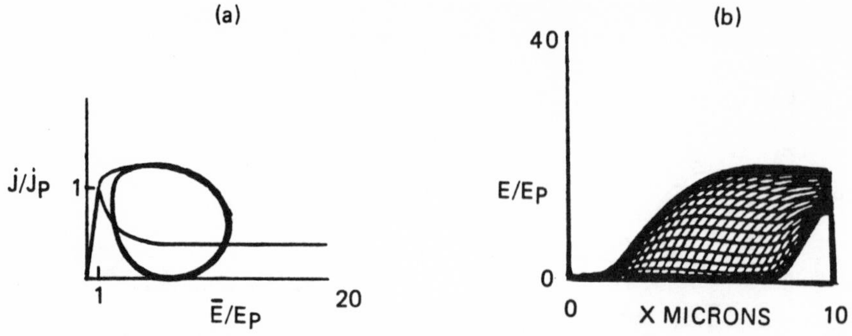

Fig. 9. Reactive circuit computations showing the presence of a permanent residual voltage drop in the vicinity of the anode contact. (a) Current density versus average electric field, (b) multiple exposures of the electric field versus distance profiles. From Grubin (1976b).

The experimental results were obtained for GaAs samples cut fro bulk, doped Monsanto n-GaAs wafers and homegrown epitaxial specimen The Monsanto wafers had quoted Hall mobilities usually between 4000 5000 cm^2/V-sec and carrier concentrations between $5x10^{14}$ and 10^{16}/cr The contacts were either tin/nickel/tin or gold/germanium/nickel.

Figure 10 shows experimental current voltage and voltage versu distance profiles for a sample whose active region boundary field wa below the threshold field for negative differential mobility (Solomo et al., 1975). To assure this we took the expedient approach of removing the active region from the influence of the metal contact b substantially reducing the cross sectional area of a large region between the contacts. This sample and samples of this type, when in resistive circuit, exhibited oscillatory characteristics that were more closely related to doping inhomogenieties than to the transit-time between the contacts.

Figure 11 shows experimental results for a device that sustaine transit-time oscillation. We see the enhanced sublinearity in the current versus voltage relation. This sublinearity, prior to the instability is due to the increased voltage drop at the cathode as the bias is increased. This increasing voltage drop is revealed in the probe measurements. We estimate a cathode field prior to an instability of approximately 6.2kV/cm.

In Figure 12 we show measurements for a device that did not exhib any time dependent behavior. Here one notices extreme saturation in

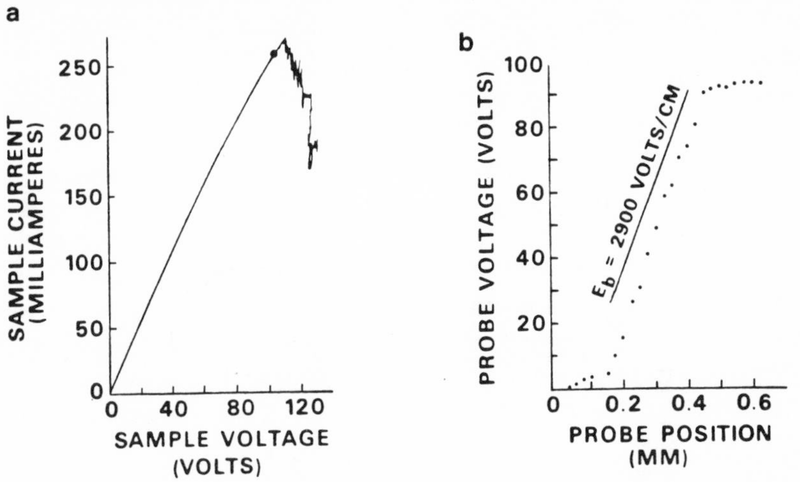

Fig. 10. (a) Current versus voltage for a low cathode boundary field GaAs sample. The probing point is indicated by the closed circle, (b) the probed voltage versus distance measurements. From Solomon et al. (1975).

current at high bias levels, while at low bias levels the current voltage relation is linear. The voltage versus distance measurements show a very large cathode voltage drop and a cathode field which we estimate to be greater than 60 kV/cm. We show measurements in both polarities to offer evidence that the large voltage drop is not a consequence of a high resistance cathode region. In both polarities we see that there is not a large anode voltage drop.

The results of the cathode boundary dependence of two terminal negative differential mobility devices is summarized in Fig. 13. Here for the same device parameters we plot current density versus average electric field (sample voltage/sample length) for different fixed values of cathode field. We superimpose the equilibrium drift velocity versus electric field relation. The properties of the instability fall into three categories. At low E_c (curve A), the j- E curve is essentially linear up to the field at peak velocity. At this point several things may occur. The possibilities depend in a detailed way on the characteristics of the GaAs and the circuit. Domains may be nucleated at large doping fluctuations in the bulk and lead to transit-time oscillations. A more interesting possibility is that after a domain is nucleated and reaches the anode it may remain there in a stationary field configuration and the current will saturate. In a resonant circuit the field may oscillate at the resonant frequency. For E_c within the shaded region, the current

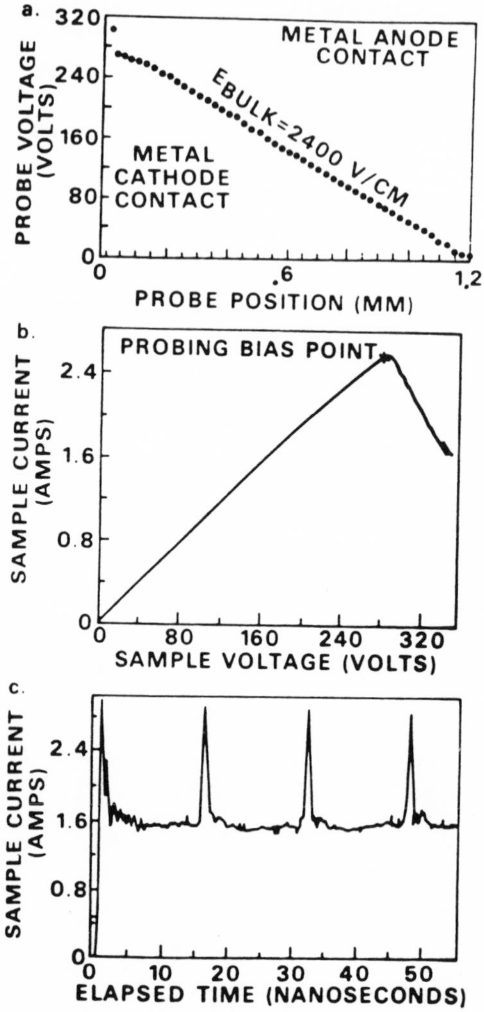

Fig. 11. Probed sample voltage versus distance for and intermediate cathode boundary field, at the bias point indicated in (b). (b) Current versus voltage, (c) instability current versus time response. From Solomon et al. (1975).

shows a departure from linearity due to the appearance of a large cathode voltage drop. At threshold the current switches along the load line. Threshold is controlled by the cathode field and not by the threshold field for negative differential mobility. Threshold occurs before the bulk enters the region of negative differential

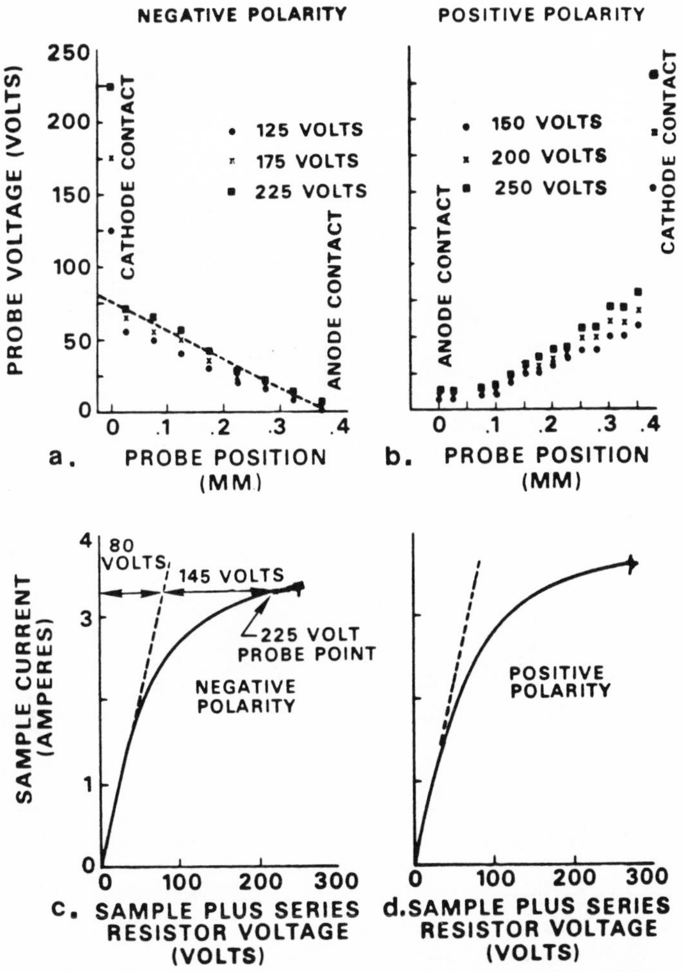

Fig. 12. Probed sample voltage versus distance for a high cathode field sample in (a) negative and (b) positive polarities. Current versus voltage curves for (c) negative and (d) positive polarities. From Solomon et al. (1975).

mobility. For high values of cathode field the current voltage characteristic is linear only at low bias. At high bias the curve saturates at the current density $j_s = N_0ev_s$. The departure from linearity corresponds to the appearance of a large cathode voltage drop and transit-time oscillations in a resistive circuit do not occur. In a resonant circuit the sample may sustain some oscillatory activity for a certain range of boundary field. We note that in all cases where an oscillation occurs, stable domain propagation requires

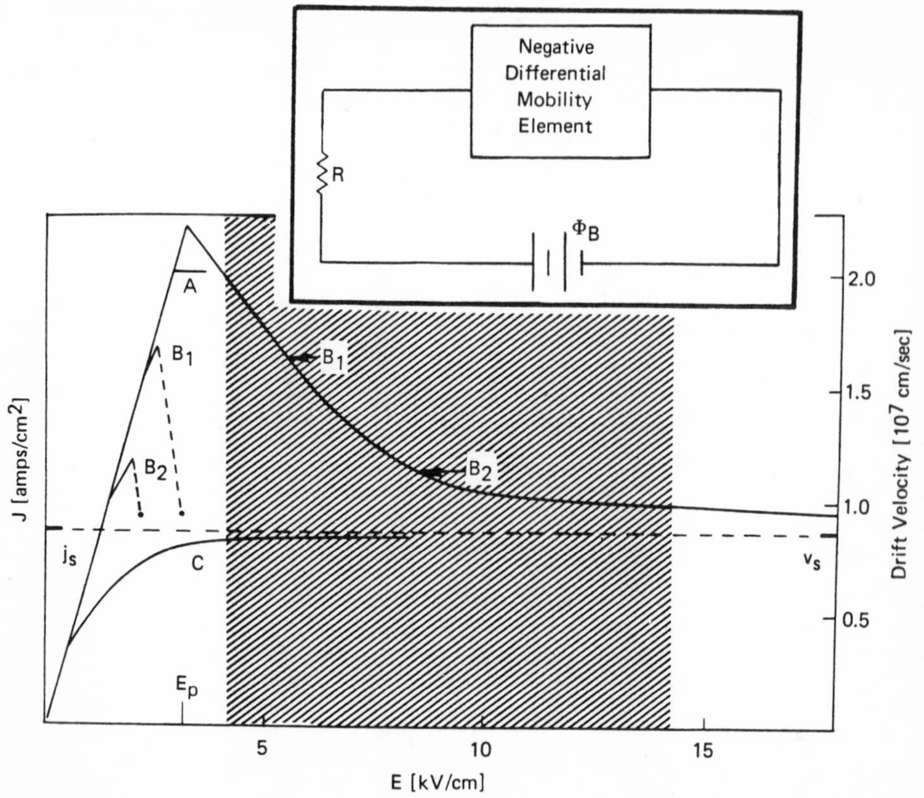

Fig. 13. The equilibrium velocity electric field curve and the computer simulated current density versus average electric field curve for various fixed values of cathode boundary field E_c. For the simulation the NDM element is in the circuit shown in the inset. For curve A, E_c = 0.0. The values for curve B_1 and B_2 are indicated by the arrows. for curve C, E_c = 24kv/cm. The simulated element had an active region length of 100 microns and a doping level of $10^{15}/cm^3$. The right and left hand ordinates are related by $j_n = N_0ev$ and $v_s = 0.8 \times 10^7$ cm/sec. From Shaw et al. (1969).

that the current level exceed the saturated drift current, sometimes referred to as the sustaining current.

The above discussion appears to present a rather complete phenomenological description of the contact and circuit dependence of gallium arsenide two terminal devices. The model has also been applied to the indium phosphide device. Here the presently known contact behavior is somewhat richer than that for gallium arsenide and while (2) and (3) can be used to simulate the InP behavior, the fixed cathode field approximation is in some cases inadequate. In particular, while a number of classes of devices can be explained with contact curves represented by Figs. 3 through 5, the appearance of very high efficiency oscillations (Colliver et al., 1973) with prethreshold current voltage curves similar to that of Fig. 8, requires $j_c(E_c)$ curves similar to that of #4 in Fig. 2 for its simulation. This curve is qualitatively different from the curves discussed earlier, for as the range of current oscillations the cathode conduction current is approximately constant, while the cathode field can almost freely move about. A plot of the electric field profiles during the course of the oscillation is shown in Fig. 14. We note that immediately downstream from the cathode contact the electric field profile is approximately uniform. We now turn to the subject of three terminal devices.

THREE TERMINAL DEVICES

We are interested in the hot electron contributions to the operation of gallium arsenide field effect transistors (FET). The FET in its simplest form is a semiconductor slab with three terminals. Two of these are usually low resistance contacts, while the third is either a Schottky contact or a PN junction with an accompanying region of charge depletion. For a unipolar conduction device, operation is based on modulation of the depletion region which is usually accomplished by changes in the gate bias. Small and large signal (Colliver et al., 1973) gain are possible. Figure 15 is a sketch of the device and connecting lumped elements, whose operation we have simulated. Our (two dimensional) numerical simulations and experiments (Grubin et al., 1979) demonstrate that the GaAs FET can be placed in one of two groups as determined by the ratio

$$K = \frac{\text{Gate voltage at cutoff}}{\text{Drain voltage at the onset of current saturation for zero gate voltage}} \quad . \tag{7}$$

Devices with K greater than unity sustain current oscillations, whose origin lies in the presence of negative differential mobility in the semiconductor. Those with K approximately equal or less than unity are electrically stable. We point out that in the original FET discussion, as given by Shockley (1952), K was necessarily equal to unity. The situation with K greater than unity is summarized in Fig. 16a, while that for K approximately equal to unity is summarized in Fig. 16b (Grubin, 1977). In Fig. 16 we sketch the current voltage relation for a gallium arsenide FET with the parameters listed in the caption. For reference we have drawn the velocity electric field

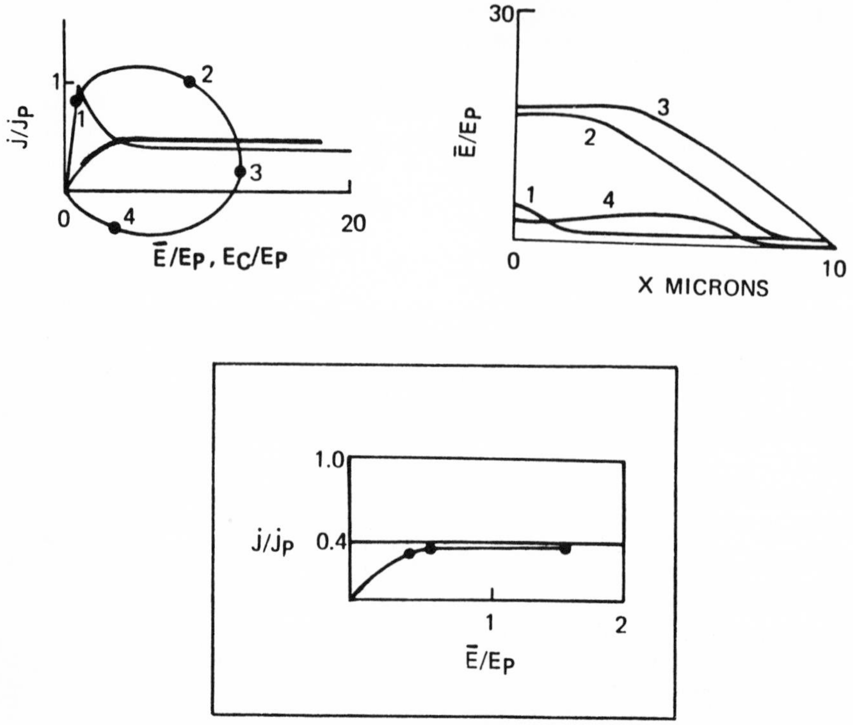

Fig. 14. Reactive circuit calculations for the cathode conduction current curve #4 of Fig. 2. The inset shows the pre-instability current density versus average electric field curves for the element in a resistive circuit. From Shaw et al. (1979).

relation for GaAs scaled to the current and voltage parameters. The first point to note is that the current levels do not approach the peak current associated with GaAs. This is a consequence of the additional resistance supplied by the gate region as well as velocity limitations. We recall from our earlier discussion of two terminal GaAs devices that sublinearity in the GaAs I-V relation is often accompanied by a current instability. For the wider channel device shown in Fig. 16 similar behavior occurs. The instability is represented by the dashed part of the I-V curves of Fig. 16a. The x's in the diagram represent average current and voltage values for the instability, and the presence of negative conductance is due to the dynamic propagating domain. The closed circles in Fig. 16 represent stationary, time-independent points, and we note that when there is an instability it is surrounded by regions of nonzero gate and drain

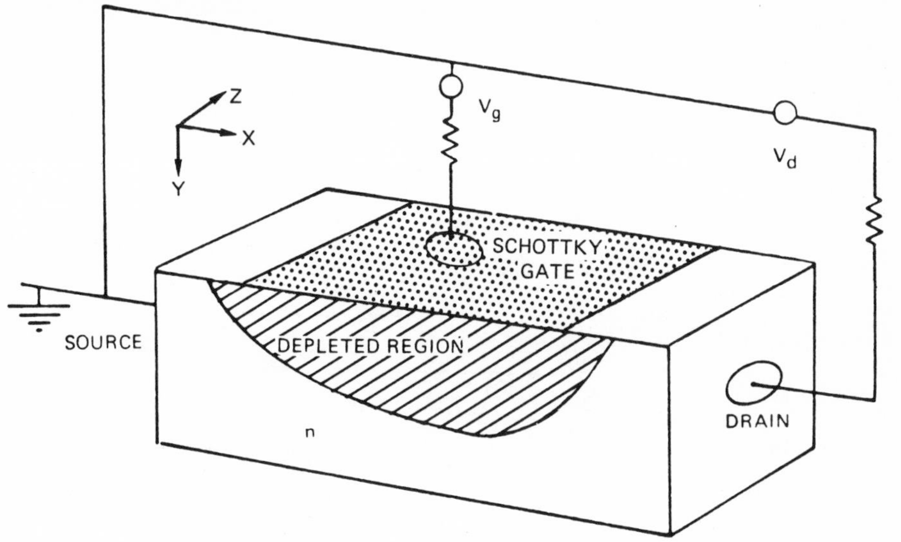

Fig. 15. Schematic representation of a Schottky - gate field effect transistor.

bias for which there is no time-dependent behavior. We now illustrate the space charge distribution associated with the x's and closed circles.

The internal distribution of charge and current associated with the current and potential levels '1' and '2' of Fig. 16a are shown in Fig. 17 (Grubin and McHugh, 1978). Column a shows a set of current density streamlines through the device. The length of each streamline is proportional to the magnitude of the vector current density at the point in question. The maximum length of the individual x and y components before overlap is $j_p = N_0 e v_p$ where v_p is the peak carrier velocity. We see in both cases that the current density is greatest under the gate contact as required by current continuity. For the higher bias case, the current density under the gate region is at least as great as j_p and velocity limitation introduces carrier accumulation. The density of charge particles in the FET is generally nonuniform and column b of Fig. 17 is a projection of this distribution as it may relate to the current density profile of column a. We point out that the particle density increases in the downward direction. The first frame shows a region of charge depletion directly under the gate contact. The second frame shows the formation of a weak stationary dipole layer under the gate contact. We see from these diagrams that the x-component of electric field has reached the NDM threshold field value at the drain edge of the gate contact. We point out that an analysis of the potential distribution

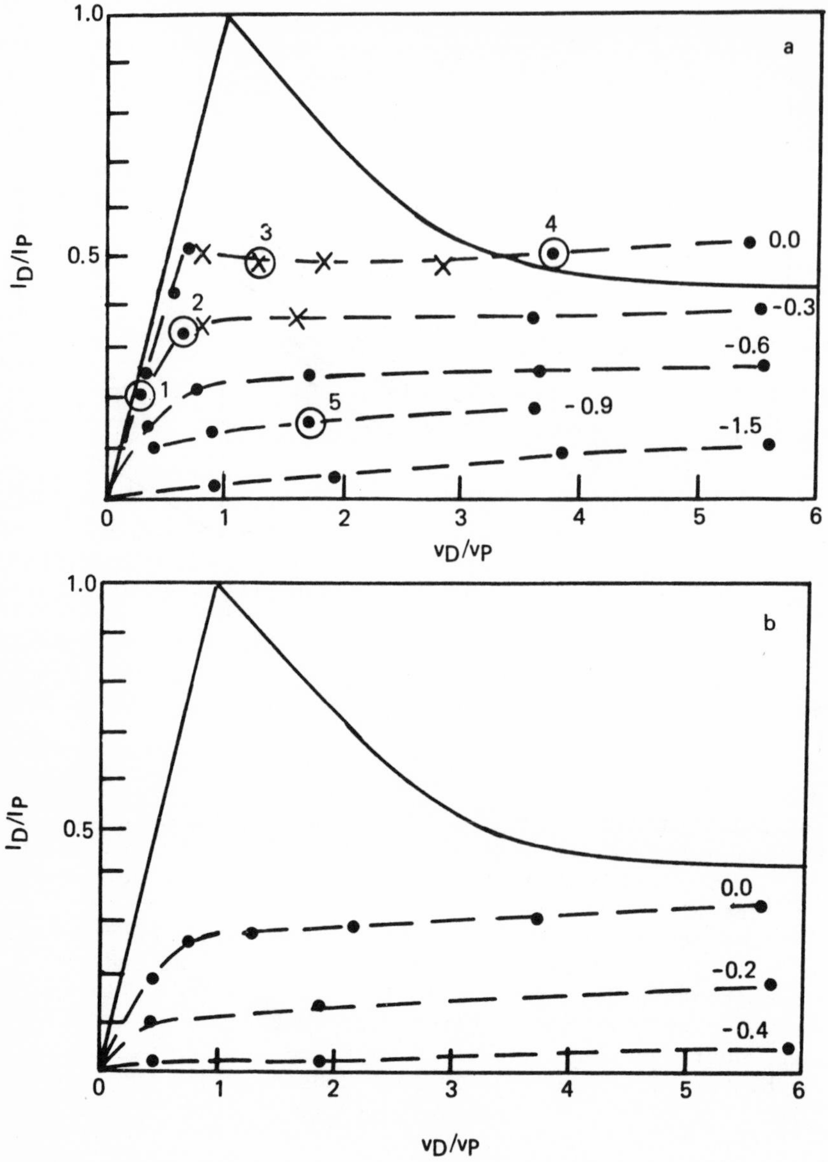

Fig. 16. Drain current - drain voltage relation. Circles denote computed points, x's denote averages of the current during an oscillation. For the calculations the doping of the active region was $10^{15}/cm^3$. Thus a wide channel device as represented in (a) had a channel height of 2.24 microns. The narrow channel device as represented by (b) had a channel height of 1.22 microns. From Grubin (1977). Note: I_p $N_O e v_p A$, $v_p = E_p L$.

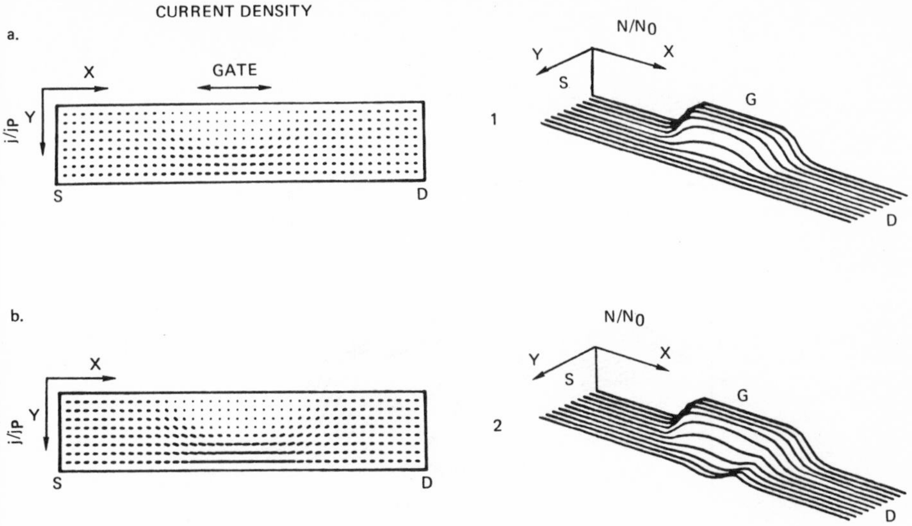

Fig. 17. The internal distribution of current and charge corresponding to the current and voltage levels represented by numbers '1' and '2' of Fig. 16. Note that the particle density surrounding the nonuniform distribution is uniformly distributed within the source-gate region and the gate-drain region. From Grubin and McHugh (1978).

shows that at the high bias levels most of the potential drop is under the gate contact.

We now consider the presence of an instability (Grubin and McHugh, 1978). We recall that for two terminal devices the instability was determined by the value of electric field at the cathode boundary region and that the threshold current density for the instability could be anywhere between j_S and j_P, where j_S is the current density associated with the high-field saturated drift velocity. In three terminal devices, the initiation of a domain instability generally occurs under the gate contact. The instability occurs at a value of current density approximately equal to j_ρ. Figure 18 provides a dramatic representation of the FET instability, where we see the nucleation, propagation and recycling of a high field domain. The sequence of events associated with the instability is as follows: Domain growth under the gate is accompanied by an increase in potential across the device. A corresponding decrease in current occurs throughout the device and circuit, as constrained by the dc load line. As the current decreases, carriers with velocities below that of the peak velocity enter the accumulation layer which subsequently

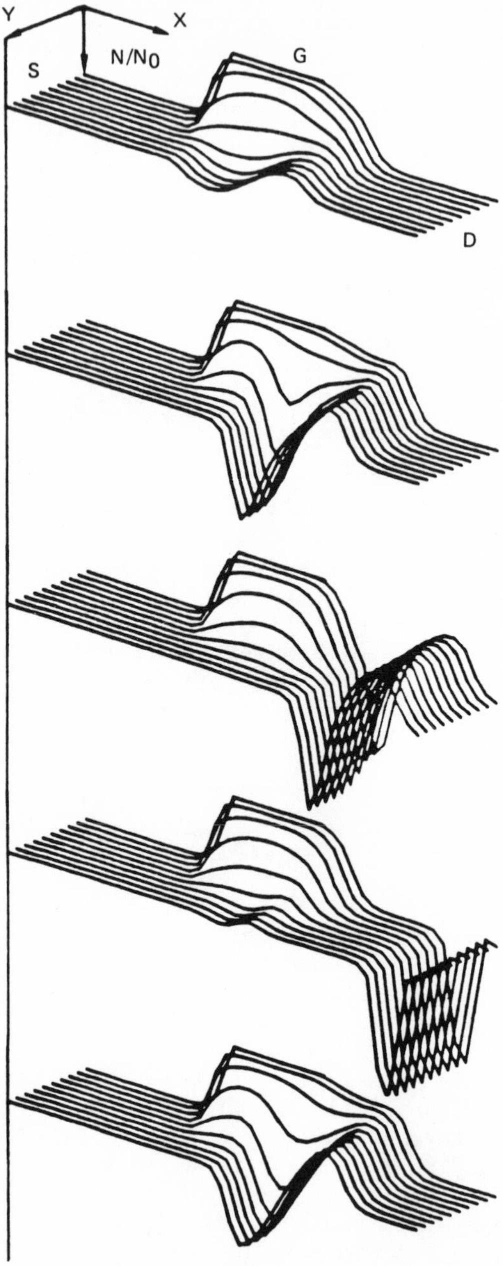

Fig. 18. Projection of the time dependent particle density when an instability occurs. The parameters at which this occurs are represented by the number '3' in Fig. 16. From Grubin and McHugh (1978).

begins to detach. The domain spreads as it leaves the gate region and it settles into a value of current density somewhat in excess of that associated with the saturated drift velocity of the electrons. Prior to reaching the drain contact the domain dynamics appear to be one dimensional.

The numerical situation at high drain bias levels and zero gate bias is similar to that of two terminal devices. Namely the oscillations cease and are accompanied by the presence of an accumulation layer extending from the gate to the drain contact. This is illustrated in Fig. 19. The region of space charge accumulation is accompanied by a large nonuniform potential drop in the gate to drain region with resulting high values of electric field and electrons traveling at their saturated drift values. This region of space charge accumulation is qualitatively similar to the anode-adjacent domain discussed in connection with Fig. 9.

Our discussion of two terminal devices indicates that large values of cathode boundary fields limit the downstream carrier velocity to values below the saturated drift value. Instabilities, if present, are damped. An analogous situation occurs with three terminal devices when a large negative bias is applied to the gate contact. For this case the depletion layer moves toward the bottom of the channel and the source-drain current is low (Grubin and McHugh, 1978). A large potential drop is present under the gate contact resulting in the formation of a high field domain. As seen in Fig. 20, the domain consists of an accumulation region surrounded by a depletion zone.

The experimental situation with regard to hot electron contributions to the FET may be summarized by turning to Fig. 21 (Grubin and McHugh,1978) where we show the dc current voltage characteristics for a device whose parameters are listed in the caption and for which $K > 1$. The cross hatched region shows the range over which current oscillations were observed as detected by a spectrum analyzer. Oscillations were detected over a range of 6 to 40 GHz. In all of our experiments in which oscillations were observed, the instability occurred beyond the 'knee' of the current voltage relation. Prior to the oscillation the drift velocity in the gate to drain region was approximately equal to or greater than the saturated drift velocity of the carriers. Saturation in current occurred when the average field within the conducting channel under the gate contact was approximately equal to the threshold field for negative differential mobility (Grubin, 1977). The instability was suppressed either by going to a sufficiently high drain bias or by going to a sufficiently large and negative gate bias, such that in the latter case the gate to drain channel current density was below j_s.

The above discussion demonstrates that in wide channel devices hot electron effects in three terminal devices are similar to that of

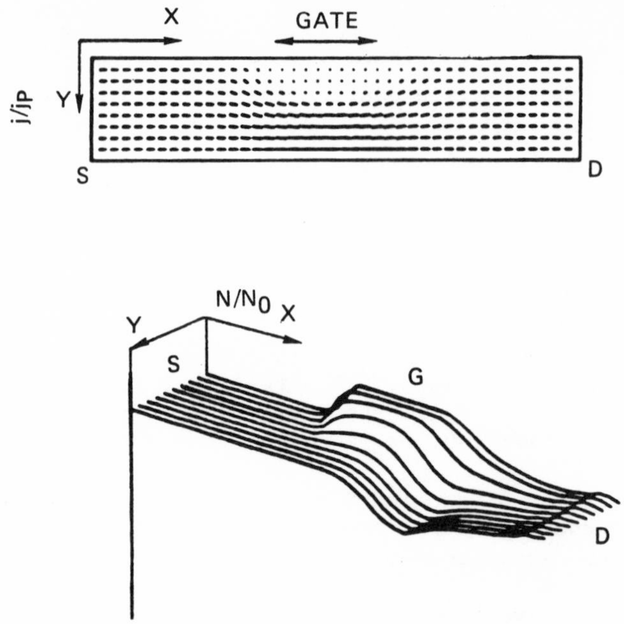

Fig. 19. Distribution of current and particle density for the case represented by '4' of Fig. 16. The instability has ceased. Note that the current density within the gate to drain region, where there is particle accumulation, is accompanied by current density levels that are approximately equal to the saturated drift current. From Grubin and McHugh (1978).

two terminal devices. A simple intuitive picture of this commonality arises after making a one-to-one correspondence between the potential drop in that portion of the conducting channel that is under the gate contact and a phenomenological cathode boundary for two dimensional devices.

The situation with narrow channel devices is somewhat less dramatic from the point of view of domain instabilities. Generally because of the nonlinear distribution of potential under the gate contact, very thin devices sustain current densities significantly below j_S in the region between the gate-to-drain contact. As in two terminal devices, this is sufficient to prevent an instability.

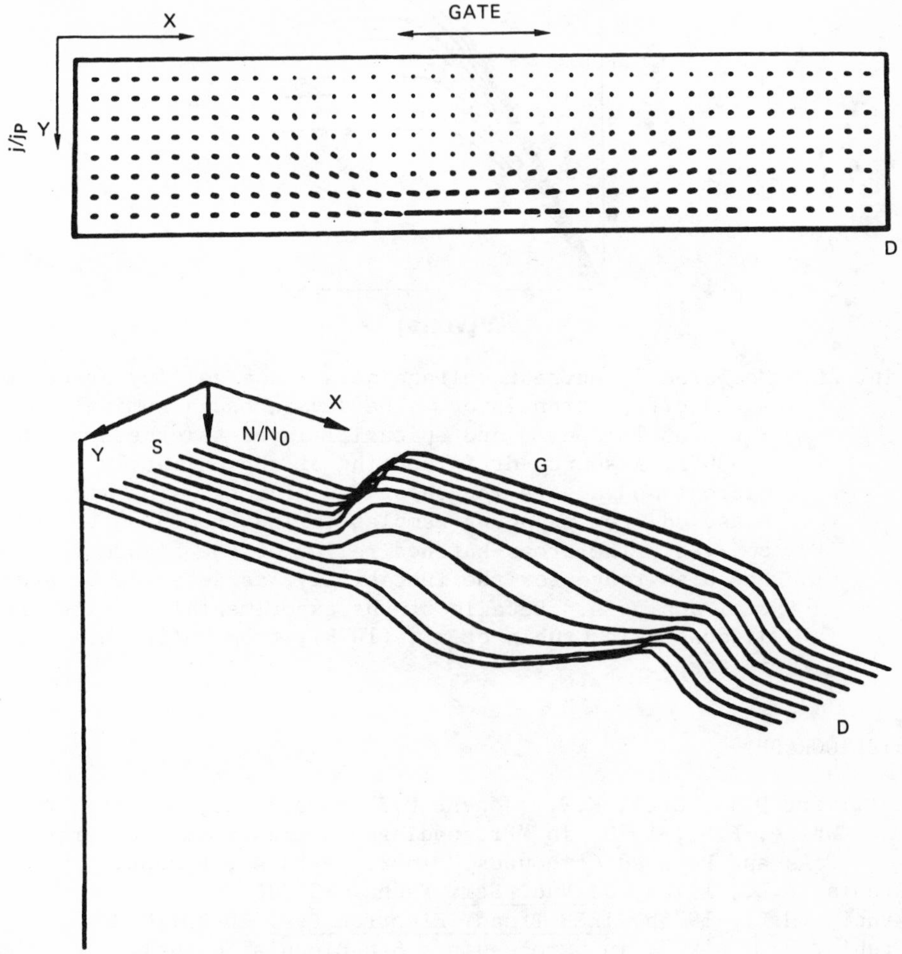

Fig. 20. Distribution of current and particle density for the case represented by '5' in Fig. 16. The instability has ceased. Note the large dipole layer within the vicinity of the gate contact, and the fact that the current density between the gate and drain region is very small. From Grubin and McHugh (1978).

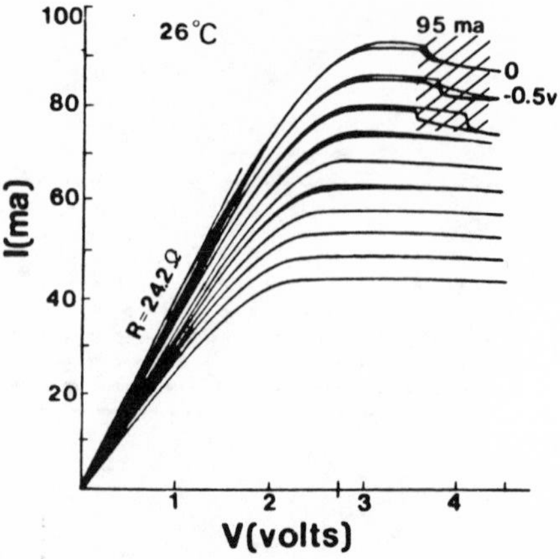

Fig. 21. Measured dc current voltage data for a gallium arsenide field effect transistor. The device had a nominal doping level of $10^{17}/cm^3$, and epitaxial layer thickness of 3000 ± 500°A, a source-drain spacing of approximately 8.5 microns and a gate length of 3.0 microns. The mobility measured from adjacent samples varied from 3000 to 4000 cm^2/v-s. The cross-hatched region of the figure denotes the bias range for the instability, as detected by spectral analysis. Details of the experimental results are discussed in Grubin et al. (1979), from which this data is taken.

BIBLIOGRAPHY

Colliver, D.J., Gray, K.W., Jones, D.J. Rees, H.D., Gibbons, G., and White, P.M., 1973, in "Proceedings International Conference on GaAs and Related Compounds," Inst. of Phys., London.
Grubin, H.L., 1976a, J. Vac. Sci. Techn. 13:786.
Grubin, H.L., 1976b, IEEE Trnas. Electron Dev. ED-23:1012.
Grubin, H.L., 1977, in "Proceedings 6th Biennial Cornell Electrical Engineering Conference," Cornell University.
Grubin, H.L., 1978, IEEE Trans. Electron Dev. ED-25:511.
Grubin, H.L., 1979, this volume.
Grubin, H.L., and McHugh, T.M., 1978, Sol.-State Electr. 21:69.
Grubin, H.L., Ferry, D.K., and Gleason, K.R., 1979, Sol.-State Electr in press.
Gunn, H.B., 1964, IBM J. Res. Develop. 8:141.
Kroemer, H., 1968, IEEE Trans. Electron Dev. ED-15:819.
Rideout, V.L., 1975, Sol.-State Electr. 18:541.
Shaw, M.P., Solomon, P.R., and Grubin, H.L., 1969, IBM J. Res. Develop. 13:587.
Shaw, M.P., Grubin, H.L., and Solomon, P.R., 1979, "The Gunn-Hilsum Effect," Academic Press, New York.

Solomon, P.R., Shaw, M.P., Grubin, H.L., and Kaul, R.D., 1975, IEEE Trans. Electron Dev. ED-22:127.
Shockley, W., 1952, Proc. IRE 40:1365.
Shockley, W., 1954, Bell Sys. Tech. J. 33:799.
Thim, H.W., 1968, J. Appl. Phys. 39:3897.

MODELING OF CARRIER TRANSPORT IN THE FINITE COLLISION DURATION REGIME: EFFECTS IN SUBMICRON SEMICONDUCTOR DEVICES

D. K. Ferry

Department of Electrical Engineering
Colorado State University
Fort Collins, Colorado 80523

INTRODUCTION

Over the last two decades, the electronics industry has responded to market forces by producing less expensive, but more complex and sophisticated, integrated circuits by large-scale-integration (LSI). The growth of LSI has in fact been phenomenal, and the application of special integrated circuits and microprocessors has blossomed. In fact, the complexity of these circuits, where complexity is defined here by the number of individual devices on an integrated circuit chip, has approximately doubled each year over this time span. There are, of course, several factors which contribute to this increase in complexity, including major effects arising from increased die size, increased circuit cleverness, and reduced device size (Moore, 1975). This latter factor, reduction of the individual feature size in a device, is of paramount importance and dimensions are currently down to the 1-2 micrometer range. Progress in the micro-electronics industry is strongly coupled with the ability to continue to make ever increasing numbers of smaller, and more clever, devices on a single chip; i.e., the move toward very-large-scale-integration (VLSI) will be of paramount importance to this continued progress. It is apparent that extrapolation of today's technology will produce individual devices whose dimensions are of the order of 0.2-0.5 micrometers (Ballantyne, 1978).

On the other hand, the advent of high-resolution electron, X-ray, molecular, and ion-beam lithographic techniques is leading us toward an era in which individual feature sizes might well be fabricated on the molecular scale of 10-20 nanometers. In fact, individual metallization structures already have been made on this scale by electron lithographic techniques (Broers, et al., 1978). It becomes feasible then to conceive of very small device structure that are so small that the bulk properties of the host semiconducto material may be significantly less important than size and environment related effects.

From the above discussion, it is apparent that one can reasonably ask just how far can we expect device size to be reduced, and, more importantly, do we understand the physical principles that wil govern device behavior in that region of smallness? Previously, Hoeneisen and Mead (1972a, 1972b) discussed minimal device size for MOS and bipolar devices based upon extrapolation and scaling of present day device structures. They considered a general study of the static limitations imposed by oxide breakdown, source-drain punchthrough, ionization in the channel, etc., on typical MOS structures, and similar effects in the bipolars. Although they suggeste a minimal channel length of 0.11 micrometer for the MOS, for exampl it was found that a more typical device would have a channel length of 0.24 micrometer. Others have approached this topic in a similar manner and also have considered the role of doping fluctuations, power dissipation, noise, and inter-connections in limiting packing density of these devices. And yet, for all of these studies, it is evident that the emphasis has been on other questions than whether or not our current understanding of device physics is adequate to handle devices on a smaller spatial scale.

It is readily evident that technological momentum is pushing to ever smaller devices, and the technology is there to prepare really very small devices. It, moreover, becomes obvious that we must now ask whether our physical understanding of devices and thei operation can indeed be extrapolated down to very small space and time scales, or do the underlying quantum electronic principles prevent a down-scaling of the essential semiclassical concepts upon which our current understanding is based? More importantly, do the underlying quantum electronic principles offer new options for devi operation? In earlier work, a quantum-mechanical study of the dynamics of single and multi-device operation on such small spatial (and temporal) scales was made and the foundation laid for understanding the physics of device operation and possible synergetic features on these small spatial scales (Barker and Ferry, 1979a, Ferry and Barker, 1979a). Moreover, we have laid a conceptual fram work for examining the detailed physics and modeling necessary to characterize these small devices. These foundations are based upon fully quantum mechanical system dynamics including nonlinear electronic processes. Some of these nonlinear electronic processes have already been illustrated to be of importance in present day devices; i.e., effects related to high field transport in light-mass semiconductors (Barker, 1978) and to dielectric breakdown in oxides (Ferry, 1979a).

The traditional approach to the modeling problem of semiconductor device operation is to divide the device into several areas for which several different, but linear, approximations hold, and to solve the one-dimensional transport problem consistently within these areas. However, it is now well-recognized that such approximate approaches are inappropriate for devices on the small scale, even for 1-5μm channel length devices. To gain the proper insight and understanding of device behavior for the small scale device, more general and accurate two-dimensional (at a minimum) solutions are required which are generally only obtainable by numerical techniques. Suitable computer codes for these solutions are thus becoming an important part of the device designers' repetoire. As we pointed out above, the solution to the two-dimensional Poisson's equation represents no conceptual difficulties. The major physical effects however, are dominantly tied up in the manner in which the charge fluctuations and current response are coupled to the local electric field, formally related through the continuity equation. To accurately determine the current response, one must solve an appropriate transport equation, as discussed above, and it is in these transport equations that many of the major modifications arising from quantum electronics appear.

In the following sections, we examine the major deviation from the Boltzmann transport equation due to the intracollisional field effect (ICFE). We then turn to the retarded transport equations and show how this ICFE leads to a retardation of the collisional interactions.

THE INTRACOLLISIONAL FIELD EFFECT

The Boltzmann transport equation (BTE) has long been the basis for semi-classical transport studies in semiconductors and other materials. Its utility also stems from the fact that it is readily transformable into a path variable form which can be adapted to numerical solutions for complicated energy-dependent scattering processes (Ferry, 1979b). In this form, the BTE is often referred to as the Chambers-Rees path integral equation, and serves as the basis for Monte Carlo and iterative calculations of transport. However, the BTE is valid only in the weak coupling limit under the assumptions that the electric field is weak and slowly varying at most, the collisions are independent, and the collisions occur instantaneously in space and time. Each of these approximations can be expected to be violated in future sub-micron dimensioned semiconductor devices. We have previously shown that in such devices, the time scales are such that collision durations are no longer negligible when compared to the relevant time scale upon which transport through the device occurs. In this situation, even for time-independent fields, the quantum kinetic equations are nonlocal in time and momentum. It may be recalled that the BTE can be

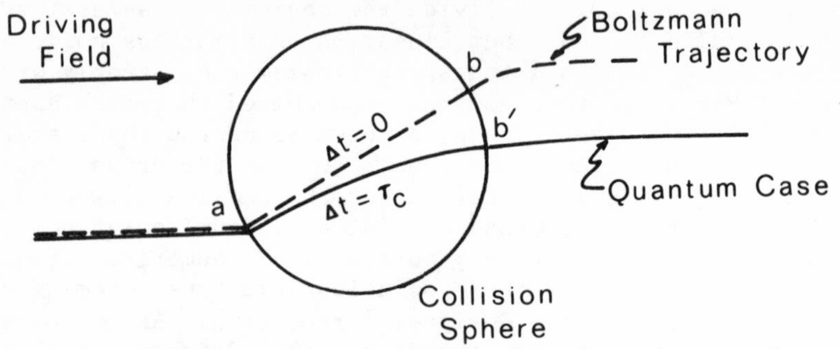

Fig. 1. Naive picture of a collision event. The intra-collisional field effect modifies the trajectory through the collision thus affecting the collision dynamics.

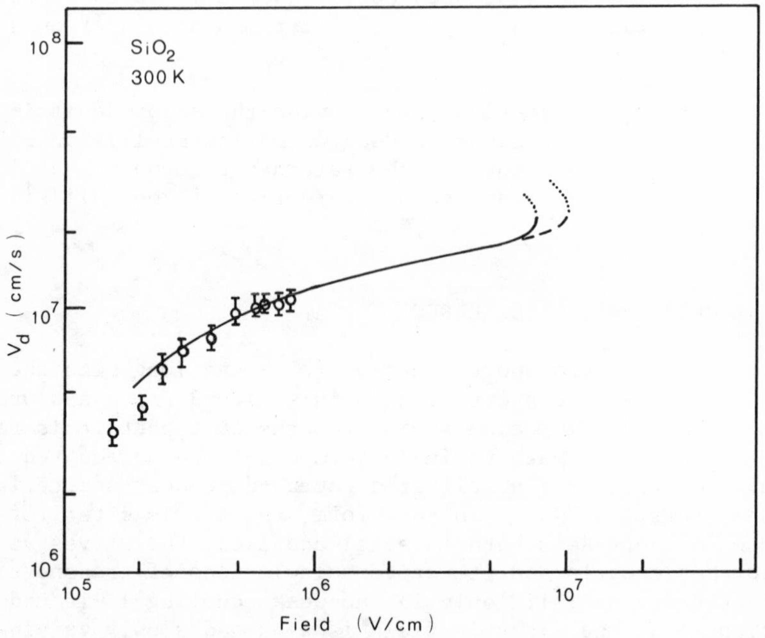

Fig. 2. Velocity-field relationship for high-field transport in SiO_2. The subthreshold emission of LO phonons leads to a lowering of the breakdown field (solid curve) by some 30% over the value found without the ICFE (dashed curve). Data points are due to Hughes (1975, 1978). The dotted portion of the curves are unstable.

rigorously derived from the density matrix Liouville equation formulation of quantum transport (Kohn and Luttinger, 1957, Barker, 1973). In the BTE approach, the collision terms are derived under the assumption that the collisions occur instantaneously, which is a reasonable approximation when the mean-time between collisions is large. At high fields, such as will occur in very small devices, the collision duration is significant and correction terms must be generated (for the BTE) to account for the non-zero time duration of each collision. If the instantaneous collision approximation that leads to the BTE is relaxed, an additional field contribution appears as a differential super-operater term (see, e.g., the discussion in Barker and Ferry, 1979a) in the collision integrals evaluated in the momentum representation, resulting in an intra-collisional field effect (Barker, 1973, 1978; Thornber, 1978).

The intracollisional field effect (ICFE) can be partially understood by referring to Fig. 1. In the Boltzmann case, the collision occurs instantaneously, so that the carrier enters the collision sphere at a and instantaneously exits at b. However, the collision does not occur instantaneously, but requires a non-zero collision duration τ_c. In this case, it can now be accelerated by the field *during the collision.* Thus it exits not at b, but at b' some time $\Delta t = \tau_c$ later. When τ_c begins to become comparable to τ, the mean-time between collision, this ICFE will have a significant effect on the transport dynamics, particularly in the transient response region.

The mathematical details of the ICFE have previously been given by Barker (1979), so we shall not go into these details here. Rather, we shall give some of the supportive evidence for the observability of the effect. In very large fields, such as can occur in SiO_2 near breakdown, the ICFE can indeed by very significant (Ferry, 1979a). Two major modifications of the scattering integral occur as a result of this intracollisional process. First, the total energy-conserving δ-function is broadened by the presence of the electric field. Second, the threshold energy required for the emission of an optical phonon is modified, which causes a shift (in energy) of the δ-function. This latter process is easily understood in physical terms. In the emission of an optical phonon, where the electron is scattered against the electric field, the field will absorb a portion of the electron's energy during the collision, and hence a reduction in energy loss to the lattice will be favored. The opposite effect, an enhancement in energy loss to the lattice, occurs for emission along the electric field. These effects can be incorporated into the appropriate scattering integrals in the iterative technique, and this has been carried out. In Fig. 2, the velocity-field relationship is shown for the case when the field-induced modifications of the scattering parameters are included. The dotted curve represents the two LO scattering results above without

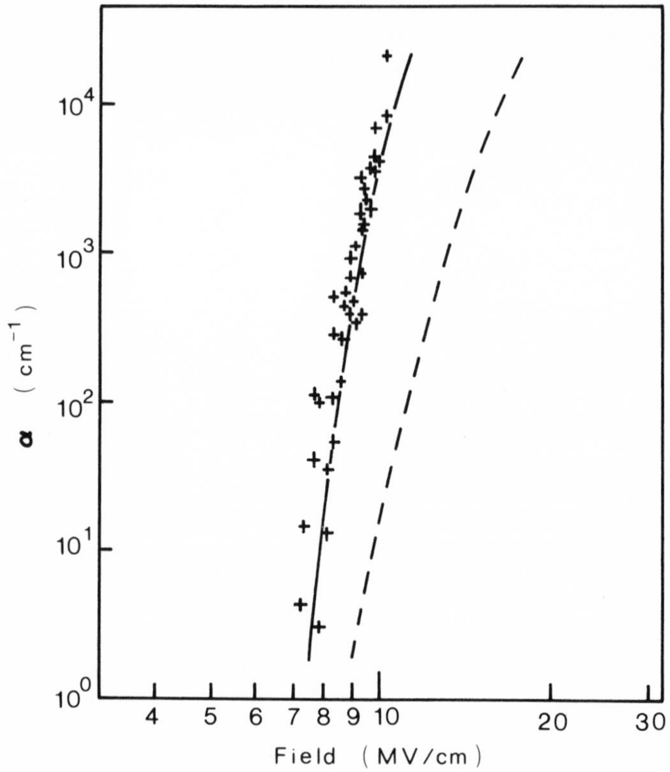

Fig. 3. Ionization rates in SiO_2 as a function of field. The dash curve is for $m_h/m_e = 100$ and ignores the subthreshold phon emission. The solid curve takes into account subthreshold emission, for $m_h \sim 2.9m_0$. Data is due to Solomon and Klei (1975). This is fairly substantive evidence for the presence of the ICFE.

the field-induced modifications. For electric fields above (5-6) x 10^6 V/cm, the broadening and shift of the scattering resonances produce a noticeable effect upon the velocity-field relationship. The dashed portion of the curve represents a region where the distribution function is unstable. This reduction in threshold can be further observed in the impact ionization rates in SiO_2. In Fig. 3, we show these rates as measured by Solomon and Klein (1975) and as calculated earlier (Ferry, 1979a). It is necessary to include effects arising from the ICFE to adequately fit theory to experiment.

The second major effect of the ICFE arises for scattering between non-equivalent sets of valleys (Barker, 1979). There, the ICFE causes a virtual lowering of the final state valleys with respect to the initial state valleys. In GaAs, for example, at

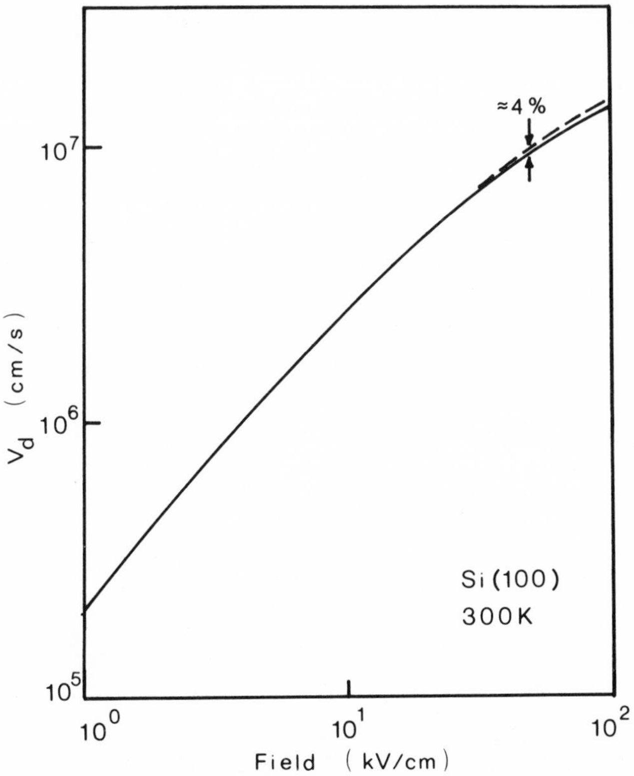

Fig. 4. Velocity-field curve for electrons in a quantized inversion layer at the surface of Si in a MOSFET. The virtual lowering of the higher lying subbands by the intra-collisional field effect results in the difference between the solid and dashed curves. Here, we assume $n_s = 10^{13}/cm^2$ and $N_{depl.} = 4 \times 10^{10}/cm^2$.

100 kV/cm, the L-valleys are reduced during the collision by 36 meV. However, at 100 kV/cm, the electrons scatter very rapidly anyway and this effect is essentially unobservable. However, in quasi-two-dimensional inversion layers in Si MOSFETS, the subband spacing can be comparable to the virtual lowering. Transport in these systems has been treated earlier (Stern and Howard, 1967; Stern, 1974; Fang and Fowler, 1968, 1970; Ferry, 1976, 1978). In Fig. 4, we show the velocity-field curves in a Si inversion layer taking into account the virtual lowering of the upper subbands. Here the maximum effect is about 4%, probably not able to be delineated by current experimental techniques.

THE RETARDED EQUATIONS

When the ICFE is included as a modification of the lowest orde kinetic equations, a high field quantum kinetic equation, which replaces the BTE, is found as

$$\frac{\partial f(\vec{p},t)}{\partial t} + e\vec{E}(t) \cdot \nabla_p f(\vec{p},t) = \int_0^t dt' \sum_{\vec{p}'} \{S(\vec{p},\vec{p}';t,t')\, f(\vec{p}',t')$$

$$- S(\vec{p}',\vec{p};t,t')\, f(\vec{p},t')\}, \tag{1}$$

where the momenta $\vec{p},\vec{p}'$ are explicit functions of the retarded time t' on the right-hand-side through the relationship

$$\vec{p}(t') = \vec{p} - \int_{t'}^{t} e\vec{E}(t'')dt'', \tag{2a}$$

$$\vec{p}'(t') = \vec{p}' - \int_{t'}^{t} e\vec{E}(t'')dt'', \tag{2b}$$

and the transition terms S take the form, for inelastic phonon scattering,

$$S(p,p';t,t') = \mathrm{Re}\, \frac{2\pi}{\hbar} \sum_{\vec{q}} \left(\frac{1}{\pi\hbar}\right) \exp\left(-\frac{t-t'}{\tau_\Gamma}\right) \times \tag{3}$$

$$\times\ (N_q + \tfrac{1}{2} + \tfrac{1}{2}\eta)\ |V(q)|^2\ \delta_{\vec{p},\vec{p}' + \eta\vec{q}}\ \exp\left[-i \int_t^{t'} \frac{dt''}{\hbar}\, \beta(\vec{p},\vec{p}';t'')\right],$$

where

$$\beta(\vec{p},\vec{p}';t'') \equiv \varepsilon[\vec{p}(t'')] - \varepsilon[\vec{p}'(t'')] + \eta\hbar\omega_{\vec{q}}. \tag{4}$$

The two exponential factors in (3) are related to the joint spectra density function, which reduces to an energy conserving δ-function the instantaneous collision, low-field limit, $\varepsilon(\vec{p})$ is the quasi-particle renormalized energy, $\hbar/\tau_\Gamma$ is the joint linewidth due to collisional broadening of the initial and final states, and η takes the values +1, -1 for phonon emission or absorption, respectively, in the in-scattering term. For the out-scattering term, the role o $\vec{p}$, $\vec{p}'$ are inter-changed, although this does not upset detailed balance in the equilibrium sense.

In small semiconductor devices, where the dimensional scale is of the order of 1.0 μm or less, the carrier concentration will in general be relatively high. We use the drifted Maxwellian approach to developing a set of coupled balance equations, using (1) instead of the BTE. With this approach, a hierarchy of moment equations can be generated, from which the various parameters can be determined. In this work, we extend these moment equations to include first-order effects arising from the non-zero time duration of the collisions. Starting from the density-matrix developed form of the BTE, we show that collision terms derived in the normal case, but modified for intra-collisional field effects, must be convolved with a decay effect over an effective collision duration. Thus, the balance equations are modified in a straight-forward fashion, although the details are much more complicated. This latter follows from the role of the intra-collisional field effects, which both broaden and shift the resonances, effectively lengthening the effective collision duration and weakening the effect of the collision itself. If (1) is Laplace Transformed, the moment equations can be developed by multiplying by an arbitrary function $\phi(\vec{p})$, integrating over the momentum, so that the moment equations are developed in the transform-domain, and then retransforming. This yields (Ferry and Barker, 1979b).

$$< \frac{\partial\phi}{\partial t} > + e\vec{E} \cdot <\nabla_p \phi> = \frac{1}{\tau_c} \int_0^t \exp(-t'/\tau_c) <\dot{\phi}_c(t-t')> dt', \qquad (5)$$

where $\dot{\phi}_c$ is the time-rate of change of $\phi(\vec{p})$ due to collisions. The result in (5) is particularly interesting, in that it allows $<\dot{\phi}_c>$ to be evaluated in the case of instantaneous scattering, and this result to be averaged over an effective collision duration τ_c (Barker and Ferry, 1979b, Ferry and Barker, 1979c), weighted by the function $\exp(-t/\tau_c)$. If, as is the case at low fields, $<\dot{\phi}_c>$ does not change during a collision duration, the normal result ($t >> \tau_c$) is obtained. However, in large fields, where the intra-collisional field effect is important, the variation of $<\dot{\phi}_c>$ during the collision becomes important. These effects will also be important in high-frequency transport where the collision duration becomes comparable to the relaxation times and the period of the wave.

Calculations of the transient response have been carried out for GaAs, considering Γ-L-X ordering. By following these to a point in time in which a steady-state is reached, the v-E curve for long times is found. The parameters used agree closely with those adopted elsewhere. We take $\varepsilon_{\Gamma-L} = 0.3$ eV and the central valley parameters as normally adopted. The L-valleys are taken to be Ge-like, with $m_d = 0.22m_o$ per valley and $m_c = 0.15m_o$. Intervalley phonon coupling constants of 1×10^9 eV/cm are taken for the Γ-L and L-L intervalley phonons. In Fig. 5 is shown the effective collision-duration τ_c, as

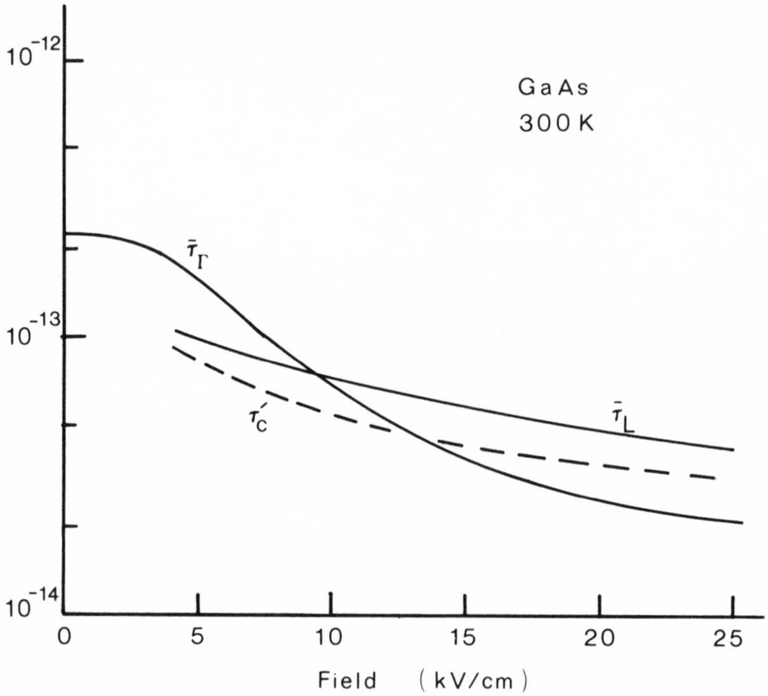

Fig. 5. Variation of the effective collision duration and mean-time-between collisions for the Γ and L valleys of GaAs. The latter quantities are calculated from the effective mobilities from $\bar{\tau} = m^*\mu/e$.

well as the average mean-time-between-collisions for the Γ and L valleys. From this, it can readily be seen that the collision duration must be considered in GaAs.

In Fig. 6, the transient response for a steady, homogeneous field of 10 kV/cm, applied at $t = 0$, is shown. The response for a retarded collisional interaction rises quicker and settles faster than that of the unretarded case. The quicker rise follows from the retarded momentum relaxation effects, while the faster settling occurs due to retarded energy relaxation effects which causes an overshoot to occur in the temperature as well.

The collisional retardation speeds up the transient process primarily due to the effect of slowing down changes in the effective momentum and energy via collisional relaxation as well. This is especially sensitive to the population in the various valleys. Th

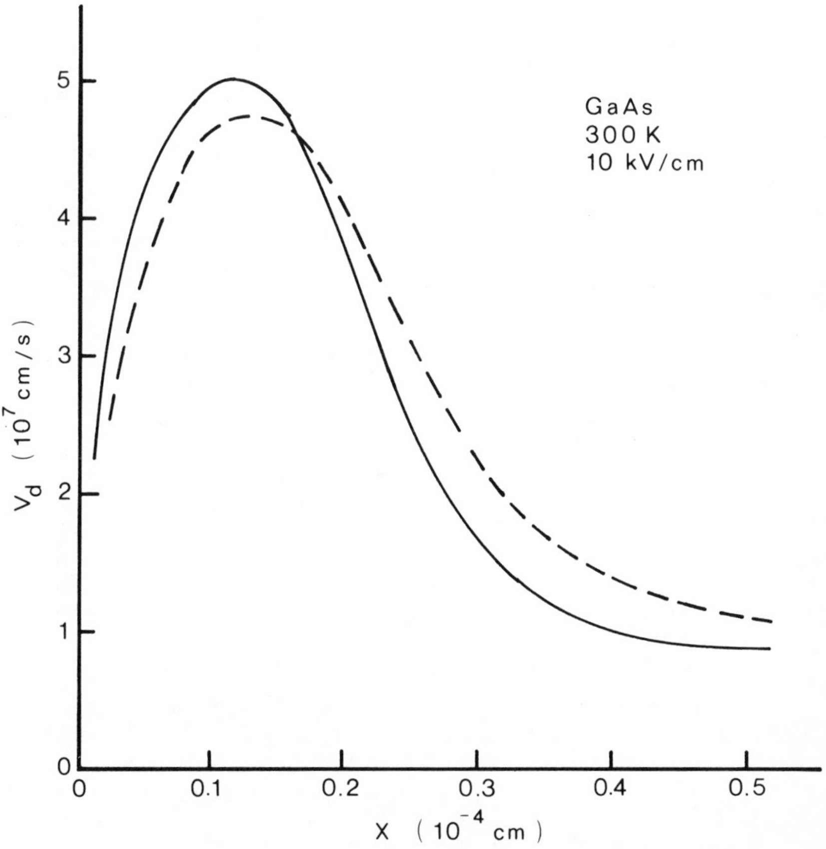

Fig. 6. Velocity as a function of distance for electrons in GaAs. It is assumed that the carriers enter the homogeneous, high electric field at t = 0, x = 0. The solid curve includes the effects of retardation of the collisional interaction due to the finite-collision duration.

is significant, as in the repopulation region of the velocity-field curve, the net velocity is largely dominated by the repopulation itself and the small signal a.c. mobility is extremely sensitive to the energy distribution function, so it is extremely important that any simulation technique be very efficient in yielding this portion of the distribution function.

DISCUSSION

The role of retardation of the collisional interaction speeds up the process of transient response of the velocity. Although it retards the collisional relaxation and allows a faster rise of the velocity response, it allows a faster rise of the effective temperature. This latter results in a faster settling of the response to the steady-state values, and has a number of important consequences for sub-micron semiconductor devices. First, the frequency of mobility roll-off should be increased resulting in better high frequency response. Secondly, the general quickening of the transient results in the transient being important over a smaller distance than first thought. With one significant exception, the role of retardation could be included by simply scaling a device to a somewhat smaller size. The exception, however, is an important one and deals with the effective velocity achieved in small structures. Here, the much larger and faster overshoot response will lead to different effective velocity-field curves.

BIBLIOGRAPHY

Ballantyne, J. M., Editor, 1978, "Proceedings of the NSF Workshop o Opportunities for Microstructures Science, Engineering, and Technology," Cornell University Press, Ithaca, New York.

Barker, J. R., 1973, J. Phys. C 6:2663.

Barker, J. R., 1978, Sol.-State Electr. 21:267.

Barker, J. R., 1979, previous chapter in this volume.

Barker, J. R., and Ferry, D. K., 1979a, Sol.-State Electr., in pres

Barker, J. R., and Ferry, D. K., 1979b, Phys. Rev. Letters 42:1779.

Fang, F., and Fowler, A. B., 1968, Phys. Rev. 169:619.

Fang, F., and Fowler, A. B., 1970, J. Appl. Phys. 41:1825.

Ferry, D. K., 1976, Phys. Rev. B 14:5364.

Ferry, D. K., 1978, Sol.-State Electr. 21:115.

Ferry, D. K., 1979a, J. Appl. Phys. 50:1422.

Ferry, D. K., 1979b, previous chapter in this volume.

Ferry, D. K., and Barker, J. R., 1979a, Sol.-State Electr., in pres

Ferry, D. K., and Barker, J. R., 1979b, Sol.-State Commun., 30:361.

Ferry, D. K., and Barker, J. R., 1979c, Sol.-State Electr., in pres

Hoeneisen, B., and Mead, C. A., 1972a, Sol.-State Electr. 15:819.

Hoeneisen, B., and Mead, C. A., 1972b, Sol.-State Electr. 15:891.

Hughes, R. C., 1975, Phys. Rev. Letters 35:449.

Hughes, R. C., 1978, Sol.-State Electr. 21:251.

Kohn, W., and Luttinger, J. M., 1957, Phys. Rev. 108:590.

Moore, G., 1975, in "Proc. Intern. Electron Device Mtg., 1975" IEEE Press, New York, p. 11.

Solomon, P., and Klein, N., 1975, Sol.-State Commun. 17:1317.

Stern, F., 1974, Crit. Rev. Sol. State Science 4:499.

Stern, F., and Howard, W. E., 1967, Phys. Rev. 163:816.

Thornber, K. K., 1978, Sol.-State Electr. 21:259.

ON THE PHYSICS OF SUB-MICRON SEMICONDUCTOR DEVICES

J. R. Barker

Physics Department
University of Warwick
Coventry CV4 7AL, U.K.

INTRODUCTION

Considerable evidence now exists that we are on the threshold of a second industrial revolution ushered in by the recent development and massive proliferation of modern semiconductor micro-electronics. The impetus from economic, defense and technological interests is pushing towards ever increasing micro-miniaturization (Ballantyne, 1979, Einspruch, 1979). So far the achievement of large-scale-integration (LSI) has been sustained by a relatively sound base of reliable scientific knowledge in the fields of materials science, fabrication processes, microelectronics and semiconductor physics. The advent of high resolution electron, X-ray, molecular and ion-beam lithographic techniques is currently opening prospects for very-large-scale-integration (VLSI) in which individual feature sizes down to the molecular scale of 10 - 20 nanometers are not inconceivable. Indeed, simple experimental structures of 0.1 to 0.01 micrometers have already been achieved (Broers et al., 1978). However, these new developments are considerably hampered by a large gap in our understanding of non-equilibrium semiconductor heterostructures on scales intermediate to the true atomic scale ($\lesssim$ 10 Angstroms) and the bulk solid-state macro-scale ($\gtrsim$ 1μm = 10,000 Angstroms). It is already apparent that simple down-scaling of processing, device-function and performance, bulk physics, etc., is not adequate; nor indeed, is a straight-forward up-scaling of known atomic-scale phenomena. Sub-micron device physics thus appears as part of a new interdisciplinary science: microstructure science.

Some of the physical constraints which dominate sub-micron devices have been discussed in recent studies (Barker, 1979a, Barker and Ferry, 1979a, 1979b, Ferry and Barker, 1979a, 1979b). They are (1) short spatial scales; (2) fast temporal scales; (3) very high driving fields (which arise from the limitations imposed on down-scaling voltages by thermal fluctuations and the discrete value of the electronic charge); (4) finite (non-bulk) device volume; (5) hi doping/carrier densities; (6) non-isolation from the device environment (the device boundaries and environment of contact regions, interfaces, interconnects, and other devices strongly influence intra-device electronic processes and provide additional mechanisms for inter-device coupling over and above the normal capacitative coupling and communication via currents in the interconnect block); at the highest packing densities the full system architecture may become more important than the underlying crystal structure of the host material.

At the smallest scale - feature sizes down to a few hundred Angstroms - the constraints lead to electronic processes which are not amenable to semi-classical device modeling, and which require concepts inherent in inhomogeneous many-body non-linear quantum response theory. It is not known to what extent conventional effective mass theory, band theory, dielectric response theory, and statistical averaging procedures can be applied at this level. For large devices it is usual to describe intra-device processes by bul theory; the current injection/extraction, and environmental influences are modelled by boundary conditions on the currents and fields alone, and are imposed at the finite device surfaces. Any spatial inhomogeneity is then a secondary feature imposed on the carrier distributions by the local self-consistent electric fields. For sub-micron devices, it may instead be essential to consider the full inhomogeneous device heterostructure and extend the effective device boundaries out beyond the range of the true device-environme coupling. Since strong inter-device coupling may lead to cooperati effects, it may well prove necessary to develop an analogue of quas particle theory to extract from strongly-coupled device arrays an effective renormalized device array in which the individual quasi-devices are weakly-linked, and so amenable to simple analysis.

In the following we describe some of the recently developed theoretical ideas on sub-micron device physics and hint at new phys cal phenomena, devices and device systems which may be forthcoming from the realization of very-large-scale-integration.

PHYSICAL SCALES AND PHENOMENA

The reduction of any problem in condensed matter physics to manageable proportions depends on the physicist's abilities to isolate those physical scales and associated processes which are "dominant" and which processes can be treated as "perturbations".

Down to scales of about 0.25μm the valid conceptual base is still (with a few special reservations) that of conventional solid-state electronics: effective mass and band theory and Boltzmann transport. Scaling laws can take us down to this level to provide physical and material conditions for micro-electronic applications. Thus, for device channel lengths much greater than about 0.25μm (depending on the host material) the device transit time τ_d, mean-free-time τ and elementary collisions duration τ_c satisfy the inequality

$$\tau_c << \tau << \tau_d. \qquad (1)$$

As we have discussed in the quantum transport theory lectures (Barker, these proceedings) this scale is adequately covered by Boltzmann transport phenomenology, except for very high frequency applications, very high fields and certain transient phenomena. At sub-micron scales the inter-relation between different physical scales changes and new phenomena become possible.

At the middle scale (which defines the medium small device, or MSD), where channel lengths are reduced to a range around 0.1-0.25μm, the mean free time may become a significant fraction of the transit time, and more importantly may become comparable to the collision duration:

$$\tau_c \lesssim \tau < \tau_d. \qquad (2)$$

This situation arises partially from the previously mentioned impracticality of down-scaling applied voltages (and temperatures): the subsequent higher operating fields lead to reductions in τ due to velocity saturation, or, more physically, due to the required increase in power dissipation. A detailed quantum transport study of this regime (Barker, 1979a, Barker and Ferry, 1979a, 1979b, Ferry and Barker, 1979a, 1979b) suggests that MSD transport physics may still be based in a Boltzmann-like picture for a Wigner carrier distribution which incorporates the perturbative effects of retarded/ partially completed scattering events, multiple scattering, intra-collisional field effects (manifest also in the dynamic self-consistent non-equilibrium screening) and the perturbative influence of scattering on the acceleration process (memory effect). Effects which are second-order in bulk devices, e.g. over-shoot and Jones-Rees relaxation phenomena are anticipated to be first order effects in this regime (Barker and Ferry, 1979a, Ferry and Barker, 1979b). Surface/interface scattering and scattering by phonons in the immediate device environs become important but should still be perturbative. Strong quantum effects are possible, but depend on what type of device/device-configuration is under scrutiny. Indeed, we should recall that the function and performance of any device is also controlled by other characteristic scales; for example: spatial scales associated with minority carrier diffusion length, depletion length, or tunneling length control the respective physical phenomer of minority carrier injection, punch-through, and tunneling. The latter are responsible in turn for the operation of bi-polar transistors, JFET, CCD, MOS, and tunnel diodes. In MSDs, we encounter new scales associated with non-equilibrium screening lengths (field induced de-screening of scattering processes and phonon softening may both be possible), collision lengths, the range of environmental scattering processes, over-shoot scales and size quantization (de Broglie wavelength). The latter is particularly relevent to MOS structures.

At the smallest scales so far conceived with channel lengths $\lesssim 250$Å (25 nanometers), which defines the very small device (or VSD) the concept of a mean-free-time becomes meaningless. This occurs because at the still higher operating fields (~ 600 kV/cm) and with the very short channel length, the transit time becomes shorter thar τ and indeed comparable to the collision duration τ_c: $\tau_c \lesssim \tau_d$. Her neither the fields nor the influence of the environment can be treated perturbatively and the Boltzmann picture fails entirely. Coherent and tunneling transport processes become possible (such as already occurs in superconducting systems because of the macroscopic phase coherence of the superconducting wave function). Devices on this scale are unavoidably "box-like": three-dimensional and finite the subsequent size-quantization (which will be highly sensitive to the local fields) could lead to energy spacings in the range 1 meV to 1 eV. Room temperature quantum effects are thus possible. This size-quantization may force a drastic change in the way we picture transport; for example, it might be best viewed as field-induced tunneling between different inhomogeneous size-localized electron states. It is worth noting that VSD systems on the 20 - 200Å scale are comparable in size to "small-particle" or grain sizes. These are known to have esoteric properties; for example, the magnetic susceptibility of a very small system having an even number of electrons behaves like exp($-\varepsilon$/kT) at low temperatures and as 1/kT for an odd number of electrons. Current injection/extraction is problematic on VSD scales because of the discrete nature of the carriers and size effects. Thus if the lateral dimensions of the metallized interconnects reaches about 100Å it may become appropriat to consider transport as de Broglie wave-guided transport rather tha controlled by classical point-particle trajectories.

At VSD scales the carrier wavelengths extend into the environment of contacts, interfaces, interconnects, and other devices. The spatial inhomogeneity of the total system cannot therefore be ignored. Indeed charge-density waves and de-Broglie waveguides have been suggested as controllable features of VSD arrays. One-dimensional superlattices in materials prepared by molecular beam epitaxy and several types of intercalates have already hinted at new phenomena to be expected from replicated arrays of small devices

(Dingle et al., 1978; Sai-Halasz, 1978). Dingle et al., (1978) has recently demonstrated how the inhomogeneity may be turned to advantage: impurities localized in one layer-type of an MBE material release free carriers which become trapped in the other layer-type, thereby producing a two-dimensional electron gas with high mobility. Super-lattice effects arising from VSD replication may be of considerable importance in determining intra-device carrier dynamics over and above the influence of the host crystal structure. At VSD scales in VLSI structures we encounter a domain which in its feature sizes, interactions, and complexity is approaching that of biological systems and other natural systems which can support phase transitions and other non-equilibrium cooperative processes. *Cooperative behaviour in VLSI systems may be possible provided the appropriate order parameters and coherence lengths can be properly identified and controlled.*

ON MEDIUM SMALL DEVICES

Let us consider some general features expected in a specific MSD: the MOSFET. A detailed discussion of the corresponding transport physics is not presented as this is covered in Ferry's seminar (these proceedings) - see also the studies by Barker and Ferry (1979a, 1979b) and Ferry and Barker (1979a, 1979b). The MSD scale is important for MOS devices because it is presently believed to represent the minimum scale for MOSFET operation. The problem of estimating the minimal size for MOS and bipolar devices has been extensively discussed by Honeisen and Mead (1972) using extrapolative and scaling arguments based on large-scale device structures. They performed a general study of the static limitations imposed by oxide breakdown, source-drain punchthrough, ionization in the channel, etc., on typical MOS structures, and similar effects in the bipolar devices. Although these considerations suggest a minimal channel length of 0.11 micrometer for the MOSFET, it was found that a more typical device would have a channel length of 0.24 micrometers (which interestingly coincides with our own estimates of the threshold for the breakdown of Boltzmann phenomenology [Barker, 1979a, Barker and Ferry, 1979a]. Other studies have arrived at similar conclusions, but these were based upon assessment of the influence of doping fluctuations, power dissipation, noise and interconnections in limiting the packing density of these devices. These considerations are also supported by our own more recent studies, which use physical concepts and estimators based on a deeper non-linear quantum transport theory of MSD processes.

Figure 1 illustrates a typical medium small device of MOS structure which (Barker and Ferry, 1979a) is scaled to appear similar to that of Honeisen and Mead (1972). Total dimensions are 2500 x 2500 x 100 Angstroms, where the latter is set by the channel

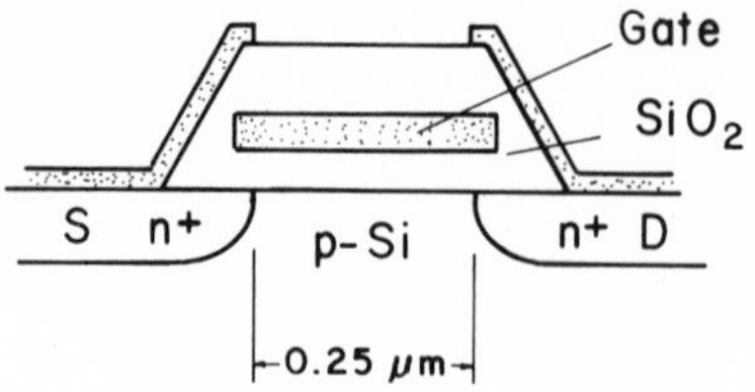

Fig. 1. A model medium small MOS device.

dimension, not by the depletion depth. Limitations on this device are primarily set by oxide breakdown and source-to-drain punchthrough. As the device size is shrunk to the MSD level, substrate doping (actually channel region doping) must increase to control source-drain punchthrough. But this requires higher oxide fields to generate the inversion layer, so that oxide breakdown becomes important. The model MSD is characterized by channel region doping of 3×10^{17} cm^{-3} and oxide thickness of 140Å. At fields just below oxide breakdown (~ 7×10^6 V/cm) the inversion layer surface carrier density is 1.0×10^{13} cm^{-2}, although typical operation would be at lower gate voltages and surface concentrations. A maximum operating effective field of 100 kV/cm is anticipated near the drain end. The small device volume evidently imposes considerable problems with respect to fringing fields (Ferry and Barker, 1979b).

The model MSD is readily characterized as being in strong inversion whenever the gate voltage exceeds the turn-on voltage. In this case the bands are strongly bent and the potential well formed by the insulator-semiconductor surface and the electrostatic potential in the semiconductor can be sufficiently narrow that quantum mechanical effects become important. At the large inversion densities discussed here the electrostatic potential must be found self-consistently taking into account the large extra charge density. This leads to a very thin channel extent, in some cases comparable to or smaller than the mean electron wavelength. Size quantization will occur since the electrons are confined to the potential well, and will lead to widely separated sub-bands: a quasi-two dimensional electron gas is thereby created at the semiconductor-insulator interface. Such effects are known in current large MOS devices at very low temperatures, but in the MSD case sub-band separations of order 180 meV should be possible <u>at room temperature</u> and will significantly alter, for example, velocity-field characteristics.

Such effects as velocity overshoot and Jones-Rees relaxation are expected to be important in the MSD. Device performance is also likely to be significantly affected by the host of high-field effects discussed in the quantum transport lectures (Barker, 1979c). Barker

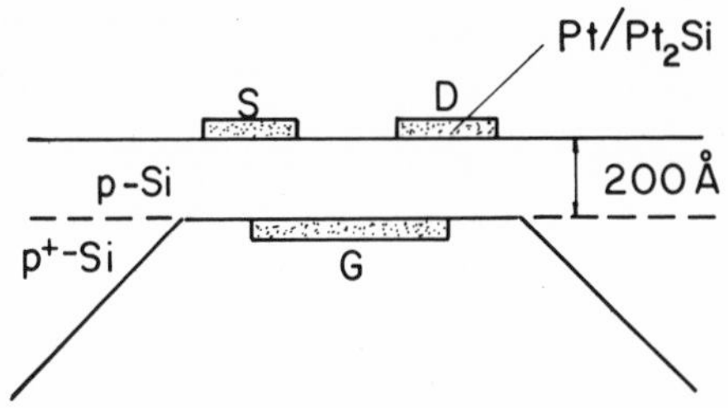

Fig. 2. A model very small MESFET device.

and Ferry (1979a) have recently derived and solved a generalised quantum kinetic equation which replaces the Boltzmann equation in the MSD regime. The transport equation is non-local in space, time and momentum because of the non-negligible effect of the finite, field dependent collision duration. Both Chambers-Rees path variable methods and the moment balance hot electron methodology may be extended into the MSD regime with appropriate (but subtle) modifications (Ferry and Barker, 1979a, Barker and Ferry, 1979b).

VERY SMALL SCALE DEVICES

A Model VSD

In Fig. 2, we sketch a model VSD of conventional function, that might be fabricated by two-side silicon wafer processing using existing technologies (Ferry and Barker, 1979c). This device is a junction p-channel FET with a reversed-biased Shottky gate. Several processing steps are required. First, a highly doped, thin p^+ -Si layer is formed with Boron doping ($> 7 \times 10^{19}$ cm^{-3}) either by ion implantation or diffusion at the top surface. After the surface oxide and boron-silicon alloy layers are removed, a thin (300Å) layer can be grown using solid-phase epitaxy via a Pt_2Si transport layer with the over-layer put down either by evaporation or molecular beam epitaxy. This layer, doped p-type to about 6-7 x 10^{18} cm^{-3} will be the active layer. Then the backside is thinned by preferential etching, and the thin p^+ layer removed by oxidation/stripping or anodic etching of the backside. Finally, a silicide such as Pt_2Si can be formed on the backside and both front and back silicides patterned using the STEM system. The gate remains on the backside of the silicon wafer in the thinned region.

There are generally complications due to the low Shottky barrier heights on p-type material. However, this is no problem in the VSD, because of the low operating voltages, and barrier heights of 0.2 eV for Pt_2Si on p-Si can be obtained. In standby operation, with zero gate bias, the built in potential of the gate Schottky barrier depletes the thin layer about 60Å into the layer from the backside. Thus pinch-off is obtained for a gate bias of 1.5 V and this sets the operating voltage of the device. The average field in the channel is then high: about 600 kV/cm.

It should be noted that the contact regions were not diffused into the layer to form "ohmic" contacts. The reason is due to quantum mechanical considerations: if p^+-contact regions are encountered in a very small device, the channel region will lie some few tenths of a volt higher in carrier energy, due both to doping and the size-quantization effects. This would be sufficient, in the material considered here, to completely deplete the channel region regardless of gate voltage. Thus to contain carriers in the channel region, the contact must present barriers to the carriers: blocking contacts. This has an added advantage, in that carriers can be brought into the channel region at an elevated kinetic energy, which thus reduces the spatial extent over which the temporal and spatial relaxation is significant. There may also be a significant fraction of the carriers in localized states in the channel: bringing the carriers in hot from the contacts will reduce the effect of these states as well. One should note that the thermionic current in such a device is negligible, and the current is brought out of the contacts by tunneling.

The strong interest in VSD structures is easily understood if we examine figures of merit for the p-channel MESFET, assuming that it worked like a normal device. For a doping density of $5 \times 10^{18} cm^{-3}$, and 2 V operating voltage, we may estimate a transit time $\tau_d \sim 0.25$ picoseconds, a rise time of 1 - 2 picoseconds and a speed-power product of $2 - 3 \times 10^{-17}$ Joules: comparable to Josephson logic. An estimated 5-10 μmhos transconductance leads to an estimated switching time of 2-3 picoseconds. The VSD is thus very fast and low power: provided the VSD performed conventionally (which is questionable).

Transport Physics

At present there is no consensus on what type of transport theory is applicable to VSDs. Until recently there has been some pessimism; perhaps VSDs will not work at all: the very high operating fields might precipitate breakdown, avalanching and serious metal migration to name but a few catastrophes. This school of thought has based its ideas on extrapolation of semiclassical ideas in conventional bulk transport theory. Such arguments can be quite misleading, for example a semiclassical calculation of the current extraction from the blocking contacts in the model MESFET above gives almost zero thermionic emission. Actually one needs a quantum calculation and the current due to tunneling is quite sufficient for device operation. A second school of thought ignores the possibility of catastrophic effects and argues that when the transit time is shorter than the bulk mean free time the carriers will freely accelerate down the channel suffering negligible collisions. This is the classical concept of ballistic transport and is enticingly reminiscent of vacuum tube electronics. Our own view is that semiclassical bulk concepts are meaningless on the small VSD scale, rather one should use quantum transport concepts coupled to a proper incorporation of the influence (proximity effect) of the device environment on the dimensionally constrained finite device. We shall take up this view in the following section. Although only exploratory steps have so far been taken, the expected transport physics of VSDs is more varied and richly structured than the two semiclassical schools would suggest.

Let us recall that the VSD is characterised by very short space scales, very fast temporal response, very high fields ($\lesssim$ 600 kV/cm), high carrier densities ($\lesssim 10^{18}cm^{-3}$) but small total carrier number, strong size-related effects: coupling to contacts, interfaces, surfaces, interconnects, and other devices. The ultra-fast transit times ($\lesssim$ 1-2 picoseconds) plus the quenching effect of the high fields on the intra-device collisions (Barker, 1979c) preclude any significant dissipation of energy within the device volume. However, a detailed study of the device density matrix equations of motion which take into account the coupling to the environment shows that dissipation will predominantly occur over the extended region in the vicinity of the device. This is an interesting mechanism because it also admits a second-order device-device correlation/interaction, provided sufficient carrier and phonon wavefunction overlap occurs (indeed for metallized interconnect regions of the order of 100Å, de Broglie waveguide modes should dominate the current flow and sufficient coherence in this flow could lead to very long-range coupling). By using projection calculus methods on the full system density matrix equations we have been able to show (Ferry and Barker, 1979c) that the interaction of a particular device with its host VLSI system may be classified into coherent (time reversible) and incoherent (dissipative, irreversible) components. The coherent effects include state renormalization arising from the short-range interaction with the regular part of the finite device boundaries: the effect is enhanced by the very high inhomogeneous electric field within the device volume. There is also a possible coherent long-range interaction with the environment which stems from effects of device replication (long-range order in the total device array) and gives rise for example to super-lattice phenomena. The coherent processes may over-ride both the bulk intra-device carrier and scatterer (phonons and impurities) dynamics. The incoherent processes include surface/interface roughness scattering, surface/interface phonon scattering, interface plasmon scattering, long-range

(≲ 1.0 micrometer) device-electron: insulator-phonon scattering, and phonon mediated electron-electron interdevice scattering.

The above features suggest that many modes of transport will be possible in VSD structures, depending on the degree of isolation fro surrounding devices and the precise device configuration and operating conditions. In Table 1, we list some theoretical lines of attac on VSD transport physics.

Quantum transport effects which are merely perturbative in MSDs are likely to be strong in VSDs. For example, the retardation of th intra-device scattering in space, time and quasi-momentum, memory effects (as the intra-device flow becomes increasingly coherent), intra-collisional field effects and so on (Barker, 1979c). The shor transit times and the quenching intra-collisional field effect are likely to severely inhibit some classically predicted effects such a intra-device avalanching and breakdown, although this could still pose problems in the region surrounding a device.

Ballistic transport is possible in VSDs (Ferry and Barker, 1979 1979c) when the bulk mean-free-path is longer than the channel length. However, this ballistic mode will not be free-carrier-like in the classical sense because of the coherent proximity effects, an indeed the high carrier densities imply significant plasma effects. In this connection, recent work by Yanson in the USSR (described in Einspruch, 1979) has reported electron-phonon spectroscopy using randomly fabricated metal-metal contacts with dimensions on the orde of 100Å. The contacts arise from "pin-holes" through an insulating film between two metal films. Electron flow through the contact is reportedly ballistic, with electrons dropping through 1 Volt within one mean-free-path, with subsequent non-linear emission of phonons. Yanson has claimed observation of electron de Broglie waveguide mode in some contacts.

A semi-ballistic stochastic transport mode is possible if the dissipative interaction with the environment is sufficiently strong that carriers stream through the device under continuous dissipative interaction with the environs. A distribution function based kineti approach is then impractical, and a better physical picture is found by treating the transport as strongly driven pseudo-Brownian motion governed by a non-linear retarded Langevin equation (LE of Table 1) for the current response. Such a formalism has been constructed for VSDs (Ferry and Barker, 1979c), and is suitable for handling the dominating influence of fluctuations in this type of response.

The high carrier densities predicted for VSDs suggest that carrier-carrier interactions will be important. Correlation effects shielding of the other interactions and the fields, phonon softening etc., must therefore be properly accounted for. Some progress in

Table 1. Lines of Attack in the Quantum Transport Theory of Very Small Devices.

The density matrix

Liouville equation + environmental modifications

Heisenberg picture — Direct approach

- Stochastic method (LE) for each transport parameter memory functionals generalized Langevin theory (B,D)
- Plasma code simulation (B) (C?)

Schrodinger picture — Indirect approach

- Electron density matrix + projection calculus + super-resolvent techniques.
- Quasi-Boltzmann equation (A)
- Quasi-Boltzmann + stochastic environmental interaction (A)
- Full quantum mechanical transport equations for generalized Wigner distributions (B,C)
- Parameterized density matrix techniques (F)
- Feynman Path integral methodology (E)

Key:

A Dominant dissipation in device

B Dominant dissipation in environment

C Dissipation in device and environment

D Continuous dissipation in environment

E Very strong electron-phonon coupling at very high fields

F Very strong carrier-carrier scattering

SPECIAL FEATURES

Intra collisional field effect
Spatio-temporal retardation
Non-equilibrium screening
Super-lattice effects
Size quantization
Environmental coupling
Memory effects
Many body effects
Phonon sub system response
Coupling to Maxwell equations
Inhomogeneous transport

providing a rigorous non-equilibrium shielding theory for inhomogeneous transport has been recently reported (Barker, 1979b, 1979c), but is not yet adequate for VSDs (although it should cover MSD cases). The strong carrier-carrier interactions do however provide some fresh theoretical advantages. Thus, if carrier-carrier scattering is faster than any other scattering process, it is possible to construct an analogue of displaced Maxwellian theory in classical hot electron physics, but based on a parameterised density matrix (Ferry and Barker, 1979c). Here one parameterizes the density matrix via the effective space/time dependent chemical potential, carrier temperature, and drift velocity, which in turn are computed as non-linear quantum correlation functions obtained as solutions to the micro-continuity equations for carrier number, momentum and energy. The method is an extension of Poincare's method of integral invariants and was first advocated for hot electron physics by Zubarev and Kalashnikov.

SYNERGETIC EFFECTS AND NEW DEVICE CONCEPTS

So far we have examined local aspects of sub-micron VLSI: the behaviour of a particular device in a fixed system matrix. We have also tacitly dealt with conventional concepts of device function. Looking further ahead, however, it is easily seen that there are global aspects of VLSI which have exciting possibilities for new physical phenomena, radically new device and device system concepts, and for a possible link-up with biological problems. These new, but at this stage still ill-defined, possibilities depend on the multi-element character of VLSI and on the existence of inter-element coupling on the atomic scale (provided of course that we relax the conventional design ethic of completely isolating each device in our thinking). To understand what is involved we turn now to a brief discussion of the concepts of synergetic phenomena (Haken, 1978) and modern ideas on so-called dissipative structures (Turner, 1974).

Synergetic phenomena are the joint actions of many sub-systems so as to produce structure and functioning on the full system scale. These are well known in physical science and are characteristic of nonlinear systems. Systems which contain mechanisms for spontaneous symmetry breaking and self-organization are particularly interesting for example, soft and hard mode instabilities, mode locking in laser systems, switching phenomena, and even the Gunn effect. Self-organizing systems have the following general characteristics: a given sub-system is in dynamical interaction with an environment (which may contain other sub-systems); the dynamical evolution of the sub-system is slaved to (controlled by) the environment via order-parameters, which in the simplest cases may be just external forces. The slaved sub-system then reacts back on the environment. In synergetic theory, it is usually assumed that the slaving is adiabatic, i.e. instantaneous coupling occurs. Complex systems may, as a result, operate in different modes which are well defined by the behaviour of the order parameters. Often, very complex systems may show well regulated behaviour which is controlled by a small number of order parameters. In some cases the non-linearity of the order parameter equations of motion may lead to bifurcation phenomena in which continuous adjustment of an order parameter leads to spontaneous symmetry breaking in which a particular state evolves to a pair (or more) of stable and unstable states. This often appears as jump phenomena in very nonlinear systems. Fluctuation phenomena then become of crucial importance in determining to which state the system eventually stabilizes.

Phase transitions provide a familiar illustration of cooperative behaviour in many-body systems. For example, a thermodynamic equilibrium system of elementary excitations coupled via residual, weak short-range forces may under certain conditions (i.e. at transition points) undergo instabilities in which the short-range forces cooperate to generate long-range correlations which spontaneously precipitate a transition from one ordered macro-state to another. Once such a new phase is formed its maintenance does not require any interaction with the outside world: in a constant environment, equilibrium structures are self-sustaining.

In recent years considerable attention has been turned towards a different class of ordered structures: the so-called dissipative structures. In these, order may be created spontaneously in open systems far from equilibrium and which obey specific non-linear kinetic laws. For such non-equilibrium structures stability is not self-sustaining but is maintained by a continuous exchange of energy and matter with the surroundings. Biological systems are an example, and we may also cite chemically reacting mixtures under open system conditions and Benard cell formation.

The author (1979a), and others (notably Krumhansl, 1979) have speculated that comparably complex signal processing systems may also support synergetic phenomena as the scale and complexity approaches that of natural systems. Thus, both many-body considerations and synergetic phenomenology lead us to expect strong qualitative differences in stability between sequential and concurrent (array) processing signal systems. The argument is simple: we know from model studies that one-dimensional systems cannot support phase-transitions, but as soon as two- and three-dimensional cross-interactions are introduced, the system can condense or lock-into ordered macrostructures and make transitions between them. Multi-dimensionality has a strong equivalency to concurrency. Multi-dimensionality is also a prerequisite for cooperative phenomena which are stable against local fluctuations or control.

Preliminary theoretical studies of the quantum mechanical operation of an array of very small devices suggest that synergetic effects are possible when the spatial scale of the individual device is less than a few hundred angstroms (Barker and Ferry, 1979c). In a previous section were listed the important device-environment interactions which are expected to dominate transport and device performance in VSDs. By environment, we mean other devices, interconnects, surfaces and interfaces, etc. Of course, device-environment interactions are not unknown even in present scale device systems. For example, gated logic systems and programmable logic arrays have device-environment control exercised via the interconnection matrix. A true device-device interaction also appears in large MOS memory chips (near-cell interference in read/write situations) which is currently treated as a reliability problem rather than an effect to be capitalized on. And of course, capacitative coupling between devices is a very familiar phenomenon. The true device-environment interactions we envisage here are only possible on a very small scale for which super-lattice effects and dissipative coupling become possible. A major difficulty facing the theoretical study of device-device and device-interconnect coupling is the problem of determining the correct dielectric response (especially in the screening of the interactions) in finite, small sub-systems with a limited total number of carriers.

Our detailed studies suggest that sufficient inter-device slaving features may exist on the very short space-time scales of sub-micron VLSI for a wide range of options for novel device and device-system design and function. Of these, we may single out the concept of a passive holistic device whose function and performance might be controlled via renormalization of the intra-device dynamics by altering global features; e.g. super-lattice voltage levels in the surrounding device array. A further possibility, of relevance to non-von Neumann microcomputer architectures is the deliberate fabrication of partially isolated device arrays which may undergo self-organizing transitions induced by the modulation of a few externally imposed control parameters. Possible applications might include self-healing microcomputers (against radiation damage, for example) and programmable structures (rather than fixed architectures).

The main theoretical tools to explore synergetic VLSI already exist. But applications will require knowledge of the intended function and skeletal VLSI structure, characterization of the relevant inter-device coupling, and selection of the appropriate control fields and currents. This knowledge does not yet exist and must be the target for future research. However, let us sketch a derivation of the basic device equations of motion with the aim of establishing the basic features pre-requisite to synergetic phenomena. (An alternative approach is described in Ferry and Barker [1979c]).

FORMAL THEORY OF COUPLED DEVICE ARRAYS

At the macro-level, we exercise control over an N-element device system via applied fields (or voltages) and input and feedback currents. We label these F^i_{ext} (i = 1 . . . N; F^i_{ext} is a set of applied controls acting on the ith element: for compactness the label i refers to devices and any other spatially delineated structure in the system, e.g. contacts, interconnects, etc.). The applied generalized forces will give rise to local applied forces F^i which are the self-consistent solutions of appropriate macro-equations of motion, e.g. Poisson's equation and current continuity equations, coupled self-consistently to the dynamical variables of each device. The latter are completely specified as quantum statistical expectation values by the individual device density matrices $\rho(i;t)$. In the absence of inter-device coupling we have simply the Liouville-von Neumann equations of motion ($\hbar = 1$):

$$i\partial_t \rho(i;t) = \hat{H}(i)\rho(i;t), \qquad (3)$$

where $\hat{H} \equiv [H, \ldots]$ is the commutator generating super-Hamiltonian. $H(i) \equiv H(i,F^i(t))$ is the Hamiltonian for device i and is assumed to be time-dependent through the coupling to the generalized time-dependent forces F^i.

In the coupled N-element system, we write the full Hamiltonian as

$$H(1, 2, \ldots N) = \sum_{i=1}^{N} H(i,F^i) + \sum_{i>j=1}^{N} H'(i,j,F^k), \qquad (4)$$

where labels {i} refer to complete sets of dynamical variables for the ith device; $H'(i,j,F^k)$ is the coupling between devices i and j (which is assumed here to be instantaneous and pairwise for simplicity, although more general interactions most certainly exist). The system density matrix $\rho(1, 2, \ldots N;t)$ satisfies

$$i\partial_t \rho(1 \ldots N;t) = \hat{H}(1, \ldots N;t)\rho(1 \ldots N;t). \qquad (5)$$

From ρ we may define <u>reduced</u> density matrices as

$$\rho(1, 2, \ldots M;t) \equiv \Omega^M \underset{(M+1, \ldots N)}{\mathrm{TR}} \rho(1, 2, \ldots N),$$

$$(1 \le M \le N), \qquad (6)$$

where Ω is the system volume and the partial trace is over the residual set {M + 1, . . . N}. Using (4) - (6), and exploiting the cyclic properties of the trace, we arrive at the N-coupled individua device density matrix equations

$$i\partial_t\rho(i) = \hat{H}(i)\rho(i) + \Omega^{-1} \sum_{i\neq j=1\,(j)}^{N} \mathrm{TR}\,\{\hat{H}'(i,j)\rho(i,j)\}, \tag{7}$$

which differ from (3) by the coupling to $\rho(i,j)$, the <u>two-device</u> density matrix. Similarly we may generate a hierarchy of equations for $\rho(i,j)$, $\rho(i,j,k)$, etc. Let us next introduce inter-device corre lation operators $g(ij)$, $g(ijk)$, etc., which vanish if there are no device-device correlations, as

$$g(ij) \equiv \rho(j)\rho(i) - \rho(ij), \tag{8}$$

$$g(ijk) \equiv \rho(ijk) - g(jk)\rho(i) - g(ik)\rho(j) - g(ij)\rho(k) - \rho(i)\rho(j)\rho(k). \tag{9}$$

The coupled hierarchy may be closed under suitable assumptions on the correlation operators, just as in many-body theory (and with similar problems!). Thus if three-device and higher-order correlations are negligible, we may truncate the hierarchy to obtain

$$i\partial_t\rho(i) = \{\hat{H}(i) + \Omega^{-1} \sum_{j=1\,(j)}^{N}{}' \mathrm{TR}\,\hat{H}'(i,j)\rho(j)\}\rho(i) - \Omega^{-1} \sum_{j=1\,(j)}^{N}{}' \mathrm{TR}\,\hat{H}'(ij)g(ij), \tag{10}$$

$$i\partial_t g(ij) = \hat{H}'(i,j)\rho(j)\rho(i) + \hat{H}(ij)g(ij), \tag{11}$$

where $H(ij) \equiv H(i) + H(j) + H'(ij)$.

Equation (11) may be formally solved using causal boundary conditions initialized at $t = 0$. This gives

$$g(t) = g^o + \left(\frac{1}{i}\right) \int_o^t d\tau\, e^{-i\int_{t-\tau}^{t} d\tau'\hat{H}(ij,\tau')}\, \hat{H}'(ij,t-\tau)\rho(j,t-\tau)\rho(i,t-\tau) \tag{12}$$

Here g^o is the memory/transient term. Defining

$$\Omega^{-1} \sum_j{}' \underset{(j)}{TR} \ldots \rho(j;t) \equiv \langle \ldots \rangle_t ,$$

we insert (12) into (10) to find

$$i\partial_t \rho(i) = \{\hat{H}(i) + \langle \hat{H}'(i,j)\rangle_t\}\rho(i) - \langle \hat{H}'(i,j) g^o(ij;t)\rangle_t$$

$$+ i\int_o^t d\tau \langle \hat{H}'(i,j;t) e^{-i\int_{t-\tau}^{t} d\tau' \hat{H}(ij;\tau')} \hat{H}'(ij;t-\tau)\rangle_{t-\tau} \rho(i,t-\tau). \quad (13)$$

The set of equations in (13) are examples of the fundamental equations necessary for ab initio theories of synergetic multi-device operation. Compared with the set of equations (3) of isolated-device dynamics, they are extremely rich in non-linear, non-local structure and contain the seeds for cooperative phenomena. Let us point out three features of importance: (1) the individual device density matrices are driven by the normal device Hamiltonian H(i) but the dynamics are renormalized by the average instantaneous interaction $\langle H'(ij)\rangle_t$ with other devices: thus if $\sum_j H'(ij)$ has periodic components, the dynamics described by H(i) are over-ridden by superlattice effects; (2) the term in $g^o(ij)$ describes memory of the initializing state of the full system; (3) the third term in (13) describes generalized, retarded inter-device "scattering". This term contains the seeds for dissipative inter-device coupling on appropriate time scales. For example, if the time variation of the generalized forces is slow on the time scale of variation of $\rho(i;t)$, we may form the asymptotic equation of motion ($t \to \infty$)

$$i\partial_t \rho(i) = \langle \hat{H}(i) + \hat{H}'(i,j)\rangle_t \rho(i) - \langle \hat{H}' g^o\rangle_t + i\hat{c}\rho(i), \quad (14)$$

where

$$\hat{c} \equiv i\langle \hat{H}'(ij;t)\{\hat{H}(i,j,t) - io^+\}^{-1}\hat{H}'(i,j;t)\rangle_t . \quad (15)$$

Provided the spectrum of H(i,j) has a continuous part, the expression for the "collision" operator c has both hermitian and non-hermitian components:

$$i\hat{c} \equiv -\langle \hat{H}' P(\tfrac{1}{\hat{H}})\hat{H}'\rangle_t - i\pi\langle \hat{H}'\delta(\hat{H})\hat{H}'\rangle_t , \quad (16)$$

where $\delta(\hat{H})$ describes energy-conservation for dissipative scattering processes in the full system and P denotes the Cauchy Principal Part.

The above rudimentary arguments illustrate that sufficient non-linear slaving features (both adiabatic and retarded) may exist in VLSI for the contemplation of synergetic behaviour. We would strongly advocate formal developments of this type to under-pin (and suggest) any future phenomenology for synergetic VLSI.

REFERENCES

Ballantyne, J. M., Ed., 1979, "Proceedings of the NSF Workshop on Opportunities for Microstructure Science, Engineering, and Technology," Cornell University Press, Ithaca, New York.

Barker, J. R., 1979a, in "Proceedings of the NSF Workshop on Opportunities for Microstructure Science, Engineering, and Technology," J. M. Ballantyne, Editor, Cornell University Press, Ithaca, New York.

Barker, J. R., 1979b, to be published.

Barker, J. R., 1979c, Chapter 5 these proceedings.

Barker, J. R., and Ferry, D. K., 1979a, Solid-State Electr. in press

Barker, J. R., and Ferry, D. K., 1979b, Phys. Rev. Letters 42:1779.

Barker, J. R., and Ferry, D. K., 1979c, in "Proceedings of the 1979 IEEE Conference on Cybernetics and Society," IEEE Press, in press.

Broers, A. N., Harper, J. M. E., and Molzen, W. W., 1978, Appl. Phys. Letters 33:392.

Dingle, R., Stormer, H. L., Grossard, A. C., and Wiegmann, W., 1978, Appl. Phys. Letters 33:665.

Einspruch, E., Editor, 1979, "Microstructure Science, Engineering, and Technology," National Academy of Sciences, Washington, DC.

Ferry, D. K., and Barker, J. R., 1979a, Solid State Commun. 30:301.

Ferry, D. K., and Barker, J. R., 1979b, in "Proceedings of the University/Government/Industry Microelectronics Symposium," IEEE Press, New York, p. 88.

Ferry, D. K., and Barker, J. R., 1979c, Solid-State Electr. in press

Haken, H., 1978, "Synergetics", Springer-Verlag, Berlin.

Hoeneisen, B., and Mead, C. A., 1972, Solid-State Electr. 15:819; 15:891.

Krumhansl, J. A., 1979, in "Proceedings of the NSF Workshop on Opportunities for Microstructure Science, Engineering, and Technology," J. M. Ballantyne, Editor, Cornell University Press, Ithaca, New York.

Sai-Halasz, G. A., 1978, in "Proceedings of the 14th International Conference on Physics of Semiconductors," Inst. of Physics, London, p. 21.

Turner, J. S., 1974, "Lecture Notes in Physcis 28: Statistical Physics," Springer-Verlag, New York, p. 248.

LIST OF PARTICIPANTS

Akers, L.
Dept. of Electrical Engineering
University of Nebraska
Lincoln, Nebraska, USA

Al-Mass'ari, M. A. S.
Physics Department
University of Rijad
Rijad, Saudi Arabia

Apsley, N.
Royal Signals and Radar Est.
Great Malvern, Worco., U.K.

Baglee, D. A.
Dept. of Electrical Engineering
Colorado State University
Fort Collins, Colorado, USA

Barker, J. R.
Physics Department
Warwick University
Coventry, U.K.

Bauer, G.
University of Leoben
Leoben, Austria

Birgül, G.
Nuclear Research and Training
Center
Ankara, Turkey

Blakey, P. A.
Dept. of Electrical Engineering
University of Michigan
Ann Arbor, Michigan, USA

Bosi, S.
Institute of Physics
University of Modena
Modena, Italy

Büget, U.
Atomic Energy Commission
Ankara, Turkey

Calandra, C.
Institute of Physics
University of Modena
Modena, Italy

Calecki, D.
Solid State Physics
L'Ecole Normale Superieure
Paris, France

Convert, G.
Thomson-CSF
Orsay, France

Cook, R. K.
Phillips Hall
Cornell University
Ithaca, New York, USA

Cooper, L. R.
Office of Naval Research
800 N. Quincy
Arlington, Virginia, USA

Cristoloveanu, S.
CNRS - Grenoble
Grenoble, France

Dambkes, H.
Gesamthochschule Duisberg
Duisberg, Germany

Dandekar, S.
Phillips Hall
Cornell University
Ithaca, New York, USA

Daugaard, S.
Colorado State University
Fort Collins, Colorado, USA

Debney, B.
Plessey Research Centre
Caswell, Towcester, U.K.

Del Giudice, M.
Thomson-CSF
Orsay, France

Elta, M.
MIT-Lincoln Laboratory
Lexington, Massachusetts, USA

Ferry, D. K.
Dept. of Electrical Engineering
Colorado State University
Fort Collins, Colorado, USA

Frensley, W. R.
Texas Instruments, Inc.
Central Research Lab.
Dallas, Texas, USA

Gasquet, D.
Universite Science et Techn.
du Languedoc
Montpelier, France

Grave, T.
Physics Institute
Universität Heidelberg
Heidelberg, Germany

Greene, R.
Naval Research Laboratory
Washington, D.C., USA

Grondin, R.
Dept. of Electrical Engineering
University of Michigan
Ann Arbor, Michigan, USA

Grubin, H.
United Technologies Research Center
East Hartford, Connecticut, USA

Günal, I.
Middle East Technical University
Ankara, Turkey

Hearn, C. J.
Marine Sciences Laboratory
Gwynedd, U.K.

Hess, K.
Dept. of Electrical Engineering
University of Illinois
Urbana, Illinois, USA

Inoue, M.
Phillips Hall
Cornell University
Ithaca, New York, USA

Jauho, A.
Physics Department
Cornell University
Ithaca, New York, USA

Jacoboni, C.
Institute of Physics
University of Modena
Modena, Italy

Kahlert, H.
Boltzmann Institute
Kopernikusgasse 15
Vienna, Austria

Karabiyik, H.
Nuclear Research Training Center
Ankara, Turkey

Kaufmann, I.
Office of Naval Research (London)
London, England

Kocevar, P.
Physics Institute
University of Graz
Graz, Austria

Koch, F.
Physics Department
Technische Universität Munchen
Garching, Germany

Kratzer, S. G.
Phillips Hall
Cornell University
Ithaca, New York, USA

Laude, L. D.
Université de L'Etat a Mons
Mons, Belgium

Leburton, J. P.
Université de Liege
Liege, Belgium

Lovato, M.
Université de L'Etat a Mons
Mons, Belgium

Miller, D. A. B.
Heriot-Watt University
Edinburgh-Riccarton, U.K.

Moore, B. T.
Dept. of Electrical Engineering
Colorado State University
Fort Collins, Colorado, USA

Nardinocchi, G.
Institute of Physics
University of Modena
Modena, Italy

Nicholas, R.
Clarendon Laboratory
Oxford, U.K.

Nordbotten, A.
Defence Research Establishment
Kjeller, Norway

Nougier, J. P.
Université Science et Tech.
du Languedoc
Montpelier, France

Perrier, P.
Université de Toulouse
Toulouse, France

Reggiani, N.
Institute of Physics
University of Modena
Modena, Italy

Reich, R. K.
Dept. of Electrical Engineering
Colorado State University
Fort Collins, Colorado, USA

Seiler, D. G.
Physics Dept.
North Texas State University
Denton, Texas, USA

Smirl, A.
Physics Dept.
North Texas State University
Denton, Texas, USA

Türek, D.
Nuclear Research Training Center
Ankara, Turkey

Ulbrich, R.
Institute of Physics
Universität Dortmund
Dortmund, Germany

Vaissiere, J. C.
Université Science et Tech.
du Languedoc
Montpelier, France

Van Welzenis, R. G.
Eindhoven University of Technology
Eindhoven, The Netherlands

Vogl, P.
Physics Institute
University of Graz
Graz, Austria

Williams, C. K.
Dept. of Electrical Engineering
North Carolina State University
Raleigh, North Carolina, USA

Windhorn, T.
Dept. of Electrical Engineering
University of Illinois
Urbana, Illinois, USA

Wirkner, J.
Institute fur Angewandte Physik
Erlangen, Germany

Wood, J.
University of Leeds
Leeds, U.K.

Zeller, R. T.
Physics Department
Brown University
Providence, Rhode Island, USA

Zimmerman, J.
Université de Lille
Villeneuve d'Ascq, France

INDEX